Discrete Mathematics and Theoretical Computer Science

Springer

London
Berlin
Heidelberg
New York
Barcelona
Hong Kong
Milan
Paris
Singapore
Tokyo

Sergiu Rudeanu

Lattice Functions and Equations

 Springer

Professor Sergiu Rudeanu, PhD
Faculty of Mathematics, University of Bucharest, Str. Academiei 14, 70109 Bucharest, Romania

British Library Cataloguing in Publication Data
Rudeanu, Sergiu
    Lattice functions and equations. - (Discrete mathematics
    and theoretical computer science)
    1. Algebra, Boolean
    I. Title
    511.3'24
ISBN 1852332662

Library of Congress Cataloging-in-Publication Data
Rudeanu, Sergiu
    Lattice functions and equations / Sergiu Rudeanu.
      p. cm. -- (Discrete mathematics and theoretical computer science, ISSN 1439-9911)
    Includes bibliographical references and index.
    ISBN 1-85233-266-2 (acid-free paper)
    1. Lattice theory. I. Title. II. Series.
QA171.5.R75 2001
511.3'3—dc21   2001034202

Discrete Mathematics and Theoretical Computer Science Series ISSN 1439-9911
ISBN 1-85233-266-2 Springer-Verlag London Berlin Heidelberg
a member of BertelsmannSpringer Science+Business Media GmbH
http://www.springer.co.uk

Typesetting: Camera ready by author
Printed and bound at the Athenæum Press Ltd., Gateshead, Tyne & Wear
34/3830-543210 Printed on acid-free paper SPIN 10755314

Preface

The field of lattice functions and equations started *avant la lettre*: it was Boole himself who initiated the study of Boolean functions and equations, which has been extensively studied in the literature, both from a theoretical standpoint and in view of applications; see our monograph Rudeanu [1974a], herein after referred to as BFE.

The aim of the present book is twofold: to update BFE and to survey lattice functions and equations (LFE) in classes of lattices more general than Boolean algebras. Yet the book is self-contained and it seems appropriate to explain its title now.

A *lattice (Boolean) function* is a function f with arguments and values in a lattice L (Boolean algebra B) and which can be expressed by a formula built up from variables and constants of L (of B) by superpositions of the basic operations of L (of B). In particular if the formula does involve variables and basic operations but no constants, then f is said to be a *simple lattice (Boolean) function*. An equation $f = g$ over L (over B), where f and g are lattice (Boolean) functions or simple lattice (Boolean) functions, is called a *lattice (Boolean) equation*, or a *simple lattice (Boolean) equation*, respectively. We also refer to functions $f : \{0,1\}^n \longrightarrow \{0,1\}$ and to equations between such functions as *truth functions* and *truth equations*, or equivalently, we call them *switching functions* and *switching equations*, respectively. The latter terminology was used by switching theorists in the sixties; nowadays it seems that "switching" has been forgotten and replaced by "Boolean". Since, as is well known, every truth ≡ switching function is a simple Boolean function, our terminology does not contradict the current usage, but is just a refinement of it. BFE and LFE are pleadings for the interest of our approach.

Here are the contents of the present monograph.

Chapter 1 deals with equations in a very general sense. Reproductive solutions of such equations are studied in some detail. Chapter 2 provides the prerequisites of universal algebra necessary for a better understanding of certain points in the subsequent chapters. In particular the parallel concepts of Boolean function and simple Boolean function, as well as those of Post function and simple Post function, turn out to be specializations of the universal-algebra concepts of algebraic function and term function, respectively. Chapter 3 provides all lattice-theoretical prerequisites and studies LFE in lattices, bounded distributive lattices, pseudocomplemented distributive lattices, Heyting algebras and Stone algebras. Chapter 4 is devoted to equational compactness in semilattices and Boolean algebras; equational compactness is the condition stating that an infinite system of algebraic equations has a solution whenever all of its finite subsystems are consistent. As shown in Chapter 5, the theory of LFE in Post algebras is very close to BFE. Chapter 6, having the title "A revision of Boolean fundamentals", studies linear Boolean equations and suggests generalizations of several basic concepts and techniques of BFE: minterms, prime implicants, reproductive solutions and tree-like soutions. Generalized systems of Boolean equations are also introduced. Chapter 7 is devoted to certain closure operators in the algebra of Boolean functions (isotone, monotone, independent and decomposition closures), while Chapter 8 studies Boolean transformations, including the range, injectivity and fixed points, as well as functional dependence of Boolean functions. Boolean equations with unique solution, quadratic truth equations and Boolean equations on computers are some of the topics dealt with in Chapter 9. Boolean differential calculus and the decomposition of Boolean functions, already presented in BFE, are resumed and developed in Chapters 10 and 11, respectively. Boolean differential calculus is a field of Boolean-based mathematics; by this term we mean the attempt to establish analogues of several mathematical concepts and theorems when their universe of discourse is replaced by a Boolean algebra. Other branches of Boolean-based and Post-based mathematics are dealt with in Chapter 12; among them, Boolean geometry is a rather developed field. Chapter 13 collects several results of various natures. The last chapter is devoted to applications in graph theory, automata theory, synthesis of circuits, fault detection, databases, marketing and others. There are also three Appendices, in particular a list of open problems.

See Di Nola, Sessa, Pedrycz and Sanchez [1989] for the related field of fuzzy relation equations, which has numerous applications and might be viewed as a study of matrix equations over a Brouwerian lattice. In Appendix 3 we suggest it would be of interest to have a few other companion monographs to the present book.

Within each chapter we number separately theorems, propositions, lemmas, corollaries and remarks in the form n.p ; for instance, Theorem n.p means theorem p in section n. Statements from another chapter are referred to in the form m.n.p, where m denotes the chapter. The same rules apply to displayed formulas. The notation Author [year]* designates items which we could not find and which we quote from other sources, mainly from Mathematical Reviews.

Acknowledgements. For many years I have had the privilege of the constant friendship of Frank M. Brown, Cristian S. Calude, Peter L. Hammer, Afrodita Iorgulescu, Robert A. Melter, Ivo G. Rosenberg, Michel Serfati and Dan A. Simovici. They have brilliantly proved the theorem "A friend in need is a friend, indeed", and in particular I felt it quite recently, in connection with this book. Ever since Cris Calude invited me to write it, I have benefited from his permanent support. The LATEX assistance of Laurenţiu Leuştean and Gheorghe

Ştefănescu was crucial to me. Many thanks are also due to Daniel Drăgulici, Cristina Purcărea, Nicolae Ţăndăreanu and Elena Zaitseva for their precious help.

It is my pleasure to acknowledge the kind editorial assistance of Springer.

I wish finally to express my deep love and gratitude to my wife Delia and my son Răzvan. This book and my entire scientific activity would not have been possible without their steadfast understanding and support.

Sergiu Rudeanu
March 2001

Table of contents

1. Exotic equations

There are many kinds of equations studied in mathematics, so that at first glance it seems dificult, if not impossible, to say something non-trivial about equations in general, with no specification at all. Yet we are going to show that the fundamental concepts of reproductive solution and general reproductive solution, extensively studied in the theory of Boolean equations, apply to the most general concept of equation as well, yielding actually non-trivial results.

In §1 we study the concepts of general solution and reproductive general solution for equations in the most general sense, while in §2 we assume that the equations are defined over finite sets satisfying certain mild conditions.

1 An abstract theory of equations

The first task is to specify the framework, i.e., to find an appropriate definition for "the most general concept of equation".

Let T be a non-empty "total" set. Roughly speaking, an equation over T is the problem of finding all the elements of T that have a certain property. If the intuitive concept of property is modelled by a function $\pi : T \longrightarrow \{0,1\}$, then the equation becomes

(1.1) $$\pi(x) = 1$$

and $S = \{x \in T \mid \pi(x) = 1\}$ is the set of solutions or the *solution set* of equation (1.1). Conversely, each subset $S \subseteq T$ is the solution set of the equation $\chi_S(x) = 1$, where $\chi_S : T \longrightarrow \{0,1\}$ is the characteristic function of the set S. We thus obtain a bijection between the equations of the form (1.1) and the subsets of T, so that we may identify the former with the latter.

On the other hand, the intuitive concept of property defined for the elements of T is modelled more directly by a unary relation on T: some elements satisfy the unary relation, others do not. Yet a unary relation on T is nothing but a subset of T (just as a binary relation is a subset of $T \times T$ etc.).

The above remarks justify Definition 1.1 below, which in particular *includes all the equations that will be studied in this book.*

Definition 1.1. By an *equation* over a non-empty set T we mean a subset $S \subseteq T$; the elements of S are called the *solutions* of the equation. An equation $S \neq \emptyset$ is said to be *consistent*. The equation T is called the *identity equation.*□

Definition 1.1 of the general concept of equation is due to Prešić [1973], while Banković [1983] uses the unary-relation form.

The concept of general (or parametric) solution of an equation is well known in various contexts. Schröder [1890-1905], vol.1, introduced the reproductive general solution of Boolean equations, which were extensively studied by Löwenheim [1908], [1910] and his followers; the term "reproductive" was introduced by Löwenheim [1910]; cf. BFE, page 57. Definitions 1.2 and 1.3 below are due to Prešić [1968].

Definition 1.2. A *general solution* of a consistent equation $S \subseteq T$ is a function $f : T \longrightarrow T$ such that $f(T) = S$. $\qquad\qquad\square$

In other words, this means that for every $x \in T$,

$$(1.2) \qquad\qquad x \in S \Longleftrightarrow \exists t \; x = f(t) \,.$$

Definition 1.3. A *reproductive general solution* or simply a *reproductive solution* of a consistent equation $S \subseteq T$ is a general solution f of S such that for every $x \in T$,
$$(1.3) \qquad\qquad x \in S \Longrightarrow x = f(x) \,.$$

$\qquad\qquad\square$

Proposition 1.1 below and its Corollaries 1.1 and 1.2 collect results from Prešić [1968], [1972].

Proposition 1.1. *The following conditions are equivalent for $\emptyset \neq S \subseteq T$ and $f : T \longrightarrow T$:*

(i) f is a reproductive solution of S ;

(ii) f is a general solution of S and $f|_S = 1_S$;

(iii) $f(T) \subseteq S$ and $\forall x \, (x \in S \Longrightarrow x = f(x))$;

(iv) $\forall x \, (x \in S \Longrightarrow x = f(x))$ and $\forall x \, (x \notin S \Longrightarrow f(x) \in S)$;

(v) $f^2 = f$ and $\forall x \, (x \in S \Longleftrightarrow x = f(x))$.

COMMENT: The conditions $f|_S = 1_S$ and $f^2 = f$ are known as f being a *retraction* and f being *idempotent*, respectively.

PROOF: (i) $\Longleftrightarrow$ (ii) $\Longleftrightarrow$ (iii) $\Longleftrightarrow$ (iv): Obvious.

(iii) $\Longrightarrow$ (v): Since $f(x) \in S$ it follows that $f(x) = f(f(x))$. Further, if $x = f(x)$ then $x \in f(T)$, hence $x \in S$.

(v) $\Longrightarrow$ (iii): Since $f(x) = f(f(x))$, it follows that $f(x) \in S$.

(iii) $\Longrightarrow$ (ii): The inclusion $S \subseteq f(T)$ follows from $x \in S \Longrightarrow x = f(x) \Longrightarrow x \in f(T)$. $\qquad\qquad\square$

Corollary 1.1. *Every idempotent map $f : T \longrightarrow T$ is a reproductive solution of the set $f(T)$ of its fixed points.*

PROOF: Suppose $f : T \longrightarrow T$ is idempotent and take $S = \{x \in T \mid x = f(x)\} = \{f(x) \mid x \in T\}$. Then $S \neq \emptyset$ and condition (v) in Proposition 1.1 is fulfilled. $\qquad\square$

Corollary 1.2. *Every consistent equation S has reproductive solutions: they are all the maps $f : T \longrightarrow T$ of the form*

(1.4) $$f(x) = x \ if \ x \in S, \ else \ h(x) \,,$$

where $h : T \backslash S \longrightarrow S$.

PROOF: Immediate from condition (iii) in Proposition 1.1. □

Corollary 1.3. *The correspondence $f \mapsto f(T)$ establishes a surjection from the set of all idempotent maps $f : T \longrightarrow T$ onto the set of all non-empty subsets of T.*

PROOF: Corollary 1.2 implies $f(T) = S$. □

The next proposition uses a concept borrowed from matrix algebra (cf. Penrose).

Definition 1.4. By a *generalized inverse* of a function $f : A \longrightarrow B$ is meant a function $g : B \longrightarrow A$ such that $fgf = f$. □

If $A \neq \emptyset$ then $B \neq \emptyset$ and a generalized inverse can be constructed by choosing, for each $y \in f(A)$, an element x such that $f(x) = y$ and by setting $g(y) = x$, while $g(y)$ is arbitrarily chosen for $y \in B \backslash f(A)$. If $A = B = \emptyset$ then $g = f =$ the empty function.

Proposition 1.2. (Kečkić and Prešić [1984]. *Let f be a general solution of an equation S and and g a generalized inverse of f. Then fg is a reproductive soluion of S.*

PROOF: Condition (iii) in Proposition 1.1 is easily checked: $fg(T) \subseteq f(T) = S$ and if $x \in S$ then $x = f(t)$ for some t, hence $fg(x) = fgf(t) = f(t) = x$. □

Božić [1975] constructed all general solutions and all reproductive solutions of a consistent equation starting from a given one. Propositions 1.3-1.6 and Lemmas 1.1, 1.2 below represent an improvement of Božić's results, obtained in Rudeanu [1978b].

Proposition 1.3. *Let f be a general solution of an equation S and $g : T \longrightarrow T$. Then g is a general solution of S if and only if $g(T) \subseteq S$ and $f = gh$ for some $h : T \longrightarrow T$.*

PROOF: If g is a general solution then $g(T) = S = f(T)$, hence for every $t \in T$ we can choose $x \in T$ such that $f(t) = g(x)$ and set $h(t) = x$, which implies $gh(t) = f(t)$. Conversely, if $g(t) \subseteq S$ and $f = gh$, then for every $s \in S$ there is $t \in T$ such that $s = f(t) = gh(t) \in g(T)$; it follows that $S \subseteq g(T)$, hence $S = g(t)$. □

Lemma 1.1. *If f is a general solution and g a reproductive solution of an equation, then $f = gf$.*

PROOF: For every $t \in T$ we have $f(t) \in S$, hence $f(t) = g(f(t))$.   □

Proposition 1.4. *Let f be a reproductive solution of an equation S and g : $T \longrightarrow T$. Then g is a reproductive solution of S if and only if $g(T) \subseteq S$ and $f = gf$.*

PROOF: Necessity follows from Lemma 1.1. To prove sufficiency note first that g is a general solution by Proposition 1.3. If $s \in S$ then $s = f(s) = gf(s) = g(s)$.   □

It remains to solve the functional equation in Proposition 1.3 in order to obtain an explicit form of the corresponding general solution of S (the same idea applied to Proposition 1.4 yields nothing but Corollary 1.2 of Proposition 1.1).

Recall that with each function $f : A \longrightarrow B$ is associated the equivalence $\ker f$ on A defined by $x \ker f\, x' \iff f(x) = f(x')$.

Lemma 1.2. *Let $f, g : T \longrightarrow T$. Then:*

(i) The equation $f = gh$ is consistent if and only if

$$(1.5) \qquad\qquad \ker h \subseteq \ker f .$$

(ii) When condition (i) is fulfilled, the solutions g are given by

$$(1.6) \qquad\qquad g(t) = f(x) \text{ if } t = h(x) , \text{ else arbitrary} .$$

PROOF: Clearly $f = gh$ implies (1.5). Conversely, suppose condition (1.5) is fulfilled. Then formula (1.6) unambiguously defines a map $g : T \longrightarrow T$ such that $g(h(x)) = f(x)$ and conversely, if $gh = f$ then (1.6) holds.   □

Proposition 1.5. *Let f be a general solution of an equation S and $g : T \longrightarrow T$. Then g is a general solution of S if and only if it is of the form*

$$(1.7) \qquad\qquad g(t) = f(x) \text{ if } t = h(x) , \text{ else } k(x) ,$$

for some $h : T \longrightarrow T$ fulfilling (1.5) and some $k : T \backslash h(T) \longrightarrow S$.

PROOF: Immediate from Proposition 1.3 and Lemma 1.2.   □

We can supplement Proposition 1.5 by an explicit description of all the functions $h : T \longrightarrow T$ that fulfil (1.5).

Proposition 1.6. *Let $f : T \longrightarrow T$. Then there is a bijection between the functions $h : T \longrightarrow T$ that fulfil (1.5) and the pairs $(\equiv, j)$ where $\equiv$ is an equivalence relation on T included in $\ker f$ and $j : T/\!\equiv \longrightarrow T$ is an injection.*

PROOF: We associate with each function h satisfying (1.5) the pair $(\ker h, \hat{h})$, where $\hat{h} : T/\ker h \longrightarrow T$ is the injection defined by $\hat{h}(\hat{x}) = h(x)$. Conversely, we associate with each pair $(\equiv, j)$ satisfying the above conditions, the function $h : T \longrightarrow T$ defined by $h(x) = j(\hat{x})$. Then

$$(1.8) \qquad\qquad x \ker h\, y \iff \hat{x} \ker j\, \hat{y} \iff x \equiv y \implies x \ker f\, y .$$

Further, the function h_1 associated with the pair $(\ker h, \hat{h})$ coincides with h because $h_1(x) = \hat{h}(\hat{x}) = h(x)$. Finally if h is the function associated with a pair $(\equiv, j)$ then the pair $(\ker h, \hat{h})$ associated with h coincides with $(\equiv, j)$. For $\equiv$ is $\ker h$ as shown in (1.8) and $j(\hat{x}) = h(x) = \hat{h}(\hat{x})$. □

Proposition 1.7. (Banković [1979]). *Let f be a general solution of an equation S and $g : T \longrightarrow T$. Then g is a reproductive solution of S if and only if it is of the form $g = fh$ for some $h : T \longrightarrow T$ such that*
$$fh(s) = s \text{ for all } s \in S.$$

PROOF: Let g be a reproductive solution of S. For every $t \in T$ we have $g(t) \in S = f(T)$; choose $x \in T$ such that $g(t) = f(x)$ and set $h(t) = x$. Then $fh(t) = f(x) = g(t)$ and if $s \in S$ then $fh(s) = g(s) = s$. The converse is immediate.   □

The next proposition is an abstract version of the substitution method for solving a system of equations.

Proposition 1.8. (Banković [1983]). *Let S_1 and S_2 be two equations and f a general (reproductive) solution of S_1. Set $S' = \{x \in T \mid f(x) \in S_2\}$. Then:*

(i) The system $S_1 \cap S_2$ is consistent if and only if S' is consistent.

(ii) If g is a general (reproductive) solution of S', then fg is a general (reproductive) solution of $S_1 \cap S_2$.

PROOF: (i) If $s \in S_1 \cap S_2$ then $s = f(x)$ for some x and $s \in S_2$, hence $x \in S'$. Therefore condition $S' \neq \varnothing$ is necessary, while sufficiency will follow from (ii).

(ii) Set $fg = h$. Then $h(T) \subseteq f(T) = S$ and for every $t \in T$ we have $g(T) \in S'$, therefore $h(t) = f(g(T)) \in S_2$. Thus $h(T) \subseteq S_1 \cap S_2 \neq \varnothing$. To prove the converse inclusion take $s \in S_1 \cap S_2$. Then $s = f(t)$ for some t and $f(t) \in S_2$, hence $t \in S'$, therefore $t = g(u)$ for some u, which implies $s = f(g(u)) = h(u) \in h(T)$.

If the solutions f and g are reproductive, then we can take $t := s$ and $u := t$, hence $s = f(s) = f(g(s)) = h(s)$.   □

The study of general and reproductive solutions of arbitrary equations is continued by Chvalina [1987] in terms of semigroup theory. He characterizes those general solutions f of an equation S for which $f|_S$ is a permutation such that if $s_1, s_2 \in S$ then $f^m(s_1) = f^n(s_2)$ for some $m, n \in \mathbb{N}$, as well as those general solutions g of S for which $g|_S = f|_S$. He also devises a topological characterization of reproductive solutions within general solutions. Prešić [2000] generalizes the concept of reproductive solutions to equations whose unknowns are subject to constraints.

2 Equations over finite sets

This section deals with a consistent equation S over a finite set T, say $= \{t_0, t_1, \ldots, t_m\}$. It was Prešić [1971] who initiated this study and based it on two ideas. The first one is slightly generalized in the following result:

Proposition 2.1. *Suppose S is a consistent equation and with each element $x \in T$ is associated a linear order L_x on T for which x is the first element. Then the function $f : T \longrightarrow T$ such that for any $x \in T$, $f(x)$ is the first element of S with respect to L_x, is a reproductive solution of S.*

PROOF: Clearly $f(T) \subseteq S$ and $f(s) = s$ for every $s \in S$.   □

In fact Prešić associated with each $x \in T$ a cyclic permutation p_x of T and his linear order was

(2.1) $$x < p_x(x) < p_x^2(x) < \ldots < p_x^m(x) \,,$$

where the powers p^i are defined by $p^0(x) = x$, $p^{i+1}(x) = p(p^i(x))$.

Banković [1985] noted that cyclic permutations can be replaced by any function $p : T \longrightarrow T$ such that

(2.2) $$\{x, p(x), p^2(x), \ldots, p^m(x)\} = T \qquad (\forall x \in T)$$

and he used the linear order defined by $x < p(x) < \ldots < p^m(x)$.

Lemma 2.1. *Suppose S is a consistent equation. Let $p : T \longrightarrow T$ be any function satisfying (2.2) and define $b : T \longrightarrow T$ by*

(2.3) $$b(x) = x \text{ if } x \in S, \text{ else } p(x) \,.$$

Then for each $i \in \{1, \ldots, m + 1\}$, $b^i(x)$ is the first term in the sequence $x, p(x), \ldots, p^{i-1}(x)$ that belongs to S if the sequence does contain elements of S, otherwise $b^i(x) = p^i(x)$.

PROOF: The property holds for $i := 1$. Now we suppose the property true for i and prove it for $i + 1$.

If $k \leq i - 1$ is the first exponent such that $p^k(x) \in S$, then $b^i(x) = p^k(x) \in S$, hence $b^{i+1}(x) = b(b^i(x)) = b^i(x) = p^k(x)$ is the first term of the sequence $x, p(x), \ldots, p^i(x)$ that belongs to S.

If none of the elements $x, p(x), \ldots, p^{i-1}(x)$ belongs to S, then $b^i(x) = p^i(x)$. If $p^i(x) \in S$, then $b^{i+1}(x) = b^i(x) = p^i(x)$ is the first term of the sequence $x, p(x), \ldots, p^i(x)$ that belongs to S; otherwise $b^{i+1}(x) = p(b^i(x)) = p^{i+1}(x)$.   □

Proposition 2.2. (Banković [1985]). *Under the hypotheses of Lemma* 2.1, b^{m+1} *is a reproductive solution of S.*

PROOF: It follows from Lemma 2.1 via (2.2) and $S \neq \emptyset$ that $b^{m+1}(x)$ is the first term of the sequence $x, p(x), \ldots, p^m(x)$ that belongs to S, whence the conclusion follows by Proposition 2.1.   □

As a matter of fact, Prešić and Banković worked with equations of the form (1.1), while Ghilezan [1970] noted that this restriction is not necessary.

The following result will be needed in the sequel:

Lemma 2.2. (Banković [1990]). *A function $f : T \longrightarrow T$ is a general solution of an equation S if and only if it fulfils*

(i) $f(T) \subseteq S$, *and*

(ii) there is a permutation β of $\{0, 1, \ldots, m\}$ such that

$$(2.4) \qquad t_{\beta(k)} \in S \Longrightarrow f(t_k) = t_{\beta(k)} \qquad (k = 0, 1, \ldots, m) \,.$$

PROOF: Sufficiency is obvious. Conversely, suppose f is a general solution. Define a map $\varphi_0 : S \longrightarrow T$ by choosing, for each element $t \in S$, an element $\varphi_0(t)$ such that $t = f(\varphi_0(t))$. Then φ_0 is obviously injective and it follows immediately that φ_0 can be extended to a bijection $\varphi : T \longrightarrow T$. The defining property of φ_0 implies $t_h = f(\varphi(t_h))$ for all $t_h \in S$. The last property can be written in the form (2.4) for the permutation β of $\{0, 1, \ldots, m\}$ defined by $\beta(k) = h \Longleftrightarrow \varphi(t_h) = t_k$. □

The second main idea in Prešić [1971] was the introduction of an algebraic structure of the set T. Namely, Prešić took two elements 0,1 which may be outside T or may be two distinguished elements of T and he defined two binary partial operations $+$ and $\cdot$ on $T \cup \{0, 1\}$ by the following rules:

$$(2.5) \qquad x + 0 = 0 + x = x, \; x \cdot 0 = 0 \cdot x = 0, \; x \cdot 1 = 1 \cdot x = x \,,$$

for every $x \in T \cup \{0, 1\}$, the operation $\cdot$ being ususally denoted by concatenation. He also defined

$$(2.6) \qquad x^y = 1 \text{ if } x = y \,, \text{ else } 0 \,,$$

for every $x, y \in \{0, 1\}$, as well as a family $\sum_{i=0}^{n}$ of partial operations given by

$$(2.7) \qquad \sum_{i=0}^{0} x_i = x_0, \; \sum_{i=0}^{n+1} x_i = \left(\sum_{i=0}^{n} x_i\right) + x_{n+1} \qquad (n \in \mathbf{N}) \,.$$

As a matter of fact, Prešić was seemingly unaware that this algebraic structure was first discovered by Post [1921]; cf. Proposition 13.2.1.

Proposition 2.3. *Every function $f : T \longrightarrow T$ can be written in the form*

$$(2.8) \qquad f(x) = \sum_{i=0}^{m} f_i x^{t_i} \qquad (\forall x \in T) \,,$$

where the coefficients are uniquely determined by

$$(2.9) \qquad f_k = f(t_k) \qquad (k = 0, 1, \ldots, m) \,.$$

PROOF: For each $k = 0, 1, \ldots, m$, the equality

$$(2.10) \qquad f(t_k) = \sum_{i=0}^{m} f(t_k) t_k^{t_i}$$

holds because the right-hand side of (2.10) contains the term $f(t_k) t_k^{t_k} = f(t_k)$, while the other terms are 0. Conversely, (2.8) implies, for each $k = 0, 1, \ldots, m$,

$$f(t_k) = \sum_{i=0}^{m} f_i t_k^{t_i} = f_k t_k^{t_k} = f_k.$$

□

Prešić [1988] initiated the study of the equation

(2.11) $$a_0 x^{t_0} + a_1 x^{t_1} + \ldots + a_m x^{t_m} = 0,$$

where $a_0, a_1, \ldots, a_m \in \{0,1\}$ and the unknown $x \in T$. Equation (2.11) generalizes certain Boolean equations and Post equations.

Let us denote by $S(a_0, \ldots, a_m)$ the (possibly empty!) set of solutions to equation (2.11); according to the discussion in §1, $S(a_0, \ldots, a_m)$ may be identified with equation (2.11) itself.

Remark 2.1. (Prešić [1988]). For $x := t_k$ the left member of equation (2.11) reduces to a_k, hence

(2.12) $$t_k \in S(a_0, \ldots, a_m) \Longleftrightarrow a_k = 0 \qquad (k = 0, 1, \ldots, m)$$

therefore equation (2.11) is consistent if and only if $a_k = 0$ for some k. In other words, taking also into account the associativity of the operation $\cdot$ on $\{0,1\}$, we see that *the consistency condition for equation* (2.11) *is*

(2.13) $$a_0 a_1 \ldots a_m = 0.$$

□

Unlike the previous results, those which we are going to report below seem to be helpful in the actual construction of a general or of a reproductive solution of equation (2.11).

Prešić [1988] described all reproductive solutions of equation (2.11), while Banković [1992b] gave a simpler proof of that result. Two characterizations of the general solution of equation (2.11) were obtained by Banković [1990], [1993b]. Propositions 2.4 and 2.5 below provide simpler forms of the general solutions and reproductive solutions, respectively; cf. Rudeanu [1998b].

Lemma 2.3. *In the case of equation* (2.11) *the implication* (2.4) *in Lemma 2.2 is equivalent to*

(2.14) $$f(t_k) = t_{\beta(k)} a^0_{\beta(k)} + f(t_k) a^1_{\beta(k)} \qquad (k = 0, 1, \ldots, m).$$

PROOF: It follows from (2.14) that

(2.15) $$a_{\beta(k)} = 0 \Longrightarrow f(t_k) = t_{\beta(k)} \qquad (k = 0, 1, \ldots, m);$$

but (2.15) is equivalent to (2.4) by (2.12). Conversely, suppose (2.4) holds. Then (2.14) is checked immediately by considering the cases $a_{\beta(k)} = 1$ and $a_{\beta(k)} = 0$ and using again (2.12). □

Proposition 2.4. *A function* $f : T \longrightarrow T$ *is a general solution of equation* (2.11) *if and only if it is of the form*

$$(2.16) \qquad f(x) = \sum_{k=0}^{m} (t_{\beta(k)} a_{\beta(k)}^0 + f_k a_{\beta(k)}^1) x^{t_k} \qquad (\forall x \in T),$$

where β *is a permutation of* $\{0, 1, \ldots, m\}$ *and*

$$(2.17) \qquad f_k \in S(a_0, \ldots, a_m) \qquad (k = 0, 1, \ldots, m).$$

PROOF: Suppose (2.16) and (2.17) hold. Consider an arbitrary but fixed $k \in \{0, 1, \ldots, m\}$. Taking into account Proposition 2.3, it follows that if $a_{\beta(k)} = 1$ then $f(t_k) = f_k \in S(a_0, \ldots, a_m)$, while

$$t_{\beta(k)} \in S(a_0, \ldots, a_m) \Longleftrightarrow a_{\beta(k)} = 0 \Longrightarrow f(t_k) = t_{\beta(k)}.$$

Therefore f is a general solution by Lemma 2.2. Conversely, necessity follows from Proposition 2.3 and Lemmas 2.2 and 2.3.   □

Proposition 2.5. *A function* $f : T \longrightarrow T$ *is a reproductive solution of equation* (2.11) *if and only if it is of the form*

$$(2.18) \qquad f(x) = \sum_{i=0}^{m} (t_k a_k^0 + f_k a_k^1) x^{t_k} \qquad (\forall x \in T),$$

where

$$(2.17) \qquad f_k \in S(a_0, \ldots, a_m) \qquad (k = 0, 1, \ldots, m).$$

PROOF: It follows from Propositions 2.4 and 2.3 that any function satisfying (2.18) and (2.17) is a general solution of equation (2.11) and

$$(2.19) \qquad f(t_k) = t_k a_k^0 + f_k a_k^1 \qquad (k = 0, 1, \ldots, m)$$

therefore if $t_k \in S(a_0, \ldots, a_m)$ then $f(t_k) = t_k$ because $a_k = 0$. Conversely, suppose f is a reproductive solution of equation (2.11). Then f satisfies conditions (i) and (ii) in Lemma 2.2 with the identity in the rôle of β. Taking also into account Lemma 2.3 we see that relations (2.17) and (2.19) hold for $f_k := f(t_k)$. The proof is completed by Proposition 2.3.   □

   The results in Prešić [1988], Banković [1990], [1993b], as well as Propositions 2.4 and 2.5, enable us to actually construct a general or a reproductive solution of equation (2.11) provided we know a priori at least a solution of the equation (for instance, we may take all f_k occurring in Proposition 2.4 or 2.5 equal to that particular solution). Proposition 2.6 below does not require the knowledge of a particular solution, the only hypothesis being the consistency of the equation.

Lemma 2.4. (Banković [1995a]). *For any vector* $(b_0, \ldots, b_m) \in \{0, 1\}^{m+1}$, *the vector* $(c_0, \ldots, c_m)$ *defined by*

(2.20) $c_i = b_0 \ldots b_{i-1} b_i^0$ $(i = 0, 1, \ldots, m-1)$, $c_m = b_0 \ldots b_{m-1} b_m$

has m components equal to 0 and one component equal to 1.

PROOF: If $b_0 = \ldots = b_m = 1$, then $c_0 = \ldots = c_{m-1} = 0$ and $c_m = 1$. Otherwise let k be the first index such that $b_k = 0$. Then $b_0 = \ldots = b_{k-1} = 1$, hence $c_k = 1$ and the terms $c_0, \ldots, c_{k-1}$ are 0 because their last factor is 0, while the terms $c_{k+1}, \ldots, c_m$ are 0 because they contain the factor b_k. $\square$

Definition 2.1. A vector $(b_0, \ldots, b_m) \in \{0,1\}^{m+1}$ and a vector $(s_0, \ldots, s_m) \in T^{m+1}$ will be called *equivalent* provided there is a permutation β of $\{0, 1, \ldots, m\}$ such that

(2.21) $b_k = a_{\beta(k)}$ *and* $s_k = t_{\beta(k)}$ $(k = 0, 1, \ldots, m)$.

$\square$

Proposition 2.6. (Banković [1995a]). *Suppose equation (2.11) is consistent. A function $f : T \longrightarrow T$ is a general (reproductive) solution of equation (2.11) if and only if it is of the form*

$$f(x) = \sum_{k=0}^{m} (a_{k0}^0 t_{k0} + a_{k0} a_{k1}^0 t_{k1} + \ldots +$$
(2.22) $$a_{k0} a_{k1} \ldots a_{k,m-2} a_{k,m-1}^0 t_{k,m-1}$$
$$+ a_{k0} a_{k1} \ldots a_{k,m-2} a_{k,m-1} t_{km}) x^{t_k} \quad (\forall x \in T) ,$$

where for each $k = 0, 1, \ldots, m$, the vectors $(a_{k0}, \ldots, a_{km}) \in \{0,1\}^{m+1}$ and $(t_{k0}, \ldots, t_{km}) \in T^{m+1}$ are equivalent, and where $(a_{00}, a_{10}, \ldots, a_{m0})$ is a permutation of $(a_0, \ldots, a_m)$ (and where $(a_{00}, a_{10}, \ldots, a_{m0}) = (a_0, \ldots, a_m)$).

PROOF: Suppose f is represented in the form (2.22) and satisfies the first variant of the above conditions. Take an arbitrary but fixed $k \in \{0, 1, \ldots, m\}$; we know by Proposition 2.3 that the coefficient of x^{t_k} in (2.22) is $f(t_k)$. Since $a_0 \ldots a_m = 0$ by Remark 2.1 and $(a_{k0}, \ldots, a_{km})$ is a permutation of $(a_0, \ldots, a_m)$, there is a first index h such that $a_{kh} = 0$. Then $a_{k0} \ldots a_{k,h-1} a_{kh}^0 t_{kh} = t_{kh}$ and it follows from Lemma 2.4 that the other terms in the coefficient of x^{t_k} are 0, hence $f(t_k) = t_{kh}$. But the equivalence of the vectors $(a_{k0}, \ldots, a_{km})$ and $(t_{k0}, \ldots, t_{km})$ implies that $a_{kh} (= 0)$ is the coefficient of $x^{t_{kh}}$ in equation (2.11), therefore $t_{kh} \in S(a_0, \ldots, a_m)$. We have thus proved that $f(T) \subseteq S(a_0, \ldots, a_m)$.

Further, let $t_h \in S(a_0, \ldots, a_m)$. Since $(a_{00}, \ldots, a_{m0})$ is a permutation of $(a_0, \ldots, a_m)$, there is an index k such that $a_h = a_{k0}$, which implies $t_h = t_{k0}$. Since t_h is a solution of equation (2.11), it follows that $a_h = 0$, therefore $f(t_k) = a_{k0}^0 t_{k0}$ again by Lemma 2.4 and Proposition 2.3. So $f(t_k) = t_h$.

If the stronger hypothesis $(a_{00}, \ldots, a_{m0}) = (a_o, \ldots, a_m)$ is fulfilled, then in the above argument we have $k = h$, hence $f(t_h) = t_h$, showing that f is a reproductive solution.

Now suppose f is a general solution. Let β be the permutation provided by Lemma 2.2. Take again an arbitrary but fixed $k \in \{0, 1, \ldots, m\}$. Set $a_{k0} = a_{\beta(k)}$ and $t_{k0} = t_{\beta(k)}$. If $a_{\beta(k)} = 0$, complete the permutation $(t_{k0}, t_{k1}, \ldots, t_{km})$ arbitrarily and let $(a_{k0}, \ldots, a_{km})$ be the vector of $\{0,1\}^{m+1}$ equivalent to

$(t_{k0}, \ldots, t_{km})$. Note that $t_{\beta(k)} \in S(a_0, \ldots, a_m)$ and $a_{k0}^0 t_{k0} = t_{\beta(k)} = f(t_k)$ by Lemma 2.2. If $a_{\beta(k)} = 1$, take the index h defined by $f(t_k) = t_h$ and set $a_{k1} = a_h$, $t_{k1} = t_h$, complete the permutation $(t_{k0}, t_{k1}, \ldots, t_{km})$ arbitrarily and let $(a_{k0}, \ldots, a_{km})$ be the vector of $\{0, 1\}^{m+1}$ equivalent to $(t_{k0}, \ldots, t_{km})$. Note that $a_{k0} a_{k1}^0 t_{k1} = t_h = f(t_k)$.

So we have constructed pairs of equivalent vectors

$$(a_{k0}, \ldots, a_{km}), \; (t_{k0}, \ldots, t_{km})$$

for $k = 0, 1, \ldots, m$. Moreover, $(a_{00}, a_{10}, \ldots, a_{m0}) = (a_{\beta(0)}, a_{\beta(1)}, \ldots, a_{\beta(m)})$ is a permutation of $(a_0, \ldots, a_m)$. Let $g(x)$ be the right-hand side of equation (2.22). Taking into account the above discussion and Lemma 2.4, we see that for each $k = 0, 1, \ldots, m$, if $a_{\beta(k)} = 0$ then $g(t_k) = a_{k0}^0 t_{k0} = f(t_k)$, while if $a_{\beta(k)} = 1$ then $g(t_k) = a_{k0} a_{k1}^0 t_{k1} = f(t_k)$. Thus $g = f$.

Finally suppose f is a reproductive solution of equation (2.11). Then in the above construction we take the identity in the rôle of the permutation β from Lemma 2.2 and obtain $a_{k0} = a_k$ ($k = 0, 1, \ldots, m$), that is, $(a_{00}, a_{10}, \ldots, a_{m0}) = (a_0, \ldots, a_m)$. $\square$

Banković [1997a] generalized Proposition 2.6 for the n-unknown analogue of equation (2.11).

The next proposition and its corollary do not refer to finite sets, but still use the partial algebraic structure (2.5),(2.6).

Proposition 2.7. (Banković [????]). *Let T be a set equipped with the partial operations (2.5),(2.6), $r : T \longrightarrow \{0, 1\}$ and $g : T \longrightarrow T$ a general solution of the equation $r(x) = 1$. Then a function $f : T \longrightarrow T$ is a general solution of the equation $r(x) = 1$ if and only if there exist $\alpha, \beta : T \longrightarrow T$, α surjective, such that*

$$(2.23) \qquad f(t) = r(\alpha(t)) \cdot \alpha(t) + r'(\alpha(t)) \cdot g(\beta(t)) \qquad (\forall t \in T) .$$

Moreover, one can choose α to be a bijection.

PROOF: Let S be the solution set.

Suppose the identity (2.23) holds for some α, β. Take $y \in T$. If $\alpha(y) \in S$ then $f(y) = \alpha(y) \in S$, otherwise $f(y) = g(\beta(y)) \in S$. Now take $x \in S$. Since $x = \alpha(t)$ for some $t \in T$, it follows that

$$f(t) = r(x) \cdot x + r'(x) \cdot g(\beta(t)) = x .$$

Conversely, suppose f is a general solution.

Then each $x \in S$ is of the form $x = f(t)$ for some $t \in T$, therefore the axiom of choice implies the existence of a function $h : S \longrightarrow T$ such that $x = f(h(x))$ for all $x \in S$. Clearly h is an injection, so that we can extend it to obtain a bijection $\varphi : T \longrightarrow T$. Set $\alpha = \varphi^{-1}$. Similarly, for each $t \in T$, since $f(t) \in S$, it follows that $f(t) = g(u)$ for some $u \in T$, which yields a function $\beta : T \longrightarrow T$ such that $f(t) = g(\beta(t))$ for all $t \in T$.

Now for any $t \in T$, if $\alpha(t) = x \in S$ then

$$r(\alpha(t))\alpha(t) + r'(\alpha(t))g(\beta(t)) = \alpha(t) = x = f(h(x)) = f(\varphi(x)) = f(t) \,,$$

while if $\alpha(t) \notin S$ then

$$r(\alpha(t))\alpha(t) + r'(\alpha(t))g(\beta(t)) = g(\beta(t)) = f(t) \,.$$

□

Corollary 2.1. (Prešić [1972]; see also Proposition 5.3.11). *Same hypotheses on T, r and g. Then a function $f : T \longrightarrow T$ is a reproductive solution of the equation $r(x) = 1$ if and only if there is a function $\beta : T \longrightarrow T$ such that*

(2.24) $$f(t) = r(t) \cdot t + r'(t) \cdot g(\beta(t)) \qquad (\forall t \in T) \,.$$

PROOF: If the identity (2.24) holds, then f is a general solution by the previous proposition. Besides, $f(x) = x$ for every solution x.

Conversely, suppose f is a reproductive solution. Then in the proof of the proposition we can take h to be the inclusion of S in T and extend it to $\varphi := 1|_S$, therefore $\alpha = 1|_S$ as well. □

2. Universal algebra

In this chapter we recall the elements of universal algebra that will be used in this book. We insist on the universal-algebraic concepts of polynomial and algebraic function, because they generalize lattice functions. Partial algebras and multi-sorted algebras are disregarded, as being not relevant to the subject matter of this book. For the proofs that are omitted the reader is referred to any standard book on universal algebra, such as Burris and Sankappanavar [1981], Cohn [1965], Grätzer [1979], Chapter 0 in Henkin, Monk and Tarski [1971] or Pierce [1968].

1 First concepts and subdirect decompositions

In this section we recall the most elementary concepts and results of universal algebra and we conclude with a representation theorem.

The abstract concept of *algebra* is defined as follows: a pair $\mathbf{A} = (A; F)$, where A is a set called the *support* of the algebra and $F = (f_i)_{i \in I}$ is a family of *operations* $f_i : A^{n(i)} \longrightarrow A$, $i \in I$. For each $i \in I$, the natural number $n(i)$ is called the *arity* of the operation f_i. In particular for some $i \in I$ we may have $n(i) = 0$, in which case f_i is a *zero-ary* operation (it depends on no arguments) and can be identified with an element $f_i \in A$. The family $\Sigma = (n(i))_{i \in I}$ of arities is called the *type* or the *signature* of the algebra $\mathbf{A}$ and we say that $\mathbf{A}$ is an *algebra of type* Σ or a $\Sigma - algebra$. Two algebras of the same type are said to be *similar*.

Clearly the conventional structures studied in algebra, such as monoids, groups, rings, lattices etc., fall within the concept of algebra defined above and it is customary to represent the family of operations and the type as lists corresponding to an arbitrary but fixed linear ordering of the set I. For instance, a group can be viewed as an algebra $(A; \cdot, ^{-1}, e)$ of type $(2, 1, 0)$ and also as an algebra $(A; \cdot)$ of type (2).

If $\mathbf{A} = (A; F)$ is an algebra and $S \subseteq A$, then S is said to be *closed* provided

$$(1.1) \qquad \forall i \in I \; \forall x_1, \ldots, x_{n(i)} \in S \; \; f_i(x_1 \ldots, x_{n(i)}) \in S \,,$$

which implies in particular

(1.1′) $\forall i \in I \quad n(i) = 0 \Longrightarrow f_i \in S$.

A closed set S endowed with the restrictions of the operations f_i to $S^{n(i)}$, $i \in I$, becomes an algebra $(S; (f_i|_{S^{n(i)}})_{i \in I})$ of the same type as $\mathbf{A}$ and which is referred to as a *subalgebra* of $\mathbf{A}$.

It is a frequent and beneficial abuse of notation to use the same symbol A for an algebra $\mathbf{A} = (A; F)$ and for its support A. It is also customary and recommended to use the term *subalgebra* both for the closed set S and for the associated algebra having the support S.

For instance, if A is a group, the subalgebras of the algebra $(A; \cdot, ^{-1}, e)$ coincide with the subgroups of A, every subgroup of A is also a subalgebra of the algebra $(A; \cdot)$, but the converse does not hold. Consider e.g. the additive group $\mathbf{Z}$; then $\mathbf{N}$ and $\emptyset$ are subalgebras of the algebra $(\mathbf{Z}; +)$ but not subgroups of $\mathbf{Z}$.

Clearly any intersection of subalgebras of an algebra A is also a subalgebra of A. This implies that for any subset X of an algebra A, the intersection of all the subalgebras that include X is the least subalgebra that includes X. This intersection is called the *subalgebra generated by* X and will be denoted by $< X >_A$ or simply by $< X >$.

An alternative *recursive definition* for $< X >$ is the following:

(i) the elements of X are in $< X >$;

(ii) for every $i \in I$, if $x_1, \ldots, x_{n(i)} \in < X >$, then $f_i(x_1, \ldots, x_{n(i)}) \in < X >$;

(iii) every element of $< X >$ is obtained by applying rules (i) and (ii) finitely many times.

An important consequence of this recursive definition is the following *compactness property*: for every element $z \in < X >$, there exists a finite subset $X_z \subseteq X$ such that $z \in < X_z >$.

The recursive definition of $< X >$ also yields a method for proving that a certain property holds for every element of $< X >$. This method is known as *algebraic induction* and it comprises two steps:

(j) one proves the property for the elements of X, and

(jj) for each $i \in I$, supposing that the property holds for the elements $x_1, \ldots, x_{n(i)} \in < X >$, one proves that the element $f_i(x_1, \ldots, x_{n(i)})$ has the property, too.

A *homomorphism* $\varphi : A \longrightarrow B$ from the algebra $(A; F)$ to the similar algebra $(B; G)$ is a map $\varphi : A \longrightarrow B$ such that for every $i \in I$ and every $x_1, \ldots, x_{n(i)} \in A$,

(1.2) $\varphi(f_i(x_1, \ldots, x_{n(i)})) = g_i(\varphi(x_1), \ldots, \varphi(x_{n(i)}))$,

which implies in particular

(1.2′) $\forall i \in I \quad n(i) = 0 \Longrightarrow \varphi(f_i) = g_i$.

Let $\mathrm{Hom}_{\Sigma}(\mathbf{A}, \mathbf{B})$ or simply $\mathrm{Hom}(A, B)$ denote the set of all homomorphisms from A to B.

If $n(i) = 2$ and the customary infixed notation is used, say $f_i(x_1, x_2) = x_1 * x_2$ and $g_i(y_1, y_2) = y_1 \bullet y_2$, then the corresponding condition (1.2) becomes the familiar condition $\varphi(x_1 * x_2) = \varphi(x_1) \bullet \varphi(x_2)$.

An *isomorphism* is a bijective homomorphism. It is easily seen that if $\varphi : A \longrightarrow B$ is an isomorphism then $\varphi^{-1} : B \longrightarrow A$ is also an isomorphism. We indicate this situation by $A \cong B$ and we say that the algebras A and B are *isomorphic*. A homomorphism (an isomorphism) $\varphi : A \longrightarrow A$ is called an *endomorphism* (*automorphism*) of A. The composite $\psi\varphi$ of two homomorphisms (isomorphisms), provided it exists, is a homomorphism (an isomorphism) as well. The *identity mapping* $1_A : A \longrightarrow A$ of any algebra is an automorphism.

It is easy to check that if A, B are Σ-algebras and $\varphi : A \longrightarrow B$ is a homomorphism, then

(1.3) $\qquad\qquad$ S is a subalgebra of $A \Longrightarrow \varphi(S)$ is a subalgebra of B ,

(1.4) $\qquad\qquad$ T is a subalgebra of $B \Longrightarrow \varphi^{-1}(T)$ is a subalgebra of A .

This implies that if $X \subseteq A$ then

(1.5) $\qquad\qquad\qquad$ $\varphi(< X >_A) = < \varphi(X) >_B$.

For $\varphi(< X >_A)$ is a subalgebra of B by (1.3), $\varphi(X) \subseteq \varphi(< X >_A)$ because $X \subseteq < X >_A$, and if T is a subalgebra of B such that $\varphi(X) \subseteq T$, then $\varphi(< X >_A) \subseteq T$ because $X \subseteq \varphi^{-1}(T)$ which is a subalgebra of A by (1.4), hence $< X >_A \subseteq \varphi^{-1}(T)$, therefore $\varphi(< X >_A) \subseteq \varphi(\varphi^{-1}(T)) \subseteq T$.

A *congruence* of an algebra A is an equivalence relation $\sim$ on A such that

(1.6) $\quad$ $\begin{aligned} &\forall i \in I\ n(i) > 0 \Longrightarrow \forall x_k, y_k \in A\ (k = 1, \ldots, n(i)) : \\ &x_k \sim y_k \quad (k = 1, \ldots, n(i)) \implies f_i(x_1, \ldots, x_{n(i)}) \sim \\ &f_i(y_1, \ldots, y_{n(i)}) . \end{aligned}$

Every algebra A has at least two congruences: the *equality* on A, also known as the *diagonal* $\Delta_A = \{(x, x) \mid x \in A\}$ of A, and the *universal congruence* $\omega_A = A \times A$, which identifies every two elements of A.

The *kernel* of a homomorphism $\varphi : A \longrightarrow B$, i.e., the relation $\ker \varphi$ defined on A by

(1.7) $\qquad\qquad\qquad$ $x \ker \varphi\ y \Longleftrightarrow \varphi(x) = \varphi(y)$

is a congruence of the algebra A and it will be seen below that every congruence is of this form.

If $\sim$ is a congruence of the algebra $(A; F)$, then the quotient set $\hat{A} = A/\sim$ can be made into an algebra $\hat{A} = A/\sim = (\hat{A}, \hat{F})$ of the same type, where the operations from $\hat{F} = (\hat{f}_i)_{i \in I}$ are defined as follows: for every $i \in I$ and every $[x_1], \ldots, [x_{n(i)}] \in \hat{A}$,

(1.8) $\qquad\qquad$ $\hat{f}_i([x_1], \ldots, [x_{n(i)}]) = [f_i(x_1, \ldots, x_{n(i)})]$,

where $[x]$ stands for the coset of the element $x \in A$ modulo $\sim$. Property (1.6) ensures that the operations (1.8) are well defined. The algebra $A/\sim$ is known as the *quotient algebra* of A modulo $\sim$.

Note that definition (1.8) expresses the fact that the *natural surjection*

(1.9) $\mathrm{nat} \sim : A \longrightarrow A/\sim$, $\mathrm{nat} \sim (x) = [x]\ \forall x \in A$,

is a homomorphism. Note also that

(1.10) $\sim = \ker \mathrm{nat} \sim$,

which shows that every congruence is the kernel of a homomorphism.

A typical example of a quotient algebra is the ring $\mathbf{Z}_n = \mathbf{Z}/(\mathrm{mod}\, n)$ of integers modulo n. In group theory congruences appear as normal subgroups. More exactly, the assignments

$$H \mapsto \sim_H ,\ \forall x,y \in G\ (x \sim_H y \Longleftrightarrow xy^{-1} \in H) ,$$

$$\sim \mapsto H_\sim = \{x \in G \mid x \sim e\} ,$$

establish a bijection between the normal subgroups H of a group $(G; \cdot, ^{-1}, e)$ and its congruences $\sim$ in the sense (1.6). The quotient group G/H is in fact the quotient algebra $G/\sim_H$ in the sense (1.8). Similarly, in ring theory congruences appear as two-sided ideals.

The *homomorphism theorem* (also called the *first isomorphism theorem*) states that if $\varphi : A \longrightarrow B$ is a surjective homomorphism, then B is isomorphic to $A/\ker \varphi$:

(1.11) $\varphi \in \mathrm{Hom}(A, B)$ & φ surjection $\Longrightarrow B \cong A/\ker \varphi$.

To be specific, the isomorphism $\alpha : A \longrightarrow A/\ker \varphi$ satisfies $\alpha \varphi = \mathrm{nat}\ \ker \varphi$, i.e.,

(1.12) $\alpha(\varphi(x)) = [x]\ (\forall x \in A)$.

The following definition generalizes direct products of groups, of rings etc. Given a family of algebras $(A_t; F^t)$, $F^t = (f_i^t)_{i \in I}$, $t \in T$, their *direct product* is the algebra $(\Pi_{t \in T} A_t; (f_i)_{i \in I})$, where

(1.13)
$$\forall i \in I\ f_i : (\Pi_{t \in T} A_t)^{n(i)} \longrightarrow \Pi_{t \in T} A_t ,$$
$$f_i((x_{1t})_{t \in T}, \ldots, (x_{n(i)t})_{t \in T}) = (f_i^t(x_{1t}, \ldots, x_{n(i)t}))_{t \in T}$$

for every $(x_{1t})_{t \in T}, \ldots, (x_{n(i)t})_{t \in T} \in \Pi_{t \in T} A_t$.

A *subdirect product* of a family $(A_t)_{t \in T}$ of Σ-algebras is a subalgebra S of $\Pi_{t \in T} A_t$ such that for each $t \in T$ the homomorphism $p_t \iota$ is surjective, where ι and p_t denote inclusion and canonical projection, respectively:

$$S \xrightarrow{\iota} \Pi_{t \in T} A_t \xrightarrow{p_t} A_t .$$

A *decomposition* of an algebra A *as a subdirect product* of a family $(A_t)_{t \in T}$ of algebras, or a *subdirect decomposition* of A for short, is an isomorphism φ from A to a subdirect product of the family $(A_t)_{t \in T}$. In other words, $A \cong \varphi(A)$, where $\varphi : A \longrightarrow \Pi_{t \in T} A_t$ is an injective homomorphism such that $p_t \varphi$ are surjections for all $t \in T$. The decomposition is called *proper* if none of the homomorphisms

$p_t\varphi$, $t \in T$, is an isomorphism. An algebra is said to be *subdirectly irreducible* if it has n o proper subdirect decomposition.

If $\varphi : A \longrightarrow \Pi_{t \in T} A_t$ is a subdirect decomposition, then $\bigcap_{t \in T} \ker(p_t\varphi) = \Delta_A$. Conversely, with every family of congruences $(\theta_t)_{t \in T}$ of A such that $\bigcap_{t \in T} \theta_t = \Delta_A$ is associated a subdirect decomposition $\varphi : A \longrightarrow \Pi_{t \in T}(A/\theta_t)$ for which $\theta_t = \ker(p_t\varphi) \; \forall t \in T$.

An algebra A is subdirectly irreducible iff either A is a singleton or the intersection of all its *proper congruences*, i.e., of all congruences distinct from Δ_A, is itself a proper congruence. In other words, *an algebra A is subdirectly irreducible iff it is not a singleton and every two distinct elements are separated by a proper congruence*, i.e., if $a, b \in A$ and $a \neq b$ then there is a congruence $\theta \neq \Delta_A$ such that $(a, b) \notin \theta$.

A theorem due to Birkhoff states that *every algebra has a subdirect decomposition into subdirectly irreducible algebras*.

2 Term algebra, identities and polynomials

In this section we present the essentials of the universal algebra viewpoint on the familiar concept of identity.

Let $\Sigma = (n(i))_{i \in I}$ and let $X = \{x_j \mid j \in J\}$ be a set of elements called *variables*; we assume that $j \neq j' \Longrightarrow x_j \neq x_{j'}$. Unless otherwise stated, Σ and X are fixed in this and the subsequent section.

Recall the notation Z^* for the set of all the *words* over an alphabet Z; in particular λ will denote the *empty word*. Let $F = \{F_i \mid i \in I\}$ be a set of elements called *operator symbols*; we assume that $i \neq i' \Longrightarrow F_i \neq F_{i'}$ and $X \cap F^* = X^* \cap F = \emptyset$. Set further

$$(2.1) \qquad Z = X \cup F = \{x_j \mid j \in J\} \cup \{F_i \mid i \in I\}$$

and associate with each $i \in I$ an operation of arity $n(i)$ on Z^*, which will be denoted by the same symbol F_i, as follows:

$$(2.2) \qquad F_i : (Z^*)^{n(i)} \longrightarrow Z^* \; , \; F_i(w_1, \ldots, w_{n(i)}) = F_i w_1 \ldots w_{n(i)} \; ,$$

for every $w_1, \ldots, w_{n(i)} \in Z^*$, where $F_i w_1 \ldots w_{n(i)}$ means the concatenation of the letter F_i with the words $w_1, \ldots, w_{n(i)}$. This makes Z^* into a Σ−algebra $(Z^*; (F_i)_{i \in I})$.

The *term algebra $T_\Sigma(X)$ of type Σ over the variables X* is by definition the subalgebra of Z^* generated by X, i.e.,

$$(2.3) \qquad T_\Sigma(X) = <X>_{Z^*} \; ,$$

which makes sense because the set Z (and hence X) can be viewed as a subset of Z^*. The elements of $T_\Sigma(X)$ are called *terms of type Σ in the variables X*, or *well-formed expressions*.

Remark 2.1. The recursive definition of $< X >$ given in §1 yields the following *alternative definition of terms*:
  (i) the variables are terms ;
  (ii) for every $i \in I$, if $w_1, \ldots, w_{n(i)}$ are terms, then $F_i w_1 \ldots w_{n(i)}$ is a term ;
  (iii) every term is obtained by applying rules (i),(ii) finitely many times. □

Remark 2.2. The trivial property that *every term w involves finitely many variables (possibly none)*, namely the letters from X in the word w, is a particular case of the compactness property of subalgebras (cf. §1). □

The concept of term is the rigorous formulation of the intuitive idea of expression for the elements of an algebraic struture. Consider, for instance, the type (2,1,0). Then any three-element set can be chosen in the rôle of I. Instead of the impersonal notation F_i let us choose the symbols M,D,E, reminiscent of multiplication, (unary) division and the unit e, respectively, in a group. Take $x, y, z \in X$; then Mxy, E, Dz, $MxDMyz$ are examples of terms reminiscent of the elements $x \cdot y$, e , z^{-1} , $x \cdot (y \cdot z)^{-1}$, respectively, in a group $(G; \cdot, ^{-1}, e)$, where $x, y, z \in G$.

The term algebra $T_\Sigma(X)$ has the fundamental property of being *absolutely free*. This means that for every Σ−algebra A and every map $\varphi : X \longrightarrow A$ there exists a unique homomorphism $\overline{\varphi} : T_\Sigma(X) \longrightarrow A$ such that $\overline{\varphi}|_X = \varphi$. In particular $T_\Sigma(\emptyset)$ has the stronger property of being the *initial algebra* of type Σ, which means that for every Σ−algebra A there exists a unique homomorphism $\Phi : T_\Sigma(\emptyset) \longrightarrow A$.

Let A be a Σ−algebra. By an *identity of A in the variables X* (or simply an *identity of A*) is meant a pair $(w, w') \in (T_\Sigma(X))^2$ such that $\varphi(w) = \varphi(w')$ for every $\varphi \in \mathrm{Hom}(T_\Sigma(X), A)$. Let

(2.4)
$$\mathrm{Id}_{\Sigma, X}(A) = \bigcap\{\ker \varphi \mid \varphi \in \mathrm{Hom}(T_\Sigma(X), A)\}$$
$$= \{(w, w') \in (T_\Sigma(X))^2 \mid \varphi(w) = \varphi(w') \;\forall \varphi \in \mathrm{Hom}(T_\Sigma(X), A)\}$$

or simply IdA, be the set of all the identities of A.

Let further **K** be a class of Σ−algebras. By an *identity of* **K** is meant an identity common to all the algebras of **K**. The set

(2.5)
$$\mathrm{Id}_{\Sigma, X}(\mathbf{K}) = \bigcap_{A \in \mathbf{K}} \mathrm{Id}_{\Sigma, X}(A)$$
$$= \bigcap\{\ker \varphi \mid \varphi \in \mathrm{Hom}(T_\Sigma(X), A) , \; A \in \mathbf{K}\}$$

of all identities of **K** is also denoted simply by Id**K**.

For instance, consider again the type (2,1,0) and the terms $w = MMxyz$ and $w' = MxMyz$, where $x, y, z \in X$. If $(G; \cdot, ^{-1}, e)$ is a group and $\varphi \in \mathrm{Hom}(T_\Sigma(X), G)$, then $\varphi(w) = \varphi(Mxy) \cdot \varphi(z) = (\varphi(x) \cdot \varphi(y)) \cdot \varphi(z) = \varphi(x) \cdot (\varphi(y) \cdot \varphi(z)) = \varphi(x) \cdot \varphi(Myz) = \varphi(w')$, showing that $(w, w') \in \mathrm{Id}G$. In this way any identity of the group G, in the usual sense of the word, is associated with an identity in the sense of the above definition. We are going to get a deeper insight of this connection.

Coming back to the general case, the elements of the set

(2.6) $$A^X = \{v \mid v : X \longrightarrow A\}$$

will be called the *valuations* of the set X in A and we will represent them in the form

(2.7) $$v = (\ldots, a_j, \ldots) \,, \text{ where } a_j = v(x_j) \; (\forall j \in J) \,.$$

Now we make the set

(2.8) $$A(X) = A^{A^X} = \{g \mid g : A^X \longrightarrow A\}$$

into a Σ−algebra $(A(X); (\mathbf{f}_i)_{i \in I})$ defined as follows: for every $i \in I$, $g_1, \ldots, g_{n(i)} \in A(X)$ and $v \in A^X$,

(2.9) $$\mathbf{f}_i(g_1, \ldots, g_{n(i)})(v) = f_i(g_1(v), \ldots, g_{n(i)}(v)) \,.$$

The *A-projections* associated with the variables x_j are defined by

(2.10) $$x_{jA} : A^X \longrightarrow A \,, \; x_{jA}(\ldots, a_j, \ldots) = a_j \; (\forall j \in J) \,,$$

or equivalently,

(2.10') $$x_{jA}(v) = v(x_j) \quad (\forall v \in A^X) \; (\forall j \in J) \,.$$

The correspondence $x_j \mapsto x_{jA}$ defines a map from X to $A(X)$ and the latter has a unique homomorphic extension $\pi_A \in \mathrm{Hom}(T_\Sigma(X), A(X))$, which is sometimes called the *canonical projection*. So

(2.11) $$\pi_A(x_j) = x_{jA} \quad (\forall j \in J) \,,$$

which suggests introducing the notation

(2.12) $$\pi_A(w) = w_A \quad (\forall w \in T_\Sigma(X)) \,.$$

The function $w_A : A^X \longrightarrow A$ is said to be the *polynomial generated by* w in the algebra A and by a *polynomial, in the variables* X , *of the algebra* A we mean any function from A^X to A which is generated by some $w \in T_\Sigma(X)$.

To make explicit the above definition, note that the homomorphism condition states that for every $i \in I$ and every $w_1, \ldots, w_{n(i)} \in T_\Sigma(X)$,

(2.13) $$\pi_A(F_i w_1 \ldots w_{n(i)}) = f_i(\pi_A(w_1), \ldots, \pi_A(w_{n(i)}))$$

and this can be written in the form

(2.14) $$(F_i w_1 \ldots w_{n(i)})_A = f_i(w_{1A}, \ldots, w_{n(i)A}) \,,$$

or else, taking into account (2.9),

(2.15) $$(F_i w_1 \ldots w_{n(i)})_A(v) = f_i(w_{1A}(v), \ldots, w_{n(i)A}(v)) \; (\forall v \in A^X)$$

and in view of (2.7) this can also be written in the form

(2.15') $$\begin{aligned} &(F_i w_1 \ldots w_{n(i)})_A(\ldots, a_j, \ldots) \\ &= f_i(w_{1A}(\ldots, a_j, \ldots), \ldots, w_{n(i)A}(\ldots, a_j, \ldots)) \end{aligned}$$

for every family $(\ldots, a_j, \ldots)$ of elements $a_j \in A$.

Now let us introduce the set $A[X]$ of polynomials (cf. (2.12)). So

(2.16) $$A[X] = \pi_A(T_\Sigma(X))$$

and a further insight of this set is provided by

Proposition 2.1. $A[X] = < \{x_{jA} \mid j \in J\} >_{A(X)}$.

PROOF: Follows from (2.16), (1.5), (2.3) and (2.11):

$$\pi_A(T_\Sigma(X)) = \pi_A(< X >_{Z\cdot}) = < \pi_A(X) >_{A(X)} = < \{\pi_A(x_j) \mid j \in J\} >_{A(X)} .$$

$\square$

Stated in words: *the algebra $A[X]$ of polynomials is the subalgebra of $A(X)$ generated by the projection functions x_{jA}.*

Remark 2.3. Proposition 2.1 yields the following alternative *recursive definition of polynomials* (cf. the recursive definition of $< X >$ in §1):

(i) the projection functions are polynomials ;

(ii) for every $i \in I$, if $g_1, \ldots, g_{n(i)}$ are polynomials, then $f_i(g_1, \ldots, g_{n(i)})$ is a polynomial ;

(iii) every polynomial is obtained by applying rules (i) and (ii) finitely many times . $\square$

In view of Remarks 2.1 and 2.3, certain properties of terms and of polynomials can be proved by algebraic induction (cf. §1).

Lemma 2.1. *Let $w \in T_\Sigma(X)$ and let X_0 be the finite set of all the variables that appear in w (cf. Remark 2.2). If $v, v' \in A(X)$ are such that $v|_{X_0} = v'|_{X_0}$, then $w_A(v) = w_A(v')$.*

PROOF: By algebraic induction. If $w = x_j$, $j \in J$, the property follows from (2.10'). If $w = F_i w_1 \ldots w_{n(i)}$ where $w_1, \ldots, w_{n(i)}$ satisfy the property, then (2.15) implies

$$w_A(v) = f_i(w_{1A}(v), \ldots, w_{n(i)}(v)) = f_i(w_{1A}(v'), \ldots, w_{n(i)}(v')) = w_A(v').$$

$\square$

In view of Lemma 2.1 the following construction makes sense. Let $w \in T_\Sigma(X)$ and let $x_1, \ldots, x_n$ be a set of variables containing all the variables that appear in w (the notation $x_1, \ldots, x_n$ is conventional, these variables are not supposed to have any privileged position in the set X). Then we associate with w the function

$$g : A^n \longrightarrow A , \ g(a_1, \ldots, a_n) = w_A(v)$$

(2.17) $\forall a_1, \ldots, a_n \in A$, $\forall v \in A^X$ such that $v(x_k) = a_k$ ($k = 1, \ldots, n$) .

In practice we can denote the polynomial generated by w and the associated function (2.17) by the same symbol w_A. So $w_A : A^X \longrightarrow A$ or $w_A : A^n \longrightarrow A$ according to the context and we can denote the values of w_A by $w_A(a_1, \ldots, a_n)$ in both cases, so that identities (2.10) and (2.15') can be written as follows:

(2.10'')
$$x_{jA}(a_1,\ldots,a_n) = a_j \qquad (j = 1,\ldots,n) ,$$

(2.15'')
$$(F_i w_1 \ldots w_{n(i)})_A(a_1,\ldots,a_n)$$
$$= f_i(w_{1A}(a_1,\ldots,a_n),\ldots,w_{n(i)A}(a_1,\ldots,a_n)) .$$

In other words, the definition of w_A can be roughly paraphrased as follows: *the polynomial generated by a term is obtained by interpreting the variables* $x_1,\ldots,x_n$ *as variables in the usual sense* (i.e., as projection functions) *and the operator symbols* F_i *as the operations* f_i *of the algebra,* $i \in I$.

For instance, consider again the type (2,1,0), the terms $w_1 = Mxy$, $w_2 = E$, $w_3 = Dz$, $w_4 = MxDMyz$ and a group $(G; \cdot,^{-1}, e)$. Then $w_{1G}(x,y) = x \cdot y$, $w_{2G} = e$, $w_{3G}(x) = x^{-1}$ and $w_{4G}(x,y,z) = x \cdot (y \cdot z)^{-1}$, where x, y, z are (usual) variables running in G. This also illustrates the important fact that *distinct terms may generate the same polynomial*, e.g. $w_{4G} = w_{5G}$, where $w_{5G} = MxMDzDy$.

We conclude this section with a characterization of identities in terms of polynomials.

Lemma 2.2. *Let* $w \in T_\Sigma(X)$ *and* $v \in A^X$. *Let* $\bar{v} \in \mathrm{Hom}(T_\Sigma(X), A)$ *be the unique homomorphic extension of* $v : X \longrightarrow A$ $(T_\Sigma(X)$ *is absolutely free; cf.* §1*). Then*

(2.18)
$$\bar{v}(w) = w_A(v) .$$

PROOF: If $w = x_j$ then (2.18) follows from (2.10') since $\bar{v}(x_j) = v(x_j) = x_{jA}(v)$. If $w = F_i w_1 \ldots w_{n(i)}$ where $w_1,\ldots,w_{n(i)}$ fulfil (2.18), then (2.15) implies

$$\bar{v}(w) = f_i(\bar{v}(w_1),\ldots,\bar{v}(w_{n(i)})) = f_i(w_{1A}(v),\ldots,w_{n(i)A}(v)) = w_A(v).$$

□

Theorem 2.1. *The following conditions are equivalent for* $w_1, w_2 \in T_\Sigma(X)$ *and a* $\Sigma-$*algebra A:*

(i) $(w_1, w_2) \in \mathrm{Id}_{\Sigma,X}(A)$;

(ii) $w_{1A} = w_{2A}$.

PROOF: (i) $\Longrightarrow$ (ii) by (2.12), since $\pi_A \in \mathrm{Hom}(T_\Sigma(X), A)$.

(ii) $\Longrightarrow$ (i): Take $\varphi \in \mathrm{Hom}(T_\Sigma(X), A)$; in view of (2.4) we have to prove that $(w_1, w_2) \in \ker \varphi$. Set $v = \varphi|_X$. Then $v \in A^X$, hence $\bar{v} = \varphi$ by the uniqueness of $\bar{v}$, therefore $\varphi(w_1) = w_{1A}(v) = w_{2A}(v) = \varphi(w_2)$.   □

3 Polynomials, identities (continued) and algebraic functions

Polynomials can also be interpreted as finitary operations on the set A, generalizing the basic operations f_i of the algebra A. Other topics in this section are the transfer of identities from certain algebras to others and the introduction of the larger class of algebraic functions.

An alternative construction of polynomials over a $\Sigma-$algebra is the following.

Let n be an integer, $n \geq 1$. The *algebra of polynomials of n variables over the algebra A* is the specialization of the algebra $A[X]$ from §2 to the case $X = \{1, \ldots, n\}$. We use the special notation

(3.1)
$$A^n = A^{\{1,\ldots,n\}} = \{v \mid v : \{1,\ldots,n\} \longrightarrow A\} \,,$$

(3.2)
$$v = (a_1, \ldots, a_n) \,, \text{ representation of } v \in A^n \,,$$

(3.3)
$$O_n(A) = A^{A^n} = \{g \mid g : A^n \longrightarrow A\} \,,$$

(3.4)
$$P_n(A) = a[\{1,\ldots,n\}] = \pi_A(T_\Sigma(\{1,\ldots,n\})) \,,$$

for the *n-ary valuations* on A, the *algebra of n-ary operations* of A and the *algebra of polynomials of n variables* of A, respectively. Then we define the set of all *finitary operations* of A and the set of all *polynomials* of A by

(3.5)
$$O(A) = \bigcup_{n \in \mathbf{N}} O_n(A) \,,$$

(3.6)
$$P(A) = \bigcup_{n \in \mathbf{N}} P_n(A) \,,$$

respectively.

An operation $g \in O_n(A)$ can be identified with a function $g \in A(\mathbf{N})$ such that $g(v) = g(v')$ whenever $v, v' \in A^{\mathbf{N}}$ satisfy

(3.7)
$$v|_{\{1,\ldots,n\}} = v'|_{\{1,\ldots,n\}} \,.$$

Thus $O(A)$ can be identified with a subset of $A(\mathbf{N})$. Under this intepretation the following result holds:

Theorem 3.1. *The set $A[\mathbf{N}]$ of polynomials in the sense (2.16) coincides with the set $P(A)$ of polynomials defined by (3.6).*

PROOF: Take $w \in T_\Sigma(\mathbf{N})$. Then there is a positive integer n such that the variables occurring in w belong to the set $\{1, \ldots, n\}$. If $v, v' \in A^{\mathbf{N}}$ satisfy (3.7) then $w_A(v) = w_A(v')$ by Lemma 2.1, therefore $w_A \in O_n(A)$ under the identification defined above. We prove by algebraic induction that $w_a \in P_n(A) = A[\{1, \ldots, n\}]$. If $w = x_j$, $j \in \{1, \ldots, n\}$, then $w_A = x_{jA} \in A[\{1, \ldots, n\}]$. If $w = F_i w_1 \ldots w_{n(i)}$ where $w_{kA} \in A[\{1, \ldots, n\}]$ $(k = 1 \ldots, n(i))$, then again $w_A \in A[\{1, \ldots, n\}]$.

Conversely, take $g \in P(A)$. Fix a positive integer n such that $g \in P_n(A)$. Then $g = w_A$ for some $w \in T_\Sigma(\{1, \ldots, n\})$. Now g is identified with an element of $A(\mathbf{N})$ as was indicated above and on the other hand the word w also represents a term $w \in T_\Sigma(\mathbf{N})$ for which we still have $g = w_A$. Thus $g \in A[\mathbf{N}]$. □

Theorem 3.1 shows that the $A(X)$−approach and the $O(A)$−approach are essentially equivalent. The former approach, adopted in §2, in the remaining of this section and in Ch.4, is generally speaking suitable for algebraic logic and some topics in universal algebra, including equational compactness, to be studied in Ch.4. The $O(A)$−approach will be more convenient in most chapters of this book.

Consider again an arbitrary set X.

An important property of polynomials is that they behave with respect to homomorphisms and congruences exactly as the basic operations f_i and g_i of the algebras $(A; (f_i)_{i \in I})$ and $(B; (g_i)_{i \in I})$.

Proposition 3.1. *For every $w \in T_\Sigma(X)$ and every $\varphi \in \mathrm{Hom}(A, B)$,*

$$(3.8) \qquad \varphi(w_A(\ldots, a_j, \ldots)) = w_B(\ldots, \varphi(a_j), \ldots) \; (\forall (\ldots, a_j, \ldots) \in A^X) \,.$$

PROOF: By algebraic induction. The starting point is

$$\varphi(x_{jA}(\ldots, a_j, \ldots)) = \varphi(a_j) = x_{jB}(\ldots, \varphi(a_j), \ldots) \,,$$

while the inductive step is based on (2.15′):

$$\varphi((F_i w_1 \ldots w_{n(i)})_A(\ldots, a_j, \ldots))$$
$$= \varphi(f_i(w_{1A}(\ldots, a_j, \ldots), \ldots, w_{n(i)A}(\ldots, a_j, \ldots)))$$
$$= g_i(\varphi(w_{1A}(\ldots, a_j, \ldots)), \ldots, \varphi(w_{n(i)A}(\ldots, a_j, \ldots)))$$
$$= g_i(w_{1B}(\ldots, \varphi(a_j), \ldots), \ldots, w_{n(i)B}(\ldots, \varphi(a_j), \ldots))$$
$$= (F_i w_1 \ldots w_{n(i)})_B(\ldots, \varphi(a_j), \ldots) \,.$$

□

Corollary 3.1. *If A is a subalgebra of B then $w_A(v) = w_B(\iota v)$ for every $v \in A^X$, where ι is the inclusion $\iota : A \longrightarrow B$. In other words, $w_B|_{A^X} = w_A$.* □

Corollary 3.2. *If A is a subalgebra of B and every polynomial of B is uniquely determined by its restriction on A^X, then the identities of B coincide with those of A.*

PROOF: By Corollary 3.1: $w_{1B} = w_{2B} \iff w_{1B}|_{A^X} = w_{2B}|_{A^X} \iff w_{1A} = w_{2A}$. □

Corollary 3.2 yields in particular the Müller-Löwenheim Verification Theorem for Boolean algebras; cf. BFE, Chapter 2, §6.

Corollary 3.3. *Suppose* **K** *is a class of similar algebras which contains an algebra A which is isomorphic to a subalgebra of B for every algebra B ∈ **K**. If for every B ∈ **K** the polynomials of B are uniquely determined by their restrictions to A^X, then the identities common to all algebras of **K** coincide with the identities of A.*   □

Corollary 3.4. *For every $w \in T_\Sigma(X)$, every congruence $\sim$ of an algebra A and every $(\dots, a_j, \dots), (\dots, b_j \dots) \in A^X$,*

$$(1.6') \qquad a_j \sim b_j \ (\forall j \in J) \implies w_A(\dots, a_j, \dots) \sim w_A(\dots, b_j, \dots) \ .$$

COMMENT: Property (1.6′) is known as the *substitution property*; it extends to polynomials property (1.6) in the definition of congruences.
PROOF: If $a_j \sim b_j \ (\forall j \in J)$, then property (3.8) applied to the homomorphism nat$\sim$ yields

$$[w_A(\dots, a_j, \dots)] = w_{A/\sim}(\dots, [a_j], \dots) = w_{A/\sim}(\dots, [b_j], \dots) = [w_A(\dots, b_j, \dots)] \ .$$

□

Polynomials can be also used to describe the subalgebra generated by a set. Let $C \subseteq A$ and let $\iota : C \longrightarrow A$ be the inclusion. Take C in the rôle of X and denote by w^C the elements of $T_\Sigma(C)$. Then:

Proposition 3.2. *With the above notation,*

$$(3.9) \qquad \begin{aligned} < C > &= \{(w^C)_A(\iota) \mid w^C \in T_\Sigma(C)\} \\ &= \{(w^C)_A(c_1, \dots, c_n) \mid c_1, \dots, c_n \in C \ , \ n \in \mathbf{N} \ , \ w^C \in \\ & \quad T_\Sigma(C)\} \ . \end{aligned}$$

PROOF: Denote by R the right-hand side of (3.9). If $c \in C$ then (2.11) yields

$$(c^C)_A(\iota) = (c^C)_A(\dots, \iota(c), \dots) = \iota(c) = c \ ,$$

showing that $C \subseteq R$.
If $i \in I$ and $(w_1)^C, \dots, (w_{n(i)})^C \in T_\Sigma(C)$, then (2.15′) yields

$$f_i((w_1^C)_A(\iota), \dots, (w_{n(i)}^C)_A(\iota)) = (F_i w_1^C \dots w_{n(i)}^C)_A(\iota) \in R \ ,$$

showing that R is a subalgebra.
Finally if S is a subalgebra of A and $C \subseteq S$, we prove by algebraic induction that $R \subseteq S$. First, as shown above, if $c \in C$ then $(c^C)_A(\iota) = c \in S$. Further, if $i \in I$ and $(w_k^C)_A \in S \ (k = 1, \dots, n(i))$, then

$$(F_i w_1^C \dots w_{n(i)}^C)_A(\iota) = f_i((w_1^C)_A(\iota), \dots, (w_{n(i)}^C)_A(\iota)) \in S \ .$$

□

The characterization $w_{1A} = w_{2A}$ of identities provided by Theorem 2.1 turns out to be the key tool in studying the transfer of identities from certain algebras to others.

Proposition 3.3. *Identities are preserved by taking subalgebras, homomorphisms and direct products.*

PROOF: Let $w_1, w_2 \in T_\Sigma(X)$ and let A and A_t, $t \in T$, be Σ-algebras.

(i) We prove that if S is a subalgebra of A and $w_{1A} = w_{2A}$, then $w_{1S} = w_{2S}$. If $v \in S^X$ then $w_{1S}(v) = w_{1A}(\iota v) = w_{2A}(\iota v) = w_{2S}(v)$ by Corollary 3.1 of Proposition 3.1.

(ii) We prove that if B is a homomorphic image of A and $w_{1A} = w_{2A}$, then $w_{1B} = w_{2B}$. Let φ be a surjective homomorphism from A to B. If $v = (\ldots, b_j, \ldots) \in B^X$, for each $j \in J$ choose an element $a_j \in A$ such that $\varphi(a_j) = b_j$. Then Proposition 3.1 implies

$$w_{1B}(\ldots, b_j, \ldots) = w_{1B}(\ldots, \varphi(a_j), \ldots) = \varphi(w_{1A}(\ldots, a_j, \ldots))$$

$$= \varphi(w_{2A}(\ldots, a_j, \ldots)) = w_{2B}(\ldots, \varphi(a_j), \ldots) = w_{2B}(\ldots, b_j, \ldots).$$

(iii) We prove that if $C = \Pi_{t \in T} A_t$ and $w_{1A_t} = w_{2A_t}$ for every $t \in T$, then $w_{1A} = w_{2A}$. By applying Proposition 3.1 to the canonical projections $p_t : C \longrightarrow A_t$ we obtain

$$p_t(w_{1A}(\ldots, (a_t)_{t \in T}, \ldots)) = w_{1A_t}(\ldots, a_t, \ldots)$$

$$= w_{2A_t}(\ldots, a_t, \ldots) = p_t(w_{2A}(\ldots, (a_t)_{t \in T}, \ldots))$$

for all $t \in T$, therefore $w_{1A}(\ldots, (a_t)_{t \in T}, \ldots) = w_{2A}(\ldots, (a_t)_{t \in T}, \ldots)$, that is, $w_{1A} = w_{2A}$. □

Corollary 3.5. *Identities are preserved by taking isomorphic images of subdirect products.* □

Corollary 3.6. *If every algebra of a class* **K** *is isomorphic to a subdirect power of an algebra* $A \in$ **K** *(i.e., a subdirect product of copies of* A*), then* $\mathrm{Id}_{\Sigma, X}(\mathbf{K}) = \mathrm{Id}_{\Sigma, X}(A)$ *(cf.(2.6)).*

PROOF: The inclusion $\mathrm{Id}_{\Sigma, X}(A) \subseteq \mathrm{Id}_{\Sigma, X}(\mathbf{K})$ follows from Corollary 3.5, while the converse inclusion is trivial. □

A class **K** of similar algebras is called a *variety* provided it is closed under formation of homomorphic images, subalgebras and direct products. A fundamental theorem due to Birkhoff establishes that a class **K** is a variety if and only if it is *equational*. The latter term designates a class **K** of Σ-algebras for which there is a set E of identities such that $A \in \mathbf{K} \Longleftrightarrow E \subseteq \mathrm{Id}A$ for every Σ-algebra A. The usual classes of algebras are equational, e.g. the monoids, groups, rings, fields, lattices, Boolean algebras etc. (provided the signature is appropriately chosen: for instance, groups should be viewed as algebras $(G : \cdot, ^{-1}, e)$ of type $(2,1,0)$).

Now we introduce a generalization of the concept of polynomial. First we associate with each element $a \in A$ the function

(3.10) $f_a \in A(X)$, $f_a(v) = a$, $(\forall v \in A^X)$.

Further, having in mind Proposition 2.1, we define

(3.11) $$\overline{A}[X] =< \{x_{jA} \mid j \in J\} \cup \{f_a \mid a \in A\} >_{A(X)} ;$$

the elements of the subalgebra $\overline{A}[X]$ of $A(X)$ will be called *algebraic functions*, following Grätzer [1979] (first edition, 1968).

At this point we mention that *polynomials* are also known as *term functions*, while Lausch and Nöbauer [1973] reserve the term *polynomials* for the above defined algebraic functions. Grätzer's terminology for terms and polynomials is *polynomials* and *polynomial functions*, respectively. As a matter of fact the Lausch-Nöbauer terminology seems more appropriate because, unlike the elements of $A[X]$, the elements of $\overline{A}[X]$ use constants in their constructions, exactly as the usual ring polynomials. However the terminology polynomials/algebraic functions seems to be better known.

Note the following immediate consequence of Proposition 2.1 and (3.11):

(3.12) $$\overline{A}[X] =< a[X] \cup \{f_a \mid a \in A\} >_{A(X)} .$$

To obtain an explicit description of algebraic functions consider the set

(3.13) $$G = \{f_a \mid a \in A\} \cup \{x_{jA} \mid j \in J\} ,$$

which is included both in $T_\Sigma(G)$ and $A(X)$. Denote the elements of $T_\Sigma(G)$ by $w^G, w_1^G, \ldots$ etc.

Lemma 3.1. *For every subalgebra S of $A(X)$, every $w^G \in T_\Sigma(G)$ and every $(\ldots, g_k, \ldots) \in S^G$,*

(3.14) $$(w^G)_S(\ldots, g_k, \ldots)(v) = (w^G)_A(\ldots, g_k(v), \ldots) \quad (\forall v \in A^X) .$$

PROOF: It follows from (2.11) that

$$(x_k^G)_S(\ldots, g_k, \ldots)(v) = g_k(v) = (x_k^G)_A(\ldots, g_k(v), \ldots),$$

while the inductive step uses (2.16) and (2.10):

$$(F_i w_1^G \ldots w_{n(i)}^G)_S(\ldots, g_k, \ldots)(v)$$

$$= f_i((w_1^G)_S(\ldots, g_k, \ldots), \ldots, (w_{n(i)}^G)_S(\ldots, g_k, \ldots))(v)$$

$$= f_i((w_1^G)_S(\ldots, g_k, \ldots)(v), \ldots, (w_{n(i)}^G)_S(\ldots, g_k, \ldots))(v)$$

$$= f_i((w_1^G)_A(\ldots, g_k(v), \ldots), \ldots, (w_{n(i)}^G)_A(\ldots, g_k(v), \ldots))$$

$$= (F_i w_1^G \ldots w_{n(i)}^G)_A(\ldots, g_k(v), \ldots) .$$

$\square$

Corollary 3.7. *If there is a bijection between G and X, then for every subalgebra S of $A(X)$, every $w \in T_\Sigma(X)$ and every $(\ldots, g_j, \ldots) \in S^X$,*

(3.15) $$w_S(\ldots, g_j, \ldots) = w_A(\ldots, g_j(v), \ldots) \quad (\forall v \in A^X) .$$

$\square$

Proposition 3.4. *The algebraic functions* $g \in \overline{A}[X]$ *are the functions of the form*

(3.16) $$g = (w^G)_{A(X)}(\ldots, f_a, \ldots; \ldots, x_{jA}, \ldots),$$

where $w^G \in T_\Sigma(X)$ *and for every* $(\ldots, a_j, \ldots) \in A^X,$

(3.17) $$g(\ldots, a_j, \ldots) = (w^G)_A(\ldots, a, \ldots; \ldots, a_j, \ldots).$$

PROOF: Formula (3.16) follows from Proposition 3.2 with $A := A(X)$ and with $C := G$ defined by (3.13), while (3.17) is obtained from (3.16) via Lemma 3.1. □

COMMENT: Since only finitely many a_j's occur in (3.17) and the a_j's are arbitrary elements of A, i.e. *variables* in the usual sense of the word, Proposition 3.4 can be paraphrased in the following form:

Proposition 3.5. *The algebraic functions* $g : A^n \longrightarrow A$ *are the functions of the form*

(3.18) $$g(x_1, \ldots, x_n) = p(a_1, \ldots, a_m, x_1, \ldots, x_n) \qquad (\forall x_1, \ldots, x_n \in A),$$

where $m \geq 0$, $a_1, \ldots, a_m \in A$ *and* $p : A_{m+n} \longrightarrow A$ *is a polynomial.* □

Corollary 3.8. *Any function obtained from an algebraic function by fixing certain variables is also an algebraic function.* □

Proposition 3.6. *Every algebraic function satisfies the substitution property.*

PROOF: From Proposition 3.5 and Corollary 3.4 of Proposition 3.1. □

Proposition 3.7. *If* $g \in A(X)$ *is an algebraic function (a polynomial) and* $\sigma : X \longrightarrow X$, *then the function* $h \in A(X)$ *defined by*

(3.19) $$h(\ldots, x_j, \ldots) = g(\ldots, \sigma(x_j), \ldots)$$

is also an algebraic function (a polynomial).

PROOF: If $g = f_a$ then $h = g$. If $g = x_{jA}$ then $h = x_{kA}$ where $k = \sigma(x_j)$. The inductive step follows from the fact that if $g = f_i(g_1, \ldots, x_{n(i)})$ then $h = f_i(h_1, \ldots, h_{n(i)})$, where h_k are the functions associated with g by (3.19) $(k = 1, \ldots, n(i))$. □

Remark 3.1. In the $O(A)$-approach one defines, for every integer $n \geq 1$, the set

(3.20) $$AF_n(A) = \overline{A}[\{1, \ldots, n\}]$$

of *algebraic functions of n variables* of A and the set

(3.21) $$AF(A) = \bigcup_{n \in \mathbf{N}} AF_n(A)$$

of all *algebraic functions* of A. □

Remark 3.2. The *projection functions* x_{jA} corresponding to the case $X = \{1, \ldots, n\}$ (cf. (2.10″)) are usually denoted by π^n_{jA}, i.e.,

$$(3.22) \qquad \pi^n_{jA}(a_1, \ldots, a_n) = a_j \ (\forall a_1, \ldots, a_n \in A) \qquad (\forall j \in \{1, \ldots, n\}) \,.$$

The recursive definition of *polynomials of n variables* is obtained from Remark 2.3 by changing everywhere "polynomials" by "polynomials of n variables". The recursive definition of *algebraic functions of n variables* is obtained from the latter definition by changing everywhere "polynomials" to "algebraic functions" and by adding the starting rule "the constant functions of n variables are algebraic functions of n variables". □

An *algebraic equation* over an algebra A is an equation (cf. Ch.1, §1) of the form $f(X) = g(X)$, where $f, g \in AF_n(A)$ for some $n \in \mathbf{N}$. The main subject of this book is the study of algebraic equations in certain classes of lattices, mainly in Boolean and Post algebras.

The next lemmas will be needed in subsequent chapters.

Lemma 3.2. *If* $g \in \overline{A}[X]$ *and* $\varphi \in \operatorname{Hom}(\overline{A}[X], A)$ *satisfies* $\varphi(f_a) = a$ *for all* $a \in A$, *then*

$$(3.23) \qquad \qquad \varphi(g) = g(\ldots, \varphi(x_{jA}), \ldots) \,.$$

PROOF: It follows from (2.11) and (3.10) that

$$x_{jA}(\ldots, \varphi(x_{jA}), \ldots) = \varphi(x_{jA}) \,,$$

$$f_a(\ldots, \varphi(x_{jA}), \ldots) = a = \varphi(f_a) \,,$$

while the inductive step uses (2.9):

$$\mathbf{f}_i(g_1, \ldots, g_{n(i)})(\ldots, \varphi(x_j), \ldots)$$

$$= f_i(g_1(\ldots, \varphi(x_j), \ldots), \ldots, g_{n(i)}(\ldots, \varphi(x_j), \ldots))$$

$$= f_i(\varphi(g_1), \ldots, \varphi(g_{n(i)})) = \varphi(\mathbf{f}_i(g_1, \ldots, g_{n(i)})) \,.$$

□

Lemma 3.3. (Grätzer [1962], Beazer [1974c]). *Suppose* $(A; (f_i)_{i \in I})$ *and* $(B; (g_i)_{i \in I})$ *are similar algebras,* $\varphi : A \longrightarrow B$ *is a surjective homomorphism and* $f : A^n \longrightarrow A$ *is a function with the substitution property. Define*

$$(3.25) \qquad g(\varphi(x_1), \ldots, \varphi(x_n)) = \varphi(f(x_1, \ldots, x_n)) \qquad (\forall x_1, \ldots, x_n \in A) \,;$$

then $g : B^n \longrightarrow B$ *is a function with the substitution property* (cf. Corollary 3.4 of Proposition 3.1).

PROOF: Note first that g is well defined, because if $\varphi(x_h) = \varphi(y_h)$ $(h = 1, \ldots, n)$, then, since $\ker \varphi$ is a congruence, it follows that $\varphi(f(x_1, \ldots, x_n)) = \varphi(f(y_1, \ldots, y_n))$.

Further, let θ be a congruence on B.

Define $x \tau x' \iff \varphi(x) \theta \varphi(x')$, where $x, x' \in A$. Then τ is a congruence of A, because it is clearly an equivalence relation and

$$x_k \tau x_k' \ (k = 1, \ldots, n(i))$$

$$\iff \varphi(x_k) \theta \varphi(x_k') \ (k = 1, \ldots, n(i))$$

$$\implies g_i(\varphi(x_1), \ldots, \varphi(x_{n(i)})) \theta g_i(\varphi(x_1'), \ldots, \varphi(x_{n(i)}'))$$

$$\iff \varphi(f_i(x_1, \ldots, x_{n(i)})) \theta \varphi(f_i(x_1', \ldots, x_{n(i)}'))$$

$$\iff f_i(x_1, \ldots, x_{n(i)}) \tau f_i(x_1', \ldots, x_{n(i)}') \ .$$

Finally

$$\varphi(x_h) \theta \varphi(x_h') \ (h = 1, \ldots, n)$$

$$\iff x_h \tau x_h' \ (h = 1, \ldots, n) \implies f(x_1, \ldots, x_n) \tau f(x_1', \ldots, x_n) \tau f(x_1', \ldots, x_n')$$

$$\iff \varphi(f(x_1, \ldots, x_n)) \theta \varphi(f(x_1', \ldots, x_n'))$$

$$\iff g(\varphi(x_1), \ldots, \varphi(x_n)) \theta g(\varphi(x_1'), \ldots, \varphi(x_n')) \ .$$

$\square$

The next lemma is a generalization of two intermediate results obtained in Beazer [1974c].

Lemma 3.4. *Suppose A is an algebra, B is a subalgebra of A and $\psi : A \longrightarrow B$ is a surjective homomorphism such that $\psi(b) = b$ for all $b \in B$. If $f, g : A^n \longrightarrow A$ are functions with the substitution property such that $f|_B = g|_B$ then $\psi f = \psi g$.*

PROOF: Let $f', g' : B^n \longrightarrow B$ be the functions associated with f and g, respectively, by the construction (3.25) in Lemma 3.3. Then for every $b_1, \ldots, b_n \in B$,

$$f'(b_1, \ldots, b_n) = f'(\psi(b_1), \ldots, \psi(b_n)) = \psi(f(b_1, \ldots, b_n))$$

$$= \psi(g(b_1, \ldots, b_n)) = \ldots = g'(b_1, \ldots, b_n) \ ,$$

that is, $f' = g'$. Hence for every $x_1, \ldots, x_n \in A$,

$$\psi(f(x_1, \ldots, x_n)) = f'(\psi(x_1), \ldots, \psi(x_n))$$

$$= g'(\psi(x_1), \ldots, \psi(x_n)) = \psi(g(x_1, \ldots, x_n)) \ .$$

$\square$

Lemma 3.5. *The set of polynomials (of algebraic functions) is closed under composition of functions.*

PROOF: For the convenience of notation consider functions of n variables. We will prove by algebraic induction (cf. Remarks 2.3 and 3.2) that every polynomial (algebraic function) h has the property that for any polynomials (algebraic functions) $g_1, \ldots, g_n$, the composite function $h(g_1, \ldots, g_n)$ is a polynomial (is algebraic). The initial step of the proof is clear. Now take $i \in I$ and suppose the functions $h_1, \ldots, h_{n(i)}$ have the desired property. Then the definition of composition and (2.9) yield

$$((\mathbf{f}_i(h_1, \ldots, h_{n(i)}))(g_1, \ldots, g_n))(v)$$

$$= (\mathbf{f}_i(h_1, \ldots, h_{n(i)}))(g_1(v), \ldots, g_n(v))$$

$$= f_i(h_1(g_1(v), \ldots, g_n(v)), \ldots, h_{n(i)}(g_1(v), \ldots, g_n(v)))$$

$$= f_i(h_1(g_1, \ldots, g_n)(v), \ldots, h_{n(i)}(g_1, \ldots, g_n))(v))$$

$$= \mathbf{f}_i(h_1(g_1, \ldots, g_n), \ldots, h_{n(i)}(g_1, \ldots, g_n))(v) \, ,$$

therefore $(\mathbf{f}_i(h_1, \ldots, h_{n(i)}))(g_1, \ldots, g_n)$ is in fact the polynomial (algebraic function) $\mathbf{f}_i(h_1(g_1, \ldots, g_n), \ldots, h_{n(i)}(g_1, \ldots, g_n))$; cf. the inductive hypothesis.   □

3. Lattices

The main concern of this book is the study of algebraic functions and algebraic equations, i.e., of equations expressed by algebraic functions, in the class of bounded distributive lattices and in certain subclasses, mainly (relatively) pseudocomplemented lattices, Post algebras and Boolean algebras. The aim of the present chapter is to carry out this program for bounded distributive lattices, pseudocomplemented distributive lattices, relatively pseudocomplemented lattices, Stone algebras and Heyting algebras.

The lattice-theoretical prerequisites are established in §§ 1,2. In §3 we study algebraic functions and algebraic equations, as well as the more general family of functions with the substitution property and equations expressed in terms of the latter functions, in the classes of lattices mentioned above. We present several generalizations of well-known properties of Boolean functions and equations. In particular a generalization of the Verification Theorem yields information about the identities of those classes of lattices.

For missing proofs and more details (in particular for more examples of members of the various classes of lattices) the reader is referred e.g. to Balbes and Dwinger [1974], Birkhoff [1967], Crawley and Dilworth [1973], Davey and Priestley [1990] or Grätzer [1978].

1 Posets and distributive lattices

This section is an informal introduction to partially ordered sets, lattices (both as partially ordered sets and algebras), (bounded) distributive lattices and semilattices. The focus is on computation rules.

We first recall that a *partially ordered set* or *poset* for short is a pair $(P; \leq)$, where P is a set and $\leq$ is a relation of *partial order* on P, which means that the following properties hold: for every $x, y, z \in P$,

(1.1) $$x \leq x \qquad (reflexivity),$$

(1.2) $$x \leq y \ \& \ y \leq x \Longrightarrow x = y \qquad (antisymmetry),$$

(1.3) $$x \leq y \, \& \, y \leq z \Longrightarrow x \leq z \qquad (transitivity).$$

Thus $(\mathbf{N}; \leq)$, $(\mathbf{Z}; \leq)$, $(\mathbf{Q}; \leq)$, $(\mathbf{R}; \leq)$, where $\leq$ has its usual meaning, $(\mathbf{N}; |)$, where $|$ stands for the divisibility relation, and $(\mathcal{P}(E); \subseteq)$, where E is any set, are posets. If $(P; \leq)$ is a poset, then any subset $A \subseteq P$, endowed with the restriction of $\leq$ to the elements of A, is a poset as well; the order of A is said to be *induced* by that of P.

In any poset $(P; \leq)$ it is useful to work also with the following relations:

(1.4) $$x < y \Longleftrightarrow x \leq y \, \& \, x \neq y \, ,$$

(1.5) $$x \geq y \Longleftrightarrow y \leq x, \ x > y \Longleftrightarrow y < x \, .$$

The relation $<$ is called the *strict partial order* associated with $\leq$ and it satisfies the following properties: for every $x, y, z \in P$,

(1.6) $$x \not< x \, ,$$

(1.7) $$x < y \Longrightarrow y \not< x \, ,$$

(1.8) $$x < y \, \& \, y < z \Longrightarrow x < z \, .$$

The relation $\geq$ is a partial order on P, called the *dual* of $\leq$.

Let $(P; \leq)$ be a poset and $A \subseteq P$. An element x of P is said to be: a *lower bound* (an *upper bound*) of A iff $x \leq a$ ($x \geq a$) for all $a \in A$; if, moreover, $x \in A$, then x is known as the *first* $\equiv$ the *least* $\equiv$ the *smallest* (the *last* $\equiv$ the *greatest*) element of A. The least (greatest) element of A, provided it exists, is unique; it may be denoted by $\min A$ ($\max A$). In particular, if $a, b \in P$, $a \leq b$ and A is the *interval*

$$[a, b] = \{x \in P \mid a \leq x \leq b\} \, ,$$

then $\min A = a$ and $\max A = b$. By $\inf A \equiv \text{g.l.b.}A$ ($\sup A \equiv \text{l.u.b.}A$) we mean the *greatest lower bound* (*least upper bound*) of A. So $\inf A$ ($\sup A$), provided it exists, is unique. A *minimal* (*maximal*) element of A is an element $x \in A$ such that $a \not< x$ ($x \not< a$) for all $a \in A$, or equivalently, such that $a \leq x \Longrightarrow a = x$ ($x \leq a \Longrightarrow x = a$) for all $a \in A$. A set A may have several minimal (maximal) elements or a single one or none, but if A has least (greatest) element, this is the single minimal (maximal) element of A.

A *totally ordered set* or a *chain* is a poset $(T; \leq)$ in which every two elements are *comparable*, i.e., $x \leq y$ or $y \leq x$. For instance, every subset $Y \subseteq \mathbf{R}$ is a chain with respect to the usual order $\leq$. In a totally ordered set the concepts of minimal (maximal) element and least (greatest) element coincide.

It is important to note that the concepts introduced so far occur in pairs: $(\leq, \geq)$, $(<, >)$, (lower bound, upper bound), (least, greatest), (inf, sup) and (minimal, maximal). For each of these pairs (α, β), concept α with respect to the poset $(P; \leq)$ coincides with concept β with respect to the *dual poset* $(P; \geq)$, and conversely. For instance, an element x is a lower bound of a set A with respect to the order $\leq$ if and only if it is an upper bound of A with respect to the order $\geq$; etc. The concepts α and β occurring in the pairs (α, β) mentioned above are said to be *dual* to each other. The above remarks yield the following *duality principle*: from each theorem about posets one obtains another theorem by replacing each concept occurring in the former theorem by its dual. These two theorems are said to be *dual* to each other. For instance, the theorem stating the uniqueness of the greatest (least) element is the dual of the theorem stating the uniqueness of the least (greatest) element.

A *lattice* (as defined by Ore) is a poset $(L; \leq)$ such that for every $x, y \in L$ there exists $\inf\{x, y\}$ and $\sup\{x, y\}$. For instance, every chain $(C, \leq)$ is a lattice in which $\inf\{x, y\} = \min\{x, y\}$ and $\sup\{x, y\} = \max\{x, y\}$. The poset $(\mathbf{N}, |)$ is a lattice in which $\inf\{m, n\} = \text{g.c.d.}\{m, n\}$ and $\sup\{m, n\} = \text{l.c.m.}\{m, n\}$. The poset $(\mathcal{P}(E); \subseteq)$ is a lattice in which $\inf\{X, Y\} = X \cap Y$ and $\sup\{X, Y\} = X \cup Y$. A *complete lattice* is a poset $(L; \leq)$, $L \neq \emptyset$, such that for every subset $A \subseteq L$ there exist $\inf A$ and $\sup A$. This definition includes the case $A := \emptyset$; note that $\inf \emptyset = \sup L$ and $\sup \emptyset = \inf L$, these elements being the greatest element of L and the least element of L, respectively. For instance, every finite non-empty lattice is a complete lattice, $(\mathbf{N}, |)$ and $(\mathcal{P}(E); \subseteq)$ are complete lattices.

More generally, whenever a poset P (not a subset of it!) has least/greatest element, this element will be denoted by 0/by 1 and called the *zero/one* of P.

Let $(L; \leq)$ be a lattice. For every $x, y \in L$, let us introduce the notation

$$(1.9) \qquad x \wedge y = \inf\{x, y\} , \; x \vee y = \sup\{x, y\} ;$$

then $x \wedge y$ and $x \vee y$ are elements of L uniquely determined by the elements x, y. In other words, we have defined two binary operations $\wedge$ and $\vee$ on L and it is easily seen that they verify the following identities:

$$(1.10) \qquad x \wedge y = y \wedge x , \; x \vee y = y \vee x \quad \text{(commutativity)},$$

$$(1.11) \qquad (x \wedge y) \wedge z = x \wedge (y \wedge z) , \; (x \vee y) \vee z = x \vee (y \vee z) \quad \text{(associativity)},$$

$$(1.12) \qquad x \wedge (x \vee y) = x , \; x \vee (x \wedge y) = x \quad \text{(absorption)}.$$

We are thus led to another description of the concept of lattice, which is in fact the original definition, due to Dedekind. By a *lattice* is meant an algebra $(L; \wedge, \vee)$ of type $(2,2)$, whose operations $\wedge$ and $\vee$, called *meet* or *conjunction*, and *join* or *disjunction*, respectively, satisfy the identities (1.10)-(1.12). So any Ore lattice can be made into a lattice in the sense of Dedekind.

Conversely, let $(L; \wedge, \vee)$ be a lattice as defined by Dedekind. It is easy to prove, using (1.12) and (1.10), that the following properties hold: for every $x, y \in L$,

$$(1.13) \qquad x \wedge x = x , \; x \vee x = x \quad \text{(idempotency)},$$

$$(1.14) \qquad x \wedge y = x \Longleftrightarrow x \vee y = y .$$

Then the relation $\leq$ defined by

(1.15) $x \leq y \iff x \wedge y = x \; (\iff x \vee y = y)$

is reflexive by (1.13), antisymmetric by (1.10) and transitive by (1.11). Moreover, it follows from (1.10), (1.11), (1.13) and (1.15) that properties (1.9) also hold, so that $(L; \leq)$ is a lattice in the sense of Ore.

Finally one proves that the above constructions establish a bijection between the lattices $(L; \leq)$ and the lattices $(L; \wedge, \vee)$ that exist on a set L. (As a matter of fact, this can be extended to an isomorphism of categories). We will take advantage of this fact by making no distinction between the two options, i.e., by working both with the relation $\leq$ and the operations $\wedge, \vee$ of the lattice.

The *duality principle* also holds for lattices, the list of dual concepts being enriched by the pair $(\wedge, \vee)$.

Note also the following properties, which are immediate:

(1.16) $u \leq x \wedge y \iff u \leq x \; \& \; u \leq y \, , \; x \vee y \leq v \iff x \leq v \; \& \; y \leq v \, ,$

(1.17) $x \leq y \implies x \wedge z \leq y \wedge z \; \& \; x \vee z \leq y \vee z \, ,$

(1.18) $x \leq y \; \& \; t \leq v \implies x \wedge t \leq y \wedge v \; \& \; x \vee t \leq y \vee v \, .$

A lattice satisfying the identities

(1.19) $x \wedge (y \vee z) = (x \wedge y) \vee (x \wedge z), \; x \vee (y \wedge z) = (x \vee y) \wedge (x \vee z) \, ,$

is said to be *distributive* (as a matter of fact, each of the two identities (1.19) implies the other). For instance, any totally ordered set, $(\mathbf{N}; |)$ and $(\mathcal{P}(E); \subseteq)$ are distributive lattices. The set of normal subgroups of a group, endowed with $\subseteq$, is a non-distributive lattice.

If a lattice has first element 0 (zero) and last element 1 (one), then for every $x, y \in L$,

(1.20) $0 \leq x \, , \; x \leq 1 \, ,$

(1.21) $x \wedge 0 = 0 \, , \; x \vee 1 = 1 \, ,$

(1.22) $x \wedge 1 = x \, , \; x \vee 0 = x \, ,$

(1.23) $x \wedge y = 1 \iff x = y = 1 \, , \; x \vee y = 0 \iff x = y = 0 \, .$

A *bounded (distributive) lattice* is an algebra $(L; \wedge, \vee, 0, 1)$ of type (2,2,0,0) such that $(L; \wedge, \vee)$ is a (distributive) lattice, while 0 and 1 are its zero and one, respectively. By an abuse of notation, $(L; \wedge, \vee, 0, 1)$ may be denoted by $(L; \leq)$, where $\leq$ is the associated order, or even simpler by its support L.

For example, $(\mathbf{N}; |)$ and $(\mathcal{P}(E); \subseteq)$ are bounded distributive lattices, while the distributive lattices $\mathbf{N}, \mathbf{Z}, \mathbf{Q}, \mathbf{R}$, with the usual order, are not bounded. We have already noted that any complete lattice is bounded; $([0, 1] \cap \mathbf{Q}; \leq)$, where $\leq$ is the usual order, is a bounded distributive lattice that is not complete.

Note that the dual of a bounded (distributive) lattice $(L; \wedge, \vee, 0, 1)$ is the bounded (distributive) lattice $(L; \vee, \wedge, 1, 0)$, therefore the *duality principle* holds for bounded (distributive) lattices.

While in the next sections we will deal with several specializations of bounded distributive lattices, we conclude this section with a concept which is intermediate between posets and lattices.

A *join semilattice* (*meet semilattice*) is a poset $(S; \leq)$ such that for every $x, y \in S$ there exists $\sup\{x, y\}$ ($\inf\{x, y\}$). So the concepts of join semilattice and meet semilattice are dual to each other.

Trivially a lattice is both a join semilattice and a meet semilattice. If E is an infinite set, then $(\mathcal{P}_i(E); \subseteq)$, where $\mathcal{P}_i(E)$ stands for the family of all infinite subsets of E, is a join semilattice but not a lattice.

As we have done for lattices, if $(S; \leq)$ is a join semilattice we can define $x \vee y = \sup\{x, y\}$ for every $x, y \in S$ and obtain an algebra $(S; \vee)$ of type (2) whose operation is commutative, associative and idempotent (cf. the second halves of (1.10), (1.11) and (1.13)). We arrive at the same result if we start from a meet semilattice and define $x \wedge y = \inf\{x, y\}$. Conversely, let $(S; \circ)$ be a *semilattice*, i.e., an algebra of type (2) whose operation is commutative, associative and idempotent. If we define $x \leq y \iff x \circ y = x$ then $(S; \leq)$ is a meet semilattice, while the definition $x \leq y \iff x \circ y = y$ yields a join semilattice $S; \leq)$. In practice we denote a semilattice in the algebraic sense by $(S; \wedge)$ or $(S; \vee)$ according as we intend to regard it as a meet semilattice or as a join semilattice.

It is easy to prove by algebraic induction that the *polynomials* of a (join) semilattice S are of the form

$$(1.24) \qquad p(x_1, \ldots, x_n) = x_{k_1} \vee \ldots \vee x_{k_m} ,$$

where $\emptyset \neq \{k_1, \ldots, k_m\} \subseteq \{1, \ldots, n\}$, while the *algebraic functions* are the polynomials, the constant functions and the functions of the form

$$(1.25) \qquad f(x_1, \ldots, x_n) = a \vee p(x_1, \ldots, x_n) ,$$

where $a \in S$ and p is a polynomial.

2 Classes of (relatively) (pseudo)complemented lattices

In this section we introduce pseudocomplemented lattices, Stone algebras, relatively pseudocomplemented lattices, Heyting algebras and Boolean algebras. We focus on computation rules; functions and identities in these classes of lattices will be studied in the next section.

Definition 2.1. Let x be an element of a lattice L with 0. If the element

$$(2.1) \qquad x^* = \max\{y \in L \mid x \wedge y = 0\}$$

exists (in which case it is unique), then x^* is called the *pseudocomplement* of x

and we say that x is *pseudocomplemented*. The lattice L is said to be *pseudo-complemented* provided all of its elements are pseudocomplemented. The *dual pseudocomplement* x^+ of an element x of a lattice with 1 is the element

$$(2.1')\qquad\qquad x^+ = \min\{y \in L \mid x \vee y = 1\},$$

provided it exists, in which case (x^+ is unique and) x is said to be *dually pseudocomplemented*. The lattice L is called *dually pseudocomplemented* provided all of its elements are dually pseudocomplemented. □

For instance, any bounded chain is pseudocomplemented, with $0^* = 1$ and $x^* = 0$ for every $x \neq 0$. The lattice of open subsets of a topological space X is pseudocomplemented, with $U^* =$ the interior of $X \backslash U$. These two lattices are also dually pseudocomplemented. Another example is the algebraic system associated with the intuitionistic propositional calculus.

Remark 2.1. Every pseudocomplemented lattice L is bounded: we have $0^* = 1$ because $\{y \in L \mid 0 \wedge y = 0\} = L$. □

Proposition 2.1. *The following properties hold in a pseudocomplemented lattice: for every $x, y \in L$,*

$$(2.2)\qquad\qquad x \wedge y = 0 \Longleftrightarrow y \leq x^*,$$

$$(2.3)\qquad\qquad x \wedge x^* = 0,$$

$$(2.4)\qquad\qquad x^* = 1 \Longleftrightarrow x = 0,$$

$$(2.5)\qquad\qquad 0^* = 1 \;\&\; 1^* = 0,$$

$$(2.6)\qquad\qquad x \leq x^{**},$$

$$(2.7)\qquad\qquad x^{***} = x^*,$$

$$(2.8)\qquad\qquad x \leq y \Longrightarrow y^* \leq x^*,$$

$$(2.9)\qquad\qquad x \wedge y = 0 \Longleftrightarrow x^{**} \wedge y = 0,$$

$$(2.10)\qquad\qquad x = x^{**} \Longleftrightarrow \exists y \; x = y^*,$$

$$(2.11)\qquad\qquad (x \wedge y)^{**} = x^{**} \wedge y^{**}.$$

PROOF: Property (2.2) is a paraphrase of Definition 2.1, while (2.3) and (2.4) are obtained from (2.2) by taking $y := x^*$ and $y := 1$, respectively. From (2.4) and Remark 2.1 we obtain (2.5). Taking into account (2.3) in property (2.2) written for $x := x^*$ and $y := x$ we infer (2.6). If $x \leq y$ then since $y \wedge y^* = 0$ by (2.3), it follows that $x \leq y^* = 0$, which implies $y^* \leq x^*$ by (2.2), so property

(2.8) holds. Now from (2.6) and (2.8) we deduce $x^{***} \leq x^*$, while (2.6) written for $x := x^*$ yields $x^* \leq x^{***}$, therefore (2.7) is valid. If $x \wedge y = 0$ then $y \leq x^*$, hence $x^{**} \wedge y \leq x^{**} \wedge x^* = 0$, while the converse implication from (2.9) is a consequence of (2.6). To prove (2.10) note first that

$$x \wedge y = 0 \iff y \leq x^* = x^{***} \iff x^{**} \wedge y = 0$$

by (2.2), (2.7) and (2.2). Now $x = y^* \implies x = y^{***} = x^{**}$, while the converse implication from (2.10) is obvious. Finally from $x \wedge y \leq x$ and a double application of (2.8) we infer $(x \wedge y)^{**} \leq x^{**}$ and similarly $(x \wedge y)^{**} \leq y^{**}$, hence $(x \wedge y)^{**} \leq x^{**} \wedge y^{**}$. To prove the converse inequality from (2.11) note that $x \wedge y \wedge (x \wedge y)^* = 0$ and apply (2.9) twice, which yields $x^{**} \wedge y^{**} \wedge (x \wedge y)^* = 0$. □

Corollary 2.1. *The map $x \mapsto x^{**}$ is a closure operator (i.e., $x \leq y \implies x^{**} \leq y^{**}$, $x \leq x^{**}$ and $x^{****} = x^{**}$) satisfying also $0^{**} = 0$.* □

Proposition 2.2. *The following identities hold in a pseudocomplemented distributive lattice :*

(2.12) $$(x \vee x^*)^* = 0 ,$$

(2.13) $$(x \vee y)^* = x^* \wedge y^* ,$$

(2.14) $$(x \vee y)^{**} = (x^{**} \vee y^{**})^{**} .$$

PROOF: It follows from (1.19), (1.23), (2.2) and (1.16) that

$$z \wedge (x \vee y) = 0 \iff (z \wedge x) \vee (z \wedge y) = 0 \iff z \wedge x = z \wedge y = 0$$

$$\iff z \leq x^* \; \& \; z \leq y^* \iff z \leq x^* \wedge y^* ,$$

which establishes (2.13) by (2.2). Then (2.13) implies (2.12) because $(x \vee x^*)^* = x^* \wedge x^{**} = 0$ by (2.3) and also (2.14) because

$$(x^{**} \vee y^{**})^{**} = (x^{***} \wedge y^{***})^* = (x^* \wedge y^*)^* = (x \vee y)^{**}$$

by (2.7). □

Proposition 2.3. *The following identities are equivalent in a pseudocomplemented distributive lattice:*

(2.15) $$x^* \vee x^{**} = 1 ,$$

(2.16) $$(x \wedge y)^* = x^* \vee y^* ,$$

(2.17) $$(x \vee y)^{**} = x^{**} \vee y^{**} .$$

PROOF: $(2.15) \Longrightarrow (2.16)$: Using in turn (2.3), (2.9), (2.2), (1.17) and (2.15) we obtain

$$z = (x \wedge y)^* \Longrightarrow x \wedge y \wedge z = 0 \Longrightarrow x^{**} \wedge y \wedge z = 0$$
$$\Longrightarrow x^{**} \wedge z \leq y^* \Longrightarrow x^{**} \wedge z \leq y^* \wedge z \Longrightarrow z = z \wedge (x^* \vee x^{**})$$
$$= (z \wedge x^*) \vee (z \wedge x^{**}) \leq (z \wedge x^*) \vee (y^* \wedge z) \leq x^* \vee y^* \,;$$

the converse inequality follows from

$$x \wedge y \wedge (x^* \vee t^*) = (x \wedge y \wedge x^*) \vee (x \wedge y \wedge y^*) = 0 \,.$$

$(2.16) \Longrightarrow (2.17)$: It follows from (2.16) and (2.13) that

$$x^{**} \vee y^{**} = (x^* \wedge y^*)^* = (x \vee y)^{**} \,.$$

$(2.17) \Longrightarrow (2.15)$: Applying in turn (2.7), (2.17), (2.12) and (2.5) we deduce that

$$x^* \vee x^{**} = x^{***} \vee x^{**} = (x^* \vee x)^{**} = 0^* = 1 \,.$$

$\square$

Definition 2.2. A pseudocomplemented distributive lattice satisfying any, hence all of the equivalent conditions in Proposition 2.3, is called a *Stone algebra*. A lattice which is both a Stone algebra and a dual Stone algebra is termed a *double Stone algebra*.
$\square$

For instance, a bounded chain is a double Stone algebra.

Definition 2.3. Let x, y be elements of a lattice L. If the element

(2.18) $$x \rightarrow y = \max\{z \in L \mid x \wedge z \leq y\}$$

exists (in which case it is unique), then $x \rightarrow y$ is called the *relative pseudocomplement* of x in y. The lattice L is said to be *relatively pseudocomplemented* if $x \rightarrow y$ exists for every $x, y \in L$. If, moreover, the lattice L has 0, then it is called a *Heyting algebra* or a *pseudo-Boolean algebra*. The *dual relative pseudocomplement* $y - x$ of x in y is the element

(2.18′) $$y - x = \min\{z \in L \mid x \vee z \geq y\}$$

provided it exists (in which case it is unique). The lattice L is called *dual relatively pseudocomplemented* provided $y - x$ exists for every $x, y \in L$. If, moreover, the lattice L has 1 then it is called a *Brouwerian algebra*.
$\square$

For instance, a bounded chain is a Heyting algebra with $x \rightarrow y = 1$ if $x \leq y$ and $x \rightarrow y = y$ if $y < x$ and it is a Brouwerian algebra with $y - x = 0$ if $y \leq x$ and $y - x = y$ if $y > x$. The algebraic system associated with the intuitionistic propositional calculus is a Heyting algebra .

Proposition 2.4. *Let L be a relatively pseudocomplemented lattice. For every $a \in L$ and every $X \subseteq L$,*

(2.19) $$a \wedge \bigvee_{x \in X} x = \bigvee_{x \in X} (a \wedge x) \,,$$

to the effect that whenever $\bigvee_{x \in X} x$ exists, the right side exists and the equality holds (where $\bigvee$ denotes sup; cf. Definition 3.1).

PROOF: Let $b = \bigvee_{x \in X} x$ and $Y = \{a \wedge x \mid x \in X\}$. If $X = \emptyset$ then $Y = \emptyset$, therefore both suprema are 0 and the equality holds. Now suppose $X \neq \emptyset$ and prove that $a \wedge b = \sup Y$. If $x \in X$ then $a \wedge x \leq a \wedge b$. If y is an upper bound of Y then for every $x \in X$ we have $a \wedge x \leq y$, that is $x \leq a \to y$. This implies $b \leq a \to y$, that is $a \wedge b \leq y$.   □

Corollary 2.2. *Every relatively pseudocomplemented lattice is distributive.*   □

Proposition 2.5. *Let L be a complete lattice. Then L is a Heyting algebra if and only if it satisfies identity (2.19).*

PROOF: Necessity follows from Proposition 2.4. Conversely, suppose L is a complete lattice satisfying (2.19) and take $a, b \in L$. Set $X = \{x \in L \mid a \wedge x \leq b\}$ and $c = \sup X$. Then $0 \in X \neq \emptyset$ and $a \wedge c = \bigvee_{x \in X}(a \wedge x) \leq b$, while if $a \wedge x \leq b$ then $x \in X$, hence $x \leq c$. So $c = a \to b$.   □

In the sequel we adopt the convention that $\wedge$ and $\vee$ connect stronger than $\to$.

Proposition 2.6. *The following properties hold in a relatively pseudocomplemented lattice L: for every $x, y, z \in L$,*

(2.20)
$$x \wedge z \leq y \iff z \leq x \to y \,,$$

(2.21)
$$x \wedge (x \to y) \leq y \,,$$

(2.22)
$$y \leq x \to y \,,$$

(2.23)
$$x \to x = 1 \,,$$

(2.24)
$$x \leq y \iff x \to y = 1 \,,$$

(2.25)
$$x \leq y \implies z \to x \leq z \to y \;\&\; y \to z \leq x \to z \,,$$

(2.26)
$$x \wedge (x \to y) = x \wedge y \,,$$

(2.27)
$$x \to y \wedge z = (x \to y) \wedge (x \to z) \,,$$

(2.28)
$$x \wedge y \to z = x \to (y \to z) \,,$$

(2.29)
$$x \vee y \to z = (x \to z) \wedge (y \to z) \,.$$

PROOF: Property (2.20) is a paraphrase of Definition 2.3, while (2.21), (2.22) and (2.23) are obtained from (2.20) by taking $z := x \to y$, $z := y$ and $y := x$, respectively. Now (2.24) follows from (2.20) with $z := 1$. Also, (2.26) is obtained from (2.21) and (2.22) by taking meet with x. Property (2.25) follows from (2.21) and (2.20) because if $x \leq y$ then $z \wedge (z \to x) \leq x \leq y$ and $x \wedge (y \to z) \leq y \wedge (y \to z) \leq z$. The proofs of (2.27)-(2.29) are based on (2.20), (2.25) and (2.26):

For (2.27) we start from $y \wedge z \leq y$ and obtain $x \to y \wedge z \leq x \to y$ and similarly $x \to y \wedge z \leq x \to z$, therefore $x \to y \wedge z \leq (x \to y) \wedge (x \to z)$. The converse inequality follows from

$$x \wedge (x \to y) \wedge (x \to z) = x \wedge y \wedge (x \to z) = x \wedge y \wedge z \leq y \wedge z.$$

For (2.28) we compute

$$(x \wedge y) \wedge (x \to (y \to z)) = x \wedge y \wedge (y \to z) = x \wedge y \wedge z \leq z,$$

which implies $x \to (y \to z) \leq x \wedge y \to z$. To prove the converse inequality we obtain in turn

$$x \wedge y \wedge (x \wedge y \to z) = x \wedge y \wedge z \leq z,$$

$$x \wedge (x \wedge y \to z) \leq y \to z,$$

$$x \wedge y \to z \leq x \to (y \to z).$$

For (2.29) we start from $x \leq x \vee y$ and obtain $x \vee y \to z \leq x \to z$ and similarly $x \vee y \to z \leq y \to z$, therefore $x \vee y \to z \leq (x \to z) \wedge (y \to z)$. The converse inequality follows from

$$(x \vee y) \wedge (x \to z) \wedge (y \to z)$$

$$= (x \wedge (x \to z) \wedge (y \to z)) \vee (y \wedge (x \to z) \wedge (y \to z))$$

$$= (x \wedge z \wedge (y \to z)) \vee (y \wedge z \wedge (x \to z)) \leq z.$$

□

Proposition 2.7. *Every Heyting algebra is a pseudocomplemented distributive lattice satisfying the identities*

(2.30) $$x^* = x \to 0,$$

(2.31) $$(x \to y)^* = x^{**} \wedge y^*.$$

PROOF: Property (2.30) follows from (2.1) and (2.18). Further from $0 \leq y$ we obtain $x^* = x \to 0 \leq x \to y$ by (2.25), hence $(x \to y)^* \leq x^{**}$ by (2.8), while (2.22) implies $(x \to y)^* \leq y^*$, therefore $(x \to y)^* \leq x^{**} \wedge y^*$. To prove the converse inequality note that $x \wedge y^* \wedge (x \to y) = x \wedge y \wedge y^* = 0$ by (2.26), hence $x^{**} \wedge y^* \wedge (x \to y) = 0$ by (2.9). □

Definition 2.4. A lattice which is both a Heyting algebra and a Brouwerian algebra is called a *double Heyting algebra* or a *bi-Brouwerian lattice* (the latter term was suggested by Beazer [1974a]). □

As already remarked, a bounded chain is a double Heyting algebra.

So a double Heyting algebra satisfies properties (2.1)–(2.31) as well as their duals (2.1′)–(2.31′). Here are some of them:

(2.2′) $$x \vee y = 1 \iff x^+ \leq y ,$$

(2.3′) $$x \vee x^+ = 1 ,$$

(2.6′) $$x^{++} \leq x ,$$

(2.8′) $$x \leq y \implies y^+ \leq x^+ ,$$

(2.15′) $$x^+ \wedge x^{++} = 0 ,$$

(2.19′) $$a \vee \bigwedge_{x \in X} x = \bigwedge_{x \in X} (a \vee x) ,$$

(2.20′) $$x \vee z \geq t \iff z \geq y - x ,$$

(2.22′) $$y \leq x \vee (y - x) ,$$

(2.23′) $$x - x = 0 ,$$

(2.24′) $$x \leq y \iff x - y = 0 ,$$

(2.25′) $$x \leq y \implies z - y \leq z - x \ \& \ x - z \leq y - z ,$$

(2.26′) $$x \vee (y - x) = x \vee y ,$$

(2.29′) $$z - x \wedge y = (z - x) \vee (z - y) ,$$

(2.30′) $$x^+ = 1 - x ,$$

(2.31′) $$(y - x)^+ = x^{++} \vee y^+ .$$

$\square$

Proposition 2.8. (Beazer [1974a]). *In a double Heyting algebra the operations · and + defined by*

(2.32) $$x \cdot y = (x \to y) \wedge (y \to x) ,$$

(2.32′) $$x + y = (x - y) \vee (y - x) ,$$

satisfy

(2.33) $$x = y \iff x \cdot y = 1 \iff x + y = 0 .$$

PROOF: In view of duality it suffices to prove the first equivalence. But $x \cdot x = 1$ by (2.23). Conversely, if $x \cdot y = 1$ then $x \to y = y \to x = 1$, hence $x = y$ by (2.24). $\qquad\square$

We now introduce Boolean algebras as a further specialization of the above classes of lattices.

Definition 2.5. An element x of a bounded lattice L is said to be *complemented* or *Boolean* if there is an element $x' \in L$ such that

(2.34) $$x \wedge x' = 0 \ \& \ x \vee x' = 1 \,,$$

in which case x' is called a *complement* of x. The lattice L is termed *complemented* provided all of its elements are complemented. $\qquad\square$

An element of a bounded lattice may have a single complement, several complements or none. Consider, for instance, the *diamond* and the *pentagon*, which are bounded lattices defined on the five-element set $\{0, a, b, c, d, 1\}$, 0 and 1 being the least and greatest element, respectively, in both lattices; in the diamond the elements a, b, c are not comparable, while in the pentagon $a < b$ and c is not comparable to a and b. So in the diamond each of the elements a, b, c is the complement of the two others, while in the pentagon a and b are complements of c and each of the elements a and b has the unique complement c. As a further example note that in a bounded chain the elements distinct from 0 and 1 have no complements.

Remark 2.2. In every bounded lattice, each of the elements 0 and 1 is the unique complement of the other. $\qquad\square$

Proposition 2.9. *In a bounded distributive lattice each complemented element has a unique complement.*

PROOF: Let y_1 and y_2 be complements of x. Then

$$y_1 = y_1 \wedge 1 = y_1 \wedge (x \vee y_2) = (y_1 \wedge x) \vee (y_1 \wedge y_2) = 0 \vee (y_1 \wedge y_2)$$

$$= (x \wedge y_2) \vee (y_1 \wedge y_2) = (x \vee y_1) \wedge y_2 = 1 \wedge y_2 = y_2 \,.$$

$\qquad\square$

Definition 2.6. A *Boolean algebra* is a complemented distributive lattice with $0 \neq 1$. The (unique!) complement of an element x will be denoted by x'. Unless otherwise stated, Boolean algebras will be regarded as algebras $(B; \wedge, \vee, ', 0, 1)$ of type (2,2,1,0,0). $\qquad\square$

As a matter of fact there are numerous systems of axioms for Boolean algebras, as well as for several classes of lattices; see Rudeanu [1963].

Remark 2.3. The principle of duality holds for double Stone algebras, double Heyting algebras and Boolean algebras. $\qquad\square$

Proposition 2.10. *Every Boolean algebra B is a double Stone algebra and a double Heyting algebra, where for every* $x, y \in B$,

$$(2.35) \qquad\qquad x^* = x^+ = x' \,,$$

$$(2.36) \qquad\qquad x \to y = x' \vee y \ \& \ y - x = x' \wedge y \,.$$

PROOF: In view of duality it suffices to prove that B is a Stone algebra and a Heyting algebra. Since B contains 0, the latter assertion amounts to proving that $x' \vee y$ is the relative pseudocomplement of x in y. But

$$x \wedge z \leq y \Longrightarrow z = z \wedge (x \vee x') = (z \wedge x) \vee (z \wedge x') \leq y \vee x' \,,$$

$$z \leq x' \vee y \Longrightarrow x \wedge z \leq x \wedge (x' \vee y) = (x \wedge x') \vee (x \wedge y) = x \wedge y \leq y \,.$$

Now Proposition 2.7 implies that B is a pseudocomplemented distributive lattice with $x^* = x \to 0 = x' \vee 0 = x'$. Besides, properties (2.34) show that x is the complement of x', that is, $x'' = x$, hence $x' \vee x'' = x' \vee x = 1$, therefore B is a Stone algebra. $\qquad\square$

Proposition 2.11. *The following properties hold in a Boolean algebra B: for every* $a, x, y, z \in B$,

$$(2.34) \qquad\qquad x \wedge x' = 0 \ \& \ x \vee x' = 1 \,,$$

$$(2.37) \qquad\qquad 0' = 1 \ \& \ 1' = 0 \,,$$

$$(2.38) \qquad\qquad x'' = x \,,$$

$$(2.39) \qquad\qquad (x \wedge y)' = x' \vee y' \ \& \ (x \vee y)' = x' \wedge y' \,,$$

$$(2.40) \qquad\qquad x \wedge (x' \vee y) = x \wedge y \ \& \ x \vee (x' \wedge y) = x \vee y \,,$$

$$(2.41) \qquad\qquad x \leq y \Longrightarrow y' \leq x' \,,$$

$$(2.42) \qquad\qquad x \wedge z \leq y \Longleftrightarrow z \leq x' \vee y \ \& \ y \leq x \vee z \Longleftrightarrow x' \wedge y \leq z \,,$$

$$(2.43) \qquad\qquad x \leq y \Longleftrightarrow x \wedge y' = 0 \Longleftrightarrow x' \vee y = 1 \,,$$

$$(2.44) \qquad\qquad x = y \Longleftrightarrow (x' \vee y) \wedge (x \vee y') = 1 \Longleftrightarrow (x \wedge y') \vee (x' \wedge y) = 0 \,,$$

$$(2.45) \qquad\qquad a \wedge \bigvee_{x \in X} x = \bigvee_{x \in X} (a \wedge x) \ \& \ a \vee \bigwedge_{x \in X} x = \bigwedge_{x \in X} (a \vee x)$$

$$\text{whenever the left sides exist} \,.$$

COMMENT: Properties (2.38) and (2.39) are known as the law of *double negation* and the *De Morgan* laws, respectively. It seems appropriate to use the term

Boolean absorption for the laws (2.40); cf. BFE, Comment to Theorem 1.2.

PROOF: Property (2.34) holds by definition and we have just noted in the proof of Proposition 2.10 that (2.34) implies (2.38). Property (2.37) is a paraphrase of Remark 2.2. Property (2.42) holds by Proposition 2.10 and the same is true for (2.43), because

$$x \wedge y = 0 \Longleftrightarrow x \wedge y'' = 0 \Longleftrightarrow x \leq y''' = y'$$

by (2.38). Then (2.39), (2.40), (2.41), (2.44) and (2.45) hold by (2.13) & (2.16), (2.26) & (2.26'), (2.8) & (2.38), (2.32) & (2.32') and (2.19) & (2.19'), respectively, again via Proposition 2.10.   □

Remark 2.4. Properties (2.40), (2.42) and (2.43) are valid in the following more general framework: x, y, z are elements of a bounded distributive lattice and only x is supposed to be complemented (only y in the second equivalence (2.43)). For the first equivalence (2.42) this was shown in the proof of Proposition 2.10; the other proofs are left to the reader.   □

Computation in a Boolean algebra is facilitated by the use of *symmetric difference* or *ring sum* operation $+$, defined by

(2.46) $x + y = (x \wedge y') \vee (x' \wedge y) \,,$

which is so called because every Boolean algebra $(B; \wedge, \vee,', 0, 1)$ can also be viewed as a ring $(B; +, \cdot, 0, 1)$, where $+$ is defined by (2.46) and $x \cdot y = x \wedge y$. This ring is said to be a *Boolean ring with unit*, to the effect that it is commutative, idempotent (i.e., $x^2 = x$ for all x), of characteristic 2 (i.e., $x + x = 0$ for all x) and has unit 1. Conversely, every Boolean ring with unit $(B; +, \cdot, 0, 1)$ can be made into a Boolean algebra $(B; \wedge, \vee,', 0, 1)$, where $x \wedge y = x \cdot y$,

(2.47) $x \vee y = x + y + x \cdot y \,,$

(2.48) $x' = x + 1 \,.$

Note also that

(2.49) $x + y = x \vee y \Longleftrightarrow x \wedge y = 0 \,,$

(2.50) $x = y \Longleftrightarrow x + y = 0 \,,$

(2.51) $x + x' \cdot y = x \vee y \,.$

See e.g. BFE, Chapter 1, §3 and in particular Theorem 1.12, stating that the category of Boolean algebras is isomorphic to the category of Boolean rings with unit.

The rest of this chapter is based on the fact that various kinds of lattices can be regarded as algebras of various types. Let us explain this point.

Lattices are algebras $(L; \wedge, \vee)$ of type (2,2) and the classes of lattices introduced so far may be viewed just as lattices with special properties. Alternatively, bounded (distributive) lattices may be regarded as algebras $(L; \wedge, \vee, 0, 1)$ of type

(2,2,0,0). The various subclasses of bounded distributive lattices can be viewed either as consisting of bounded distributive lattices with supplementary properties or as classes of algebras of richer types: pseudocomplemented distributive lattices and in particular Stone algebras are algebras $(L; \wedge, \vee, ^*, 0, 1)$ of type (2,2,1,0,0), Heyting algebras can be regarded as algebras $(L; \wedge, \vee, \rightarrow, 0, 1)$ of type (2,2,2,0,0) etc. Therefore the following definitions are in order.

Definition 2.7. Unless otherwise stated, from now on bounded (distributive) lattices will be viewed as algebras of type (2,2,0,0), pseudocomplemented distributive lattices and in particular Stone algebras, as well as their duals, as algebras of type (2,2,1,0,0), relatively pseudocomplemented lattices and their duals as algebras of type (2,2,2,0,0). Double Stone algebras are algebras of type (2,2,1,1,0,0), while double Heyting algebras are regarded as algebras of type (2,2,2,2,0,0). □

Definition 2.8. Consider a class of lattices, whose members are called *c lattices; c* is missing for the class of all lattices. Then the subalgebras, homomorphisms, congruences, algebraic functions and polynomials in this class will be called *sub-c-lattices, c-lattice homomorphisms, c-lattice congruences, c-lattice functions* and *simple c-lattice functions*, respectively. The name *c-lattice equations* will designate equations expressed in terms of *c*-lattice functions. If the members of the class are called *C algebras*, then the above concepts will be termed *sub-C-algebras, C homomorphisms, C congruences, C functions, simple C functions* and *C equations*, respectively. □

Remark 2.5. Bounded-lattice functions coincide with lattice functions, while bounded-lattice congruences coincide with lattice congruences. The simple bounded-lattice functions are the simple lattice functions and the constant functions 0 and 1. □

Remark 2.6. Taking into account Proposition 2.4, Boolean algebras can be viewed as algebras of various types, e.g.: $(B; \wedge, \vee, 0, 1)$ of type (2,2,0,0), $(B; \wedge, \vee, ', 0, 1)$ of type (2,2,1,0,0), $(B; \wedge, \vee, \rightarrow, 0, 1)$ of type (2,2,2,0,0), $(B; \wedge, \vee, ^*, ^+, 0, 1)$ of type (2,2,1,1,0,0) or $(B; , \wedge, \vee, \rightarrow, -, 0, 1)$ of type (2,2,2,2,0,0). Yet the concepts introduced according to Definition 2.8, i.e., *sub-Boolean algebra, Boolean homomorphism, Boolean congruence, Boolean function* and *simple Boolean function*, are the same for all of the above types. For instance, the concepts corresponding to algebras $(B; \wedge, \vee, ', 0, 1)$ and $(B, \wedge, \vee, \rightarrow, 0, 1)$ are the same because $x \rightarrow y = x' \vee y$ and $x' = x \rightarrow 0$, which implies e.g. that a subset closed with respect to $\rightarrow$ is also closed with respect to ' and conversely. The other proofs are left to the reader. □

Having in mind the above specifications, we conclude this section with a study of Boolean elements in relatively pseudocomplemented lattices.

Definition 2.9. The set of *complemented* or *Boolean* elements of a bounded lattice L will be denoted by $B(L)$. □

Proposition 2.12. *If L is a bounded distributive lattice, then $B(L)$ is a sub-bounded-lattice and if $0 \neq 1$ then it is also a Boolean algebra.*

PROOF: For the first statement note that $0, 1 \in B(L)$ by Remark 2.1 and that if $x, y \in B(L)$ then

$$(x \wedge y) \wedge (x' \vee y') = (x \wedge y \wedge x') \vee (x \wedge y \wedge y') = 0 \vee 0 = 0 \,,$$

$$(x \wedge y) \vee (x' \vee y') = (x \vee x' \vee y') \wedge (y \vee x' \vee y') = 1 \wedge 1 = 1 \,,$$

showing that $x' \vee y'$ is the complement of $x \wedge y$, so that $x \wedge y \in B(L)$, and similarly $x \vee y \in B(L)$.

The subalgebra $B(L)$ is a bounded distributive lattice by Proposition 2.3.3 and it is complemented by definition. □

Definition 2.10. An element x of a pseudocomplemented lattice L is said to be *regular* provided $x^{**} = x$. The set of regular elements of L will be denoted by $R(L)$. □

Remark 2.7. The element x is regular iff $x = y^*$ for some $y \in L$ (because if $x = y^*$ then $x^{**} = y^{***} = y^* = x$, while the converse is obvious). □

Theorem 2.1. (Glivenko). *Let $(L; \wedge, \vee, ^*, 0, 1)$ be a pseudocomplemented distributive lattice with $0 \neq 1$ and define $x \overset{\circ}{\vee} y = (x \vee y)^{**}$. Then:*

(i) $(R(L); \wedge, \overset{\circ}{\vee}, ^*, 0, 1)$ is a Boolean algebra;

(ii) the map $\varphi(x) = x^{**}$ is a surjective pseudocomplemented-lattice homomorphism $\varphi : L \to R(L)$;

(iii) $B(L)$ is a sub-bounded-lattice of $R(L)$.

PROOF: (i) It follows from (2.11), (2.14), (2.7), $0^{**} = 0$ and $1^{**} = 1$ (cf.(2.5)) that $R(L)$ is closed with respect to $\wedge, \overset{\circ}{\vee}, 0$ and 1, respectively. In the poset $R(L)$ clearly $x \wedge y = \inf\{x, y\}$ and also $x \overset{\circ}{\vee} y = \sup\{x, y\}$, because $x, y \leq x \vee y \leq (x \vee y)^{**}$ and if $x, y \leq z \in R(L)$ then $(x \vee y)^{**} \leq z^{**} = z$; cf. Corollary 2.1 of Proposition 2.1. Therefore $R(L)$ is a bounded lattice. Distributivity follows from (2.11):

$$x \overset{\circ}{\vee} (y \wedge z) = (x \vee (y \wedge z))^{**} = ((x \vee y) \wedge (x \vee z))^{**}$$

$$= (x \vee y)^{**} \wedge (x \vee z)^{**} = (x \overset{\circ}{\vee} y) \wedge (x \overset{\circ}{\vee} z) \,.$$

Finally $x \wedge x^* = 0$ by (2.3) and $x \overset{\circ}{\vee} x^* = (x \vee x^*)^{**} = 0^* = 1$ by (2.12) and (2.5), which proves that if $x \in R(L)$ then x^* is the complement of x in $R(L)$.

(ii) φ is a surjection by Remark 2.7 and $\varphi(x^*) = x^{***} = (\varphi(x))^*$. Then $\varphi(x \wedge y) = \varphi(x) \wedge \varphi(y)$ by (2.11). Finally Corollary 2.1 of Proposition 2.1 yields $\varphi(0) = 0$, $\varphi(1) = 1$ and together with (2.14) it implies

$$\varphi(x \overset{\circ}{\vee} y) = (x \vee y)^{****} = (x \vee y)^{**} = \varphi(x) \overset{\circ}{\vee} \varphi(y) \,.$$

(iii) Let x' be the complement of an element $x \in B(L)$. Then $x \wedge x' = 0$ implies $x' \leq x^*$, hence $1 = x \vee x' \leq x \vee x^*$. From $x \vee x^* = 1$ we infer

$$x^{**} = x^{**} \wedge (x \vee x^*) = x^{**} \wedge x \leq x \leq x^{**},$$

therefore $x = x^{**} \in R(L)$.

Thus $B(L) \subseteq R(L)$. In particular if $x, y \in B(L)$ then $x \vee y \in R(L)$ by Proposition 2.12, therefore $x \vee y = (x \vee y)^{**} = x \overset{\circ}{\vee} y$. $\qquad \square$

Proposition 2.13. *If L is a Stone algebra, then $B(L) = R(L)$ and $x' = x^*$ for every $x \in B(L)$.*

PROOF: If $x \in R(L)$ then (2.3) and (2.15) yield $x \wedge x^* = 0$ and $x \vee x^* = 1$, showing that $x \in B(L)$ with $x' = x^*$. $\qquad \square$

Recall (cf. Proposition 2.7) that every Heyting algebra is a pseudocomplemented lattice with $x^* = x \to 0$.

Theorem 2.2. (Glivenko). *Let $(L; \wedge, \vee, \to, 0)$ be a Heyting algebra with $0 \neq 0^*$ and define $x \overset{\circ}{\vee} y = (x \vee y)^{**}$, $x \overset{\circ}{\to} y = x^{**} \overset{\circ}{\vee} y$. Then:*

(i) $(R(L); \wedge, \overset{\circ}{\vee}, \overset{\circ}{\to}, 0)$ is a Boolean algebra with $x' = x \overset{\circ}{\to} 0$ and $1 = 0^*$;

(ii) the map $\varphi(x) = x^{**}$ is a surjective Heyting homomorphism $\varphi : L \longrightarrow R(L)$;

(iii) $B(L)$ is a sub-bounded-lattice of $R(L)$.

PROOF: $R(L)$ is closed with respect to $\overset{\circ}{\to}$ by Remark 2.7. Now repeatedly apply Theorem 2.1. Consider the Boolean algebra $(R(L); \wedge, \overset{\circ}{\vee}, ^*, 0, 1)$. We have $1 = 0^*$ by (2.5), while $x^* = x^* \overset{\circ}{\vee} 0 = x \overset{\circ}{\to} 0$. This completes the proof of (i). Besides, φ is a psudocomplemented-lattice homomorphism and (iii) holds. Finally (2.31) and (2.13) imply

$$\varphi(x \to y) = (x \to y)^{**} = (x^{**} \wedge y^*)^* = (x^* \vee y)^{**} = \varphi(x^* \vee y)$$

$$= \varphi(x^*) \overset{\circ}{\vee} \varphi(y) = (\varphi(x))^* \overset{\circ}{\vee} \varphi(y) = \varphi(x) \overset{\circ}{\to} \varphi(y) .$$

$\qquad \square$

3 Functions and equations

This section sketches the beginning of a possible theory of functions and equations in the classes of lattices introduced in the previous section.

First we characterize lattice functions and study lattice equations in bounded distributive lattices. Then we study identities and equations expressed in terms of functions with the substitution property, in pseudocomplemented distributive lattices, Stone algebras, Heyting algebras and double Heyting lattices. (Recall that the class of functions with the substitution property is larger than the class of algebraic functions.)

Proposition 3.1. *Let L be an arbitrary lattice and $n \in \mathbf{N}$. Then every lattice function $f : L^n \longrightarrow L$ is isotone, i.e.,*

$$(3.1) \qquad x_h \leq y_h \ (h = 1, \ldots, n) \Longrightarrow f(x_1, \ldots, x_n) \leq f(y_1, \ldots, y_n) \ .$$

PROOF: by algebraic induction. The constant functions f_a, $a \in L$, and the projection functions π_{hL}^n, $(h = 1, \ldots, n)$ are clearly isotone. If the functions $f, g : L^n \longrightarrow L$ are isotone, it is plain that so are the functions $f \wedge g$ and $f \vee g$ defined by

$$(3.2) \qquad (f \wedge g)(x_1, \ldots, x_n) = f(x_1, \ldots, x_n) \wedge g(x_1, \ldots, x_n) \ ,$$

$$(3.3) \qquad (f \vee g)(x_1, \ldots, x_n) = f(x_1, \ldots, x_n) \vee g(x_1, \ldots, x_n) \ .$$

$\square$

Definition 3.1. The symbols $\bigwedge$ and $\bigvee$ denote infima and suprema, as follows: $\bigwedge_{a \in A} a$ or $\bigwedge A$ ($\bigvee_{a \in A} a$ or $\bigvee A$) is the infimum (supremum) of the set $A \subseteq L$, whenever it exists. In particular $\bigwedge \emptyset = 1$ and $\bigvee \emptyset = 0$. Another particular notation is

$$\bigwedge_{h=1}^{m} a_h = a_1 \wedge \ldots \wedge a_m, \quad \bigvee_{h=1}^{m} a_h = a_1 \vee \ldots \vee a_m \ .$$

$\square$

Lemma 3.1. (Goodstein [1967]). *If $(L; \wedge, \vee, 0, 1)$ is a bounded distributive lattice, $n \in \mathbf{N}$ and $N = \{1, \ldots, n\}$, then every lattice function $f : L^n \longrightarrow L$ can be represented in the form*

$$(3.4) \qquad f(x_1, \ldots, x_n) = \bigvee_{S \subseteq N} (f(\delta_{S1}, \ldots, \delta_{Sn}) \wedge \bigwedge_{h \in S} x_h) \ (\forall x_1, \ldots, x_n \in L) \ ,$$

where for every $S \subseteq N$,
$$(3.5) \qquad \delta_{Sh} = 1 \ if \ h \in S, \ else \ 0 \qquad (h = 1, \ldots, n) \ .$$

PROOF: by induction on n. For $n := 1$ we prove

$$(3.6) \qquad f(x) = (f(1) \wedge x) \vee f(0) \qquad (\forall x \in L)$$

by algebraic induction. The identities $a = (a \wedge x) \vee a$ and $x = (1 \wedge x) \vee 0$ establish the first step. Then from (3.6) and $g(x) = (g(1) \wedge x) \vee g(0)$ ($\forall x \in L$) we deduce, using (3.2), distributivity, $f(0) \leq f(1)$ (cf. Proposition 3.1) and (3.3), that

$$(f \wedge g)(x) = f(x) \wedge g(x)$$

$$= (f(1) \wedge g(1) \wedge x) \vee (f(1) \wedge g(0) \wedge x) \vee (f(0) \wedge g(1) \wedge x) \vee (f(0) \wedge g(0))$$

$$= (f(1) \wedge g(1) \wedge x) \vee (f(0) \wedge g(0)) = ((f \wedge g)(1) \wedge x) \vee (f \wedge g)(0) ,$$

$$(f \vee g)(x) = ((f(1) \vee g(1)) \wedge x) \vee f(0) \vee g(0) = ((f \vee g)(1) \wedge x) \vee (f \vee g)(0) .$$

Now suppose the property is true for $n-1$. In view of Corollary 2.3.8 of Proposition 2.3.5, for each $\alpha \in \{0,1\}$, the function $f_\alpha : L^{n-1} \longrightarrow L$ defined by $f_\alpha(x_1, \ldots, x_{n-1}) = f(x_1, \ldots, x_{n-1}, \alpha)$ is algebraic and for each $(a_1, \ldots, a_{n-1}) \in L^{n-1}$, the function $g_{a_1 \ldots a_{n-1}} : L \longrightarrow L$ defined by $g_{a_1, \ldots, a_{n-1}}(x) = f(a_1, \ldots, a_{n-1}, x)$ is algebraic. Then for every $(a_1, \ldots, a_n) \in L^n$, setting $N' = \{1, \ldots, n-1\}$, we get

$$f(a_1, \ldots, a_n) = g_{a_1 \ldots a_{n-1}}(a_n)$$

$$= (g_{a_1 \ldots a_{n-1}}(1) \wedge a_n) \vee g_{a_1 \ldots a_{n-1}}(0)$$

$$= (f(a_1, \ldots, a_{n-1}, 1) \wedge a_n) \vee f(a_1, \ldots, a_{n-1}, 0)$$

$$= (f_1(a_1, \ldots, a_{n-1}) \wedge a_n) \vee f_0(a_1, \ldots, a_{n-1})$$

$$= ((\bigvee_{S' \subseteq N'} f_1(\delta_{S'1}, \ldots, \delta_{S',n-1}) \wedge \bigwedge_{j \in S'} a_j)) \wedge a_n) \vee$$

$$\vee (\bigvee_{S' \subseteq N'} f_0(\delta_{S'1}, \ldots, \delta_{S',n-1}) \wedge \bigwedge_{j \in S'} a_j)$$

$$= (\bigvee_{S' \subseteq N'} f(\delta_{S'1}, \ldots, \delta_{S',n-1}, 1) \wedge ((\bigwedge_{j \in S'} a_j) \wedge a_n) \vee$$

$$\vee (\bigvee_{S' \subseteq N'} f(\delta_{S'1}, \ldots, \delta_{S',n-1}, 0) \wedge \bigwedge_{s \in S'} a_j)$$

$$= \bigvee \{ f(\delta_{S1}, \ldots, \delta_{S,n-1}, \delta_{Sn}) \wedge \bigwedge_{h \in S} a_h \mid n \in S \}\} \vee$$

$$\vee \bigvee \{ f(\delta_{S1}, \ldots, \delta_{S,n-1}, \delta_{Sn}) \wedge \bigwedge_{h \in S} a_h \mid n \notin S \}\} .$$

$\square$

Theorem 3.1. (Goodstein [1967]). *Let* $(L; \wedge, \vee, 0, 1)$ *be a bounded distributive lattice and* $n \in \mathbf{N}$. *Set* $N = \{1, \ldots, n\}$. *Then:* α) *The following conditions are equivalent for a function* $f : L^n \longrightarrow L$:

(i) *f is a lattice function;*

(ii) *f can be represented in the form*

$$(3.7) \qquad f(x_1, \ldots, x_n) = \bigvee_{S \subseteq N} (a_S \wedge \bigwedge_{h \in S} x_h) \qquad (\forall x_1, \ldots, x_n \in L) ,$$

where $a_s \in L$ *for every* $S \subseteq N$;

(iii) f can be represented in the form (3.7), where

(3.8)
$$S \subseteq T \Longrightarrow a_S \leq a_T .$$

β) The coefficients a_S in the representation (iii) are uniquely determined by

(3.9)
$$a_S = f(\delta_{S1}, \ldots, \delta_{Sn}) \qquad (\forall S \subseteq N) .$$

PROOF: (i)$\Longrightarrow$(iii): The representation (3.4)-(3.5) provided by Lemma 3.1 is of the form (3.7). Besides, if $S \subseteq T$ then $\delta_{Sh} \leq \delta_{Th}$ ($h = 1, \ldots, n$), hence $f(\delta_{S1}, \ldots, \delta_{Sn}) \leq f(\delta_{T1}, \ldots, \delta_{Tn})$ by Proposition 3.1, so that condition (3.8) is fulfilled.

(iii)$\Longrightarrow$(ii): Trivial.

(ii)$\Longrightarrow$(i): Identity (3.7) expresses the following equality in the lattice $O_n(L)$ of n−ary operations of L (cf. 2.(3.3)):

$$f = \bigvee_{S \subseteq N} \left(f_{a_S} \wedge \bigwedge_{h \in S} \pi_{hL}^n \right) .$$

β) For every $T \subseteq N$, the representation (3.8)-(3.9) yields

$$f(\delta_{T1}, \ldots, \delta_{Tn}) = \bigvee \{ a_S \mid \bigwedge_{h \in S} \delta_{Th} = 1 \}$$

$$= \bigvee \{ a_S \mid h \in S \Longrightarrow \delta_{Th} = 1 \} = \bigvee \{ a_S \mid S \subseteq T \} = a_T .$$

$\square$

Corollary 3.1. *A lattice function $f : L^n \longrightarrow L$ is uniquely determined by its restriction to $\{0, 1\}^n$.*
$\square$

Corollary 3.2. *A function $\varphi : \{0, 1\}^n \longrightarrow L$ can be extended to a lattice function $f : L^n \longrightarrow L$ if and only if it is isotone, in which case the extension is unique.*

PROOF: Necessity and uniqueness follow from Proposition 3.1 and Theorem 3.1, respectively. To prove sufficiency, suppose φ is isotone and define $f : L^n \longrightarrow L$ by

$$f(x_1, \ldots, x_n) = \bigvee \{ \varphi(\alpha_1, \ldots, \alpha_n) \wedge \bigwedge_{\alpha_h = 1} x_h \mid \alpha_1, \ldots, \alpha_n \in \{0, 1\} \} .$$

Then f is a lattice function and for every $\beta_1, \ldots, \beta_n \in \{0, 1\}$,

$$f(\beta_1, \ldots, \beta_n) = \bigvee \{ \varphi(\alpha_1, \ldots, \alpha_n) \mid \alpha_1, \ldots, \alpha_n \in \{0, 1\} \ \& \bigwedge_{\alpha_h = 1} \beta_h = 1 \}$$

$$= \bigvee \{ \varphi(\alpha_1, \ldots, \alpha_n) \mid \alpha_1, \ldots, \alpha_n \in \{0, 1\} \ \& \ \alpha_h \leq \beta_h \ (h = 1, \ldots, n) \}$$

$$= \varphi(\beta_1, \ldots, \beta_n) .$$

$\square$

Remark 3.1. Every simple bounded-lattice function $f : L^n \longrightarrow L$ satisfies $f(\alpha_1, \ldots, \alpha_n) \in \{0, 1\}$ $\forall \alpha_1, \ldots, \alpha_n \in \{0, 1\}$. The easy proof is left to the reader. □

Corollary 3.3. *A function* $f : L^n \longrightarrow L$ *is a simple lattice function if and only if it can be represented as a disjunction of monomials* $x_{h_1} \wedge \ldots \wedge x_{h_m}$, $\{h_1, \ldots, h_m\} \subseteq N$.

PROOF: Sufficiency is clear. Necessity follows from Lemma 3.1 and Remark 3.1. □

Corollary 3.4. *The identities common to all bounded distributive lattices coincide with the identities of the lattice* $\{0, 1\}$.

PROOF: From Corollary 2.3.3 of Proposition 2.3.1. □

A similar result for implicative algebras was established by Drabbe [1969].

Remark 3.2. (J.C. Abbott, private communication). The above results do not hold for distributive lattices that are not bounded. More precisely: if every simple lattice function $f : L^n \longrightarrow L$ can be represented in the form $f(x) = (a \wedge x) \vee b$, then L is bounded. For in particular the representation $x = \pi_{1L}^1(x) = (a \wedge x) \vee b$ implies $b \leq x$ for all $x \in L$, that is $b = 0$, therefore $x = a \wedge x \leq a$ for all $x \in L$, showing that $a = 1$. □

The results dual to the above ones are left to the reader.

Remark 3.3. It was noted in Proposition 2.3.6 that every algebraic function has the substitution property. Grätzer [1964] studied the more general case of functions with the substitution property in a bounded distributive lattice. He showed in particular that Corollary 3.1 holds for these functions as well and that a function with the substitution property is a lattice function if and only if its restriction to $\{0, 1\}^n$ is isotone. See also BFE, Chapter 1. □

Schweigert [1975] noted that in a bounded distributive lattice every lattice function of one variable is idempotent and, moreover, this property characterizes distributivity. Schweigert's result can be refined as follows:

Proposition 3.2. *Let* L *be a bounded lattice. For every* $f : L^n \longrightarrow L$ *define* $f^1 = f$, $f^{n+1} = f \circ f^n$. *Then the following conditions are equivalent:*

(i) L *is distributive;*

(ii) every lattice function $f : L \to L$ *satisfies* $f^2 = f$;

(iii) every lattice function $f : L \to L$ *satisfies* $f^3 = f$;

(iv) every function $f : L \to L$ *of the form*

$$f(x) = ((x \vee a) \wedge b) \vee c$$

satisfies $f^3 = f$.

PROOF: (i)$\Longrightarrow$(ii): Since f is of the form $f(x) = (a \wedge x) \vee b$ by (3.6) in Lemma 3.1, it follows that

$$f^2(x) = (a \wedge ((a \wedge x) \vee b)) \vee b = (a \wedge x) \vee (a \wedge b) \vee b = f(x) .$$

(ii)$\Longrightarrow$(iii)$\Longrightarrow$(iv): Immediate.

(iv)$\Longrightarrow$(i): Suppose L is not distributive and prove that (iv) fails. But it is well known that L contains a sublattice isomorphic to the diamond or to the pentagon (cf. the examples after Definition 2.5). Therefore it suffices to prove that (iv) fails for the diamond and for the pentagon. In the case of the diamond we get $f(0) = c$, $f(c) = 1$, and $f(1) = 1$, hence $f^3(0) = 1 \neq c = f(0)$. In the case of the pentagon the function g defined by $g(x) = ((x \vee c) \wedge b) \vee a$ satisfies $g(0) = a$, $g(a) = b$ and $g(b) = b$, hence $g^3(0) = b \neq a = g(0)$.   □

Proposition 3.3. (Simovici and Reischer [1986], Reischer and Simovici [1987]). *Suppose $(L; \cdot, \vee, 0, 1)$ is a bounded distributive lattice and $h : L \longrightarrow L$ is a dual lattice automorphism. Then $(L; \cdot, \vee, h, 0, 1)$ is a Boolean algebra if and only if $f^3 = f$ for every algebraic function $f : L \longrightarrow L$.*

PROOF: Necessity is well known: if $f(x) = ax \vee bx'$ then $f^2(x) = (a \vee b)x \vee ab$ and $f^3(x) = f(x)$. To prove sufficiency, note first that the lattice L is distributive by Proposition 3.2. Moreover, since we have in particular $h^3 = h$ and h is injective, it follows that $h^2 = 1_L$. Now take an arbitrary element $a \in L$ and define $g : L \longrightarrow L$ by $g(x) = (a \vee x)h(x)$. Then

$$g^2(x) = (a \vee (a \vee x)h(x))h((a \vee x)h(x))$$

$$= (a \vee xh(x))(h(a)h(x) \vee x) = ah(a)h(x) \vee ax \vee ax \vee xh(x)$$

$$= ax \vee (ah(a) \vee x)h(x) ,$$

$$g^3(x) = a(a \vee x)h(x) \vee (ah(a) \vee (a \vee x)h(x))(h(a)h(x) \vee x)$$

$$= ah(x) \vee ah(a)x \vee xh(x) = ah(a) \vee g(x) .$$

But g is an algebraic function, hence $g^3 = g$, therefore $ah(a)x \leq g(x)$. Taking $x := 1$ we obtain $ah(a) \leq h(1) = 0$. Therefore $ah(a) = 0$, hence $a \vee h(a) = 1$ by duality, showing that $h(a) = a'$.   □

Proposition 3.4. (Simovici and Reischer [1986]). *Suppose $(L; \cdot, \vee, 0, 1)$ is a bounded lattice and $h : L \longrightarrow L$ is a dual lattice endomorphism. Then $(L; \cdot, \vee, h, 0, 1)$ is a Boolean algebra if and only if for every non-constant algebraic function $f : L \longrightarrow L$, f^2 is non-constant and it is of the form $f^2(x) = px \vee q$, with $p \geq q$.*

PROOF: Necessity is well known: if $f(x) = ax \vee bx'$ then $f^2(x) = (a \vee b)x \vee ab$ and if the function f is not a constant, then $a \neq b$, hence $a \vee b \neq ab$.

To prove sufficiency we first show that L is a distributive lattice, using the same technique as in the proof of Proposition 3.2.

Consider first the pentagon and define $f(x) = (x \vee c)b \vee a$. Then $f(0) = a$ and $f(a) = b$, showing that f is not a constant; yet f^2 is a constant:

$$f^2(x) = ((x \vee c)b \vee a = b \vee a = b\,.$$

In the case of the diamond define $f(x) = (x \vee a)b \vee c$. Then $f(0) = c$ and $f(c) = 1$, while

$$f^2(x) = ((x \vee a)b \vee c \vee a)b \vee c = b \vee c = 1\,.$$

Further the hypothesis implies $h^2(x) = px \vee q$ for some $p, q \in L$ with $p \geq q$. Then $q = h^2(0) = 0$, hence $p = h^2(1) = 1$, showing that $h^2(x) = x$ for every x.

Now we define $g : L \longrightarrow L$ by $g(x) = xh(x)$ and prove that g is a constant. Otherwise

$$g^2(x) = xh(x)h(xh(x)) = xh(x)(h(x) \vee x) = xh(x) = ax \vee b$$

for some $a, b \in L$ with $a \geq b$. Then $b = g^2(0) = g(g(0)) = g(0) = 0$, hence, taking also into account that $h(1) = 0$, we get

$$a = a \vee b = g^2(1) = g(g(1)) = g(0) = 0\,.$$

Thus $g^2(x) = 0$, a contradiction.

We have thus proved that g is a constant function. Therefore $xh(x) = g(x) = g(0) = 0$, hence $x \vee h(x) = 1$ by duality. $\qquad\square$

We now pass to the study of lattice equations.

Remark 3.4. (Goodstein [1967]). In every lattice the problem of solving an equation is equivalent to the problem of solving an inequality. For

$$f = g \Longleftrightarrow f \wedge g = f \vee g \Longleftrightarrow f \vee g \leq f \wedge g\,,$$

$$f \leq g \Longleftrightarrow f \wedge g = f \Longleftrightarrow f \vee g = g\,.$$

$\qquad\square$

In Lemma 3.2 and Theorem 3.2 below the consistency condition is due to Goodstein [1967], who also pointed out a particular solution, while the description of the set of solutions by a system of recurrent inequalities was given in Rudeanu [1968].

Lemma 3.2. *Suppose L is a bounded distributive lattice and $f, g : L \longrightarrow L$ are lattice functions. Then the inequality $f(x) \leq g(x)$ is consistent if and only if $f(0) \leq g(1)$, in which case the set of solutions is characterized by the inequalities*

(3.10) $$f(0) \leq g(0) \vee x \ \& \ f(1) \wedge x \leq g(1)\,.$$

PROOF: Taking into account Proposition 3.1, (3.6) and

$$g(x) = (g(1) \wedge x) \vee g(0) = (g(1) \vee g(0)) \wedge (x \vee g(0)) = g(1) \wedge (x \vee g(0))\,,$$

we see that the inequality $f(x) \leq g(x)$ is equivalent to the system of inequalities $f(0) \leq g(1)$, (3.10) and $f(1) \wedge x \leq x \vee g(0)$. But the last inequality holds identically. Therefore if $f(0) \leq g(1)$ then $f(x) \leq g(x)$ is equivalent to (3.10), while if $f(x) \leq g(x)$ for some x then $f(0) \leq g(1)$. $\qquad\square$

Theorem 3.2. *Suppose L is a bounded distributive lattice, $n \in \mathbb{N}$ and $f, g :$ $L^n \longrightarrow L$ are latttice functions. Then the inequality*

(3.11) $$f(x_1, \ldots, x_n) \leq g(x_1, \ldots, x_n)$$

is consistent if and only if

(3.12) $$f(0, \ldots, 0) \leq g(1, \ldots, 1),$$

in which case the set of solutions to (3.11) is characterized by the system of recurrent inequalities

(3.13.1)
$$f(0, \ldots, 0) \leq g(0, 1, \ldots, 1) \vee x_1,$$
$$f(1, 0, \ldots, 0) \wedge x_1 \leq g(1, \ldots, 1),$$

(3.13.h)
$$f(x_1, \ldots, x_{h-1}, 0, \ldots, 0) \leq g(x_1, \ldots, x_{h-1}, 0, 1, \ldots, 1) \vee x_h,$$
$$f(x_1, \ldots, x_{h-1}, 1, 0, \ldots, 0) \wedge x_h \leq g(x_1, \ldots, x_{h-1}, 1, \ldots, 1)$$

$$(2 \leq h \leq n - 1),$$

(3.13.n)
$$f(x_1, \ldots, x_{n-1}, 0) \leq g(x_1, \ldots, x_{n-1}, 0) \vee x_n,$$
$$f(x_1, \ldots, x_{n-1}, 1) \wedge x_n \leq g(x_1, \ldots, x_{n-1}, 1).$$

PROOF: If $(x_1, \ldots, x_n)$ is a solution of (3.1) then Proposition 3.1 implies

$$f(0, \ldots, 0) \leq f(x_1, \ldots, x_n) \leq g(x_1, \ldots, x_n) \leq f(1, \ldots, 1).$$

Conversely, we prove by induction on n that (3.12) implies the equivalence between (3.11) and (3.13). For $n := 1$ the result is established in Lemma 3.2. To prove the passage from $n - 1$ to n we apply the same technique as in the proof of Theorem 3.1, i.e., we use the fact that the function obtained from a lattice function by fixing certain variables is also a lattice function. First we write inequality (3.11) in the form

(3.14)
$$(f(x_1, \ldots, x_{n-1}, 1) \wedge x_n) \vee f(x_1, \ldots, x_{n-1}, 0)$$
$$\leq (g(x_1, \ldots, x_{n-1}, 1) \wedge x_n) \vee g(x_1, \ldots, x_{n-1}, 0);$$

according to Lemma 3.2, inequality (3.14) is equivalent to the system

(3.15.1) $$f(x_1, \ldots, x_{n-1}, 0) \leq g(x_1, \ldots, x_{n-1}, 1),$$

(3.15.2) $$f(x_1, \ldots, x_{n-1}, 0) \leq g(x_1, \ldots, x_{n-1}, 0) \vee x_n,$$

(3.15.3) $$f(x_1, \ldots, x_{n-1}, 1) \wedge x_n \leq g(x_1, \ldots, x_{n-1}, 1),$$

therefore (3.11) is equivalent to (3.15).

According to the inductive hypothesis, condition (3.12) implies that inequality (3.15.1) is consistent and its solutions are given by the system of recurrent inequalities (3.13.h), $h = 1, 2, \ldots, n - 1$. Since the system (3.15.2), (3.15.3) coincides with (3.13.n), it follows that system (3.15) is equivalent to (3.13). $\square$

Corollary 3.5. (Goodstein [1967]). *Every lattice function* $f : L^n \longrightarrow L$ *maps* L *onto the interval* $[f(0,\ldots,0), f(1,\ldots,1)]$.

PROOF: Take $a \in [f(0,\ldots,0), f(1,\ldots,1)]$ and prove the consistency of equation $f(x_1,\ldots,x_n) = a$. In view of Remark 3.4, the equation is equivalent to

$$f(x_1,\ldots,x_n) \vee a \leq f(x_1,\ldots,x_n) \wedge a ,$$

whose consistency condition is

$$f(0,\ldots,0) \vee a \leq f(1,\ldots,1) \wedge a$$

by Theorem 3.2. But the latter inequality holds because both sides equal a. $\square$

Corollaries 3.6 and 3.7 below are taken from Rudeanu [1968].

Corollary 3.6. *If condition* (3.12) *holds, then every vector* $(x_1,\ldots,x_n)$ *satisfying*

(3.16.1) $$f(0,\ldots,0) \leq x_1 \leq g(1,\ldots,1) ,$$

(3.16.h)
$$f(x_1,\ldots,x_{h-1},0,\ldots,0) \leq x_h \leq g(x_1,\ldots,x_{h-1},1,\ldots,1)$$
$$(2 \leq h \leq n) ,$$

is a solution of (3.11). $\square$

Corollary 3.7. *If* L *is a double Heyting algebra and condition* (3.12) *holds, then the solutions of* (3.11) *are characterized by the system of recurrent inequalities*

(3.17.1) $$f(0,\ldots,0)-g(0,1,\ldots,1) \leq x_1 \leq f(1,0,\ldots,0) \to g(1,\ldots,1) ,$$

(3.17.h)
$$f(x_1,\ldots,x_{h-1},0,\ldots,0) - g(x_1,\ldots,x_{h-1},0,1,\ldots,1) \leq x_h$$
$$\leq f(x_1,\ldots,x_{h-1},1,0,\ldots,0) \to g(x_1,\ldots,x_{h-1},1,\ldots,1)$$
$$(2 \leq h \leq n-1) ,$$

(3.17.n)
$$f(x_1,\ldots,x_{n-1},0) - g(x_1,\ldots,x_{n-1},0) \leq x_n \leq$$
$$\leq f(x_1,\ldots,x_{n-1},1) \to g(x_1,\ldots,x_{n-1},1) .$$
$\square$

For an attempt to extend some of the above results to distributive lattices that are not bounded see Rudeanu [1968].

Theorem 3.3 below is due to Beazer [1974c]. It generalizes the corresponding result for pseudocomplemented-lattice functions obtained by Goodstein [1967].

Theorem 3.3. *Let* L *be a pseudocomplemented distributive lattice or a Heyting algebra, with* $0 \neq 1$, *and* $n \in \mathbf{N}$. *Suppose* $f : L^n \longrightarrow L$ *has the substitution property and define* $g(x_1^{**},\ldots,x_n^{**}) = (f(x_1,\ldots,x_n))^{**}$ *for every* $x_1,\ldots,x_n \in L$. *Then:*

(i) $g : (R(L))^n \longrightarrow R(L)$ *is a Boolean function ;*

(ii) $f(x_1,\ldots,x_n) = 0 \Longleftrightarrow g(x_1^{**},\ldots,x_n^{**}) = 0$;

(iii) $f(x_1,\ldots,x_n) = 0$ *holds identically if and only if* $f(\alpha_1,\ldots,\alpha_n) = 0$ *for every* $\alpha_1,\ldots,\alpha_n \in \{0,1\}$;

(iv) *the equation* $f(x_1,\ldots,x_n) = 0$ *has solutions if and only if*

(3.18)
$$\bigwedge\{f(\alpha_1,\ldots,\alpha_n) \mid \alpha_1,\ldots,\alpha_n \in \{0,1\}\} = 0 .$$

PROOF: (i) In view of Theorem 2.1 or of Theorem 2.2, $R(L)$ is a Boolean algebra and the map $\varphi : L \longrightarrow R(L)$, $\varphi(x) = x^{**}$, is a surjective homomorphism. Therefore $g : (R(L))^n \longrightarrow R(L)$ is well defined and has the substitution property by Lemma 2.3.3. But in a Boolean algebra functions with the substitution property coincide with Boolean functions; cf. Grätzer [1962]; see also BFE, Theorem 1.13.

(ii) follows from the fact that $0^{**} = 0$, hence $x = 0 \Longleftrightarrow x^{**} = 0$.

To prove (iii) and (iv) note first that $\alpha^{**} = \alpha$ for $\alpha \in \{0,1\}$.

(iii) It follows from (ii) that f is identically zero if and only if g is so. But the latter condition is equivalent to

(3.19)
$$g(\alpha_1,\ldots,\alpha_n) = 0 \qquad (\forall \alpha_1,\ldots,\alpha_n \in \{0,1\}) ;$$

see e.g. BFE, Theorem 2.13. Finally (ii) implies that (3.19) is equivalent to $f(\alpha_1,\ldots,\alpha_n) = 0 \ \forall \alpha_1,\ldots,\alpha_n \in \{0,1\}$.

(iv) It follows from (ii) that the equation $f(x_1,\ldots,x_n) = 0$ is consistent if and only if the equation $g(x_1^{**},\ldots,x_n^{**}) = 0$ is so. But the latter condition is equivalent to

(3.20)
$$\bigwedge\{g(\alpha_1,\ldots,\alpha_n) \mid \alpha_1,\ldots,\alpha_n \in \{0,1\}\} = 0 ;$$

see e.g. BFE, Theorem 2.3. But

$$\bigwedge\{g(\alpha_1,\ldots,\alpha_n) \mid \alpha_1,\ldots,\alpha_n \in \{0,1\}\}$$

$$= \bigwedge\{(f(\alpha_1,\ldots,\alpha_n))^{**} \mid \alpha_1,\ldots,\alpha_n \in \{0,1\}\}$$

$$= (\bigwedge\{f(\alpha_1,\ldots,\alpha_n) \mid \alpha_1,\ldots,\alpha_n \in \{0,1\}\})^{**}$$

by (2.11), therefore (3.20) is equivalent to (3.18).   □

At this point we recall that the basic concepts of *general solution* and *reproductive solution* of an equation are given in Definitions 1.1.2 and 1.1.3, respectively. However, instead of saying that a function $f : T \longrightarrow T$ is a general (reproductive) solution, we will rather say, as in BFE, pp.56-57, that formula $x = f(t)$ *defines the* general (reproductive) solution.

The next corollary is a generalization of a theorem due to Löwenheim; see e.g. BFE, Theorem 2.11.

Corollary 3.8. (Beazer [1974c]). *Let* L *be a Stone algebra and* $f : L^n \longrightarrow L$ *a function with the substitution property. If* $f(\xi_1,\ldots,\xi_n) = 0$ *for some* $(\xi_1,\ldots,\xi_n)$

$\in L^n$, then formulas

(3.21)
$$x_k = (\xi_k \wedge f(t_1, \ldots, t_n)) \vee (t_k \wedge (f(t_1, \ldots, t_n))^*)$$
$$(k = 1, \ldots, n)$$

define the reproductive general solution of equation $f(x_1, \ldots, x_n) = 0$.

PROOF: The function g associated with f in Theorem 3.3 satisfies $g(\xi_1^{**}, \ldots, \xi_n^{**}) = 0$. Taking into account (2.11) and (2.17), we see that (3.21) implies

$$x_k^{**} = (\xi_k^{**} \wedge g(t_1^{**}, \ldots, t_n^{**})) \vee (t_k^{**} \wedge (g(t_1^{**}, \ldots, t_n^{**}))^*)$$

$(k = 1, \ldots, n)$, hence from Löwenheim's result BFE, Theorem 2.11, we infer $g(x_1^{**}, \ldots, x_n^{**}) = 0$, therefore $f(x_1, \ldots, x_n) = 0$ by Theorem 3.3. Conversely, if $f(t_1, \ldots, t_n) = 0$, then (3.21) reduces to $x_k = t_k$ ($k = 1, \ldots, n$). $\qquad\square$

Theorem 3.4. (Beazer [1974c]). *Suppose L is a Stone algebra having the smallest dense element d (i.e., $d = \min\{x \in L \mid x^* = 0\}$), $n \in \mathbf{N}$ and $f, g : L^n \longrightarrow L$ are functions with the substitution property. Then $f = g$ if and only if*

(3.22) $$f(\delta_1, \ldots, \delta_n) = g(\delta_1, \ldots, \delta_n) \qquad (\forall \delta_1, \ldots, \delta_n \in \{0, d, 1\}).$$

PROOF: Suppose the functions f and g agree on the set $\{0, d, 1\}$ but

$$a = f(a_1, \ldots, a_n) \neq g(a_1, \ldots, a_n) = b$$

for some $a_h \in L$ ($h = 1, \ldots, n$). Since the subdirectly irreducible Stone algebras are the chains $\{0, 1\}$ and $\{0, 1/2, 1\}$ (cf. Balbes and Dwinger [1971], Corollary VIII.6.3), there exists a subdirect decomposition $\varphi : L \longrightarrow \prod_{t \in T} A_t$ where for each $t \in T$, $A_t = \{0, 1\}$ or $A_t = \{0, 1/2, 1\}$, the homomorphisms $p_t \varphi$ are surjective and the intersection of their kernels is Δ_L (see e.g. Ch.2, §1). Therefore there is a surjective homomorphism $\varphi_1 : L \longrightarrow \{0, 1\}$ such that $\varphi_1(a) \neq \varphi_1(b)$ or there is a surjective homomorphism $\varphi_2 : L \longrightarrow \{0, 1/2, 1\}$ such that $\varphi_2(a) \neq \varphi_2(b)$.

In the first case it follows from Lemma 2.3.4 with $A := L$, $B := \{0, 1\}$ and $\psi := \varphi_1$, that

$$\psi(a) = \psi(f(a_1, \ldots, a_n)) = \psi(g(a_1, \ldots, a_n)) = \psi(b),$$

a contradiction. In the second case we use quite a similar proof with $A := L$, $B := \{0, d, 1\}$ and $\psi := \alpha \varphi_2$, where $\alpha : \{0, 1/2, 1\} \longrightarrow B$ is the obvious isomorphism. To do so we still must prove that $\psi(d) = d$, i.e., that $\varphi_2(d) = 1/2$. But

$$(\varphi_2(d))^* = \varphi_2(d^*) = \varphi_2(0) = 0,$$

therefore $\varphi_2(d) \in \{1/2, 1\}$. So if $\varphi_2(d) \neq 1/2$ then $\varphi_2(d) = 1$. It follows that $\varphi_2(y) = 1$ for every dense element $y \in L$, because $d \leq y$. Now take an arbitrary element $x \in L$. It follows, since $x = x \wedge x^{**} = x^{**} \wedge (x \vee x^*)$ and $x \vee x^*$ is a dense element, that

$$\varphi_2(x) = \varphi_2(x^{**}) \wedge \varphi_2(x \vee x^*) = \varphi_2(x^{**}) = (\varphi_2(x))^{**},$$

therefore $\varphi_2(x) \in \{0, 1\}$, which contradicts the surjectivity of φ_2.   □

We are now going to provide a complete solution to the problem of solving a system of lattice equations in one unknown, in the case of a *double Heyting lattice*, i.e., a double Heyting algebra viewed as a bounded distributive lattice $(L; \wedge, \vee, 0, 1)$.

Note that in view of Theorem 3.1, every lattice function $f : L \longrightarrow L$ can be uniquely represented in the form $f(x) = (a \wedge x) \vee b$ with $a \geq b$, namely $a = f(1)$ and $b = f(0)$.

Lemma 3.3. (Beazer [1974a]). *In a double Heyting lattice, an equation*

(3.23) $(a \wedge x) \vee b = (c \wedge x) \vee d$, *where* $a \geq b$ & $c \geq d$,

is consistent if and only if

(3.24) $b \leq c$ & $d \leq a$,

in which case the set of solutions is the interval

(3.25) $[b + d, a \cdot c]$.

PROOF: It follows from Lemma 3.2 that if equation (3.23) is consistent, then relations (3.24) hold and the solutions x satisfy conditions

(3.26) $b \leq d \vee x$ & $a \wedge x \leq c$ & $d \leq b \vee x$ & $c \wedge x \leq a$,

which are equivalent to

(3.27) $x \geq b - d$ & $x \leq a \rightarrow c$ & $x \geq d - b$ & $x \leq c \rightarrow a$,

by (2.20) and (2.20'). According to (2.32) and (2.32'), system (3.27) is further equivalent to

(3.28) $x \geq b + d$ & $x \leq a \cdot c$,

showing that the solutions of (3.23) are in the interval (3.25). It remains to prove that if (3.24) holds, then the interval (3.25) is not empty and its elements satisfy (3.23).

Suppose (3.24) holds. Set $e = b - d = \min\{x \mid d \vee x \geq b\}$. Then $e \wedge a \leq e \leq c$ because $d \vee c = c \geq b$, hence $e \leq a \rightarrow c$. Also, $e \wedge c \leq e \leq a$ because $d \vee a = a \geq b$, hence $e \leq c \rightarrow a$, therefore $e \leq a \cdot c$ by (2.32). One proves similarly that $d - b \leq a \cdot c$, therefore $b + d \leq a \cdot c$ by (2.32'), showing that $[b + d, a \cdot c] \neq \emptyset$. Finally if x satisfies (3.28) then it satisfies the equivalent system (3.26) (cf. above). From (3.24) and (3.26) one deduces, via Lemma 3.1, that x satisfies (3.23).   □

Proposition 3.5. *In a double Heyting lattice, a system of equations of the form*

$$(3.29) \qquad (a_k \wedge x) \vee b_k = (c_k \wedge x) \vee d_k \qquad (k = 1, \ldots, m),$$

where

$$(3.30) \qquad a_k \geq b_k \ \& \ c_k \geq d_k \qquad (k = 1, \ldots, m),$$

is consistent if and only if

$$(3.31.1) \qquad b_k \leq c_k \ \& \ d_k \leq a_k \qquad (k = 1, \ldots, m),$$

$$(3.31.2) \qquad b_k + d_k \leq a_h \cdot c_h \qquad (h, k = 1, \ldots, m; h \neq k),$$

in which case the set of solutions is the interval

$$(3.32) \qquad [\bigvee_{k=1}^{m} (b_k + d_k), \bigwedge_{k=1}^{m} (a_k \cdot c_k)].$$

PROOF: If system (3.29)&(3.30) is consistent, then Lemma 3.3 implies that relations (3.31.1) hold and the set of solutions is the intersection of the solution sets

$$(3.33) \qquad [b_k + d_k, a_k \cdot c_k] \qquad (k = 1, \ldots, m)$$

of the m equations (3.29) and this intersection is precisely the interval (3.32). So the interval (3.32) is not empty, therefore relations (3.31.2) hold.

Conversely, suppose relations (3.31) hold. Then from (3.31.1) we infer, again via Lemma 3.3, that each equation (3.29) is consistent and hence their solution sets (3.33) are not empty, that is,

$$(3.34) \qquad b_k + d_k \leq a_k \cdot c_k \qquad (k = 1, \ldots, m).$$

We see from (3.31.2) and (3.34) that the intersection (3.32) of the intervals (3.33) is not empty, that is, the solution set (3.32) of the system (3.29), (3.30) is not empty. $\qquad \square$

Remark 3.5. (Beazer [1974c]). Consider again the chain $L_3 = \{0, 1/2, 1\}$, regarded either as a bounded lattice or as a Stone algebra. Then L_3 has two non-trivial lattice congruences, namely θ with $(1/2, 1) \in \theta$ and which is also a Stone congruence, and τ with $(0, 1/2) \in \tau$. It is easy to check that the function $f : L \to L$, $f(0) = f(1/2) = 1$, $f(1) = 1/2$, has the substitution property. However f is not a lattice function because it is not isotone, and it is not a Stone function either. For otherwise, using (2.7) and (2.15), f would be of the form

$$f(x) = (a \wedge x) \vee (b \wedge x^*) \vee (c \wedge x^{**}),$$

hence $1/2 = f(1) = a \vee c$ and

$$1 = f(1/2) = (a \wedge 1/2) \vee c = (a \vee c) \wedge (1/2 \vee c) = 1/2,$$

a contradiction. $\qquad \square$

We conclude this chapter by going in the opposite direction of larger classes of lattices. The following result generalizes certain equations which occur in the theory of fuzzy relation equations; cf. the monograph by Di Nola, Sessa, Pedrycz and Sanchez [1989].

Proposition 3.6. (W.-L. Liu [1990]). *Let: L be a distributive lattice, $m, n \in \mathbf{N}$,*

$$(3.35) \qquad I_k = \{k_1, k_2, \ldots, k_{n(k)}\} \subseteq \{1, \ldots, n\} \qquad (k = 1, \ldots, m) ,$$

$$(3.36) \qquad J_s = \{k \in \{1, \ldots, m\} \mid s \in I_k\} \qquad (s = 1, \ldots, n) .$$

Then a system of equations over L of the form

$$(3.37) \qquad \bigwedge_{s \in I_k} x_s = a_k \qquad (k = 1, \ldots, m)$$

is consistent if and only if

$$(3.38) \qquad \begin{aligned} & a_k = \bigvee \{ a_{k_1} \wedge a_{k_2} \wedge \ldots \wedge a_{k_{n(k)}} \mid k_i \in J_{k_i} \ (i = 1, \ldots, m) \} \\ & (k = 1, \ldots, m) , \end{aligned}$$

in which case it has the least solution $\Xi = (\xi_1, \ldots, \xi_n)$ given by

$$(3.39) \qquad \xi_s = \bigvee_{k \in J_s} a_k \qquad (s = 1, \ldots, n) .$$

PROOF: It follows from (3.39) that $\bigwedge_{s \in I_k} \xi_s$ equals the right side of (3.38), therefore the proposition can be paraphrased as follows: system (3.37) is consistent if and only if Ξ is a solution, in which case Ξ is the least solution. It remains to prove that, if the system is consistent, then Ξ is the least solution. But $s \in I_k \Longleftrightarrow k \in J_s \Longrightarrow a_k \leq \xi_s$, hence

$$(3.41) \qquad a_k \leq \bigwedge_{s \in I_k} \xi_s \qquad (k = 1, \ldots, m) .$$

Now suppose $(x_1, \ldots, x_n)$ is a solution. For every $s \in \{1, \ldots, n\}$ we have $k \in J_s \Longleftrightarrow s \in I_k \Longrightarrow a_k \leq x_s$, hence we get in turn

$$(3.42) \qquad \xi_s \leq x_s \qquad (s = 1, \ldots, n) ,$$

$$\bigwedge_{s \in I_k} \xi_s \leq \bigwedge_{s \in I_k} x_s = a_k \qquad (k = 1, \ldots, m) ,$$

$$(3.43) \qquad a_k = \bigwedge_{s \in I_k} \xi_s \qquad (k = 1, \ldots, m) .$$

$\square$

As concerns equations in the class of all lattices, the only paper we know of is Marenich [1997], in which the author counts the number of solutions of certain equations in finite lattices, with applications to combinatorics.

4. Equational compactness of lattices and Boolean algebras

While the various systems of equations studied in algebra consist in most cases of finitely many equations, in this chapter we deal with systems of infinitely many algebraic equations, where each equation comprises finitely many unknowns.

An algebra A is said to be *equationally compact* provided every system of algebraic equations over A has a solution whenever every finite subsystem of it is consistent.

In the first section we present this concept and several refinements of it, introduced by Mycielski. The next two sections study various types of equational compactness within the lattice-theoretical framework.

1 Abstract equational compactness

The concepts of algebraic function and polynomial dealt with in this section are those introduced in Ch. 2.

Definition 1.1. By an *algebraic (a polynomial) system of equations* over an algebra A is meant a system of equations of the form

$$(1.1) \qquad p_k(\ldots, x_j, \ldots) = q_k(\ldots, x_j, \ldots) \qquad (k \in K),$$

where p_k and q_k are algebraic functions (polynomials) over A, for all $k \in K$. $\square$

Definition 1.2. The algebra A will be called *equationally (2-equationally) compact* provided every algebraic system (1.1) is consistent whenever every finite subsystem (every 2-equation subsystem) of it has a solution. The concepts of *equationally (2-equationally) monocompact* algebra are obtained from the above ones by replacing systems of the form (1.1) by systems of the form

$$(1.2) \qquad p_k(x) = q_k(x) \qquad (k \in K),$$

i.e., in one unknown. The concepts of *weakly equationally (weakly 2-equationally) compact* algebra and *weakly equationally (weakly 2-equationally) monocompact* algebra are obtained from the corresponding previous concepts by replacing algebraic systems of equations by polynomial systems of equations. $\square$

Remark 1.1. The following trivial implications hold:

(1.3) $\{weakly\}\ 2 - equationally\ \{mono\}compact$
$\implies \{weakly\}\ equationally\ \{mono\}compact,$

(1.4) $\{weakly\}\ \{2-\}equationally\ compact$
$\implies \{weakly\}\ \{2-\}equationally\ monocompact,$

(1.5) $\{2-\}equationally\ \{mono\}compact$
$\implies weakly\ \{2-\}equationally\ \{mono\}compact,$

where the notation of the form $\{a\}\alpha\{b\}\beta$ stands for each of the four possibilities

$$a\alpha b\beta\ ,\ \alpha b\beta\ ,\ a\alpha\beta\ ,\ \alpha\beta\ ,$$

the same combination being taken in both sides of the implication.   □

Remark 1.2. If $(A; (f_i)_{i\in I})$ and $(A; (f_i)_{i\in H})$ are two algebras defined on the same set A and $I \subseteq H$, then each type of equational compactness with respect to the latter structure implies the same type of equational compactness with respect to the former structure.   □

The concepts of equational compactness and weak equational compactness were introduced by Mycielski [1964], starting from a definition of Łoś of the notion of algebraic compactness of Abelian groups introduced by Kaplansky. The analogy with topological compactness is striking, but in fact deeper connections have been established: every compact topological algebra is equationally compact and for certain classes of algebras every equationally compact algebra is a retract of a compact topological algebra; cf. Mycielski (op.cit.) and Węglorz [1966]. Equations in one unknown and subsystems consisting of just two equations have appeared in the work of Mycielski's followers, so that it seems natural to consider all possible combinations of these aspects, as we have done above. We will disregard the other variants studied by Mycielski: the restriction to systems with the number of equations limited by a fixed cardinal, and extensions of equational compactness to relational systems instead of algebras. The reason for doing so is that we are interested only in those types of equational compactness which have been investigated for lattices, Boolean algebras and related structures.

2 Equational monocompactness of semilattices and lattices

In this section we characterize equationally monocompact semilattices (Theorem 2.1) and 2-equationally monocompact lattices (Theorem 2.2). The results are due to Grätzer and Lakser [1969] and Beazer [1974a], respectively.

Definition 2.1. A non-empty subset D of a poset P is said to be *directed (downward directed)* provided every two elements of D have an upper bound (a lower bound) in D. P is called *join complete (meet complete)* if every non-empty subset $S \subseteq P$ has a join (a meet).   □

Theorem 2.1. *A join semilattice* $(L; \vee)$ *is equationally monocompact if and only if it satisfies the following conditions:*

(i) L *is join complete ;*

(ii) *every downward directed set has a meet ;*

(iii) *for every* $a \in L$ *and every downward directed set* $D \subseteq L$,

$$(2.1) \qquad a \vee \bigwedge_{d \in D} d = \bigwedge_{d \in D} (a \vee d).$$

PROOF: Let $(L; \vee)$ be equationally monocompact.

(i) Suppose $\varnothing \neq S \subseteq L$. The system of semilattice equations $s \vee x = x$ $(s \in S)$ is consistent, because each finite subsystem $s_i \vee x = x$ $(i = 1, \ldots, n)$ has the solution $s_1 \vee \ldots \vee s_n$. So the set U of upper bounds of S is not empty. Now we prove that the system of semilattice equations

$$(2.2) \qquad s \vee x = x \ (s \in S), \ x \vee u = u \ (u \in U),$$

has a solution. For let S_0 and U_0 be finite subsets of S and U, respectively, $S_0 \cup U_0 \neq \varnothing$. The subsystem of (2.2) corresponding to (S_0, U_0) has the solution $s_1 \vee \ldots \vee s_n$ or any element of S is a solution, according as $S_0 = \{s_1, \ldots, s_n\}$ or $S_0 = \varnothing$. Therefore system (2.2) is consistent and its unique solution is the least upper bound of S.

(ii) Suppose D is a downward directed subset of L. The system of equations $x \vee d = d$ $(d \in D)$ is consistent, because each finite subsystem $x \vee d_i = d_i$ $(i = 1, \ldots, n)$ is satisfied by any lower bound d_0 of the set $\{d_1, \ldots, d_n\}$. So the set B of lower bounds of S is not empty and the system of equations

$$(2.3) \qquad x \vee d = d \ (d \in D), \ b \vee x = x \ (b \in B),$$

is consistent by an argument similar to the one used for (2.2), but with the particular solution d_0. So the unique solution of system (2.3) is the greatest lower bound of S.

(iii) Suppose $a \in L$ and D is a downward directed subset of L. The meet b of D exists by (ii) and since the set $\{a \vee d \mid d \in D\}$ is also downward directed, its meet c exists as well. Since $a \vee b$ is a lower bound of the latter set, it follows that $a \vee b \leq c$ and we have to prove that $c \leq a \vee b$.

For every $d \in D$, it follows from $a \leq a \vee b \leq c \leq a \vee d$ that $a \vee d \leq c \vee d \leq a \vee d$, i.e., $a \vee d = c \vee d$. Therefore, reasoning as for (ii), it follows that every finite subsystem of the system

$$(2.4) \qquad x \vee d = d \ (d \in D), \ a \vee x = c \vee x,$$

has a solution. Hence system (2.4) is consistent and for every solution x of it we have $x \leq d$ for all $d \in D$, hence $x \leq b$, therefore $c \leq c \vee x = a \vee x \leq a \vee b$.

Conversely, suppose $(L; \vee)$ satisfies conditions (i)-(iii).

Note first that a semilattice function $f : L \longrightarrow L$ has one of the forms $f(x) = x$, $f(x) = a$ or $f(x) = a \vee x$, where $a \in L$ (cf. 2.(1.24), 2.(1.25)). This implies the identities

$$(2.5) \qquad f(\bigvee_{s \in S} s) = \bigvee_{s \in S} f(s) \,,$$

$$(2.6) \qquad f(\bigwedge_{d \in D} d) = \bigwedge_{d \in D} f(d) \,,$$

for any non-empty subset S of L and any downward directed subset D of L, respectively. For the joins in (2.5) and the meets in (2.6) exist by (i) and (ii), respectively, while (2.5) and (2.6) are trivial if $f(x) = x$ or $f(x) = a$, so that it remains to prove them for $f(x) = a \vee x$. But in this case (2.5) is well known, while (2.6) coincides with (iii).

An immediate consequence is the following. If

$$(2.7) \qquad f(x) = g(x)$$

is a semilattice equation in one unknown over L, $S \neq \emptyset$ is an arbitrary set of solutions of (2.7) and D is a downward directed set of solutions of (2.7), then $\bigvee_{s \in S} s$ and $\bigwedge_{d \in D} d$ satisfy (2.7) as well.

Now let

$$(2.8) \qquad f_k(x) = g_k(x) \qquad (k \in K)$$

be a semilattice system such that any finite subsystem Σ of (2.8) is consistent. So for any such subsystem the solution set $S(\Sigma)$ is not empty and $t_\Sigma = \bigvee S(\Sigma)$ is the greatest solution of Σ. The set

$$(2.9) \qquad D = \{t_\Sigma \mid \Sigma \text{ is a finite subsystem of } (2.8)\}$$

is downward directed, because if Σ_1 and Σ_2 are finite subsystems of (2.8), so is $\Sigma = \Sigma_1 \cup \Sigma_2$ and from $\emptyset \neq S(\Sigma) = S(\Sigma_1) \cap S(\Sigma_2)$ it follows that $t_\Sigma \leq t_{\Sigma_1}, t_{\Sigma_2}$. Therefore the element $t = \inf D$ exists by (ii).

Further, take an element $k \in K$ and note that the set

$$(2.10) \qquad D_k = \{t_\Sigma \in D \mid \Sigma \text{ contains equation } f_k = g_k\}$$

is also downward directed, by the same argument as before. But each $t_\Sigma \in D_k$ is a solution of system Σ and in particular of equation $f_k = g_k$, therefore $t_k = \inf D_k$ exists by (ii) and satisfies $f_k = g_k$ by the consequence pointed out above. Since $D_k \subseteq D$, it follows that $t \leq t_k$. Furthermore, for every finite subsystem Σ of (2.8) set $\Sigma' = \Sigma \cup \{f_k = g_k\}$; then $\Sigma \subseteq \Sigma'$ and $t_{\Sigma'} \in D_k$, therefore $t_k \leq t_{\Sigma'} \leq t_\Sigma$. This implies $t_k \leq t$, hence $t = t_k$ satisfies $f_k(t) = g_k(t)$. But k was arbitrary in K, which concludes the proof. $\qquad \square$

Corollary 2.1. *Every equationally monocompact lattice (and in particular every equationally compact lattice) is complete.*

PROOF: If $(L; \wedge, \vee)$ is an equationally monocompact lattice, then $(L; \vee)$ is an equationally monocompact semilattice by Remark 1.2. Therefore L is join complete by (i) and meet complete by (ii), since every non-empty subset of L is downward directed. $\qquad \square$

As a matter of fact, Grätzer and Lakser [1969] proved that equationally compact semilattices are characterized by (i), (ii′) and (iii′), where (ii′) and (iii′)

are obtained from (ii) and (iii), respectively, by using chains instead of downward directed subsets.

To go further we recall

Definition 2.2. (cf. Ch.3, §3). By a *Heyting lattice (double Heyting lattice)* we mean a Heyting algebra (double Heyting algebra) L regarded just as an algebra of type $(2,2,0,0)$, i.e., as a bounded distributive lattice $(L; \wedge, \vee, 0, 1)$.   □

Lemma 2.1. *The class of equationally monocompact distributive lattices and the class of 2-equationally monocompact lattices are included in the class of complete double Heyting lattices.*

PROOF: Let L be a lattice belonging to one of the first two classes. Then L is complete by Corollary 2.1, taking also into account (1.3) for the second class. If we succeed in proving that L is a Heyting lattice, it will follow by duality that it is a double Heyting lattice.

Let $a, b \in L$. We must prove that the set $U_{ab} = \{x \in L \mid a \wedge x \le b\}$ has greatest element. This is equivalent to the existence of a solution for the system of equations

$$(2.11) \qquad\qquad b \vee (a \wedge x) = b ,$$

$$(2.12) \qquad\qquad u \wedge x = u \qquad (u \in U_{ab}) .$$

If L is equationally monocompact and distributive, we notice that a subsystem which consists of n equations (2.12) corresponding to n elements $u_1, \ldots, u_n \in U_{ab}$ and possibly equation (2.11) has the solution $u_1 \vee \ldots \vee u_n$, while the single equation (2.11) is trivially consistent. If L is 2-equationally monocompact, we notice that a subsystem which consists of (2.11) and a single equation (2.12) corresponding to an element $u \in U_{ab}$ has the solution u, while a subsystem consisting of two equations (2.12) which correspond to two elements $u_1, u_2 \in U_{ab}$ has the solution $u_1 \wedge u_2$.   □

Lemma 2.2. *Every complete double Heyting lattice is 2-equationally monocompact in the class of (bounded) lattices.*

PROOF: Let L be a complete double Heyting lattice and consider a system of lattice equations

$$(2.13) \qquad (a_k \wedge x) \vee b_k = (c_k \wedge x) \vee d_k \qquad (k \in K),$$

where

$$(2.14) \qquad\qquad a_k \ge b_k \ \& \ c_k \ge d_k \qquad (k \in K)$$

(cf. the remark preceding Lemma 3.3.3), such that every 2-equation subsystem is consistent. Fix an index $k \in K$. In view of Proposition 3.3.3, for every $h \in K$ the subsystem of (2.13) consisting of equations h and k has the solution $x_{kh} = (b_k + d_k) \vee (b_h + d_h)$. Then, using Proposition 3.2.4, we get

$$b_k \vee \left(a_k \wedge \bigvee_{h \in K} (b_h + d_h)\right) = b_k \vee \left(a_k \wedge \bigvee_{h \in K} x_{kh}\right)$$

$$= b_k \vee \bigvee_{h \in K} (a_k \wedge x_{kh}) = \bigvee_{h \in K} (b_k \vee (a_k \wedge x_{kh}))$$

$$= \bigvee_{h \in K} (d_k \vee (c_k \wedge x_{kh})) = \ldots = d_k \vee (c_k \wedge \bigvee_{h \in K} (d_h + b_h)),$$

and since k was arbitrary in K, this shows that $\bigvee_{h \in K} (b_h + d_h)$ is a solution of system (2.13). □

Remark 2.1. In view of Remark 3.2.5, a bounded lattice L is {2-}equationally {mono}compact in the class of lattices if and only if it has this property in the class of bounded lattices. □

Theorem 2.2. *The following properties are equivalent for a lattice L:*

   (i) L is a 2-equationally monocompact lattice;

   (ii) L is an equationally monocompact distributive lattice;

   (iii) L is a complete double Heyting lattice.

PROOF: (i)$\Longrightarrow$(iii) and (ii)$\Longrightarrow$(iii): by Lemma 2.1.

   (iii)$\Longrightarrow$(i): by Lemma 2.2.

   (iii)$\Longrightarrow$(ii): L is distributive by Proposition 3.2.7 and it is 2-equationally monocompact by the previous implication, hence it is equationally monocompact by (1.3). □

   See also Kelly [1972] for a characterization of equationally compact lattices in the class of all lattices which do not contain an infinite antichain, and Beazer [1975] for equational compactness in pseudo-Post algebras.

3 Equational compactness of Boolean algebras

In this section Boolean algebras are regarded as algebras of any type for which algebraic functions and polynomials coincide with Boolean functions and simple Boolean functions, respectively; cf. Remark 3.2.6.

   As is well known, any Boolean (simple Boolean) equation, as well as any system of such equations, is equivalent to a single Boolean (simple Boolean) equation of the form $f = 1$ and the latter is also equivalent to $f' = 0$ (see e.g. BFE, Theorem 2.1). However a system of infinitely many equations cannot be reduced to a single equation unless the Boolean algebra is complete.

Theorem 3.1. (Węglorz [1966], Abian [1970b], [1976]). *Every Boolean algebra is weakly equationally compact.*

   Węglorz's proof uses model-theoretical methods. We present Abian's proof in the form of Lemma 3.1 below, which is slightly stronger than Theorem 3.1. As a matter of fact, Abian has proved even more: Theorem 3.1 is equivalent to the prime ideal theorem (see e.g. Balbes and Dwinger [1974], Theorem III.4.1).

Lemma 3.1. *If every finite subsystem of a system of simple Boolean equations is consistent, then the whole system has a solution in the subalgebra $\{0,1\}$.*

PROOF: Let B and

$$(3.1) \qquad p_k(\ldots,z_j,\ldots) = 0 \qquad (k \in K)$$

be the given Boolean algebra and system, respectively. The elements p_k $(k \in K)$ of the Boolean algebra $B[X]$ (cf. 2.(2.17)) generate an ideal P which is proper. For otherwise the characterization of the ideal generated by a set (see e.g. Balbes and Dwinger [1974], Theorem II.9.2) implies the existence of $k_1, \ldots, k_n \in K$ such that $p_{k_1} \vee \ldots \vee p_{k_n} = f_1$ (i.e., the element 1 of $B[X]$, $f_1(\ldots,b_j,\ldots) = 1$ for all $(\ldots,b_j,\ldots) \in B^X$); but according to the hypothesis there exists $(\ldots,s_j,\ldots) \in B^X$ such that

$$p_{k_h}(\ldots,s_j,\ldots) = 0 \qquad (h = 1,\ldots,n) \,,$$

a contradiction. In view of the prime ideal theorem mentioned before, the proper ideal P is included in a prime ideal M of $B[X]$. Define $h : B[X] \longrightarrow \{0,1\}$ by $h(p) = 0 \iff p \in M$. Then h is a homomorphism (see e.g. op.cit., Theorem III.3.4) and since $p_k \in M$ we have $h(p_k) = 0$ for all $k \in K$. On the other hand, for each $k \in K$ the restriction of p_k to $\{0,1\}^X$ is an element of $\{0,1\}[X] = \overline{\{0,1\}}[X]$; cf. 2.(3.11) and BFE, Theorem 1.7 and Corollary of Theorem 1.11. Since the restriction of h to $\{0,1\}[X]$ is clearly a homomorphism from $\overline{\{0,1\}}[X]$ to $\{0,1\}$, we can apply Lemma 2.3.2 and obtain $p_k(\ldots,h(x_{jB}),\ldots) = h(p_k) = 0$, showing that the element $(\ldots,h(x_{jB}),\ldots)$ of $\{0,1\}^X$ is a solution of system (3.1). □

Corollary 3.1. *Every consistent system of simple Boolean equations has a solution in the subalgebra $\{0,1\}$.* □

Theorem 3.2. (Węglorz [1966], Abian [1970a], [1976]). *A Boolean algebra is equationally compact if and only if it is complete.*

PROOF: Every equationally compact Boolean algebra is equationally compact as a lattice by Remark 1.2, therefore it is complete by Corollary 3.1.

Conversely, suppose B is a complete Boolean algebra and let

$$(3.2) \qquad g_k(\ldots,z_j,\ldots) = 0 \qquad (k \in K)$$

be a system of Boolean equations for which every finite subsystem has a solution. In view of Proposition 2.3.2, the elements of the subalgebra $< G >$ of $\overline{B}[X]$ generated by the set

$$(3.3) \qquad G = \{f_a \mid a \in A\} \cup \{g_k \mid k \in K\}$$

are of the form

$$(3.4) \qquad w^G_{\overline{B}[X]}(f_{a_1},\ldots,f_{a_m},g_{k_1},\ldots,g_{k_n}) \,,$$

where the arguments $f_{a_1},\ldots,f_{a_m},g_{k_1},\ldots,g_{k_n}$, as well as m and n, vary with w^G. We claim that the rule

$$(3.5) \qquad \varphi(w^G_{\overline{B}[X]}(f_{a_1},\ldots,f_{a_m},g_{k_1},\ldots,g_{k_n})) = w^G_B(a_1,\ldots,a_m,0,\ldots,0)$$

defines a map $\varphi :< G > \longrightarrow B$. For consider two representations of the same element, say

$$w^G_{\overline{B}[X]}(f_{a_1}, \ldots, f_{a_m}, g_{k_1}, \ldots, g_{k_n}) = (w_1)^G_{\overline{B}[X]}(f_{a_1}, \ldots, f_{a_m}, g_{k_1}, \ldots, g_{k_n}) \,,$$

where clearly taking the same set of arguments $f_{a_1}, \ldots, g_{k_n}$ does not restrict generality. According to the hypothesis, the system of equations $g_{k_h}(\ldots, z_j, \ldots) = 0$ ($h = 1, \ldots, n$) has a solution $v \in B^X$. Using Lemma 2.3.1 we obtain

$$w^G_B(a_1, \ldots, a_m, 0, \ldots, 0) = w^G_B(f_{a_1}(v), \ldots, f_{a_m}(v), g_{k_1}(v), \ldots, g_{k_n}(v))$$

$$= w^G_{\overline{B}[X]}(f_{a_1}, \ldots, f_{a_m}, g_{k_1}, \ldots, g_{k_n})(v) = (w_1)^G_{\overline{B}[X]}(f_{a_1}, \ldots, f_{a_m}, g_{k_1}, \ldots, g_{k_n})(v)$$

$$= \ldots = (w_1)^G_B(a_1, \ldots, a_m, 0, \ldots, 0).$$

Further we prove that φ is a homomorphism. Taking into account the form (3.4) of the elements of $< G >$, then 2.(2.16$'$) applied with $A := \overline{B}[X]$, $f_i := \mathbf{f}_i$, then (3.5), 2.(2.15), 2.(2.10) and again (3.5), we obtain

$$\varphi(\mathbf{f}_i((w_1)^G_{\overline{B}[X]}(f_{a_1}, \ldots, f_{a_m}, g_{k_1}, \ldots, g_{k_n}), \ldots,$$

$$(w_{n(i)})^G_{\overline{B}[X]}(f_{a_1}, \ldots, f_{a_m}, g_{k_1}, \ldots, g_{k_n}))$$

$$= \varphi((F_i(w_1)^G \ldots (w_{n(i)})^G)_{\overline{B}[X]}(f_{a_1}, \ldots, f_{a_m}, g_{k_1}, \ldots, g_{k_n}))$$

$$= (F_i(w_1)^G \ldots (w_{n(i)})^G)_B(a_1, \ldots, a_m, 0, \ldots 0)$$

$$= \mathbf{f}_i((w_1)^G_B, \ldots, (w_{n(i)})^G_B)(a_1, \ldots, a_m, 0, \ldots, 0)$$

$$= \mathbf{f}_i((w_1)^G_B(a_1, \ldots, a_m, 0, \ldots, 0), \ldots, (w_{n(i)})^G_B(a_1, \ldots, a_m, 0, \ldots, 0))$$

$$= \mathbf{f}_i(\varphi((w_1)^G_{\overline{B}[X]}(f_{a_1}, \ldots, f_{a_m}, g_{k_1}, \ldots, g_{k_n})), \ldots,$$

$$\varphi((w_{n(i)})^G_{\overline{B}[X]}(f_{a_1}, \ldots, f_{a_m}, g_{k_1}, \ldots, g_{k_n}))) \,.$$

According to a theorem due to Sikorski (see e.g. Balbes and Dwinger [1974], Theorem V.9.2), a Boolean algebra is complete if and only if it is injective. Therefore the homomorphism φ can be extended to a fomomorphism $\overline{\varphi} : \overline{B}[X] \longrightarrow B$. Now it follows from (3.5) that for all $a \in A$,

$$\overline{\varphi}(f_a) = \varphi(f_a) = \varphi((x_1)^G_{\overline{B}[X]}(f_a)) = (x_1)^G_B(a) = a \,,$$

and for all $k \in K$,

$$\overline{\varphi}(g_k) = \varphi(g_k) = \varphi((x_1)^G_{\overline{B}[X]}(g_k)) = (x_1)^G_B(0) = 0 \,,$$

therefore we can use Lemma 2.3.2 and deduce

$$g_k(\ldots, \overline{\varphi}(x_{jB}), \ldots) = \overline{\varphi}(g_k) = 0 \qquad (k \in K) \,,$$

showing that the element $(\ldots, \overline{\varphi}(x_{jB}), \ldots,)$ of B^X is a solution of system (3.2). $\qquad\square$

5. Post algebras

The first axiom system for the algebras corresponding to the many-valued logic of Post [1921] was given by Rosenbloom [1924], who called them Post algebras. Ever since then Post algebras and their generalizations have been intensively studied; see e.g. Serfati [1973b], Balbes and Dwinger [1974], Rasiowa [1974], Boicescu, Filipoiu, Georgescu and Rudeanu [1991] and the literature quoted therein.

In §1 we introduce Post algebras within the axiomatic line of Epstein [1960] and Traczyk [1963], with emphasis on the computation rules that will be used in the next two sections. The basic properties of Post functions and simple Post functions are established in §2. By a Post function (simple Post function) we mean the specialization to Post algebras of the concept of algebraic function (polynomial); cf. Ch.2, §§ 2,3. Post equations (simple Post equations), that is, equations expressed in terms of Post functions (simple Post functions), are studied in §3. The theorems on Post functions and equations parallel the theory of Boolean functions and equations; cf. BFE.

1 Basic properties of Post algebras

In this section we introduce Post algebras and establish their basic computational properties. Unless otherwise stated, the results are due to Epstein [1960]. As a matter of fact, several properties hold in more general contexts; see e.g. Boicescu, Filipoiu, Georgescu and Rudeanu [1991]. Our presentation in the sequel owes much to Serfati [1973b].

Throughout this chapter r is a fixed integer, $r \geq 2$.

Lemma 1.1. *The following conditions* (1.1) *and* (1.2) *are equivalent for* $2r - 1$ *elements* $x_1, \ldots, x_{r-1}, x^0, x^1, \ldots, x^{r-1}$ *of a Boolean algebra* $(B; \wedge, \vee, ', 0, 1)$:

(1.1.1)
$$x_1 \geq x_2 \geq \ldots \geq x_{r-1},$$

(1.1.2)
$$x^0 = x_1' , \quad x^i = x_i \wedge x_{i+1}' \ (i = 1, \ldots, r-2) , \quad x^{r-1} = x_{r-1} ,$$

and

$$(1.2.1) \qquad \bigvee_{i=0}^{r-1} x^i = 1 \, , \ x^i \wedge x^j = 0 \quad (i, j = 0, \ldots, r-1, \ i \neq j) \, ,$$

$$(1.2.2) \qquad x_k = \bigvee_{j=k}^{r-1} x^j \qquad (k = 1, \ldots, r-1) \, .$$

COMMENTS: 1) The first condition (1.2.1), the remaining conditions (1.2.1) and all of the conditions (1.2.1) are referred to by saying that $(x^0, \ldots, x^{r-1})$ is a *normal* system, an *orthogonal* system and an *orthonormal* system, respectively.

2) The notation x^i is due to Serfati [1973a], [1973b].

PROOF: (1.1)$\Longrightarrow$(1.2): It follows from (1.1) and the property 3.(2.40) of Boolean absorption that if $1 \leq k \leq r-1$ then

$$(1.3) \qquad \bigvee_{j=k}^{r-1} x^j = x_{r-1} \vee x_{r-2} \vee \ldots \vee x_k = x_k \, ,$$

hence

$$\bigvee_{i=0}^{r-1} x^i = x^0 \vee \bigvee_{j=1}^{r-1} x^j = x_1' \vee x_1 = 1$$

and if $0 \leq i < j \leq r-1$ then $x^i \wedge x^j \leq x_{i+1}' \wedge x_j = 0$ because $x_{i+1} \geq x_j$.

(1.2)$\Longrightarrow$(1.1): Properties (1.1.1) and $x^{r-1} = x_{r-1}$ are immediate. In view of (1.2.2), the first relation (1.2.1) becomes $x^0 \vee x_1 = 1$ and taking also into account the remaining relations (1.2.1) we deduce $x^0 \wedge x_1 = 0$, therefore $x^0 = x_1'$. Finally if $1 \leq i \leq r-2$ then, since for every $h \neq i$ we have $x^i \wedge x^h = 0$, or equivalently, $x^i \leq (x^h)'$, it follows that

$$(1.4) \qquad \begin{aligned} x_i \wedge x_{i+1}' &= (\bigvee_{j=i}^{r-1} x^j) \wedge \bigwedge_{h=i+1}^{r-1} (x^h)' \\ &= x^i \wedge \bigwedge_{h=i+1}^{r-1} (x^h)' = x^i \, . \end{aligned}$$

$\square$

Corollary 1.1. *Formulas (1.1.2) and (1.2.2) establish a bijection between the vectors $(x_1, \ldots, x_{r-1})$ satisfying (1.1.1) and the vectors $(x^0, x^1, \ldots, x^{r-1})$ satisfying (1.2.1).*

PROOF: Follows from (1.3) and (1.4). $\square$

Proposition 1.1. *Suppose $(L; \wedge, \vee, 0, 1)$ is a bounded distributive lattice and the elements $e_0, e_1, \ldots, e_{r-1}$ satisfy*

$$(1.5) \qquad e_0 = 0 < e_1 < e_2 < \ldots < e_{r-2} < e_{r-1} = 1 \, .$$

Then:

   α) *The following conditions are equivalent:*

(i) every element $x \in L$ has a unique representation of the form

$$(1.6) \qquad x = \bigvee_{k=1}^{r-1} (x_k \wedge e_k) \,,$$

where $x_k \in B(L)$ $(k = 1, \ldots, r-1)$ and $x_1 \geq x_2 \geq \ldots \geq x_{r-1}$;

(ii) every element $x \in L$ has a unique representation of the form

$$(1.7) \qquad x = \bigvee_{i=0}^{r-1} (x^i \wedge e_i) \,,$$

where $(x^0, x^1, \ldots, x^{r-1})$ is an orthonormal system.

β) *If the equivalent conditions (i) and (ii) hold, then the coefficients x_k and x^i occurring in the representations (i) and (ii) are related by formulas (1.1) and (1.2).*

COMMENT: Note that the elements of an orthonormal system are Boolean; to be specific,

$$(1.8) \qquad (x^i)' = \bigvee_{h=0, h \neq i}^{r-1} x^h \qquad (i = 0, 1, \ldots, r-1) \,.$$

PROOF: First we prove that (1.1) and (1.2) imply

$$(1.9) \qquad \bigvee_{k=1}^{r-1} (x_k \wedge e_k) = \bigvee_{i=0}^{r-1} (x^i \wedge e_i) \,.$$

For

$$\bigvee_{k=1}^{r-1} (x_k \wedge e_k) = \left(\left(\bigvee_{j=1}^{r-1} x^j \right) \wedge e_1 \right) \vee \left(\left(\bigvee_{j=2}^{r-1} x^j \right) \wedge e_2 \right) \vee \ldots$$

$$\vee \left(\left(\bigvee_{j=r-2}^{r-1} x^j \right) \wedge e_{r-2} \right) \vee (x^{r-1} \wedge e_{r-1})$$

$$= (x^1 \wedge e_1) \vee (x^2 \wedge (e_1 \vee e_2)) \vee \ldots \vee (x^{r-2} \wedge (e_1 \vee e_2 \vee \ldots \vee e_{r-2})) \vee$$

$$\vee (x^{r-1} \wedge (e_1 \vee e_2 \vee \ldots \vee e_{r-1})) = \bigvee_{k=1}^{r-1} (x^k \wedge e_k) = \bigvee_{i=0}^{r-1} (x^i \wedge e_i) \,.$$

(i)$\Longrightarrow$(ii): Given the element (1.6), define the elements x^i by (1.1.2). Relations (1.1) and (1.2) hold by Lemma 1.1, so that identity (1.9) shows that x has the representation (1.7). To prove uniqueness, suppose that x has another representation of the form (1.7) with orthonormal coefficients $y^0, y^1, \ldots, y^{r-1}$. Construct $y_1, y_2, \ldots, y_{r-1}$ from $y^0, y^1, \ldots, y_{r-1}$ as in (1.2.2). Then the elements y^i and y_k satisfy (1.1) and (1.2) by Lemma 1.1, so that it follows from (1.9) that x is represented in the form (1.6) with the coefficients $y_1, \ldots, y_{r-1}$. This implies $y_k = x_k$ $(k = 1, \ldots, r-1)$, therefore $y^i = x^i$ $(i = 0, \ldots, r-1)$.

(ii)$\Longrightarrow$(i): Similar proof.

Statement β) is obvious from the above discussion. $\qquad \square$

Definition 1.1. A *Post algebra of order r* or an *r-Post algebra* is an algebra $(L; \wedge, \vee, e_0 = 0, e_1, \ldots, e_{r-2}, e_{r-1} = 1)$, where $(L; \wedge, \vee, 0, 1)$ is a bounded distributive lattice, the elements e_i satisfy (1.5) and the equivalent conditions (i) and (ii) in Proposition 1.1 hold. The sequence $e_0, \ldots, e_{r-1}$ is called the *chain of constants* of L. The expansions (1.6) and (1.7) are known as the *monotone representation* and the *disjunctive representation* of x, respectively. We will refer to x_k and x^i as the *monotone components* and the *disjunctive components* of x, respectively. □

Example 1.1. In the case $r := 2$, $L = B(L)$ is a Boolean algebra: the chain of constants reduces to $0 < 1$, so that $x = x^1 = x_1$ and $x^0 = (x_1)' = x'$. □

Example 1.2. Every r–element chain $C_r = \{e_0 = 0.e_1, \ldots, e_{r-1} = 1\}$, $e_{k-1} < e_k$ $(k = 1, \ldots, r-1)$, is made into an r–Post algebra by taking, as usual, min and max as meet and join operations, respectively, and by taking C_r itself as the chain of constants. Then $B(C_r) = \{0, 1\}$, hence an orthonormal system consists of one element equal to 1, the other elements being 0. Therefore each element e_h has a unique representation of the form (1.7), namely $(e^h)^h = 1$ and $(e^h)^i = 0$ for $i \neq h$. □

Remark 1.1. More generally, in every Post algebra the monotone components and the disjunctive components of the elements e_i are given by formulas

(1.10) $(e_i)_k = 1$ if $k \leq i$, else 0 ,

for $k = 1, \ldots, r-1$, and

(1.11) $(e_i)^h = 1$ if $h = i$, else 0 ,

for $h = 0, 1, \ldots, r-1$, respectively, because the orthonormal system (1.11) provides the (unique!) representation of e_i in the form (1.17), while the elements (1.10) are obtained from (1.11) via (1.2.2). □

Example 1.3. If B is a Boolean algebra, then the set

(1.12) $\mathrm{Inc}(C_{r-1}, B) = \{f : C_{r-1} \longrightarrow B \mid f \text{ is isotone}\}$

is made into an r–Post algebra by defining the lattice operations pointwise, i.e.,

(1.13) $(f \wedge g)(e_h) = f(e_h) \wedge g(e_h)$, $(f \vee g)(e_h) = f(e_h) \vee g(e_h)$

for $h = 0, 1, \ldots, r-2$, and the chain of constants $\varepsilon_i : C_{r-1} \longrightarrow B$ $(i = 0, \ldots, r-1)$ by the condition: for $h = 0, 1, \ldots, r-2$,

(1.14) $\varepsilon_i(e_h) = 0$ if $h < r - 1 - i$, else 1 .

For it is clear that the functions ε_i are isotone and $\varepsilon_0 = 0 < \varepsilon_1 < \ldots < \varepsilon_{r-1} = 1$. Further note that $B(\mathrm{Inc}(C_{r-1}, B))$ is the set of those isotone functions $g : C_{r-1} \longrightarrow B$ for which g', which is defined pointwise, is also isotone, therefore it is the set of constant functions.

Now associate with each function $f \in \mathrm{Inc}(C_{r-1}, B)$ the constant functions $f_k : C_{r-1} \longrightarrow B$ $(k = 1, \ldots, r-1)$ defined by

(1.15) $$f_k(e_h) = f(e_{r-1-k}) \qquad (h = 0, 1, \ldots, r - 2) \,.$$

Then

$$f_k \in B(\mathrm{Inc}(C_{r-1}, B)) \ (k = 1, \ldots, r - 1) \,, \ f_1 \geq f_2 \geq \ldots \geq f_{r-1}$$

and $f = \bigvee_{k=1}^{r-1}(f_k \wedge \varepsilon_k)$ because for each e_h,

$$\bigvee_{k=1}^{r-1}(f_k(e_h) \wedge \varepsilon_k(e_h)) = \bigvee_k \{f_k(e_h) \mid h \geq r - 1 - k\} = f_{r-1-h}(e_h) = f(e_h) \,.$$

To prove the uniqueness of the representation, suppose $f = \bigvee_{k=1}^{r-1}(g_k \wedge \varepsilon_k)$, where $g_1 \geq \ldots \geq g_{r-1}$ are constant functions. Then for each h,

$$f(e_h) = \bigvee_k \{g_k \mid h \geq r - 1 - k\} = g_{r-1-h} \,.$$

$\square$

Example 1.4. It can be shown similarly that the set

(1.12') $$\mathrm{Dec}(C_{r-1}, B) = \{f : C_{r-1} \longrightarrow B \mid f \text{ is antitone}\}$$

becomes a Post algebra with the chain of constants $\delta_0 = 0 < \delta_1 < \ldots \delta_{r-1} = 1$ defined by

(1.14') $$\delta_i(e_h) = 1 \text{ if } h < i \,, \text{ else } 0$$

for $h = 0, 1, \ldots, r - 1$, while the decreasing chain of constant functions $f_{(k)} :$ $C_{r-1} \longrightarrow B \ (k = 1, \ldots, r - 1)$ associated with a function $f \in \mathrm{Dec}(C_{r-1}, B)$ is defined by

(1.15') $$f_{(k)}(e_h) = f(e_{k-1}) \qquad (h = 0, 1, \ldots, r - 1) \,.$$

$\square$

Proposition 1.2. *The dual of a Post algebra L is a Post algebra with the chain of constants*

(1.16) $$d_i = e_{r-1-i} \qquad (i = 0, \ldots, r - 1) \,,$$

the monotone components of an element $x \in L$ being

(1.17) $$y_k = x_{r-k} \qquad (k = 1, \ldots, r - 1) \,.$$

PROOF: We have $d_0 = 1 > d_1 > \ldots > d_{r-1} = 0$, while $y_1 \leq y_2 \leq \ldots \leq y_{r-1}$ and $y_k \in B(L) \ (k = 1, \ldots, r - 1)$; note that the set of Boolean elements is invariant under duality. Finally, taking into account that $e_i \wedge (x \vee e_j) = e_i$ for $i \leq j$, we get

$$\bigwedge_{k=1}^{r-1}(y_k \vee d_k) = \bigwedge_{k=1}^{r-1}(x_{r-k} \vee e_{r-k-1})$$

$$= (x_{r-1} \vee e_{r-2}) \wedge (x_{r-2} \vee e_{r-3}) \wedge \ldots \wedge (x_3 \vee e_2) \wedge (x_2 \vee e_1) \wedge x_1$$

$$= (x_1 \wedge e_1) \vee (x_2 \wedge (x_3 \vee e_2)) \wedge \ldots \wedge (x_{r-1} \vee e_{r-2}) = \ldots$$
$$= (x_1 \wedge e_1) \vee (x_2 \wedge e_2) \vee \ldots \vee (x_{r-3} \wedge e_{r-3}) \vee (x_{r-2} \wedge (x_{r-1} \vee e_{r-2}))$$
$$= \bigvee_{k=1}^{r-2} (x_k \wedge e_k) \vee x_{r-1} = x .$$

$\square$

Corollary 1.2. *The principle of duality holds for Post algebras.* $\square$

The next two propositions characterize the chain of constants and the Boolean elements of a Post algebra. In particular it turns out that relations (1.10) and (1.11) are valid in an arbitrary Post algebra.

Proposition 1.3. *Let L be a Post algebra and $i \in \{0, 1, \ldots, r-1\}$. The following conditions are equivalent for an element $x \in L$:*

(i) $x_1 = \ldots = x_i = 1$, $x_{i+1} = \ldots = x_{r-1} = 0$;

(ii) $x^i = 1$;

(iii) $x^j = 0$ for all $j \neq i$;

(iv) $x = e_i$.

PROOF: (i)$\Longleftrightarrow$(iv): The sequence (i) is actually a decreasing sequence of Boolean elements and $\bigvee_{k=1}^{r-1} (x_k \wedge e_k) = e_i$ by (1.5).

(ii)$\Longleftrightarrow$(iii): Obvious from orthonormality.

(iii)$\Longleftrightarrow$(iv): The equivalent conditions (ii) and (iii) define an orthonormal system which satisfies $\bigvee_{h=0}^{r-1} (x^h \wedge e_h) = e_i$. $\square$

Proposition 1.4. *The following conditions are equivalent for an element x of a Post algebra:*

(i) $x \in B(L)$;

(ii) $x_k = x$ for all $k \in \{1, \ldots, r-1\}$;

(iii) $x_k = x$ for some $k \in \{1, \ldots, r-1\}$;

(iv) $x^0 = x'$ & $x^i = 0$ $(i = 1, \ldots, r-2)$ & $x^{r-1} = x$;

(v) $x^i = 0$ $(i = 1, \ldots, r-2)$.

PROOF: (i)$\Longrightarrow$(ii): For $x = \bigvee_{k=1}^{r-1} (x \wedge e_k)$ is the monotone representation of x.

(ii)$\Longrightarrow$(iii)$\Longrightarrow$(i), (iv)$\Longrightarrow$(i) and (iv)$\Longrightarrow$(v): Trivial.

(ii)$\Longrightarrow$(iv) by (1.1.2).

(v)$\Longrightarrow$(iv): $x = x^{r-1}$ by (1.7) and $x^0 = (x^{r-1})'$ by (1.2.1). $\square$

It is natural to consider also expansions of the form

(1.18) $$\bigvee_{k=1}^{r-1} (a_k \wedge e_k) , \ a_k \in B(L) \qquad (k = 1, \ldots, r-1) ,$$

which generalize both the monotone and the disjunctive representation of x. The relationship bewteen an arbitrary expansion (1.18) and the canonical expansions (1.6) and (1.7) is made explicit in the next proposition.

Proposition 1.5. *The monotone components of the element x given by (1.18) are*

$$(1.19) \qquad x_k = \bigvee_{j=k}^{r-1} a_j \qquad (k = 1, \ldots, r-1) \,,$$

while its disjunctive components are

$$(1.20.0) \qquad x^0 = \bigwedge_{k=1}^{r-1} a'_k \,,$$

$$(1.20.\text{i}) \qquad x^i = a_i \wedge \bigwedge_{j=i+1}^{r-1} a'_j \qquad (k = 1, \ldots, r-2) \,,$$

$$(1.20.\text{r-1}) \qquad x^{r-1} = a_{r-1} \,.$$

PROOF: The right members of (1.19) form a decreasing sequence of Boolean elements and since for each $k = 1, \ldots, r-1$,

$$\bigvee_{j=k+1}^{r-1} (a_j \wedge e_k) \leq \bigvee_{j=k+1}^{r-1} (a_j \wedge e_j) \leq x \,,$$

it follows from (1.18) that

$$\bigvee_{k=1}^{r-1} (x_k \wedge e_k) = \bigvee_{k=1}^{r-1} \bigvee_{j=k}^{r-1} (a_j \wedge e_k)$$

$$= \bigvee_{k=1}^{r-1} ((a_k \wedge e_k) \vee \bigvee_{j=k+1}^{r-1} (a_j \wedge e_k)) = x \,.$$

Formulas (1.20) follow from (1.19) via (1.1.2).   □

Now let us determine the monotone components and the disjunctive components of $x \vee y$ and $x \wedge y$.

Proposition 1.6. *The monotone components of $x \vee y$ are*

$$(1.21) \qquad (x \vee y)_k = x_k \vee y_k \qquad (k = 1, \ldots, r-1) \,,$$

while its disjunctive components are

$$(1.22) \qquad (x \vee y)^i = (x^i \wedge \bigvee_{h=0}^{i} y^h) \vee (y^i \wedge \bigvee_{h=0}^{i} x^h) \qquad (i = 0, 1, \ldots, r-1) \,.$$

COMMENT: In particular $(x \vee y)^0 = x^0 \wedge y^0$ and $(x \vee y)^{r-1} = x^{r-1} \vee y^{r-1}$.
PROOF: The right members of (1.21) form a decreasing sequence of Boolean elements and

$$\bigvee_{k=1}^{r-1}((x_k \vee y_k) \wedge e_k) = (\bigvee_{k=1}^{r-1}(x_k \wedge e_k)) \vee (\bigvee_{k=1}^{r-1}(y_k \wedge e_k)) = x \vee y \ .$$

Then, using (1.1), (1.21), (1.2) and orthonormality, we get

$$(x \vee y)^0 = ((x \vee y)_1)' = x_1' \wedge y_1' = x^0 \wedge y^0 \ ,$$

$$(x \vee y)^i = (x \vee y)_i \wedge ((x \vee y)_{i+1})' = (x_i \vee y_i) \wedge x_{i+1}' \wedge y_{i+1}'$$

$$= (x^i \wedge y_{i+1}') \vee (x_{i+1}' \wedge y^i)$$

$$= (x^i \wedge (\bigvee_{h=i+1}^{r-1} y^h)') \vee (y^i \wedge (\bigvee_{h=i+1}^{r-1} x^h)')$$

$$= (x^i \wedge \bigvee_{h=0}^{i} y^h) \vee (y^i \wedge \bigvee_{h=0}^{i} x^h) \qquad (i = 1, \ldots, r-2) \ ,$$

$$(x \vee y)^{r-1} = (x \vee y)_{r-1} = x_{r-1} \vee y_{r-1} = x^{r-1} \vee y^{r-1} \ .$$

$\square$

Proposition 1.7. *The monotone components of $x \wedge y$ are*

(1.23) $$(x \wedge y)_k = x_k \wedge y_k \qquad (k = 1, \ldots, r-1) \ ,$$

while its disjunctive components are

(1.24) $$(x \wedge y)^i = (x^i \wedge \bigvee_{h=i}^{r-1} y^h) \vee (y^i \wedge \bigvee_{h=i}^{r-1} x^h) \qquad (i = 0, 1, \ldots, r-1) \ .$$

COMMENT: In particular $(x \wedge y)^0 = x^0 \vee y^0$ and $(x \wedge y)^{r-1} = x^{r-1} \wedge y^{r-1}$.
PROOF: The right members of (1.23) form a decreasing sequence of Boolean elements and since for every $j, k \in \{1, \ldots, r-1\}$, say $j \leq k$, we have

$$x_k \wedge e_k \wedge y_j \wedge e_j = x_k \wedge y_j \wedge e_j \leq x_j \wedge y_j \wedge e_j \ ,$$

it follows that

$$x \wedge y = (\bigvee_{k=1}^{r-1}(x_k \wedge e_k)) \cdot \vee (\bigvee_{j=1}^{r-1}(y_j \wedge e_j))$$

$$= \bigvee_{k=1}^{r-1}\bigvee_{j=1}^{r-1}(x_j \wedge y_j \wedge e_k \wedge e_j) = (\bigvee_{k=1}^{r-1}(x_k \wedge y_k \wedge e_k)) \vee$$

$$\vee (\bigvee_{j=1, j \neq k}^{r-1}(x_k \wedge y_j \wedge e_k \wedge e_j)) = \bigvee(x_k \wedge y_k \wedge e_k) \ .$$

Now the disjunctive components are obtained from (1.1), (1.23), (1.2) and orthonormality:

$$(x \wedge y)^0 = ((x \wedge y)_1)' = x_1' \vee y_1' = x^0 \vee y^0 \,,$$

$$(x \wedge y)^i = (x \wedge y)_i \vee ((x \wedge y)_{i+1})' = x_i \wedge y_i \wedge (x_{i+1}' \vee y_{i+1}')$$

$$= (x^i \wedge y_i) \vee (y^i \wedge x_i)$$

$$= (x^i \wedge \bigvee_{h=i}^{r-1} y^h) \vee (y^i \wedge \bigvee_{h=i}^{r-1} x^h) \qquad (i = 1, \ldots, r-2) \,,$$

$$(x \wedge y)^{r-1} = (x \wedge y)_{r-1} = x_{r-1} \wedge y_{r-1} = x^{r-1} \wedge y^{r-1} \,.$$

$\square$

Each of the representations (1.21) and (1.23) has the following

Corollary 1.3. *For every* $x, y \in L$,

(1.25) $$x \leq y \Longleftrightarrow x_k \leq y_k \ (k = 1, \ldots, r-1) \,.$$

$\square$

The next proposition and corollaries generalize Proposition 3.2.10, which states that a Boolean algebra is a double Stone algebra with $x^* = x^+ = x'$ and a double Heyting algebra with $x \to y = x' \vee y$ and $y - x = x' \wedge y$.

It was Rousseau [1970] who first noted that every Post algebra is a Heyting algebra. Serfati [1997] found formulas (1.26), (1.29) and (1.29'), thus proving that every Post algebra is a Brouwer algebra. Formulas (1.26'), (1.27), (1.27'), (1.28) and (1.28') seem to be new.

Proposition 1.8. *Every Post algebra L is a Heyting algebra. For every $x, y \in L$, the monotone components of the relative pseudocomplement $x \to y$ are*

(1.26) $$(x \to y)_k = \bigwedge_{j=1}^{k} (x_j' \vee y_j) \qquad (k = 1, \ldots, r-1) \,,$$

while its disjunctive components are

(1.27.0) $$(x \to y)^0 = x_1 \wedge y_1' \,,$$

(1.27.i) $$(x \to y)^i = x_{i+1} \wedge y_i \wedge y_{i+1}' \qquad (i = 1, \ldots, r-2) \,,$$

(1.27.r-1) $$(x \to y)^{r-1} = \bigwedge_{j=1}^{r-1} (x_j' \vee y_j) \,.$$

PROOF: Based on Corollary 1.3. Let z_k $(k = 1, \ldots, r-1)$ denote the right members of formulas (1.26) and $z = \bigvee_{k=1}^{r-1}(z_k \wedge e_k)$. Then

$$z_k \wedge y_k \leq (x_k' \vee y_k) \wedge x_k = y_k \wedge x_k \leq y_k \qquad (k = 1, \ldots, r-1) \,,$$

therefore $z \leq x \wedge y$. On the other hand, if $t \in L$ satisfies $t \wedge x \leq y$, then for each $j \in \{1, \ldots, r-1\}$ we have $t_j \wedge x_j \leq y_j$, hence $t_j \leq x_j' \vee y_j$ by Proposition 3.2.10, therefore

$$t_k = \bigwedge_{j=1}^{k} t_j \leq \bigwedge_{j=1}^{k} (x'_j \vee y_j) = z_k \qquad (k = 1, \ldots, r-1),$$

showing that $t \leq z$. Thus $z = x \rightarrow y$.

Further, using (1.1.2), (1.26) and the fact that $x'_j \wedge x_k = 0$ for $j < k$, we obtain

$$(x \rightarrow y)^0 = ((x \rightarrow y)_1)' = (x'_1 \vee y_1)' = x_1 \wedge y'_1,$$
$$(x \rightarrow y)^i = (x \rightarrow y)^i \wedge ((x \rightarrow y)_{i+1})'$$

$$= (\bigwedge_{j=1}^{i} (x'_j \vee y_j)) \wedge \bigvee_{m=1}^{i+1} (x_m \wedge y'_m) = (\bigwedge_{j=1}^{i} (x'_j \vee y_j)) \wedge x_{i+1} \wedge y'_m$$

$$= (\bigwedge_{j=1}^{i} (x'_j \vee y_j)) \wedge x_{i+1} \wedge y'_{i+1}$$

$$= (\bigwedge_{j=1}^{i} y_j) \wedge x_{i+1} \wedge y'_{i+1} = y_i \wedge x_{i+1} \wedge y'_{i+1} \ (i = 1, \ldots, r-2),$$

$$(x \rightarrow y)^{r-1} = (x \rightarrow y)_{r-1} = \bigwedge_{j=1}^{r-1} (x'_j \vee y_j).$$

□

Proposition 1.9. *The monotone and the disjunctive components of a relative pseudocomplement $x \rightarrow y$ are also given by*

(1.28.1) $$(x \rightarrow y)_1 = x^0 \vee \bigvee_{j=1}^{r-1} y^j,$$

(1.28.k) $$(x \rightarrow y)_k = x^0 \vee (\bigvee_{j=k}^{r-1} y^j) \vee \bigvee_{1 \leq i \leq j \leq k-1} (x^i \wedge y^j)$$
$$(k = 2, \ldots, r-1)$$

and

(1.29.i) $$(x \rightarrow y)^i = (\bigvee_{j=i+1}^{r-1} x^j) \wedge y^i \qquad (i = 0, 1, \ldots, r-2),$$

(1.29.r-1) $$(x \rightarrow y)^{r-1} = x^0 \vee y^{r-1} \vee \bigvee_{1 \leq j \leq k \leq r-2} (x^j \wedge j^k),$$

respectively.

PROOF: First we use (1.26), (1.1.2), orthonormality and absorptions:

$$(x \rightarrow y)_1 = x'_1 \vee y_1 = (\bigvee_{j=1}^{r-1} x^j)' \vee \bigvee_{j=1}^{r-1} y^j = x^0 \vee \bigvee_{j=1}^{r-1} y^j,$$

$$(x \to y)_2 = (x_1' \vee y_1) \wedge (x_2' \vee y_2)$$

$$= (x^0 \vee \bigvee_{j=1}^{r-1} y^j) \wedge (x^0 \vee x^1 \vee \bigvee_{j=2}^{r-1} y^j) = x^0 \vee (\bigvee_{j=2}^{r-1} y^j) \vee (y^1 \wedge x^1),$$

and the inductive step runs as follows:

$$(x \to y)_k = ((x \to y)_{k-1} \wedge (x_k' \vee y_k)$$

$$= (x^0 \vee (\bigvee_{j=k-1}^{r-1} y^j) \vee ((\bigvee_{1 \le i \le j \le k-2} (x^i \wedge y^j)) \wedge$$

$$\wedge ((\bigvee_{h=0}^{r-1} x^h) \vee \bigvee_{m=k}^{r-1} y^m)) = x^0 \vee (\bigvee_{m=k}^{r-1} y^m) \vee (y^{k-1} \wedge \bigvee_{h=1}^{k-1} x^h) \vee$$

$$\vee ((\bigvee_{1 \le i \le j \le k-2} (x^i \wedge y^j) \wedge \bigvee_{m=k}^{k-1} x^h)$$

$$= x^0 \vee (\bigvee_{m=k}^{r-1} y^m) \vee (\bigvee_{h=1}^{k-1} (x^h \wedge y^{k-1})) \vee (\bigvee_{1 \le i \le j \le k-2} (x^i \wedge y^j)).$$

Further, taking into account (1.1.2), Proposition 1.8 yields

$$(x \to y)^i = x_{i+1} \wedge y^i \qquad (i = 0, 1, \ldots, r-2)$$

and these relations coincide with (1.29.i) in view of (1.2.2). Finally (1.29.r-1) follows from $(x \to y)^{r-1} = (x \to y)_{r-1}$ and (1.28.r-1). $\square$

The proofs of Propositions 1.10 and 1.11 below are left to the reader. Note that the duality established in Proposition 1.2 yields immediately formulas (1.26') from (1.26). The transformation by duality of formulas (1.27)-(1.29) would need the extension of Proposition 1.2 to disjunctive components.

Proposition 1.10. *Every Post algebra L is a Brouwer algebra. For every $x, y \in L$, the monotone components of the dual relative pseudocomplement $y - x$ are*

(1.26') $$(y - x)_k = \bigvee_{j=k}^{r-1} (x_j' \wedge y_j) \qquad (k = 1, \ldots, r-1),$$

while its disjunctive components are

(1.27'.0) $$(y - x)^0 = \bigwedge_{j=1}^{r-1} (x_j \vee y_j'),$$

(1.27'.i) $$(y - x)^i = x_i' \wedge y_i \wedge y_{i+1}' \qquad (i = 1, \ldots, r-2),$$

(1.27'.r-1) $$(y - x)^{r-1} = x_{r-1}' \wedge y_{r-1}.$$

$\square$

Proposition 1.11. *The monotone and the disjunctive components of the dual relative peudocomplement* $y - x$ *are also given by*

(1.28')
$$(y - x)_k = \bigvee_{j=k}^{m-1} \bigvee_{0 \leq h < j \leq m \leq r-1} (x^h \wedge y^m) \quad (k = 1, \ldots, r-1)$$

and

(1.29'.0)
$$(y - x)^0 = x^{r-1} \vee y^0 \vee \bigvee_{1 \leq j \leq k \leq r-2} (x^k \wedge y^j) ,$$

(1.29'.k)
$$(y - x)^k = (\bigvee_{i=0}^{k-1} x^i) \wedge y^k \quad (k = 1, \ldots, r-1) .$$

$\square$

Corollary 1.4. *Every Post algebra is a double Heyting algebra.* $\square$

Proposition 1.12. *Every Post algebra is a double Stone algebra. For every* $x \in L$, *the pseudocomplement is*

(1.30)
$$x^* = x_1' = x^0$$

and its dual pseudocomplement is

(1.31)
$$x^+ = x_{r-1}' = (x^{r-1})' .$$

PROOF: In view of Proposition 3.2.7, every element $x \in L$ has the pseudocomplement $x^* = x \to 0$ and its monotone components are given by (1.26):

$$(x^*)_k = \bigwedge_{j=1}^{k} x_j' = x_1' \quad (k = 1, \ldots, r-1) ,$$

which implies (1.30) by (1.6), (1.5) and (1.1.2). So $x^* \in B(L)$ and using Proposition 1.4 we get

$$x^* \vee x^{**} = x^* \vee (x_1^*)' = x^* \vee (x^*)' = 1 ,$$

showing that L is a Stone algebra. The rest of the proof, using 3.(2.30'), is left to the reader. $\square$

Corollary 1.5. $x^{**} = x_1 = (x^0)'$ *and* $x^{++} = x_{r-1} = x^{r-1} .$ $\square$

We have seen in Proposition 3.2.6 and its dual that in every double Heyting algebra relations 3.(2.24) and 3.(2.24') hold:

(1.32)
$$x \leq y \iff x \to y = 1 \iff x - y = 0 ,$$

therefore, as noted in Proposition 3.2.8,

(1.33)
$$x = y \iff x \to y = y \to x = 1 \iff x - y = y - x = 0 .$$

Furthermore, in every Post algebra we have:

Proposition 1.13. (Bordat [1975], Serfati [1996]). *Let L be a Post algebra and $x, y \in L$. Then:*

(1.34)
$$x = y \iff \bigvee_{i=0}^{r-1} (x^i \wedge y^i) = 1 \iff \bigvee_{i=0}^{r-1} (x^i \wedge (y^i)') = 0 \,.$$

COMMENT: Since x and y can be interchanged, there is no lack of symmetry in the last condition.

PROOF: Clearly $x = y$ implies the other conditions.

If $\bigvee_{i=0}^{r-1} (x^i \wedge y^i) = 1$ holds, then taking in both sides meet with y^k yields $x^k \wedge y^k = x^k$, that is $x^k \leq y^k$, and similarly $y^k \leq x^k$. So $x^k = y^k$ for all k.

Now in the Boolean algebra $B(L)$ we apply an identity due to Löwenheim (see e.g. BFE, Theorem 4.1, or Lemma 1.2 below) and obtain

$$\bigvee_{i=0}^{r-1} (x^i \wedge y^i) = 1 \iff 0 = \left(\bigvee_{i=0}^{r-1} (x^i \wedge y^i) \right)' = \bigvee_{i=0}^{r-1} (x^i \wedge (y^i)') \,.$$

$\square$

Proposition 1.14. *Let L be a Post algebra and $x, y \in L$. Then*

(1.35)
$$x \leq y \iff \bigvee_{i=0}^{r-1} \bigvee_{h=i}^{r-1} (x^i \wedge y^h) = 1$$
$$\iff \bigvee_{i=0}^{r-2} \bigvee_{h=i+1}^{r-1} (x^h \wedge y^i) = 0 \,.$$

PROOF: It follows from Propositions 1.11 and 1.7 that

$$x \leq y \iff x \wedge y = x \iff 1 = \bigvee_{i=0}^{r-1} ((x \wedge y)^i \wedge x^i)$$

$$= \bigvee_{i=0}^{r-1} \left(\left(\left(x^i \wedge \bigvee_{h=i}^{r-1} y^h \right) \vee \left(y^i \wedge \bigvee_{h=i}^{r-1} x^h \right) \right) \wedge x^i \right)$$

$$= \bigvee_{i=0}^{r-1} \left(\left(x^i \wedge \bigvee_{h=i}^{r-1} y^h \right) \vee (x^i \wedge y^i) \right) = \bigvee_{i=0}^{r-1} \left(x^i \wedge \bigvee_{h=i}^{r-1} y^h \right) \,.$$

We obtain similarly

$$x \leq y \iff 0 = \bigvee_{i=0}^{r-1} ((x \wedge y)^i \wedge (x^i)')$$

$$= \bigvee_{i=0}^{r-1} \left(\left(y^i \wedge \bigvee_{h=i}^{r-1} x^h \right) \wedge (x^i)' \right)$$

$$= \bigvee_{i=0}^{r-2} (y^i \wedge \bigvee_{h=i+1}^{r-1} (x^h \wedge (x^i)')) = \bigvee_{i=0}^{r-2} (y^i \wedge \bigvee_{h=i+1}^{r-1} x^h) .$$

$\square$

Corollary 1.6. $x \leq y$ iff $x^h \wedge y^i = 0$ whenever $h > i$. $\square$

The next lemma will be needed in §§2,3. It generalizes well-known properties of orthonormal systems in Boolean algebras; see e.g. BFE, Theorem 4.1.

Lemma 1.2. (Serfati [1973b]). *Let* λ_i $(i = 0, 1, \ldots, s - 1)$ *be an orthonormal system in a Post algebra* L *and* $a_i, b_i \in L$ $(i = 0, 1, \ldots, s - 1)$. *Then:*

(1.36)
$$(\bigvee_{i=0}^{s-1} (a_i \wedge \lambda_i)) \vee (\bigvee_{i=0}^{s-1} (b_i \wedge \lambda_i))$$
$$= \bigvee_{i=0}^{s-1} ((a_i \vee b_i) \wedge \lambda_i) ,$$

(1.37)
$$(\bigvee_{i=0}^{s-1} (a_i \wedge \lambda_i)) \wedge (\bigvee_{i=0}^{s-1} (b_i \wedge \lambda_i))$$
$$= \bigvee_{i=0}^{s-1} ((a_i \wedge b_i) \wedge \lambda_i) ,$$

(1.38)
$$(\bigvee_{i=0}^{s-1} (a_i \wedge \lambda_i))^h = \bigvee_{i=0}^{s-1} ((a_i)^h \wedge \lambda_i) \qquad (h = 0, 1, \ldots, r-1) ,$$

(1.39)
$$(\bigvee_{i=0}^{s-1} (a_i \wedge \lambda_i))_k = \bigvee ((a_i)_k \wedge \lambda_i) \qquad (k = 1, \ldots, r-1) ,$$

where index k *stands for the monotone components,*

(1.40)
$$(\bigvee_{i=0}^{s-1} (a_i \wedge \lambda_i)) \perp (\bigvee_{i=0}^{s-1} (b_i \wedge \lambda_i))$$
$$= \bigvee_{i=0}^{s-1} ((a_i \perp b_i) \wedge \lambda_i) \qquad (\perp \in \{\rightarrow, -\}) ,$$

(1.41)
$$(\bigvee_{i=0}^{s-1} (a_i \wedge \lambda_i))^\Delta$$
$$= \bigvee_{i=0}^{s-1} ((a_i)^\Delta \wedge \lambda_i) \qquad (\Delta \in \{*, + \}) .$$

PROOF: Identities (1.36) and (1.37) are quite clear using distributivity. Further, it follows from (1.36) and (1.37) that

$$(\bigvee_{i=0}^{s-1} ((a_i)^h \wedge \lambda_i)) \wedge (\bigvee_{i=0}^{s-1} ((a_i)^m \wedge \lambda_i)) = 0 \text{ if } h \neq m ,$$

$$\bigvee_{h=0}^{r-1} \bigvee_{i=0}^{s-1} ((a_i)^h \wedge \lambda_i)) = \bigvee_{i=0}^{s-1} \bigvee_{h=0}^{r-1} ((a_i)^h \wedge \lambda_i)$$

$$= \bigvee_{i=0}^{s-1} ((\bigvee_{h=0}^{r-1} (a_i)^h) \wedge \lambda_i) = \bigvee_{i=0}^{s-1} \lambda_i = 1 ,$$

showing that the right members of (1.38) form an orthonormal system. Besides,

$$\bigvee_{h=0}^{r-1} ((\bigvee_{i=0}^{s-1} ((a_i)^h \wedge \lambda_i)) \wedge e_h) = \bigvee_{i=0}^{s-1} \bigvee_{h=0}^{r-1} ((a_i)^h \wedge \lambda_i \wedge e_h)$$

$$= \bigvee_{i=0}^{s-1} ((\bigvee_{h=0}^{r-1} ((a_i)^h \wedge e_h)) \wedge \lambda_i) = \bigvee_{i=0}^{s-1} (a_i \wedge \lambda_i) ,$$

therefore the right members of (1.38) provide an orthonormal decomposition of the element $\bigvee_{i=0}^{s-1} (a_i \wedge \lambda_i)$. But the unique orthonormal decomposition of an element is provided by its disjunctive components. This completes the proof of (1.38). One proves (1.39)−(1.41) in a similar way.   □

We refer the reader to Serfati [1973b] for the following theorem: every Post algebra becomes a unitary commutative ring of characteristic r under the operations $\oplus$ and $\odot$ defined by

$$(1.42) \qquad (x \oplus y)^i = \bigvee_{s+t \equiv i} (x^s \wedge y^t) \qquad (i = 0, 1, \ldots, r - 1) ,$$

$$(1.43) \qquad (x \odot y)^i = \bigvee_{s \cdot t \equiv i} (x^s \wedge y^t) \qquad (i = 0, 1, \ldots, r - 1) ,$$

where $\equiv$ denotes congruence modulo r, the zero and the one of the ring being e_0 and e_1, respectively, while

$$(1.44) \qquad \ominus x = (x^{r-1} \wedge e_1) \vee (x^{r-2} \wedge e_2) \vee \ldots \vee (x^1 \wedge e_{r-1}) .$$

2 Post functions

In the remainder of this chapter Post algebras are viewed as algebras $(L; \wedge, \vee, (^i)_{i=0,\ldots,r-1}, (e_i)_{i=0,\ldots,r-1})$ of type $(2, 2, (1)_r, (0)_r)$. In this section we deal with algebraic functions and polynomials over L, which we call Post functions and simple Post functions, respectively. The main results are Theorem 2.1 and Proposition 2.2, which provide the canonical form of Post functions and the characterization of simple Post functions, respectively, then the coincidence between Post functions and functions with the substitution property, established in Theorem 2.2.

Definition 2.1. By a *Post function (simple Post function)* we mean an algebraic function (a polynomial) of the Post algebra $(L; \wedge, \vee, (^i)_{i=0,\dots,r-1}, (e_i)_{i=0,\dots, r-1})$; cf. Definition 3.2.8.

To characterize Post functions, we first establish some notation similar to the one used in BFE.

Notation 1) From now on *meet* will be indicated by $\cdot$ or simply by concatenation instead of $\wedge$. So we will benefit from the usual conventions concerning omission of parentheses: $xy \vee z$ stands for $(xy) \vee z$.

2) *Iterated join* and *meet* will be denoted by $\bigvee$ and $\prod$, respectively. In particular $\bigvee_A$ will stand for the iterated join $\bigvee_{A \in C_r^n}$.

3) We will work with functions of n variables $x_1, \dots, x_n$. So subscripts will have their usual meaning, while the *monotone components* of an element x will be denoted by $x_{(1)}, x_{(2)}, \dots, x_{(r-1)}$.

4) For every $i \in \{0, 1, \dots, r-1\}$ and every $x \in L$ we will also use the alternative notation $x^i = x^{e_i}$.

5) We will use upper case letters to denote *vectors*: $X = (x_1, \dots, x_n) \in L^n, \dots, A = (\alpha_1, \dots, \alpha_n) \in C_r^n, \dots$ etc. A basic convenient notation, introduced by Serfati [1973a], [1973b], is

$$(2.0) \qquad X^A = x_1^{\alpha_1} \dots x_n^{\alpha_n} \qquad (X \in L^n, \ A \in C_r^n).$$

$\square$

Remark 2.1. It follows from (1.11) and Proposition 1.3 that if $\alpha, \beta \in C_r$, then

$$(2.1) \qquad \alpha^\beta = 1 \text{ if } \alpha = \beta, \text{ else } 0,$$

which implies the following generalization: if $A, B \in C_r^n$, then

$$(2.2) \qquad A^B = 1 \text{ if } A = B, \text{ else } 0.$$

$\square$

Lemma 2.1. *For every $X \in L^n$, the elements X^A, $A \in C_r^n$, form an orthonormal system.*

PROOF: If $A \neq B$ then $\alpha_k \neq \beta_k$ for some $k \in \{1, \dots, n\}$, therefore $x_k^{\alpha_k} \cdot x_k^{\beta_k} = 0$, implying $X^A \cdot X^B = 0$. The property $\bigvee_A X^A = 1$ is proved by induction on n.

$\square$

The well-known representation theorem for Boolean functions (see e.g. BFE, Theorem 1.6′) has a straightforward generalization to Post functions.

Theorem 2.1. (Serfati [1973b]). *Let L be a Post algebra and n a natural number. Then:*

α) *The following conditions are equivalent for a function $f : L^n \longrightarrow L$:*

(i) f is a Post function ;

(ii) f can be represented in the form

$$(2.3) \qquad f(X) = \bigvee_A c_A X^A \qquad (\forall X \in L^n),$$

where $c_A \in L$ for every $A \in C_r^n$.

β) *The coefficients c_A in the representation (2.3) are uniquely determined by*

(2.4) $$c_A = f(A) \qquad (\forall A \in C_r^n) \,.$$

COMMENTS: 1) The expansion (2.3),(2.4) is known as the *canonical disjunctive form* of the function f.

2) Reischer and Simovici [1986] have characterized Post algebras as algebras $(L; \wedge, \vee, (^i)_{i=0,\ldots,r-1}, (e_i)_{i=0,\ldots,r-1})$ of type $(2, 2, (1)_r, (0)_r)$ in which $(L; \wedge, \vee)$ is a lattice, properties (1.5) and (iv) in Proposition 3.3.2 hold and every algebraic function satisfies the above properties (ii) and β; cf. Notation 4) and (2.0).

PROOF: α) The identity (2.3) can be written in the form

(2.5) $$f = \bigvee_A c_A (\pi_{1L}^n)^{\alpha_1} (\pi_{2L}^n)^{\alpha_2} \ldots (\pi_{nL}^n)^{\alpha_n} \,,$$

where $A = (\alpha_1, \ldots, \alpha_n)$ runs over C_r^n , c_A stand for the constant functions $X \mapsto c_A \in L$, and $\pi_{kL}^n : L^n \longrightarrow L$, $\pi_{kL}^n(x_1, \ldots, x_n) = x_k$ are the projection functions (cf. Remark 3.3.2). So (2.5) shows that f is a Post functiuon.

Conversely, we prove by algebraic induction that every Post function can be written in the form (2.3). For the constant functions c (including $e_0, e_1, \ldots, e_{r-1}$) we take $c_A := c$ for all A. The projections π_{kL}^n are obtained by taking $c_A := \pi_{kL}^n(A) = \alpha_k$ because

$$\bigvee_A \alpha_k \cdot X^A = \bigvee_{i=0}^{r-1} \bigvee_{\alpha_1,\ldots,\alpha_{k-1},\alpha_{k+1},\ldots,\alpha_{r-1}} e_i x_1^{\alpha_1} \ldots x_{k-1}^{\alpha_{k-1}} x_k^{e_i} x_{k+1}^{\alpha_{k+1}} \ldots x_n^{\alpha_n}$$

$$= \bigvee_{i=0}^{r-1} (e_i \cdot x_k^{e_i}) \bigvee_{\alpha_1,\ldots,\alpha_{k-1},\alpha_{k+1},\ldots,\alpha_{r-1}}$$

$$x_1^{\alpha_1} \ldots x_{k-1}^{\alpha_{k-1}} x_{k+1}^{\alpha_{k+1}} \ldots x_n^{\alpha_n} = \bigvee_{i=0}^{r-1} e_i x_k^i = x_k \qquad (k = 1, \ldots, r-1) \,.$$

Finally if the identities (2.3) and $g(X) = \bigvee_A d_a X^A$ hold, then it follows from Lemma 1.2 applied to the orthonormal system X^A , $A \in C_r^n$ (cf. Lemma 2.1) that the functions $f \vee g$, fg and f^i $(i = 0, \ldots, r-1)$, which are defined pointwise, are of the same form.

β) Take an arbitrary but fixed $B \in C_r^n$. Then it follows from (2.3) and (2.2) that

$$f(B) = \bigvee_A c_A B^A = c_B \,.$$

$\square$

For a particular application see Ch. 12 §2.

Corollary 2.1. *Every Post function of n variables is uniquely determined by its restriction on the set C_r^n.* $\square$

Corollary 2.2. *If $f, g : L^n \longrightarrow L$ are Post functions, then*

$$f = g \Longleftrightarrow f(A) = g(A) \text{ for all } A \in C_r^n \,,$$

$$f \leq g \Longleftrightarrow f(A) \leq g(A) \text{ for all } A \in C_r^n \,.$$

COMMENT: This corollary and Propositions 3.6 and 3.7 can be viewed as a generalization of the Müller-Löwenheim Verification Theorem for Boolean functions (see e.g. BFE, Theorems 2.13 and 2.14).

PROOF: $f \leq g \Longleftrightarrow fg = g$ and fg is a Post function.   □

Corollary 2.3. *Every function $f : C_r^n \longrightarrow L$ can be uniquely extended to a Post function $f : L^n \longrightarrow L$.*   □

Corollary 2.4. *The identities and the identical inequalities common to all Post algebras coincide with those of C_r^n.*

PROOF: By Corollaries 2.3 and 2.3.3.   □

For a generalization to implications between Post functions see Propositions 3.6 and 3.7.

Proposition 2.1. (Serfati [1973b]). *The set of all Post functions $f : L^n \longrightarrow L$ is a Post algebra with respect to the following operations:*

$$(2.6) \qquad (f \circ g)(X) = \bigvee_A (f(A) \circ g(A)) X^A \qquad (\circ \in \{\wedge = \cdot, \vee\}) \,,$$

$$(2.7) \qquad e_i(X) = e_i \qquad (i = 0, 1, \ldots, r - 1) \,,$$

$$(2.8) \qquad f^i(X) = \bigvee_A (f(A))^i X^A \qquad (i = 0, 1, \ldots, r - 1) \,,$$

$$(2.9) \qquad f_{(k)}(X) = \bigvee_A (f(A))_{(k)} X^A \qquad (k = 1, \ldots, r - 1) \,,$$

and we have also

$$(2.10) \qquad (f \perp g)(X) = \bigvee_A (f(A) \perp g(A)) X^A \qquad (\perp \in \{\rightarrow, -\} \,,$$

$$(2.11) \qquad f^\triangle(X) = \bigvee_A (f(A))^\triangle X^A \qquad (^\triangle \in \{^*, ^+\}) \,.$$

PROOF: Clearly the set of (Post) functions $f : L^n \longrightarrow L$ is an algebra of the same type as L under the operations defined pointwise. The fact that the pointwise definition of the operations between Post functions amounts to (2.6) and

(2.8)-(2.11) follows from Theorem 2.1 and Lemma 1.2 applied to the orthonormal system X^A, $A \in C_r^n$ (cf. Lemma 2.1). Now there are two variants for the remainder of the proof.

1) Immediate from the fact that the class of Post algebras is equational, i.e., it can be defined by a set of identities (cf. Traczyk [1964]; see also Balbes and Dwinger [1974], Theorem X.6.1 and Corollary 2).

2) Since the class of bounded distributive lattices is equational, it follows immediately that the algebra of Post functions is a bounded distributive lattice. Also, the constant operations e_i form a strictly increasing chain. In view of Lemma 1.2, the disjunctive components $(f(X))^i$ of the element $f(X)$ are precisely the right members of (2.8); this proves (2.8). The proof of (2.9)-(2.11) is similar, using the same lemma. □

Proposition 2.1 and the general property that the algebra of all polynomials is a subalgebra of the algebra of all algebraic functions (cf. Proposition 2.2.1) justifies the following

Definition 2.2. Let L be a Post algebra and n a natural number. The *algebra* PFLn *of Post functions* (SPFLn *of simple Post functions*) of n variables is the set PFLn of all Post functions (SPFLn of all simple Post functions) $f : L^n \longrightarrow L$ endowed with the Post operations defined componentwise. □

Corollary 2.5. SPFLn *is a Post algebra and a subalgebra of* PFLn.

PROOF: The latter property was remarked above. The fact that SPFLn is a Post algebra follows from the equational definition of a Post algebra or by noting that the uniqueness of the disjunctive representation of an element is inherited by any subalgebra. □

Corollary 2.6. *The following conditions are equivalent for a Post function* $f :$ $L^n \longrightarrow L :$
  (i) $f \in B(\text{PFLn})$;
  (ii) $f(X) \in B(L)$ $\quad \forall X \in L^n$;
  (iii) $f(A) \in B(L)$ $\quad \forall A \in C_r^n$.

PROOF: (i)$\Longleftrightarrow$(ii): Immediate from the pointwise definition of operations in PFLn.
  (ii)$\Longrightarrow$(iii): Trivial.
  (iii)$\Longrightarrow$(ii): We have $f \vee f^* = 1$ because

$$(f \vee f^*)(X) = f(X) \vee f^*(X) = \bigvee_A (f(A) \vee f(A)^*)X^A = \bigvee_A X^A = 1.$$

□

Proposition 2.2. *A Post function* $f : L^n \longrightarrow L$ *is simple if and only if* $f(A) \in$ C_r *for all* $A \in C_r^n$.

PROOF: Sufficiency follows from (2.5), where $c_A = f(A) \in \{e_0, e_1, \ldots, e_{r-1}\}$ for all $A \in C_r^n$. Necessity is easily proved by algebraic induction. □

Corollary 2.7. *A simple Post function is also characterized by the property*

$$(2.12) \qquad f(X) = \bigvee_{i=0}^{r-1} e_i \bigvee \{X^A \mid f(A) = e_i\} \qquad (\forall X \in L^n) .$$

□

Corollary 2.8. *Any function obtained from a simple Post function by fixing certain variables to values in C_r is also a simple Post function.*

PROOF: The latter function is a Post function by Corollary 3.3.8, therefore it is a simple Post function by Proposition 2.2.   □

Definition 2.3. By an *r-valued truth function* of n variables is meant any function $f : C_r^n \longrightarrow C_r$.   □

Remark 2.2. Any r-valued truth function is a simple Post function.   □

Remark 2.3. If L is an r-Post algebra and n is a natural number, then there exist $|L|^{|L|^n}$ functions $f : L^n \longrightarrow L$, among which $|L|^{r^n}$ are Post functions, including r^{r^n} simple Post functions.   □

The following result is a straightforward generalization of the Corollary of Theorem 1.11 in BFE.

Proposition 2.3. *Let L be an r-Post algebra and n a natural number. Then the following conditions are equivalent:*

(i) every function $f : L^n \longrightarrow L$ is a Post function ;

(ii) every function $f : L^n \longrightarrow L$ is a simple Post function ;

(iii) every function $f : L^n \longrightarrow L$ is an r-valued truth function ;

(iv) $L = C_r$.

PROOF: (iv)$\Longrightarrow$(iii)$\Longrightarrow$(ii)$\Longrightarrow$(i): Trivial (using Remark 2.2).
       (i)$\Longrightarrow$(iv): By Remark 2.3.   □

Recall that every algebraic function has the substitution property (cf. Proposition 2.3.6) and in general the converse need not hold (cf. Remark 3.3.5). Yet certain properties of Boolean functions still hold for functions with the substitution property in pseudocomplemented distributive lattices, Heyting algebras and Stone algebras (cf. Theorems 3.3.3 and 3.3.4). In the case of Boolean functions, Grätzer [1962] (see also BFE, Theorem 1.13) proved that functions with the substitution property are the same as Boolean functions. The next theorem generalizes this result to Post algebras.

Theorem 2.2. (Beazer [1974b]). *Let L be a Post algebra. A function $f : L^n \longrightarrow L$ has the substitution property if and only if it is a Post function.*

PROOF: Suppose f has the substitution property. Define $g : L^n \longrightarrow L$ by $g(X) = \bigvee_A f(A)X^A$. To prove that $f = g$ suppose the contrary. Then there exist $a_1, \ldots, a_n \in L$ such that

$$p = f(a_1, \ldots, a_n) \neq g(a_1, \ldots, a_n) = q,$$

therefore $p_{(k)} \neq q_{(k)}$ for some $k \in \{1, \ldots, n\}$. Since $p_{(k)}, q_{(k)} \in B(L)$, a well-known property (see e.g. Boicescu, Filipoiu, Georgescu and Rudeanu [1991], Corollary 3.13 and Proposition 3.9) states the existence of a surjective Boolean homomorphism $h : B(L) \longrightarrow \{0, 1\}$ such that $h(p_{(k)}) \neq h(q_{(k)})$. Then the map $\overline{h} : L \longrightarrow C_r$ defined by

$$\overline{h}(x) = \bigvee_{j=1}^{r-1} h(x_{(j)})e_j$$

is a Post homomorphism which extends h: this follows from a theorem due to Rousseau [1970] or by a direct verification. Note that $\overline{h}(p) \neq \overline{h}(q)$. Now Lemma 2.3.3 implies the existence of the function $\overline{f} : C_r^n :\longrightarrow C_r$ defined by

$$\overline{f}(\overline{h}(x_1), \ldots, \overline{h}(x_n)) = \overline{h}(f(x_1, \ldots, x_n))$$

and similarly, setting

$$\overline{g}(\overline{h}(x_1), \ldots, \overline{h}(x_n)) = \overline{h}(g(x_1, \ldots, x_n))$$

we obtain a function $\overline{g} : C_r^n \longrightarrow C_r$. Since $f(A) = g(A)$ for all $A \in C_r^n$, it follows that

$$\overline{f}(\overline{h}(\alpha_1), \ldots, \overline{h}(\alpha_n)) = \overline{g}(\overline{h}(\alpha_1), \ldots, \overline{h}(\alpha_n))$$

for all $(\alpha_1, \ldots, \alpha_n) \in C_r^n$; but the vectors $(\overline{h}(\alpha_1), \ldots, \overline{h}(\alpha_n))$ exhaust the elements of C_r^n, therefore $\overline{h} = \overline{g}$. However this is impossible, because

$$\overline{f}(\overline{h}(a_1), \ldots, \overline{h}(a_n)) = \overline{h}(p) \neq \overline{h}(q) = \overline{g}(\overline{h}(a_1), \ldots, \overline{h}(a_n))$$

$\square$

In the next section we will need the following result, which generalizes a property of Boolean functions (see e.g. BFE, Theorem 4.6(ii)).

Lemma 2.2. (Serfati [1973b]). *Let: L be a Post algebra, λ_i $(i = 0, \ldots, s-1)$ an orthonormal system of L, and x_{ik} $(i = 0, \ldots, s-1 ; k = 1, \ldots, n)$ be elements of L. Then for every Post function $f : L^n \longrightarrow L$,*

$$(2.13) \qquad f(\bigvee_{i=0}^{s-1} x_{i1}\lambda_i, \ldots, \bigvee_{i=0}^{s-1} x_{in}\lambda_i) = \bigvee_{i=0}^{s-1} f(x_{i1}, \ldots, x_{in})\lambda_i .$$

PROOF: Based on Lemma 1.2. Setting $A = (\alpha_1, \ldots, \alpha_n)$ for $A \in C_r^n$, we have

$$f(\bigvee_{i=0}^{s-1} x_{i1}\lambda_i, \ldots, \bigvee_{i=0}^{s-1} x_{in}\lambda_i) = \bigvee_A f(A) \prod_{k=1}^{n} (\bigvee_{i=0}^{s-1} x_{ik}\lambda_i)^{\alpha_k}$$

$$= \bigvee_A f(A) \prod_{k=1}^{n} (\bigvee_{i=0}^{s-1} (x_{ik})^{\alpha_k} \lambda_i) = \bigvee_A f(A) \bigvee_{i=0}^{s-1} (\prod_{k=1}^{n} (x_{ik})^{\alpha_k}) \lambda_i$$

$$= \bigvee_A \bigvee_{i=0}^{s-1} f(A) (\prod_{k=1}^{n} (x_{ik})^{\alpha_k}) \lambda_i = \bigvee_{i=0}^{s-1} \bigvee_A f(A) (\prod_{k=1}^{n} (x_{ik})^{\alpha_k}) \lambda_i$$

$$= \bigvee_{i=0}^{s-1} (\bigvee_A f(A) (\prod_{k=1}^{n} (x_{ik})^{\alpha_k})) \lambda_i = \bigvee_{i=0}^{s-1} f(x_{i1}, \ldots, x_{in}) \lambda_i .$$

$\square$

Other properties of Post functions will be established in the next section by using the technique of Post equations; see Propositions 3.5-3.7 and 3.13. See also Reischer and Simovici [1987] for iterative properties of Post functions.

3 Post equations

This section is devoted to *Post equations*, i.e., equations expressed in terms of Post functions; cf. Definition 3.2.8. The main topics to be treated, i.e. the reduction to a single equation, the consistency condition, the method of successive eliminations of variables, the construction of the reproductive solution from a particular solution, and various forms of the general solution and of the reproductive solution, parallel the theory of Boolean equations; cf. BFE.

All of the equations studied in this section are over an arbitrary r-Post algebra L.

The results for which we give credit to Carvallo, Serfati and Bordat were obtained by these authors independently of each other and in certain cases the original result is not the one given in this section, but an equivalent form of it via the transformations pointed out in Theorem 3.1 and Remark 3.1 below. In particular Carvallo worked in the Post algebra C_r and his results published in the Comptes Rendus remained without proofs; however they are valid in an arbitrary Post algebra.

The first theorem is a straightforward generalization of the corresponding result for Boolean equations (see e.g. BFE, Theorem 2.1).

Theorem 3.1. *Every system of Post equations and/or inequalities is equivalent to a single Post equation of the form $f = 0$ and also to a single equation of the form $g = 1$.*

COMMENT: This theorem was proved by Carvallo [1968b] in a particular case, by Serfati [1973a], [1973b] using the ring structure introduced by him and by Bordat [1975], [1978] via Proposition 1.11.

PROOF: Consider a system of equations and/or inequalities

$$(3.1) \qquad f_k(X) \rho_k g_k(X) , \ \rho_k \in \{=, \leq\} , \ (k = 1, \ldots, m) ,$$

where f_k and g_k are Post functions of $X \in L^n$ $(k = 1, \ldots, m)$. As a matter of fact, since $f_k \leq g_k \iff f_k g_k = f_k$, where $f_k g_k$ is also a Post function, we can consider without loss of generality that system (3.1) consists only of equations. Then in view of Proposition 1.11, the system is equivalent to

(3.2) $$h_k(X) = 0 \qquad (k = 1, \ldots, m) ,$$

where in fact each

$$h_k = \bigvee_{i=0}^{r-1} (f_k)^i ((g_k)^i)'$$

is a Post function $(k = 1, \ldots, m)$. Finally system (3.2) is equivalent to the single equation

(3.3) $$h(X) = 0 ,$$

where $h = \bigvee_{k=1}^{m} h_k$ is a Post function, too.

The reduction to the form $g = 1$ is quite similar. $\qquad \Box$

Remark 3.1. (cf. Remark 3.2). We have $f(X) = 0 \iff f^*(X) = 1$ by Proposition 3.2.1(2.4). More precisely, in terms of the decomposition provided by Proposition 2.1(2.11),

$$\bigvee_A f(A)X^A = 0 \iff \bigvee_A f^*(A)X^A = 1 .$$

The dual property is $g(X) = 1 \iff g^+(X) = 0$. So we are faced with a lack of symmetry, to the effect that $g(X) = 1$ is not in general equivalent to $g^*(X) = 0$; we have only $g(X) = 1 \implies g^*(X) = 0 \iff g^{**}(X) = 1$. However if $g \in B(\mathrm{PFLn})$ (cf. Definition 2.2) then $g^{**} = g$, so that in this case $g(X) = 1 \iff g^*(X) = 0$. Since $f^* \in B(\mathrm{PFLn})$ for every $f \in \mathrm{PFLn}$, it follows from Theorem 3.1 that *every system of Post equations and/or inequalities is equivalent to a single equation of the form $g = 1$ where $g \in B(\mathrm{PFLn})$*. This was first noted by Bordat [1975], [1978], who then concentrated on the study of equations of this type. $\qquad \Box$

Example 3.1. (Serfati [1973b]). Let $r := 3$ and consider the following system in the unknown x:

(E1.1) $$x^1 \vee e_1 x^2 = x^1 \vee e_1 x^0 ,$$

(E1.2) $$e_1 x^0 \leq x^1 .$$

Proposition 1.11 transforms equation (E1.1) into

$$(e_0 x^0 \vee e_2 x^1 \vee e_1 x^2)^0 ((e_1 x^0 \vee e_2 x^1 \vee e_0 x^2)^0)'$$

$$\vee (e_0 x^0 \vee e_2 x^1 \vee e_1 x^2)^1 ((e_1 x^0 \vee e_2 x^1 \vee e_0 x^2)^1)'$$

$$\vee (e_0 x^0 \vee e_2 x^1 \vee e_1 x^2)^2 ((e_1 x^0 \vee e_2 x^1 \vee e_0 x^2)^2)' = 0 ,$$

which, in view of Lemma 1.2 and Remark 1.1, reduces to

$$(1x^0 \vee 0x^1 \vee 0x^2)(0x^0 \vee 0x^1 \vee 1x^2)' \vee (0x^0 \vee 0x^1 \vee 1x^2)(1x^0 \vee 0x^1 \vee 0x^2)'$$

$$\vee(0x^0 \vee 1x^1 \vee 0x^2)(0x^2 \vee 1x^1 \vee 0x^2)' = 0 ,$$

while (E1.2) can be written $e_1 x^0 \cdot x^1 = e_1 x^0$, so that (E1) becomes

(E1.1′) $x^0 \vee x^2 = 0 ,$

(E1.2′) $e_1 x^0 = 0 .$

By taking the join of (E1.1′) and (E1.2′) we recapture (E1.1′); this reflects the fact that the former equation implies the latter. Thus the original system (E1) is equivalent to the single equation (E1′).   □

Example 3.2. (cf. Examples 3.3, 3.7–3.10). Take again $r := 3$ and consider the following system of equations in the unknowns a, b, c (its significance will be seen in Ch. 12, §2):

(E2.1) $e_1 a^1 \vee ba^2 = a ,$

(E2.2) $a = e_1 a^1 \vee ca^2 ,$

(E2.3) $ac^1 \vee bc^2 = ab^0 \vee e_1 b^1 \vee cb^2 ,$

(E2.4) $ab^1 \vee b^2 = b ,$

(E2.5) $ac^1 \vee bc^2 = ba^0 \vee ca^1 \vee a^2 ,$

(E2.6) $b = cb^1 \vee b^2 ,$

(E2.7) $ab^0 \vee e_1 b^1 \vee cb^2 = ba^0 \vee ca^1 \vee a^2 ,$

(E2.8) $ac^0 \vee e_1 c^1 \vee c^2 = c ,$

(E2.9) $c = bc^0 \vee e_1 c^1 \vee c^2 .$

As in the previous example, by applying Proposition 1.11 via Lemma 1.2 we get

(E2.1′) $a^0 \vee a^1 \vee b^2 a^2 = 1 ,$

(E2.2′) $a^0 \vee a^1 \vee c^2 a^2 = 1 ,$

(E2.3′) $(c^0 \vee a^0 c^1 \vee b^0 c^2)(a^0 b^0 \vee c^0 b^2) \vee (a^1 c^1 \vee b^1 c^2)(a^1 b^0 \vee b^1 \vee c^1 b^2)$
$$\vee (a^2 c^1 \vee b^2 c^2)(a^2 b^0 \vee c^2 b^2) = 1 ,$$

(E2.4′) $b^0 \vee a^1 b^1 \vee b^2 = 1 ,$

(E2.5′) $(c^0 \vee a^0 c^1 \vee b^0 c^2)(b^0 a^0 \vee c^0 a^1) \vee (a^1 c^1 \vee b^1 c^2)(b^1 a^0 \vee c^1 a^1)$
$$\vee (a^2 c^1 \vee b^2 c^2)(b^2 a^0 \vee c^2 a^1 \vee a^2) = 1 ,$$

(E2.6′) $$b^0 \vee c^1 b^1 \vee b^2 = 1 \,,$$

(E2.7′)
$$(a^0 b^0 \vee c^0 b^2)(b^0 a^0 \vee c^0 a^1) \vee (a^1 b^0 \vee b^1 \vee c^1 b^2)(b^1 a^0 \vee c^1 a^1)$$
$$\vee (a^2 b^0 \vee c^2 b^2)(b^2 a^0 \vee c^2 a^1 \vee a^2) = 1 \,,$$

(E2.8′) $$a^0 c^0 \vee c^1 \vee c^2 = 1 \,,$$

(E2.9′) $$b^0 c^0 \vee c^1 \vee c^2 = 1 \,.$$

By multiplying equations (E2.1′) and (E2.2′), (E2.4′) and (E2.6′), (E2.8′) and (E2.9′) and by working out the other equations, we obtain the following equivalent system:

(E2.10) $$a^0 \vee a^1 \vee a^2 b^2 c^2 = 1 \,,$$

(E2.3″) $$c^0(a^0 b^0 \vee b^2) \vee c^1(a^0 b^0 \vee a^1 \vee a^2 b^0) \vee c^2(a^0 b^0 \vee b^1 \vee b^2) = 1 \,,$$

(E2.5″) $$a^0(b^0 \vee b^1 c^2 \vee b^2 c^2) \vee a^1(c^0 \vee c^1 \vee b^2 c^2) \vee a^2(c^1 \vee b^2 c^2) = 1 \,,$$

(E2.11) $$b^0 \vee b^2 \vee a^1 b^1 c^1 = 1 \,,$$

(E2.7″) $$b^0(a^0 \vee a^1 c^1 \vee a^2) \vee b^1(a^0 \vee a^1 c^1) \vee b^2(c^0 a^1 \vee c^1 a^1 \vee c^2) = 1 \,,$$

(E2.12) $$c^1 \vee c^2 \vee a^0 b^0 c^0 = 1 \,.$$

Further we multiply (E2.10) and (E2.5″), (E2.3″) and (E2.12), (E2.11) and (E2.7″) and we use the identity $x \vee x'y = x \vee y$. This yields the equivalent system

(E2.13) $$a^0(b^0 \vee (b^1 \vee b^2)c^2) \vee a^1(c^0 \vee c^1 \vee b^2) \vee a^2 b^2 c^2 = 1 \,,$$

(E2.14) $$a^0 b^0 c^0 \vee c^1((a^0 \vee a^2)b^0 \vee a^1) \vee c^2(a^0 \vee b^1 \vee b^2) = 1 \,,$$

(E2.15) $$b^0(a^0 \vee a^2 \vee c^1) \vee a^1 b^1 c^1 \vee b^2((c^0 \vee c^1)a^1 \vee c^2) = 1 \,.$$

The meet of (E2.13) and (E2.14) yields

$$a^0 b^0 c^0 \vee a^0 c^1(b^0 \vee c^2)(a^1 \vee b^0) \vee a^0 c^2(b^0 \vee c^2)(a^0 \vee b^1 \vee b^2)$$

$$\vee a^1 c^1 \vee a^1 c^2 b^2 \vee a^2 b^2 c^2 = 1$$

and the meet of this equation and (E2.15) is

$$a^0b^0c^0 \vee a^0b^0c^1 \vee a^0b^0c^2 \vee a^0b^2c^2 \vee a^1b^0c^1 \vee a^1b^1c^1$$
$$\vee a^1b^2c^1 \vee a^1b^2c^2 \vee a^2b^2c^2 = 1 \,,$$

which finally reduces to

(E2.16) $$a^0b^0 \vee a^1b^1 \vee a^2c^2 = 1 \,.$$

□

We are going to study equations of the form $f = 0$. The study of equations of the form $g = 1$ via Remark 3.1 is left in most cases to the reader.

As we did for Boolean equations, we begin our study of Post equations in n unknowns with the case $n := 1$.

Proposition 3.1. *A Post equation in one unknown*

(3.4) $$\bigvee_{i=0}^{r-1} c_i x^i = 0$$

is consistent if and only if

(3.5) $$\prod_{i=0}^{r-1} c_i = 0 \,,$$

in which case the set of solutions is a sublattice of L with least element

(3.6) $$\eta = \prod_{i=0}^{r-1} (c_i^{**} \vee e_i)$$

and greatest element

(3.7) $$\zeta = \bigvee_{i=0}^{r-1} c_i^* e_i \,.$$

COMMENTS: 1) The form (3.4) of an arbitrary Post equation in one unknown follows from Theorems 3.1 and 2.1.

2) An equation is said to be *consistent* provided it has solutions. The consistency condition (3.5) was discovered by Carvallo [1967], [1968a], [1968b], Serfati [1973a], [1973b] and Bordat [1975], [1978]. The least and greatest solutions (3.6) and (3.7), respectively, were given by Carvallo, while Bordat pointed out the lattice structure of the set of solutions, with η and ζ described by their monotone components and by their disjunctive components. Serfati noted the particular solution (3.7).

PROOF: A) Let η and ζ be defined by (3.6) and (3.7), respectively. Recall that $c^* \in B(L)$ and $(c^*)' = c^{**}$ by Proposition 3.2.13 and Remark 3.2.7. It follows from Propositions 1.6, 1.7, 1.4 and 1.3 that (3.6) and (3.7) can be written in the equivalent forms

(3.8) $$\eta_{(k)} = \prod_{i=0}^{r-1} (c_i^{**} \vee (e_i)_{(k)}) = \prod_{i=0}^{r-1} c_i^{**} \quad (k = 1, \ldots, r-1) \,,$$

$$(3.9) \qquad \zeta_{(k)} = \bigvee_{i=0}^{r-1} c_i^*(e_i)_{(k)} = \bigvee_{i=k}^{r-1} c_i^* \quad (k = 1, \ldots, r-1),$$

from which we infer by Lemma 1.1 that

$$(3.10.0) \qquad \eta^0 = (c_0^{**})' = c_0^*,$$

$$(3.10.i) \qquad \eta^i = (\prod_{h=0}^{i-1} c_h^{**}) \bigvee_{h=0}^{i} c_h^* = c_i^* \prod_{h=0}^{i-1} c_h^{**} \quad (i = 1, \ldots, r-2),$$

$$(3.10.r-1) \qquad \eta^{r-1} = \prod_{i=0}^{r-2} c_i^{**},$$

$$(3.11.0) \qquad \zeta^0 = (\bigvee_{j=1}^{r-1} c_j^*)' = \prod_{j=1}^{r-1} c_j^{**},$$

$$(3.11.i) \qquad \zeta^i = (\bigvee_{j=i}^{r-1} c_j^*) \prod_{j=i+1}^{r-1} c_j^{**} = c_i^* \prod_{j=i+1}^{r-1} c_j^{**} \quad (i = 1, \ldots, r-2),$$

$$(3.11.r-1) \qquad \zeta^{r-1} = c_{r-1}^*,$$

and this implies further

$$(3.12) \qquad \bigvee_{i=0}^{r-1} c_i \eta^i = c_{r-1} \eta^{r-1},$$

$$(3.13) \qquad \bigvee_{i=0}^{r-1} \zeta^i = c_0 \zeta^0.$$

B) Now suppose condition (3.5) holds. Then $c_0 \prod_{j=1}^{r-1} c_j = 0$ holds, hence $c_0^* \vee \bigvee_{j=i}^{r-1} c_j* = 1$, therefore $c_0^{**} \zeta^0 = 0$ and since $**$ is a closure operator by Corollary 3.2.1, the last equality implies $c_0 \zeta^0 = 0$. One proves similarly that $c_{r-1} \eta^{r-1} = 0$, therefore relations (3.12) and (3.13) show that η and ζ are solutions of equation (3.4).

Further let x and y be arbitrary solutions of the equation. Since Propositions 1.6 and 1.7 imply $(x \vee y)^i \leq x^i \vee y^i$ and $(xy)^i \leq x^i \vee y^i$, it follows that $x \vee y$ and xy satisfy equation (3.4) as well.

Besides, for any solution x and any $j \in \{1, \ldots, r-1\}$ we have $c_j x^j = 0$, hence $x^j \leq c_j^*$ and $c_j^{**} \leq (x_j)'$. Hence for any $k \in \{1, \ldots, r-1\}$ it follows that

$$x_{(k)} = \bigvee_{j=k}^{r-1} x^j \leq \bigvee_{j=k}^{r-1} c_j^* = \zeta_{(k)},$$

$$\eta_{(k)} = \prod_{i=0}^{k-1} c_i^{**} \le \prod_{i=0}^{k-1} (x^i)' = (\bigvee_{i=0}^{k-1} x^i)' = \bigvee_{j=k}^{k-1} x^j = x_{(k)} \,,$$

proving that $x \le \zeta$ and $\eta \le x$.

C) Finally suppose equation (3.4) is consistent. Take a solution x. Then $c_i \le (x^i)^*$ for each i, hence

$$\prod_{i=0}^{r-1} c_i \le \prod_{i=0}^{r-1} (x^i)^* = (\bigvee_{i=0}^{r-1} x^i)^* = 1^* = 0 \,.$$

□

Unlike what happens for Boolean equations (cf. Schröder's theorem; see e.g. BFE, Theorem 2.2), although the set of solutions of a consistent Post equation is a lattice with least and greatest elements, it need not be an interval. Thus, for instance, for each $k \in \{1, \dots, n\}$, Proposition 1.4 shows that equation $x = x_{(k)}$ has the solution set $B(L)$, so that $\eta = 0$ and $\zeta = 1$, but $[0,1] = L \ne B(L)$. This explains the interest of the following result:

Proposition 3.2. (Bordat [1975], 1978]). *Suppose equation (3.4) is consistent. Then the set of solutions is the interval $[\eta, \zeta]$ if and only if*

(3.14)
$$(\bigvee_{h=0}^{i-1} c_h^*)(\bigvee_{j=i+1}^{r-1} c_j^*) \le c_i^* \qquad (i = 1, \dots, r-2) \,.$$

PROOF: It remains to show that the interval $[\eta, \zeta]$ is included in the set of solutions if and only if relations (3.14) hold. Since the interval $[\eta, \zeta]$ has the parametric representation $x = \eta \vee t\zeta$, the desired inclusion amounts to saying that

(3.15)
$$(\eta \vee t\zeta)^i \le c_i^*$$

for all $i \in \{0, 1, \dots, r-1\}$ and all $t \in L$. Using again Lemma 1.1, we see that we have anyway

$$(\eta \vee t\zeta)^0 = (\eta_{(1)} \vee (t_{(1)}\zeta_{(1)}))' \le (\eta_{(1)})' = c_0^{***} = c_0^* \,,$$

$$(\eta \vee t\zeta)^{r-1} = (\eta \vee t\zeta)_{(r-1)} \le \eta_{(r-1)} \vee \zeta_{(r-1)}$$

$$= (\prod_{i=0}^{r-2} c_i^{**}) \vee c_{r-1}^* = (\bigvee_{i=0}^{r-2} c_i^*)^* \vee c_{r-1}^* = c_{r-1}^*$$

because (3.5) implies $c_{r-1} \le \bigvee_{i=0}^{r-2} c_i^*$. So it remains to work out condition (3.15) for an arbitrary but fixed $i \in \{1, \dots, r-2\}$. But

$$(\eta \vee t\zeta)^i = (\eta \vee t\zeta)_{(i)}((\eta \vee t\zeta)_{(i+1)})'$$

$$= (\eta_{(i)} \vee t_{(i)}\zeta_{(i)})(\eta_{(i+1)})'((t_{(i+1)})' \vee (\zeta_{(i+1)})') \,,$$

which implies that $\max_t(\eta \vee t\zeta)^i$ exists and is reached for $t_{(1)} = \dots = t_{(i)} = 1$, $t_{(i+1)} = \dots = t_{(r-1)} = 0$. Therefore

$$(3.15) \Longleftrightarrow \max_t (\eta \vee t\zeta)^i \leq c_i^* \Longleftrightarrow (\eta_{(i)} \vee \zeta_{(i)})(\eta_{(i+1)})' \leq c_i^*$$

$$\Longleftrightarrow ((\prod_{h=0}^{i-1} c_h^{**}) \vee \bigvee_{j=1}^{r-1} c_j^*) \bigvee_{h=0}^i c_h^* \leq c_i^*$$

$$\Longleftrightarrow ((\prod_{h=0}^{i-1} c_h^{**})c_i^*) \vee ((\bigvee_{j=i}^{r-1} c_j^*) \bigvee_{h=0}^i c_h^*) \leq c_i^*$$

$$\Longleftrightarrow (\bigvee_{j=i+1}^{r-1} c_j^*) \bigvee_{h=0}^{i-1} c_h^* \leq c_i^* .$$

$\square$

It is natural to ask for a parametric representation of the set of solutions. The next two propositions provide such representations.

Proposition 3.3. (Bordat [1975], [1978]). *If equation* (3.4) *is consistent, then each of the following formulas*

$$(3.16) \qquad x = \eta \vee \bigvee_{i=0}^{r-1} c_i^* t^i e_i ,$$

$$(3.17) \qquad x = \zeta \bigvee_{i=0}^{r-1} (c_i^{**} \vee t^i) e_i ,$$

where η and ζ are the solutions in Proposition 3.1, defines the reproductive general solution of equation (3.4) *(cf. Definition 1.1.3; see also the comments before Corollary 3.3.8).*

PROOF: Set $T = \bigvee_{i=0}^{r-1} c_i^* t^i e_i$ and take $j \geq 1$. Recall that if $y \in B(L)$ then y^j is y of 0 according as $j = r - 1$ or $j \neq r - 1$, hence Proposition 1.7 implies $(xy)^j = xy^j$. Now Lemma 1.2 yields

$$T^j = \bigvee_{i=0}^{r-1} (c_i^* e_i)^j t^i = \bigvee_{i=0}^{r-1} c_i^* (e_i)^j t^i = c_j^* t^j ,$$

therefore Proposition 1.6 applied to $x = \eta \vee T$ implies

$$x^0 = \eta^0 T^0 \leq \eta^0 \leq c_0^* ,$$

$$x^i \leq \eta^i \vee T^i \leq c_i^* \qquad (i = 1, \ldots, r - 1) ,$$

proving that $x = \eta \vee T$ satisfies equation (3.4).

Conversely, if x is a solution of (3.4) then for all i we have $x^i \leq c_i^*$, therefore

$$\eta \vee T = \eta \vee \bigvee_{i=0}^{r-1} x^i e_i = \eta \vee x = x .$$

We have thus proved that (3.16) is a reproductive solution and the similar proof for (3.17) is left to the reader. $\square$

Example 3.3. (cf. Examples 3.2, 3.7–3.10). Consider an element a of a 3–Post algebra and the equation

$$a^2 x^0 \vee (a^0 \vee a^2) x^1 = 0 ,$$

which is of the form (3.4). Since $c_2 = 0$, the consistency condition (3.5) is fulfilled. The least solution η and the greatest solution ζ found in Proposition 3.1 are

$$\eta = ((a^2)^{**} \vee e_0)((a^0 \vee a^2)^{**} \vee e_1)(0^{**} \vee e_2)$$

$$= a^2(a^0 \vee a^2 \vee e_1) = a^2 ,$$

$$\zeta = (a^2)^* e_0 \vee (a^0 \vee a^2)^* e_1 \vee 0^* e_2 = 1 .$$

Since $r = 3$, conditions (3.14) in Proposition 3.2 reduce to

(3.14′) $c_0^* c_2^* \le c_1^*$

and this inequality is not fulfilled for the above equation unless $a^0 = 0$, because $c_0^* c_2^* = (a^2)^* \cdot 1 = a^0 \vee a^1$, while $c_1^* = a^1$. Therefore the set of solutions is strictly included in the interval $[a^2, 1]$ and in view of Proposition 3.3 each of the following formulas defines the reproductive solution of the equation:

$$x = a^2 \vee (a^0 \vee a^1) t^0 e_0 \vee a^1 t^1 e_1 \vee 1 \cdot t^2 e_2 = a^2 \vee a^1 t^1 e_1 \vee t^2 ,$$

and

$$x = (a^2 \vee t^0) e_0 \vee (a^0 \vee a^2 \vee t^1) e_1 \vee (0 \vee t^2) e_2 = (a^0 \vee a^2 \vee t^1) e_1 \vee t^2 .$$

□

As a matter of fact, many particular solutions of a consistent equation $f(x) = 0$ can be pointed out and used in order to obtain a reproductive solution of the equation. This is shown in the next proposition.

Proposition 3.4. *Suppose equation (3.4) is consistent. Then:*
A) For each permutation $\sigma = (s_0, s_1, \ldots, s_{r-1})$ of the set $\{0, 1, \ldots, r-1\}$, the element

(3.18) $\xi_\sigma = c_{s_0}^* e_{s_0} \vee c_{s_0}^{**} c_{s_1}^* e_{s_1} \vee \ldots \vee c_{s_0}^{**} c_{s_1}^{**} \ldots c_{s_{r-2}}^{**} c_{s_{r-1}}^* e_{s_{r-1}}$

is a solution of equation (3.4).
B) For every r permutations $\sigma_i = (i_0, i_1, \ldots, i_{r-1})$ $(i = 0, 1, \ldots, r-1)$ of the set $\{0, 1, \ldots, r-1\}$ such that $(0_0, 1_0, \ldots, (r-1)_0)$ is also a permutation of this set, formula

(3.19) $$x = \bigvee_{i=0}^{r-1} \xi_{\sigma_i} t^{i_0}$$

defines the reproductive solution of equation (3.4).

COMMENT: This was first proved by Serfati [1973a], [1973b], [1996] in the particular case of the permutations $\sigma_i = (i, i-1, i-2, \ldots, 0, r-1, r-2, \ldots, i+2, i+1)$, then by Banković [1997b] for permutations σ_i subject just to the condition $\sigma_i(0) = i$ for all i. The idea of the "vertical" permutation $(0_0, 1_0, \ldots, (r-1)_0)$ is also due to Banković [1995a], [1996]; cf. Poposition 1.2.6. In a paper presented at Sympo' 99 ASE, Bucharest, Banković obtained a result very close to the present proposition, but constructing a general (not necessarily reproductive) solution. On the other hand, Banković [1998] proved that in the Post algebra C_r every reproductive solution can be given the above form.

PROOF: A) It follows immediately by a repeated application of the Boolean identity $x \vee x'y = x \vee y$, and by the consequence $\bigvee_{i=0}^{r-1} c_i^* = 1$ of the consistency condition, that the elements $c_{s_0}^*, c_{s_0}^{**} c_{s_1}^*, \ldots, c_{s_0}^{**} c_{s_1}^{**} \cdots c_{s_{r-2}}^{**} c_{s_{r-1}}^*$ form an orthonormal system. Therefore they are the disjunctive components $(\xi_\sigma)^{s_0}, (\xi_\sigma)^{s_1}, \ldots, (\xi_\sigma)^{s_{r-1}}$, respectively, of ξ_σ and since $c_{s_i} c_{s_i}^* = 0$ for all i, we get

$$\bigvee_{i=0}^{r-1} c_i(\xi_\sigma)^i = \bigvee_{h=0}^{r-1} c_{s_h}(\xi_\sigma)^{s_h} = 0.$$

B) It follows from the hypothesis that the elements $t^{0_0}, t^{1_0}, \ldots, t^{(r-1)_0}$ form an orthonormal system, therefore Lemma 2.2 and A) imply

$$f\left(\bigvee_{i=0}^{r-1} \xi_{\sigma_i} t^{i_0}\right) = \bigvee_{i=0}^{r-1} f(\xi_{\sigma_i}) t^{i_0} = 0.$$

Conversely, let x be a solution of equation (3.4). Then for each i we have $x^i \leq c_i^*$, therefore $c_i^* x^i = x^i$ and $c_i^{**} x^i = 0$, hence

$$\bigvee_{i=0}^{r-1} \xi_{\sigma_i} x^{i_0} = \bigvee_{i=0}^{r-1} (c_{i_0}^* e_{i_0} x^{i_0} \vee c_{i_0}^{**} c_{i_1}^* e_{i_1} x^{i_0} \vee \ldots$$

$$\vee c_{i_0}^{**} c_{i_1}^{**} \cdots c_{i_{r-2}}^{**} c_{i_{r-1}}^* e_{i_{r-1}} x^{i_0}) = \bigvee_{i=0}^{r-1} e_{i_0} x^{i_0} = \bigvee_{h=0}^{r-1} e_h x^h = x.$$

$\square$

Example 3.4. We have seen that if $r := 2$ then L is a Boolean algebra and the identities $x = x^1$ and $x' = x^0$ hold. In this case equation (3.4) becomes $c_0 x' \vee c_1 x = 0$. There are only two permutations (Serfati): $\sigma_0 = (0, 1)$, implying $\xi_{\sigma_0} = c_0' e_0 \vee c_0 c_1' e_1 = c_0 c_1' = c_0$, and $\sigma_1 = (1, 0)$, for which $\xi_{\sigma_1} = c_1' e_1 \vee c_1 c_0' e_0 = c_1'$, hence formula (3.19) reduces to the well-known reproductive solution due to Schröder: $x = c_0 t' \vee c_1' t$ (see e.g. BFE, Lemma 2.2). $\square$

Example 3.5. In the case $r := 3$, consider for instance the permutations $\sigma_0 = (1, 0, 2)$, $\sigma_1 = (2, 0, 1)$ and $\sigma_3 = (0, 1, 2)$. Then

$$\xi_{\sigma_0} = c_1^* e_1 \vee c_1^{**} c_0^* e_0 \vee c_1^{**} c_0^{**} c_2^* e_2 = c_1^* e_1 \vee c_1^{**} c_0^{**} c_2^*,$$

$$\xi_{\sigma_1} = c_2^* e_2 \vee c_2^{**} c_0^* e_0 \vee c_2^{**} c_0^{**} c_1^* e_1 = c_2^* \vee c_2^{**} c_0^{**} c_1^* e_1 = c_2^* \vee c_0^{**} c_1^* e_1 \,,$$

$$\xi_{\sigma_2} = c_0^* e_0 \vee c_0^{**} c_1^* e_1 \vee c_0^{**} c_1^{**} c_2^* e_2 = c_0^{**}(c_1^* e_1 \vee c_1^{**} c_2^*) \,,$$

and since we have the "vertical" permutation $(1,2,0)$, the above three particular solutions yield the reproductive solution

$$x = (c_1^* e_1 \vee c_1^{**} c_0^{**} c_2^*) t^1 \vee (c_2^* \vee c_0^{**} c_1^* e_1) t^2 \vee c_0^{**}(c_1^* e_1 \vee c_1^{**} c_2^*) t^0 \,.$$

$\square$

To be sure, certain equations can be solved in a non-algorithmic way by using the pecularities of those equations.

Example 3.6. Let us solve the equation

(E6.1) $$a \rightarrow x = c$$

in an arbitrary Post algebra. In view of Proposition 1.8, this equation is equivalent to the following system of equations in the monotone components:

(E6.2) $$\prod_{j=1}^{k}(a'_{(j)} \vee x_{(j)}) = c_{(k)} \qquad (k = 1,\ldots,r-1)\,,$$

(E6.3) $$x_{(1)} \geq x_{(2)} \geq \ldots \geq x_{(r-1)}\,,$$

which we will solve as shown e.g. in BFE.

The first equation (E6.2), i.e., $a'_{(1)} \vee x_{(1)} = c_{(1)}$, is equivalent to $a'_{(1)} c'_{(1)} \vee x_{(1)} c'_{(1)} \vee a_{(1)} x'_{(1)} c_{(1)} = 0$, so that it has the consistency condition $a'_{(1)} c'_{(1)} = 0$, or

(E6.4.1) $$a'_{(1)} \leq c_{(1)}\,,$$

and the general solution

(E6.5.1) $$a_{(1)} c_{(1)} \leq x_{(1)} \leq c_{(1)}\,.$$

Now the second equation (E6.2) becomes $c_{(1)}(a'_{(1)} \vee x_{(2)}) = c_{(2)}$, while the first inequality (E6.3) is $x'_{(1)} x_{(2)} = 0$. This new system can be decomposed into the consistency condition

(E6.4.2) $$a'_{(2)} c_{(1)} \leq c_{(2)}$$

and the equation $c_{(1)} x_{(2)} c'_{(2)} \vee (c'_{(1)} \vee a_{(2)} x'_{(2)}) c_{(2)} \vee x'_{(1)} x_{(2)} = 0$. Since $c'_{(1)} c_{(2)} = 0$, the last equation is consistent and has the general solution

(E6.5.2) $$a_{(2)} c_{(2)} \leq x_{(2)} \leq x_{(1)}(c'_{(1)} \vee c_{(2)})\,.$$

An easy induction shows that the consistency condition is (E6.4.j), where

(E6.4.j) $$a'_{(j)} c_{(1)} \ldots c_{(j-1)} \leq c_{(j)} \qquad (j = 2,\ldots,r-1)\,,$$

while the set of solutions is given by the set of recurrent inequalities (E6.5), where

(E6.5.j) $\qquad a_{(j)}c_{(j)} \leq x_{(j)} \leq x_{(j-1)}(c'_{(1)}\vee\ldots\vee c'_{(j-1)}\vee c_{(j)})$ $(j = 2,\ldots,r-1)$. $\qquad\qquad\square$

Corollary 3.1. (Serfati [1997]). *Equation* (E6.1) *is consistent if and only if* c *is a solution, in which case* c *is the greatest solution.*

PROOF: It follows from

$$c_{(j-1)}(c'_{(1)} \vee \ldots \vee c'_{(j-1)} \vee c_{(j)}) = c_{(j-1)}c_{(j)} = c_{(j)}$$

that the elements $x_{(j)} := c_{(j)}$ $(j = 1,\ldots,r-1)$ fulfil conditions (E6.5). Conversely, if the elements $x_{(j)}$ satisfy conditions (E6.5), then an easy induction shows that $x_{(j)} \leq c_{(j)}$ $(j = 1,\ldots,r-1)$. $\qquad\qquad\square$

As suggested in Remark 3.1, the study of a Post equation of the form

$$\text{(3.4')} \qquad\qquad \bigvee_{i=0}^{r-1} b_i x^i = 1$$

reduces to the study of equation (3.4) for which $c_i := b_i^+ = ((b_i)_{(r-1)})'$ $(i = 0, 1, \ldots, r-1)$; cf. Propositions 2.1 and 1.10. As stated at the beginning of this section, the reader is urged to study equation (3.4') via the above transformation. In particular the consistency condition provided by Proposition 3.1 for equation (3.4') is

$$\text{(3.20)} \qquad\qquad \prod_{i=0}^{r-1}((b_i)_{(r-1)})' = 0 \, ,$$

but a simpler equivalent condition can be obtained.

Proposition 3.1'. (Bordat [1975], [1978]). *Equation* (3.4') *is consistent if and only if*

$$\text{(3.5')} \qquad\qquad \bigvee_{i=0}^{r-1} b_i = 1 \, .$$

PROOF: Write (3.20) in the equivalent form

$$\text{(3.20')} \qquad\qquad \bigvee_{i=0}^{r-1} (b_i)_{(r-1)} = 1 \, .$$

Since Proposition 1.1 yields

$$\text{(3.21)} \qquad\qquad x_{(r-1)} \leq x \leq x_{(1)} \, ,$$

it follows that (3.20) implies (3.5'). The converse implication follows from

$$\bigvee_{i=0}^{r-1} (b_i)_{(r-1)} = \left(\bigvee_{i=0}^{r-1} b_i\right)_{(r-1)} = 1_{(r-1)} = 1 \, .$$

$\qquad\qquad\square$

Remark 3.2. It follows from Remark 3.1 and Proposition 2.1 that in the particular case when $b_i \in B(L)$ $(i = 0, \ldots, r-1)$, equation (3.4') is equivalent to

$$(3.4'') \qquad \qquad \bigvee_{i=0}^{r-1} b_i' x^i = 0 \,,$$

so that the previous results can be applied by taking $c_i := b_i'$, which implies $c_i^* = b_i$ and $c_i^{**} = b_i'$. □

As in the case of Boolean equations, the solution of a Post equation in n unknowns can be reduced to the solution of n equations in one unknown by the method of successive elimination of variables. However, whereas in the Boolean case this method has two variants, yielding the description of the set of solutions by a system of recurrent inequalities and in a parametric form, respectively (see e.g. BFE, Chapter 2, §4), in the case of Post equations only the latter variant works. This is due to the fact that, as we have seen, the set of solutions of a Post equation in one unknown need not be an interval.

The *method of succesive elimination of variables* for Post equations is in fact an algorithm for solving a Post equation, say of the form

$$(3.22) \qquad \qquad f(x_1, \ldots, x_n) = 0 \,.$$

This algorithm consists of two stages, each of them comprising (at most) n steps. It runs as follows.

Stage I. Set

$$(3.23.1) \qquad \qquad f_1 = f \,.$$

At step k, where $k \in \{1, \ldots, n\}$, we are faced with a Post equation of the form

$$(3.24.k) \qquad \qquad f_k(x_k, x_{k+1}, \ldots, x_n) = 0 \,;$$

according to Corollary 2.3.8, the function obtained from f_k by fixing the variables $x_{k+1}, \ldots, x_n$ is a Post function in the variable x_k, so that we can write (3.24.k) in the form

$$(3.25.k) \qquad \qquad \bigvee_{i=0}^{r-1} f_k(e_i, x_{k+1}, \ldots, x_n)(x_k)^i = 0$$

and we regard (3.25.k) as a Post equation in one unknown. We set

$$(3.23.k+1) \qquad f_{k+1}(x_{k+1}, \ldots, x_n) = \prod_{i=0}^{r-1} f_k(e_i, x_{k+1}, \ldots, x_n) \,.$$

Note that f_{k+1} is a Post function by the same corollary and the next equation (3.24.k+1) is the consistency condition for equation (3.25.k) by Proposition 3.1.

So at step n we have a Post equation of the form

$$(3.24.n) \qquad \qquad f_n(x_n) = 0$$

and we set

(3.23.n+1)
$$f_{n+1} = \prod_{i=0}^{r-1} f_n(e_i) .$$

If $f_{n+1} \neq 0$ then equation (3.24.n), which coincides with (3.25.n), is inconsistent. It follows immediately by reverse induction that for each $k \in \{n, n-1, \ldots, 1\}$, equation (3.25.k) is inconsistent and hence equation (3.24.k) has no solutions. In particular the given equation (3.22), which coincides with (3.24.1), is inconsistent and the algorithm stops.

If $f_{n+1} = 0$, it follows similarly that all of the equations (3.25.k) and (3.24.k) are consistent, for $k \in \{n, n-1, \ldots, 1\}$. We then pass to

Stage II. For the sake of convenience we count the steps and the corresponding formulas in reverse order: $n, n-1, \ldots, 1$.

At step "n" we take a general (reproductive) solution of equation (3.24.n), say

(3.26.n)
$$x_n = \varphi_n(t_n) ,$$

where φ_n is a Post function; such a solution exists by Proposition 3.3 and also by Proposition 3.4.

At step "k", where $k \in \{n-1, \ldots, 1\}$, we are faced with $n-k$ Post functions

$$\varphi_n : L \to L , \ \varphi_{n-1} : L^2 \to L , \ldots, \ \varphi_{k+1} : L^{n-k} \to L ,$$

where for each $j \in \{n-1, \ldots, k+1\}$ and each $(t_{j+1}, \ldots, t_n) \in L^{n-j}$,

(3.26.j)
$$x_j = \varphi_j(t_j, t_{j+1}, \ldots, t_n)$$

defines the general (reproductive) solution of the Post equation in the unknown x_j,

(3.27.j)
$$f_j(x_j, \varphi_{j+1}(t_{j+1}, \ldots, t_n), \ldots, \varphi_{n-1}(t_{n-1}, t_n), \varphi_n(t_n)) = 0 .$$

In particular for $j := k+1$ we infer that

(3.28.k+1)
$$f_{k+1}(\varphi_{k+1}(t_{k+1}, \ldots, t_n), \varphi_{k+2}(t_{k+2}, \ldots, t_n), \ldots, \varphi_n(t_n)) = 0$$

for every $(t_{k+1}, \ldots, t_n) \in L^{n-k}$, which means that the vector

$$(\varphi_{k+1}(t_{k+1}, \ldots, t_n), \varphi_{k+2}(t_{k+2}, \ldots, t_n), \ldots, \varphi_n(t_n))$$

is a solution of equation (3.24.k+1). But this equation is the consistency condition for equation (3.25.k), i.e., for the Post equation (3.24.k) in the unknown x_k. Therefore the Post equation

(3.27.k)
$$f_k(x_k, \varphi_{k+1}(t_{k+1}, \ldots, t_n), \ldots, \varphi_n(t_n)) = 0$$

in the unknown x_k, is consistent. Let

(3.26.k)
$$x_k = \varphi_k(t_k, t_{k+1}, \ldots, t_n)$$

define the general (reproductive) solution of (3.27.k), such that φ_k is a Post function in the variables

$$t_k \text{ and } f_k(e_i, \varphi_{k+1}(t_{k+1}, \ldots, t_n), \ldots, \varphi_n(t_n)) , \ (i = 0, \ldots, r-1) ;$$

such a solution exists byProposition 3.3 and also by Proposition 3.4. Taking again into account Corollary 2.3.8 and the fact that any composite of Post functions is a Post function by Lemma 2.3.5, it follows that φ_k is a Post function in the variables $t_k, t_{k+1}, \ldots, t_n$.

Theorem 3.2. (Carvallo [1967], [1968b], Serfati [1973a], [1973b], Bordat [1975], [1978]). *Equation* (3.22) *is consistent if and only if* $f_{n+1} = 0$ (cf. (3.23.n+1)), *in which case formulas* (3.26) *define the general (reproductive) solution of the equation.*

PROOF: The first statement was already established in the above discussion. Now suppose $f_{n+1} = 0$ and consider first the case when the functions φ_k are just general solutions of equations (3.28.k) (not necessarily reproductive).

Take $(t_1, \ldots, t_n) \in L^n$. Since (3.26.1) is a general solution of (3.27.1), it follows that relation (3.28.1) holds, which coincides with

$$f(\varphi_1(t_1, \ldots, t_n), \varphi_2(t_2, \ldots, t_n), \ldots, \varphi_n(t_n)) = 0 .$$

Conversely, let $(\xi_1, \ldots, \xi_n) \in L^n$ be a solution of equation (3.22). It follows immediately that $f_k(\xi_k, \xi_{k+1}, \ldots, \xi_n) = 0$ for every $k \in \{1, \ldots, n\}$. From $f_n(\xi_n) = 0$ we deduce $\xi_n = \varphi_n(\tau_n)$ for some $\tau_n \in L$. Further, take $k \in \{n, n-1, \ldots, 1\}$ such that

$$\xi_j = \varphi_j(\tau_j, \tau_{j+1}, \ldots, \tau_n) \qquad (j \in \{n, n-1, \ldots, k+1\})$$

for some $\tau_n, \tau_{n-1}, \ldots, \tau_{k+1} \in L$. But we have seen that

$$x_k = \varphi_k(t_k, \tau_{k+1}, \ldots, \tau_n)$$

defines the general solution of the equation

$$f_k(x_k, \varphi_{k+1}(\tau_{k+1}, \ldots, \tau_n), \ldots, \varphi_n(\tau_n)) = 0 ,$$

which in fact coincides with $f_k(x_k, \xi_{k+1}, \ldots, \xi_n) = 0$, so that ξ_k is a solution of it. Hence there exists $\tau_k \in L$ such that

$$\xi_k = \varphi_k(\tau_k, \tau_{k+1}, \ldots, \tau_n) .$$

Finally if the solutions φ_k are reproductive, then in the above reasoning we can take $\tau_k := \xi_k$ ($k = n, n-1, \ldots, 1$). □

Remark 3.3. The elimination of a variable x from an equation $f = 0$ ($g = 1$) can be performed without bringing that equation to the canonical form (3.4) ((3.4')) with respect to x. To be specific:

Suppose the equation is of the form

(3.22')
$$c \vee \bigvee_{i \in I} c_i x^i = 0 ,$$

where $I \subseteq \{0, \ldots, r-1\}$, while $c_i, i \in I$, and c are functions depending on variables from $\{x_1, \ldots, x_n\} \backslash \{x\}$. Since the equation is equivalent to $c = \bigvee_{i \in I} c_i x^i = 0$, the result of the elimination of x is $c = \prod_{i \in I} c_i = 0$, or

$$c \vee \prod_{i \in I} c_i = 0 \, ;$$

if $I \neq \{0, 1, \ldots, r-1\}$, this condition reduces to $c = 0$. In the second stage of the method of successive eliminations, when we introduce the parametric expressions of the variables eliminated after x into (3.22'), the equation $c = 0$ is ensured and we only have to work out equation $\bigvee_{i \in I} c_i x^i = 0$.

If the equation is of the form

(3.22'') $$b \vee \bigvee_{i \in I} b_i x^i = 1 \, ,$$

with similar hypotheses on I, b_i and b, then since equation (3.22'') can be written in the form

(3.22''') $$\bigvee_{i \in I} (b \vee b_i) x^i \vee \bigvee_{j \in J} b x^j = 1 \, ,$$

where $J = \{0, \ldots, r-1\} \backslash I$, the result of the elimination of x is

$$b \vee \bigvee_{i \in I} b_i = 1 \, .$$

$\square$

Example 3.7. (cf. Examples 3.2, 3.3, 3.8–3.10). Let us apply the successive elimination of variables to the equation

(E7.1) $$(a^1 \vee a^2 \vee b^1 \vee b^2) c^0 \vee (a^2 \vee a^0 (b^1 \vee b^2)) c^1 \vee (b^1 \vee b^0 (a^1 \vee a^2)) c^2 = 0$$

in the unknowns a, b, c over a 3–Post algebra.

By eliminating c we obtain in turn the equivalent equations

$$(a^1 \vee a^2 \vee b^1 \vee b^2)(a^2 \vee a^0 (b^1 \vee b^2))(b^1 \vee b^0 (a^1 \vee a^2)) = 0 \, ,$$

$$(a^2 \vee a^0 (b^1 \vee b^2))(b^1 \vee b^0 (a^1 \vee a^2)) = 0 \, ,$$

(E7.2) $$a^2 b^0 \vee (a^0 \vee a^2) b^1 = 0 \, .$$

By eliminating b we obtain the equation $a^2 (a^0 \vee a^2) \cdot 0 = 0$, which is identically satisfied. Thus a remains arbitrary and we pass to the second stage of the algorithm.

Equation (E7.2) has been solved in Example 3.3. We choose e.g. the reproductive solution

(E7.3) $$b = a^2 \vee a^1 t^1 e_1 \vee t^2$$

found in Example 3.3 and we introduce it into (E7.1). But

$$b = a^2 t^0 \vee (a^2 \vee a^1 e_1) t^1 \vee t^2 = a^2 t^0 \vee a t^1 \vee t^2 ,$$

$$b^0 = (a^2)' t^0 \vee a^0 t^1 , \ b^1 = a^1 t^1 ,$$

$$b^2 = a^2 t^0 \vee a^2 t^1 \vee t^2 = a^2 (t^0 \vee t^1) \vee t^2 = a^2 \vee t^2 ,$$

so that (E7.1) becomes in turn

$$(a^1 \vee a^2 \vee t^2) c^0 \vee (a^2 \vee a^0 t^2) c^1 \vee (a^1 t^1 \vee ((a^2)' t^0 \vee a^0 t^1)(a^1 \vee a^2)) c^2 = 0 ,$$

(E7.4) $$(a^1 \vee a^2 \vee t^2) c^0 \vee (a^2 \vee a^0 t^2) c^1 \vee (a^1 t^1 \vee a^1 t^0) c^2 = 0 .$$

The least solution η of equation (E7.4) is

$$\eta = (a^1 \vee a^2 \vee t^2 \vee e_o)(a^2 \vee a^0 t^2 \vee e_1) \cdot 1 = a^2 \vee (a^1 \vee t^2)(a^0 t^2 \vee e_1)$$

$$= a^2 \vee a^1 e_1 \vee t^2 (a^0 t^2 \vee e_1) = a \vee (a^0 \vee e_1) t^2 ,$$

therefore the solution (3.16) provided by Proposition 3.3 is

$$c = a \vee (a^0 \vee e_1) t^2 \vee (a^1 \vee a^2 \vee t^2)^* s^0 e_0 \vee (a^2 \vee a^0 t^2)^* s^1 e_1 \vee (a^1 (t^0 \vee t^1))^* s^2 e_2$$

$$= a \vee (a^0 \vee e_1) t^2 \vee (a^0 \vee a^1)(a^1 \vee a^2 \vee t^0 \vee t^1) s^1 e_1 \vee (a^0 \vee a^2 \vee t^2) s^2 ,$$

(E7.5) $$c = a \vee (a^0 \vee e_1) t^2 \vee (a^1 \vee a^0 (t^0 \vee t^1)) s^1 e_1 \vee (a^0 \vee t^2) s^2 ,$$

because $a^2 s^2 \le a^2 \le a$.

Formulas (E7.3) and (E7.5), where a, t and s are arbitrary parameters, define the reproductive solution of equation (E7.1).

Note also that, in view of Remark 3.2, the equation $g = 1$ equivalent to (E7.1) can be successively written in the following equivalent forms:

$$a^0 b^0 c^0 \vee (a^0 \vee a^1)(a^1 \vee a^2 \vee b^0) c^1 \vee (b^0 \vee b^2)(b^1 \vee b^2 \vee a^0) c^2 = 1 ,$$

$$a^0 b^0 c^0 \vee (a^1 \vee a^0 b^0) c^1 \vee (b^2 \vee b^0 a^0) c^2 = 1 ,$$

$$a^0 b^0 \vee a^1 c^1 \vee b^2 c^2 = 1 ,$$

i.e., it is equation (E2.6) studied in Example 3.2. We are going to resume it in the next example. $\square$

Example 3.8. (cf. Examples 3.2, 3.3, 3.7, 3.9, 3.10). Let us apply the method of successive elimination of variables to the equation in the unknowns a, b, c

(E8.1) $$a^0 b^0 \vee a^1 c^1 \vee b^2 c^2 = 1$$

obtained in Example 3.2 and solved in Example 3.7 in the equivalent form $f = 0$.

We eliminate c using Remark 3.3 and obtain

(E8.2) $$a^0 b^0 \vee a^1 \vee b^2 = 1 .$$

By eliminating b via Remark 3.3 we obtain the consistency condition $a^1 \vee a^0 \vee 1 = 1$, which is identically satisfied. Thus a remains arbitrary and we pass to the second stage of the algorithm.

Equation (E8.2) can be written in the form

$$(a^0 \vee a^1)b^0 \vee a^1 b^1 \vee b^2 = 1$$

and we can apply formula (3.17) from Proposition 3.3 as modified by Remark 3.2, the greatest solution being obviously $b := 1$. We get

$$b = (a^2 \vee t^0)e_0 \vee (a^0 \vee a^2 \vee t^1)e_1 \vee (0 \vee t^2)e_2 = (a^0 \vee a^2 \vee t^1)e_1 \vee t^2 \,,$$

(E8.3)
$$b = (a^0 \vee a^2)t^0 \vee (a^0 \vee a^2 \vee e_1)t^1 \vee t^2 \,.$$

Taking into account that if $x \in B(L)$ then Proposition 1.6 implies $(x \vee y)^0 = x'y^0$ and $(x \vee y)^{r-1} = x \vee y^{r-1}$, we obtain

$$a^0 b^0 = a^0(a^1 t^0 \vee a^1 e_1^0 t^1 \vee 0 \cdot t^2) = 0 \,,$$

$$b^2 = (a^0 \vee a^2)t^0 \vee (a^0 \vee a^2 \vee e_1^2)t^1 \vee t^2 = (a^0 \vee a^2)(t^0 \vee t^1) \vee t^2 = a^0 \vee a^2 \vee t^2 \,,$$

so that by introducing the expression (E8.3) into equation (E8.1), the result is

(E8.4)
$$a^1 c^1 \vee (a^0 \vee a^2 \vee t^2)c^2 = 1 \,.$$

Taking again into account Remark 3.2, we obtain the least solution of equation (E8.4):

$$\eta = (0^* \vee e_0)((a^1)^* \vee e_1)((a^0 \vee a^2 \vee t^2)^* \vee e_2) = a^0 \vee a^2 \vee e_1 \,,$$

so that the reproductive solution (3.16) becomes

$$c = a^0 \vee a^2 \vee e_1 \vee 0 \cdot s^0 e_0 \vee a^1 s^1 e_1 \vee (a^0 \vee a^2 \vee t^2)s^2 e_2 \,,$$

(E8.5)
$$c = a^0 \vee a^2 \vee e_1 \vee t^2 s^2 \,.$$

Formulas (E8.3) and (E8.5), where a, t and s are arbitrary parameters, define the reproductive solution of equation (E8.1). $\square$

We now pass to the direct study of a Post equation in n unknowns, i.e., without reducing it to several equations in one unknown.

Exactly as in the case of Boolean equations, the consistency condition in Proposition 3.1 has a straightforward generalization to equations in several unknowns (cf. the Boole-Schröder theorem; see e.g. BFE, Theorem 2.3).

Theorem 3.3. (Serfati [1973a], [1996], Beazer [1974c], Bordat [1975c], [1978]).
A Post equation

(3.29) $f(x_1, \ldots, x_n) = 0$

is consistent if and only if

(3.30) $$\prod_A f(A) = 0 \,.$$

PROOF: For $r := 1$ this follows from Proposition 3.1, since $c_i = f(e_i)$ by Theorem 2.1. The inductive step runs as follows:

$$\prod_A f(A) = 0 \iff \prod_{\alpha_1} \prod_{\alpha_2, \ldots, \alpha_n} f(\alpha_1, \alpha_2, \ldots, \alpha_n) = 0$$

$$\iff \exists x_1 \prod_{\alpha_2, \ldots, \alpha_n} f(x_1, \alpha_2, \ldots, \alpha_n) = 0$$

$$\iff \exists x_1 \, (\exists x_2 \ldots \exists x_n \, f(x_1, x_2, \ldots, x_n) = 0) \,.$$

$\square$

As for Boolean equations, the consistency condition enables us to study the range of a Post function.

Proposition 3.5. *The range of a Post function $f : L^n \longrightarrow L$ is a sublattice of L with least element $\prod_A f(A)$, greatest element $\bigvee_A f(A)$ and has the parametric representation*

(3.31) $$y = \prod_A f(A) \vee \bigvee_{i=0}^{r-1} (\bigvee_A f^i(A)) t^i e_i \,,$$

or equivalently,

(3.32) $$y = (\bigvee_A f(A)) \bigvee_{i=0}^{r-1} ((\bigvee_A f(A))' \vee t^i) e_i \,.$$

COMMENT: The least and greatest element of this lattice were first discovered by Serfati [1996].

PROOF: An element $y \in L$ belongs to the range of f if and only if the equation $f(X) = y$ is consistent. In view of Proposition 1.11, this equation can be written in the form

$$\bigvee_{i=0}^{r-1} f^i(X) y^i = 1$$

and its consistency condition is

$$\bigvee_A \bigvee_{i=0}^{r-1} f^i(A) y^i = 1$$

by Proposition 3.1′. The latter condition can be also written in the form

$$(3.33) \qquad \bigvee_{i=0}^{r-1} (\bigvee_A f^i(A)) y^i = 1 ,$$

or equivalently, since the coefficients are in $B(L)$,

$$(3.34) \qquad \bigvee_{i=0}^{r-1} (\bigvee_A f^i(A))' y^i = 0 .$$

So (3.33) or (3.34) is *the characteristic equation of the range* of the Post function f. We solve it via Propositions 3.1 and 3.3 (or 3.4), which provide the lattice structure and the parametric representation, respectively.

The greatest solution is

$$\zeta = \bigvee_{i=0}^{r-1} c_i^* e_i = \bigvee_{i=0}^{r-1} (\bigvee_A f^i(A)) e_i$$

$$= \bigvee_{i=0}^{r-1} \bigvee_A f^i(A) e_i = \bigvee_A \bigvee_{i=0}^{r-1} f^i(A) e_i = \bigvee_A f(A) .$$

To compute the least solution we need the conjunctive representation $x = \prod_{i=0}^{r-1} ((x^i)' \vee e_i)$ of an element $x \in L$. It can be obtained by duality from the disjunctive representation (1.17), or alternatively, by computing in turn $\prod_{i=0}^{k} ((x^i)' \vee e_i)$ for $k = 0, 1, \ldots, r-1$, which yields $\prod_{i=0}^{r-1} ((x^i)' \vee e_i) = \bigvee_{i=0}^{r-1} x^i e_i$. Therefore the least solution is

$$\eta = \prod_{i=0}^{r-1} (c_i^{**} \vee e_i) = \prod_{i=0}^{r-1} ((\bigvee_A f^i(A))' \vee e_i)$$

$$= \prod_{i=0}^{r-1} (\prod_A (f^i(A))' \vee e_i) = \prod_{i=0}^{r-1} \prod_A ((f^i(A))' \vee e_i)$$

$$= \prod_A \prod_{i=0}^{r-1} ((f^i(A))' \vee e_i) = \prod_A f(A) .$$

Finally formulas (3.31) and (3.32) are immediately obtained from Proposition 3.3. □

The fact that $\prod_A f(A)$ and $\bigvee_A f(A)$ are the least and the greatest element of the range is a straightforward generalization of the corresponding property for Boolean functions but, unlike what happens in the Boolean case (cf. Schröder's theorem; see e.g. BFE, Theorem 2.4), the range of a Post function need not be an interval; see the commnent before Proposition 3.2.

Corollary 3.2. *A Post function* $f : L^n \longrightarrow L$ *has the range* $[\prod_A f(A), \bigvee_A f(A)]$ *if and only if*

$$(3.35) \qquad (\bigvee_{h=0}^{i-1} \bigvee_A f^h(A))(\bigvee_{j=i+1}^{r-1} f^j(A)) \le \bigvee_A f^i(A) \qquad (i = 1, \ldots, r-2) .$$

PROOF: By Proposition 3.2 applied to equation (3.34).   □

It turns out that Löwenheim's formula for obtaining the reproductive solution of a Boolean equation from a particular solution of it (see e.g. BFE, Theorem 2.11) has a straightforward generalization to Post equations.

Theorem 3.4. (Carvallo [1967], [1968b], Bordat [1975], Serfati [1977]; cf. Remark 3.4 below). *Let* $\Xi = (\xi_1, \ldots, \xi_n) \in L^n$ *be a particular solution of the Post equation* $f(x_1, \ldots, x_n) = 0$. *Then formulas*

$$(3.36) \qquad x_k = \xi_k f^{**}(t_1, \ldots, t_n) \vee t^k f^*(t_1, \ldots, t_n) \qquad (k = 1, \ldots, n) ,$$

which can be written in the compact form

$$(3.37) \qquad X = f^{**}(T)\Xi \vee f^*(T)T ,$$

define the reproductive solution of the equation.

PROOF: We apply Lemma 2.2 with $s := 2$, $\lambda_0 := f(T)^{**}$, $\lambda_1 := f(T)^*$ and obtain from (3.36):

$$f(x_1, \ldots, x_n) = f(T)^{**} f(\xi_1, \ldots, \xi_n) \vee f(T)^* f(t_1, \ldots, t_n) = 0 .$$

Conversely, if $(x_1, \ldots, x_n) \in L^n$ satisfies the equation, then for each k,

$$\xi_k f(x_1, \ldots, x_n)^{**} \vee x_k f(x_1, \ldots, x_n) = \xi_k \cdot 0 \vee x_k \cdot 1 = x_k .$$

□

So in order to apply Theorem 3.4 it is necessary to obtain first a particular solution of the given equation. As was suggested in the Boolean case, this can be done by choosing at each step of the second stage of the method of successive eliminations, the greatest or the least solution of the corresponding equation in one unknown; cf. Proposition 3.1. We thus obtain the following particular solution:

$$(3.38.n) \qquad \xi_n = \bigvee_{i=0}^{r-1} f_n^*(e_i)e_i ,$$

$$(3.38.k) \qquad \xi_k = \bigvee_{i=0}^{r-1} f_k^*(e_i, \xi_{k+1}, \ldots, \xi_n)e_i \qquad (k = n-1, \ldots, 2, 1) .$$

Example 3.9. (cf. Examples 3.2, 3.3, 3.7, 3.8, 3.10). We resume the equation

$$(E7.1) \qquad (a^1 \vee a^2 \vee b^1 \vee b^2)c^0 \vee (a^2 \vee a^0(b^1 \vee b^2))c^1 \vee (b^1 \vee b^0(a^1 \vee a^2))c^2 = 0$$

solved in Example 3.7 and we are going to solve it by the method indicated above.

We need a particular solution. We can use the particular solution (3.38) determined above for any consistent equation $f = 0$, or we may try to find a simple particular solution by remarks appropriate to the particular equation we are solving. As suggested above, the discovery of such a particular solution may be facilitated by a procedure based on the idea of successive elimination of variables. Let us illustrate this.

The beginning coincides with the first stage of the method of successive elimination of variables, as in Example 3.7. So we eliminate c and obtain the equation

(E7.2) $$a^2 b^0 \vee (a^0 \vee a^2) b^1 = 0 ,$$

whose consistency condition is identically satisfied and the first stage is over.

The second stage is much simplified because we are looking only for a particular solution. Since a is arbitrary, we choose e.g. $a := 0$ and introduce this value into (E7.2), which thus reduces to $b^1 = 0$. A particular solution is $b := 0$. These two values transform equation (E7.1) into an identity. So we take e.g. $c := 0$.

The particular solution $(0,0,0)$ simplifies the reproductive solution (3.37) constructed in Theorem 3.4 to $X = f^*(T)T$ and it was noted in Example 3.7 that the (pseudo-)complement of the present function f is $f^*(a, b, c) = a^0 b^0 \vee a^1 c^1 \vee b^2 c^2$. Taking also into account the identities $x x^0 = 0$, $x x^1 = x^1 e_1$ and $x^2 \le x$, we obtain the following reproductive solution:

(E9.1) $$a = s(s^0 t^0 \vee s^1 w^1 \vee t^2 w^2) = s^1 w^1 e_1 \vee s t^2 w^2 ,$$

(E9.2) $$b = t(s^0 t^0 \vee s^1 w^1 \vee t^2 w^2) = s^1 t w^1 \vee t^2 w^2 ,$$

(E9.3) $$c = w(s^0 t^0 \vee s^1 w^1 \vee t^2 w^2) = s^0 t^0 w \vee s^1 w^1 e_1 \vee t^2 w^2 .$$

$\square$

Remark 3.4. Suppose $g(X) = 1$ is a Post equation for which $g \in B(\mathrm{PFLn})$ (cf. Definition 2.2). It is easily seen by almost the same proof as for Theorem 3.4 that if Ξ is a particular solution of the equation, then formula

(3.37') $$X = g^*(T)\Xi \vee g(T)T$$

defines the reproductive solution of the equation $g(X) = 1$. This results in a rapid construction of a reproductive solution, since (several!) particular solution(s) of an equation $g(X) = 1$ can be easily detected. $\square$

Example 3.10. (cf. Examples 3.2, 3.3, 3.7–3.9). Let us resume the equation

(E8.1) $$a^0 b^0 \vee a^1 c^1 \vee b^2 c^2 = 1$$

solved in Example 3.8. Eight particular solutions can be directly read on equation (E8.1); we choose $(0,0,0)$. Now we apply Remark 3.4. Setting $g = 1$ for equation (E8.1) and $f = 0$ for equation (E7.1), it was seen in Example 3.7 that $f^* = g$ and since $g \in B(\mathrm{PFLn})$, it follows that $g^* = f$, too. Therefore relation (3.37') for $g = 1$ coincides with (3.36) for $f = 0$. We thus recapture the reproductive solution (E9.1-3) already found in Example 3.9.   □

We are now in a position to study implications between Post functions. The results are similar to, but a bit more complicated than, those for Boolean functions (see e.g. BFE, Theorem 2.14; the analogy is completed by Corollary 2.2).

Lemma 3.1. *Let $f, g : L^n \longrightarrow L$ be Post functions. Then $f^* = g^*$ if and only if the elements $g(A)$, $A \in C_r^n$, are of the form*

$$(3.39) \qquad g(A) = f^{**}(A)(t'_{A1} \ldots t'_{A,r-2} \vee \bigvee_{k=1}^{r-2} t'_{A1} \ldots t'_{A,k-1} t_{Ak} e_k),$$

where $t_{Ak} \in B(L)$ for all $A \in C_r^n$ and all $k \in \{1, \ldots, r-2\}$.

PROOF: Since $g^* = f^* \iff g^*(A) = f^*(A) \ (\forall A \in C_r^n)$, the problem reduces to solving, for each $A \in C_r^n$, the equation $x^* = a^*$, where we have set $x := g(A)$ and $a := f(A)$. But $x^* = x^0$, so that we have to determine all the orthonormal systems $x^0, x^1, \ldots, x^{r-1}$ with $x^0 = a^*$. According to a theorem due to Löwenheim (see e.g. BFE, Theorem 4.2), the orthonormal systems of r elements are given by formulas

$$x^0 = t_0, \ , \ x^k = t'_0 t'_1 \ldots t'_{k-1} t_k \ (k = 1, \ldots, r-2) \ , \ x^{r-1} = t'_0 t'_1 \ldots t'_{r-2} \ ,$$

where $t_0, t_1, \ldots, t_{r-2}$ are arbitrary Boolean parameters. In our case t_0 is fixed to a^*.   □

Corollary 3.3. *If $r := 3$ and $f, g : L^n \longrightarrow L$ are Post functions, say $f(x) = c_0 x^0 \vee c_1 x^1 \vee c_2 x^2$, then $f^* = g^*$ if and only if g is of the form*

$$(3.40) \qquad g(x) = c_0^{**}(p_0 \vee e_1)x^0 \vee c_1^{**}(p_1 \vee e_1)x^1 \vee c_2^{**}(p_2 \vee e_1)x^2 \ ,$$

where p_0, p_1, p_2 are arbitrary in $B(L)$.

PROOF: Formulas (3.39) become $g(e_i) = c_i^{**}(t'_{e_i 1} \vee t_{e_i 1} e_1)$; but $t' \vee te = t' \vee e$.   □

Proposition 3.6. *Let $f, g : L^n \longrightarrow L$ be Post functions. If the equation $f(X) = 0$ is consistent, then the following relations hold:*

$$(3.41) \qquad (f(X) = 0 \Longrightarrow g(X) = 0) \iff g \leq f^{**} \ ,$$

$$(3.42) \qquad (f(X) = 0 \iff g(X) = 0) \iff g^* = f^* \ .$$

COMMENT: Property (3.41) and the implication $\Longrightarrow$ in (3.42) are due to Serfati [1996].

PROOF: (3.41), $\Longrightarrow$: Take $\Xi \in L^n$ such that $f(\Xi) = 0$ and note that $g(\Xi) = 0$. Then for every $X \in L^n$, Theorem 3.4 yields

$$f(f^{**}(X)\varXi \vee f^*(X)X) = 0 \,,$$

therefore using Lemma 2.2 we obtain

$$0 = g(f^{**}(X)\varXi \vee f^*(X)X) = f^{**}(X)g(\varXi) \vee f^*(X)g(X) = f^*(X)g(X) \,,$$

hence $g(X) \leq f^{**}(X)$.

(3.41), $\Longleftarrow$: $f(X) = 0 \Longrightarrow f^*(X) = 1 \Longrightarrow f^{**}(X) = 0 \Longrightarrow g(X) = 0$.

(3.42), $\Longrightarrow$: It follows from (3.41) that $f^* = f^{***} \leq g^*$ and since the hypothesis of (3.42) implies that $g(X) = 0$ is consistent as well, we can interchange f with g and obtain $g^* \leq f^*$.

(3.42), $\Longleftarrow$: Since $f^* = g^*$, it follows from Lemma 3.1 that $g \leq f^{**}$, therefore $f(X) = 0 \Longrightarrow g(X) = 0$ by (3.41). This implies that equation $g(X) = 0$ is consistent as well, so that by interchanging f and g in Lemma 3.1 we obtain $g(X) = 0 \Longrightarrow f(X) = 0$. $\qquad\square$

Proposition 3.7. (Serfati [1996]). *Let $f, g : L^n \longrightarrow L$ be Post functions. If each of the equations $f(X) = e_i$ $(i = 0, 1, \ldots, r - 2)$ is consistent, then the following relations hold:*

(3.43)
$$(f(X) = e_i \Longrightarrow g(X) = e_i) \, (i = 0, 1, \ldots, r - 2) \Longrightarrow g \leq f \,,$$

(3.44)
$$(f(X) = e_i \Longleftrightarrow g(X) = e_i) \, (i = 0, 1, \ldots, r-2) \Longleftrightarrow g = f \,.$$

PROOF: For (3.43) we use Proposition 1.11 and obtain, for each i,

$$f(X) = e_i \Longleftrightarrow \bigvee_{h=0}^{r-1} f^h(X)e_i^h = 1 \Longleftrightarrow f^i(X) = 1 \Longleftrightarrow f^{i*}(X) = 0$$

so that for each $i \in \{0, 1, \ldots, r - 2\}$ we have

$$f^{i*}(X) = 0 \Longrightarrow g^{i*}(X) = 0 \,,$$

hence $g^{i*} \leq f^{i***} = f^{i*}$, or equivalently, $(g^i)' \leq (f^i)'$. This implies

$$g_{(r-1)} = g^{r-1} = \prod_{i=0}^{r-2}(g^i)' \leq \prod_{i=0}^{r-2}(f^i)' = f^{r-1} = f_{(r-1)} \,,$$

$$g_{(1)} = (g^0)' \leq (f^0)' = f_{(1)} \,,$$

$$g_{(i)} = \bigvee_{j=i}^{r-1} g^j = (\bigvee_{h=0}^{i-1} g^h)' = \prod_{h=0}^{i-1}(g^h)'$$

$$\leq \prod_{h=0}^{i-1}(f^h)' = \ldots = f_{(i)} \qquad (i = 1, \ldots, r - 2) \,,$$

that is $g_{(k)} \leq f_{(k)}$ $(k = 1, \ldots, r-1)$, therefore $g \leq f$.

The proof of the implication $\Longrightarrow$ from (3.44) is immediate from (3.43). The converse is trivial. $\square$

The next two propositions generalize the corresponding results for Boolean equations with unique solutions, due to Whitehead, Bernstein and Parker; see e.g. BFE, Theorems 6.6 and 6.7.

Proposition 3.8. (Serfati [1996]). *Let $f : L^n \longrightarrow L$ be a Post function, $i \in \{0, 1, \ldots, r-1\}$ and $\Xi = (\xi_1, \ldots, \xi_n) \in L^n$. Then the following conditions are equivalent:*

(i) Ξ is the unique solution of equation $f(X) = e_i$;

(ii) $f^i(X) = \prod_{k=1}^{n} \bigvee_{h=0}^{r-1} x_k^h \xi_k^h$ $(\forall X \in L^n)$;

(iii) $f^i(A) = \Xi^A$ $(\forall A \in C_r^n)$.

PROOF: Since $f(X) = e_i \iff f^i(X) = 1$ by Proposition 1.3 and

$$X = \Xi \iff \prod_{k=1}^{n} \bigvee_{h=0}^{r-1} x_k^h \xi_k^h = 1$$

by Proposition 1.13, condition (i) can be written in the form

$$f^i(X) = 1 \iff \prod_{k=1}^{n} \bigvee_{h=0}^{r-1} x_k^h \xi_k^h = 1$$

or equivalently,

$$f^{i*}(X) = 0 \iff (\prod_{k=1}^{n} \bigvee_{h=0}^{r-1} x_k^h \xi_k^h)^* = 0$$

and the last condition is equivalent to (ii) by Proposition 3.6.

The equivalence between (ii) and (iii) follows by Corollary 2.2, since

$$\bigvee_{h=0}^{r-1} \alpha_k^h \xi_k^h = \xi_k^{\alpha_k} .$$

$\square$

Proposition 3.9. *Let $f : L^n \longrightarrow L$ be a Post function and $i \in \{0, 1, \ldots, r-1\}$. Then the equation $f(X) = e_i$ has a unique solution if and only if*

(3.45)
$$\prod_A (f^{i*}(A) \vee \bigvee_{B \neq A} f^i(B)) = 0 .$$

COMMENT: Condition (3.45) expresses the orthonormality of the system $f^i(A)$, $A \in C_r^n$, exactly as in the Boolean case; cf. BFE, Theorem 6.7.

PROOF: Note that the elements occurring in condition (iii) from Proposition 3.8 are Boolean, so that an element $\Xi \in L^n$ is the unique solution of equation $f(X) = e_i$ if and only if

(3.46)
$$\bigvee_A (f^i(A)(\Xi_A)' \vee (f^i(A))'\Xi^A) = 0 .$$

Therefore the equation $f(X) = e_i$ has a unique solution if and only if there exists $\Xi \in L^n$ satisfying equation (3.46). But the consistency condition for equation (3.46) is

$$0 = \prod_B \bigvee_A (f^i(A)(B^A)' \vee (f^i(A))'B^A) = \prod_B ((\bigvee_{A \neq B} f^i(A)) \vee (f^i(B))') .$$

$\square$

Corollary 3.4. *The unique solution Ξ is determined by*

(3.47)
$$\xi_k^h = \bigvee \{f^i(A) \mid A \in C_r^n \ \& \ \alpha_k = e_h\}$$
$$(h = 0, \ldots, r - 1) \ (k = 1, \ldots, n) .$$

PROOF: In view of Proposition 3.8 (iii), the right side of (3.47) equals

$$\bigvee_A \{\Xi^A \mid A \in C_r^n \ \& \ \alpha_k = e_h\} = \bigvee_{\alpha_1,\ldots,\alpha_{k-1},\alpha_{k+1},\ldots,\alpha_n}$$

$$\xi_1^{\alpha_1} \cdots \xi_{k-1}^{\alpha_{k-1}} \xi_k^{e_h} \xi_{k+1}^{\alpha_{k+1}} \cdots \xi_n^{\alpha_n}$$

$$= \xi_k^{e_h} \bigvee_{\alpha_1,\ldots,\alpha_{k-1},\alpha_{k+1},\ldots,\alpha_n} \xi_1^{\alpha_1} \cdots \xi_{k-1}^{\alpha_{k-1}} \xi_{k+1}^{\alpha_{k+1}} \cdots\cdots \xi_n^{\alpha_n} = \xi_k^h .$$

$\square$

Remark 3.5. Two variants of Propositions 3.8 and 3.9 are obtained by the following modifications: replace f by a function $g \in B(\text{PFLn})$ (cf. Definition 2.2), delete i, replace the equation $f(X) = e_i$ by $g(X) = 1$ and change f^i to g. The proofs are the same. $\square$

Example 3.11. Take again $r := 3$ and let $g \in B(\text{PFLn})$. Condition (3.45) becomes

$$(g^*0) \vee g(e_1) \vee g(1))(g^*(e_1) \vee g(0) \vee g(1))(g^*(1) \vee g(0) \vee g(e_1)) = 0$$

and is easily brought to the form

$$g(0)g(e_1) \vee g(0)g(1) \vee g(e_1)g(1) \vee g^*(0)g^*(e_1)g^*(1) = 0 ,$$

which expresses the orthonormality, as already noticed. So we apply BFE, Theorem 4.2 (Löwenheim) and obtain

$$g(1) = a , \; g(e_1) = a'b , \; g(0) = a'b' ,$$

where a, b are arbitrary parameters in $B(L)$. Therefore g is of the form

$$g(x) = a'b'x^0 \vee a'bx^1 \vee ax^2 .$$

The unique solution ξ of the equation $g(x) = 1$ is obtained from formulas (3.47), which reduce to $\xi^0 = g(0) = a'b'$, $\xi^1 = g(e_1) = a'b$, $\xi^2 = g(1) = a$, therefore

$$\xi = \xi^0 e_0 \vee \xi^1 e_1 \vee \xi^2 = a'be_1 \vee a = a \vee be_1 .$$

The reader is urged to check the above computation by using Proposition 3.1, which will yield $\eta = \zeta = a \vee be_1$, as expected.   □

Another line of research consists in looking for various forms of the reproductive (or general) solution.

The next proposition and its corollaries are due to Banković [2001]. The proposition is a straightforward generalization of the Poretski-Schröder-Itoh theorem for Boolean equations (see e.g. BFE, Theorem 3.3).

Proposition 3.10. *Let $f : L^n \longrightarrow L$ be a Post function and $H = (h_1, \ldots, h_n) : L^n \longrightarrow L^n$, where*

(3.48) $$h_k(X) = x_k f^*(X) \vee x_k^* f^{**}(X) \qquad (k = 1, \ldots, n) ,$$

or in vector form,

(3.49) $$H(X) = f^*(X)X \vee f^{**}(X)X .$$

Then the identity $f(H(X)) = 0$ holds if and only if

(3.50) $$f(A^*)f^{**}(A) = 0 \qquad (\forall A \in C_r^n) ,$$

where we have set $A^* = (\alpha_1^*, \ldots, \alpha_n^*)$, in which case (3.49) is a reproductive solution of equation $f(X) = 0$.

PROOF: The first claim follows from Corollary 2.2 and Lemma 2.2:

$$f(H(X)) = 0 \; (\forall X \in L^n) \Longleftrightarrow f(H(A)) = 0 \; (\forall A \in C_r^n)$$

$$\Longleftrightarrow f^*(A)f(A) \vee f^{**}(A)f(A^*) = 0 \; (\forall A \in C_r^n) \Longleftrightarrow (3.50) .$$

Now suppose X satisfies $f(X) = 0$. Then $H(X) = 1 \cdot X \vee 0 \cdot X^* = X$.   □

Corollary 3.5. *If $f(A) = 0$ for all $A \in \{0,1\}^n$, then (3.49) is a reproductive solution of equation $f(X) = 0$.*

PROOF: If $A \in C_r^n$ then $A^* \in \{0,1\}^n$, hence $f(A^*) = 0$ and relations (3.50) hold.   □

Corollary 3.6. *If $f : L \longrightarrow L$ is a Post function and $f(0) = 0$ then $x = tf^*(t) \vee t^* f^{**}(t)$ defines the reproductive solution of equation $f(x) = 0$.*

PROOF: For $\alpha := 0$ we have $f^{**}(\alpha) = f^{**}(0) = 0^{**} = 0$, while for $\alpha \in C_r^n \setminus \{0\}$ we have $f(\alpha^*) = f(0) = 0$, so that relations (3.50) hold.   □

The equivalence of conditions (i)-(v) in the next theorem generalizes the characterizations of reproductive solutions of Boolean equations given by Rudeanu and J.-P. Deschamps; cf. BFE, Theorem 3.5.

Theorem 3.5. *Let* $f, h_1, \ldots, h_n : L^n \longrightarrow L$ *be Post functions and* $H = (h_1, \ldots, h_n)$. *If* $f(H(X)) = 0$ *holds identically then the following conditions are equivalent:*

(i) $X = H(T)$ *defines the reproductive solution of equation* $f(X) = 0$;

(ii) $f^*(X) \leq \prod_{k=1}^n \bigvee_{i=0}^{r-1} h_k(X)^i x_k^i$ $(\forall X \in L^n)$;

(iii) $f^*(X) = \prod_{k=1}^n \bigvee_{i=0}^{r-1} h_k(X)^i x_k^i$ $(\forall X \in L^n)$;

(iv) $f^*(A) \leq H(A)^A$ $(\forall A \in C_r^n)$;

(v) $f^*(A) = H(A)^A$ $(\forall A \in C_r^n)$;

(vi) $h(X) = f^*(X)X \vee f^{**}(X)H(X)$ $(\forall X \in L^n)$.

COMMENT: The equivalence (i)$\Longleftrightarrow$(iii) and the implication (i)$\Longrightarrow$(vi) are due to Banković [2001] and [2000], respectively.

PROOF: (ii)$\Longleftrightarrow$(iv) and (iii)$\Longleftrightarrow$(v): By Corollary 2.2 because

$$\bigvee_{i=0}^{r-1} h_k(A)^i \alpha_k^i = h_k(A)^{\alpha_k}   (\forall A \in C_r^n) .$$

(iv)$\Longrightarrow$(v): The identity

$$\bigvee_A f(A) H(X)^A = f(H(X)) = 0   (\forall X \in L^n)$$

implies $f(A)H(A)^A = 0$, hence $H(A)^A \leq f^*(A)$ for every $A \in C_r^n$.

(v)$\Longrightarrow$(iv): Trivial.

(ii)$\Longleftrightarrow$(vi): Using in turn Proposition 1.13, Lemma 1.2 and the Boolean identities $xy \vee y' = x \vee y'$ and $x \leq y \Longleftrightarrow x' \vee y = 1$, we get

$$(vi) \Longleftrightarrow h_k(X) = x_k f^*(X) \vee h_k(X) f^{**}(X) \ (k = 1, \ldots, n)$$

$$\Longleftrightarrow \bigvee_{i=0}^{r-1} h_k(X)^i (x_k f^*(X) \vee h_k(X) f^{**}(X))^i = 1 \ (k = 1, \ldots, n)$$

$$\Longleftrightarrow 1 = \prod_{k=1}^n \bigvee_{i=0}^{r-1} h_k(X)^i (x_k^i f^*(X) \vee h_k(X)^i f^{**}(X)) =$$

$$= \prod_{k=1}^n \bigvee_{i=0}^{r-1} h_k(X)^i (x_k^i f^*(X) \vee f^{**}(X)) =$$

$$= \prod_{k=1}^{n} \bigvee_{i=0}^{r-1} h_k(X)^i (x_k^i \vee f^{**}(X)) =$$

$$= \prod_{k=1}^{n} (\bigvee_{i=0}^{r-1} h_k(X)^i x_k^i \vee (\bigvee_{i=0}^{r-1} h_k(X)^i) f^{**}(X)) =$$

$$= \prod_{k=1}^{n} (\bigvee_{i=0}^{r-1} h_k(X)^i x_k^i \vee f^{**}(X)) =$$

$$= (\prod_{k=1}^{n} \bigvee_{i=0}^{r-1} h_k(X)^i x_k^i) \vee f^{**}(X) \Longleftrightarrow \text{(ii)}.$$

(i)$\Longleftrightarrow$(iii): Note first that

$$X = H(X) \Longleftrightarrow x_k = h_k(X) \; (k = 1, \ldots, n)$$

$$\Longleftrightarrow \prod_{k=1}^{n} \bigvee_{i=0}^{r-1} x_k^i h_k(X)^i \Longleftrightarrow (\prod_{k=1}^{n} \bigvee_{i=0}^{r-1} x_k^i h_k(X)^i)^* = 0.$$

Since $f(H(X)) = 0$ holds by hypothesis, condition (i) reduces to $f(X) = 0 \Longleftrightarrow X = H(X)$. Taking into account the above computation, Proposition 3.6 and the fact that $x^{**} = x$ for every Boolean element, the last equivalence reduces to (iii). □

Proposition 3.11. (Banković [2000]). *Let $f, g_k, h_k : L^n \longrightarrow L \; (k = 1, \ldots, n)$ be Post functions and $G = (g_1, \ldots, g_n), H = (h_1, \ldots, h_n)$. Suppose G is a general solution of equation $f(X) = 0$. Then H is a reproductive solution of this equation if and only if there is a vector $P = (p_1, \ldots, p_n)$ of Post functions $p_k : L^n \longrightarrow L \; (k = 1, \ldots, n)$ such that*

$$(3.51) \qquad H(X) = f^*(X)X \vee f^{**}(X)G(P(X)) \qquad (\forall X \in L^n).$$

COMMENT: Notice the similarity between (3.51) and the reproductive solutions provided in Theorem 3.4, Proposition 3.10 and Theorem 3.5. As a matter of fact, Proposition 3.11 generalizes Theorem 3.4: take $\Theta \in L^n$ such that $G(\Theta) = \Xi$ and define $P(X) = \Theta$, i.e., $p_k(X) = \theta_k$ for all X and all k. See also Proposition 1.2.7 and its Corollary 1.2.1.

PROOF: As in previous proofs, (3.51) implies

$$f(H(X)) = f^*(X)f(X) \vee f^{**}(X)f(G(P(X))) = 0$$

and if x satisfies $f(X) = 0$, then $H(X) = 1 \cdot X \vee 0 \cdot G(P(X)) = X$.

Conversely, suppose H is a reproductive solution of equation $f(X) = 0$. Take $A \in C_r^n$. Since $f(H(A)) = 0$ and G is a general solution, there exists

$\Phi(A) = (\varphi_1(A), \dots, \varphi_n(A)) \in L^n$ such that $G(\Phi(A)) = H(A)$. We have thus defined n functions $\varphi_k : C_r^n \longrightarrow L$ $(k = 1, \dots, n)$ and Corollary 2.3 implies the existence of n Post functions $p_k : L^n \longrightarrow L$ that extend φ_k, for $k = 1, \dots, n$. Therefore, setting $P = (p_1, \dots, p_n)$, we have $G(P(A)) = G(\Phi(A)) = H(A)$ for all $A \in C_r^n$. On the other hand, for each $A \in C_r^n$, Theorem 3.5 yields

$$H(A) = f^*(A)A \vee f^{**}(A)H(A) = f^*(A)A \vee f^{**}(A)G(P(A)) ,$$

which establishes (3.51) by Corollary 2.2. $\square$

Proposition 3.12. (Banković [2001]). *Let* $f, h_1, \dots, h_n : L^n \longrightarrow L$ *be Post functions and* $H = (h_1, \dots, h_n)$. *Suppose equation* $f(X) = 0$ *is consistent. Then* H *is a general solution of this equation if and only if*

$$(3.52) \qquad f^*(X) = \bigvee_A \prod_{k=1}^{n} \bigvee_{i=0}^{r-1} h_k(A)^i x_k^i \qquad (\forall X \in L^n) .$$

PROOF: As in previous proofs we have

$$X = H(T) \Longleftrightarrow (\prod_{k=1}^{n} \bigvee_{i=0}^{r-1} x_k^i h_k(T)^i)^* = 0 ,$$

therefore

$$\exists T \, (X = H(T)) \Longleftrightarrow \prod_A (\prod_{k=1}^{n} \bigvee_{i=0}^{r-1} x_k^i h_k(A)^i)^* = 0 ,$$

so that the definition

$$f(X) = 0 \Longleftrightarrow \exists T \, (X = h(T))$$

of a general solution becomes

$$f^*(X) = (\prod_A (\prod_{k=1}^{n} \bigvee_{i=0}^{r-1} x_k^i h_k(A)^i)^*)^* = \bigvee_A (\prod_{k=1}^{n} \bigvee_{i=0}^{r-1} x_k^i h_k(A)^i)^{**}$$

and coincides with (3.52) because $x^{**} = x$ for $x \in B(L)$. $\square$

The following theorem has a model-theoretic character. To state and prove it we first recall some basic definitions informally.

A *first-order language* is based on an alphabet containing individual variables, propositional connectives, the quantifiers $\forall$ and $\exists$, function symbols and predicate symbols, which are also simply called predicates. The *terms* of the language are obtained by applying function symbols to individual variables. The *atomic formulas* are obtained by applying predicates to terms. The *formulas* of the language are obtained from atomic formulas by applying propositional connectives and quantifiers. A *sentence* is a closed formula, i.e., a formula in which all variables are quantified.

The *Horn formulas* of a language are defined by induction:

- the *elementary Horn formulas* are the atomic formulas and the formulas of the form $\varphi_1 \& \ldots \& \varphi_s \to \varphi$, where $\varphi_1, \ldots, \varphi_s, \varphi$ are atomic formulas;

- the Horn formulas are built up from elementary Horn formulas by applying $\&, \forall, \exists$.

In particular *the language of Post algebras* has function symbols that correspond to the algebraic type $(2, 2, (1)_r, (0)_r)$ and the predicate $\leq$. For the sake of convenience, let us identify this language with its realization in a given Post algebra. Then terms coincide with simple Post functions. The atomic formulas are the inequalities $g \leq h$ between simple Post functions. In view of Proposition 1.12 the latter inequality is equivalent to

$$\bigvee_{i=0}^{r-2} \bigvee_{k=i+1}^{r-1} g^k h^i = 0 \, ,$$

where the left side is a simple Post function. Conversely, if f is a (simple) Post function, the equality $f = 0$ is equivalent to $f \leq 0$, where the constant function 0 is a simple Post function. Therefore the *atomic formulas* can be identified with the formulas of the form $f = 0$ where f is a simple Post function. It follows by Theorem 3.1 that any conjunction of atomic formulas is also an atomic formula. Therefore the *elementary Horn formulas* are of two kinds: those of the form $f = 0$, where f is a simple Post function, and those of the form $g = 0 \to h = 0$, where g and h are simple Post functions.

The next theorem generalizes Corollary 2.4.

Theorem 3.6. (Banković [1997c]). *A Horn sentence is valid in every r−Post algebra if and only if it is valid in C_r.*

PROOF: According to a well-known theorem about the prenex form of a formula, every Horn formula can be written in the form

(3.53) $$(K_1 x_1) \ldots (K_n x_n) \bigwedge_{j=1}^{k} \varphi_j(x_1, \ldots, x_n) \, ,$$

where $K_1, \ldots, K_n \in \{\forall, \exists\}$ and φ_j are elementary Horn formulas, for $j = 1, \ldots, k$.

Consider a formula φ_j which is of the form

$$g(x_1, \ldots, x_n) = 0 \to h(x_1, \ldots, x_n) = 0 \, .$$

If the equation $g(X) = 0$ is inconsistent, then φ_j is identically true and can be omitted from (3.53). If the equation $g(X) = 0$ is consistent, then

$$\varphi_j(X) \iff g(X) \leq f^{**}(X) \iff g(X) f^*(X) = 0$$

by Proposition 3.6. Therefore we can suppose without loss of generality that each φ_j is of the form $f_j(X) = 0$, where f_j is a simple Post function, so that taking into account Theorem 3.1 we see that (3.53) reduces to

(3.54) $(K_1 x_1) \ldots (K_n x_n) f(x_1, \ldots, x_n) = 0$,

where f is a simple Post function.

Now we conclude the proof by induction on n.

For $n := 1$ we apply in a compact form Corollary 2.2 and Theorem 3.3:

$$(Kx)f(X) = 0 \Longleftrightarrow \Theta_{i=0}^{r-1} f(e_i) = 0 ,$$

where Θ is $\bigvee$ or $\prod$ according as K is $\forall$ or $\exists$. This yields the desired conclusion because $\Theta_{i=0}^{r-1} f(e_i) \in C_r$ by Proposition 2.2 and C_r is isomorphically included in any Post algebra.

The inductive step follows from the equivalence

$$(3.54) \Longleftrightarrow (K_1 x_1) \ldots (K_{n-1} x_{n-1}) \Theta_{i=0}^{r-1} f(x_1, \ldots, x_{n-1}, e_i) = 0 ,$$

where $\Theta_{i=0}^{r-1} f(x_1, \ldots, x_{n-1}, e_i)$ is a simple Post function by Corollary 2.8.   □

Theorem 3.6 is a potential tool for proving theorems on Post algebras, especially on Post functions and equations. For it generalizes a similar theorem that has actually been used in proving theorems on Boolean functions and equations; see the introduction to Ch.6, §4.

Let us give here an application to Post functions: the characterization of isotone simple Post functions $f : L^n \longrightarrow L$, where L^n is endowed with the product odering $X \leq Y \Longleftrightarrow x_k \leq y_k \ (k = 1, \ldots, n)$.

Proposition 3.13. *A Post function $f : L^n \longrightarrow L$ is isotone if and only if its restriction to C_r^n is isotone.*

PROOF: Since the property is trivial for $L := C_r$, it remains to prove that this property can be expressed by a Horn sentence.

Consider $r^n + 2n$ variables $(c_A)_{A \in C_r^n}, x_1, \ldots, x_n, y_1, \ldots, y_n$ and the sentence

(3.55) $\forall c_{(0,\ldots,0)} \ldots \forall c_{(1,\ldots,1)} \forall x_1 \ldots \forall x_n \forall y_1 \ldots \forall y_n \ P$,

where P is the following formula depending on the above variables:

$$P := (p \to q) \& (q \to p) ,$$

where

$$p := x_1 \leq y_1 \& \ldots \& x_n \leq y_n \to \bigvee_A c_A X^A \leq \bigvee_A c_A Y^A ,$$

while q is the conjunction of all the inequalities of the form $c_A \leq c_B$ for which $A \leq B$. According to the discussion which precedes Theorem 3.6, any conjunction of atomic formulas is an atomic formula. Therefore q, as well as the implication and the conclusion of p, are atomic formulas. This shows that (3.55) is indeed a Horn sentence.   □

The next lemma reveals the technique used by Banković in proving theorems about the general solution via the Vaught theorem.

Lemma 3.2. *The sentence "$X = H(T)$ defines the general solution of the Post (Boolean) equation $f(X) = 0$" is an atomic formula in the language of Post (Boolean) algebras.*

PROOF: Consider first the case of Post algebras.

In view of Proposition 3.12, the given sentence is equivalent to the identity (3.52) which, taking into account Proposition 1.1 and identities (1.30) and (1.8), can be further written in the form $h(X, T) = 0$, where h is a Post function. But h is a simple Post function in the $(n + 1)r^n$ variables $h_k(A)$ $(k = 1, \ldots, n)$, $f(A)$ $(A \in C_r^n)$.

The simpler cae of Boolean equations uses the characterization of general solutions given by J.-P. Deschamps (see e.g. BFE, Theorem 3.5).   □

Last but not least, let us say a few words about *matrix Post equations and inequalities*.

Matrix calculus over a bounded distributive lattice was initiated by Moisil [1941]. Given such a lattice $(L; \cdot, \vee, 0, 1)$, Moisil defined the product of an $m \times n$ matrix $A = \| a_{ij} \|$ and an $n \times p$ matrix $B = \| b_{jk} \|$ with entries in L by

$$(3.56) \qquad A \times B = \| c_{ik} \| , \quad c_{ik} = \bigvee_{j=1}^{n} a_{ij} b_{jk} ,$$

i.e., by analogy with the usual matrix product and which, in the case of matrices with 0-1 entries, represents the composite of relations A and B. Moisil focused on square matrices, among which those representing transitive relations, and on *inversible* square matrices, i.e., square matrices A for which there exists a matrix X such that $A \times X = X \times A = I$, where $I = \| \delta_{ij} \|$ (Kronecker delta) is the unit matrix with respect to the product (3.56). The field of lattice matrices and their applications has rapidly developed since 1952. For Boolean matrices and numerous applications to social sciences, logical design and others see Kim [1982], for matrices with entries in structures more general than lattices and applications to path problems in graphs and their practical intepretations see Cunninghame-Green [1979] and Cao, Kim and Roush [1984], and for numerous applications to fuzzy mathematics, for example fuzzy relations, fuzzy arithmetic and so on, see Di Nola, Sessa, Pedrycz and Sanchez [1989]; cf. Rudeanu [1997].

The basic results on matrix Boolean equations and inversible Boolean matrices (reported in BFE, Chapter 7, §§ 1,2) have been extended to the so-called fuzzy relation equations, which have numerous applications; cf. Di Nola, Sessa, Pedrycz and Sanchez [1989]. There is a section in Serfati [1997] dealing with matrix Post equations and inequalities, much in the same spirit as the research mentioned above. The result presented below is somewhat more specific to the case of Post algebras.

By a *Post matrix* we mean a matrix with entries in a Post algebra L; we say that the matrix is *Boolean* if all of its entries are in $B(L)$.

Proposition 3.14. (Serfati [1997]). *A square Post matrix is inversible if and only if it is an inversible Boolean matrix.*

COMMENT: Twenty conditions equivalent to the inversibility of a Boolean matrix are known; cf. BFE, Theorem 7.5.

PROOF: Sufficiency is trivial. To prove necessity suppose $A = \| a_{ij} \|$ is an inversible Post matrix of order n. Let $B = \| b_{ij} \|$ be its inverse. Then $\bigvee_{j=1}^{n} a_{ij} b_{jh} = \delta_{ih}$ for every $i, j \in \{1, \ldots, n\}$. Let $k \in \{1, \ldots, r-1\}$ and take the monotone component of order k of each side of the above equality. It follows that

$$(3.57) \qquad \bigvee_{j=1}^{n} (a_{ij})_{(k)} (b_{hj})_{(k)} = (\delta_{ih})_{(k)} = \delta_{ih}$$

by Propositions 1.6, 1.7 and 1.4. Relations (3.57) show that for each k the Boolean matrix $\| (a_{ij})_{(k)} \|$ is inversible. Therefore (see e.g. BFE, Theorem 7.5) each column j (and each row i) of the matrix is orthonormal and this implies that the complement of each element $(a_{ij})_{(k)}$ in column j equals the disjunction of the other elements of the column (cf. (1.2.1) in Lemma 1.1). Since on the other hand the monotone components form a decreasing sequence, it follows that if $k_1, k_2 \in \{1, \ldots, r-1\}$, say $k_1 \leq k_2$, then for each $i, j \in \{1, \ldots, n\}$,

$$((a_{ij})_{(k_1)})' = \bigvee_{h=1, h \neq i}^{n} (a_{ij})_{(k_1)} \geq \bigvee_{h=1, h \neq i}^{n} (a_{ij})_{(k_2)} = ((a_{ij})_{(k_2)})',$$

hence $(a_{ij})_{(k_1)} \leq (a_{ij})_{(k_2)}$ and the converse inequality is proved similarly. Thus $(a_{ij})_{(k_1)} = (a_{ij})_{(k_2)}$ for all $k_1, k_2 \in \{1, \ldots, r-1\}$, which proves that a_{ij} is a Boolean element by Proposition 1.4.   □

Corollary 3.7. *If a square Post matrix is inversible, then the inverse is its transpose.*

PROOF: This property is known in the case of Boolean matrices (see e.g. BFE, Corollary of Theorem 7.5).   □

6. A revision of Boolean fundamentals

The theory of Boolean functions and equations involve certain fundamental concepts, among which are those of minterm, prime implicant, reproductive solution and solution by recurrent inequalities. In this chapter we introduce and study certain notions which generalize the above ones. As a matter of fact, the challenge was to obtain the most general concepts which retain the main features of the concepts mentioned above.

The first section, on linear Boolean equations, constructs the prerequisites necessary for §2, but has also an intrinsic interest. In §2 we introduce the concepts of linear base, or system of generalized minterms, and interpolating systems; these are the most general concepts which behave as the set of minterms and the set $\{0,1\}^n$, respectively. The subject of §3 is an axiomatic theory of prime implicants for arbitrary (not necessarily simple) Boolean functions; the latter theory, in its turn, specializes to the classical theory of prime implicants for simple Boolean functions or truth functions. The main topic in §4 is the concept of recurrent cover, which provides all possible ways of expressing the solutions of a Boolean equation by a system of recurrent inequalities. The last section is devoted to a difficult problem which is far from being solved: the study of systems involving Boolean equations and negated Boolean equations, linked not only by logical conjunction, but also by logical disjunction.

Unless otherwise stated, throughout this chapter we work in an arbitrary Boolean algebra $(B, \cdot, \vee, ', 0, 1)$.

1 Linear Boolean equations

The term *linear Boolean equation* has been given two different meanings.

One of them coincides with the classical concept of linear equation applied to the ring structure of a Boolean algebra. Systems of linear Boolean-ring equations have been studied by Bernstein [1932], then by Parker and Bernstein [1955]; see also BFE, Chapter 6, §3.

In the sequel we deal with the other meaning of linearity in a Boolean algebra, suggested by a simple analogy and due to Löwenheim [1919].

Lemma 1.1. (Rudeanu [1977]). *The systems of inequalities*

(1.1)
$$\bigvee_{j=1}^{n} a_{ij}x_j \leq b_i \qquad (i = 1, \ldots, m)$$

is consistent and its set of solutions is given by

(1.2)
$$x_j \leq \prod_{i=1}^{m}(a'_{ij} \vee b_i) \qquad (j = 1, \ldots, n) .$$

PROOF: The following equivalences hold:

$$(1.1) \iff a_{ij}x_j \leq b_i \ (\forall j)(\forall i)$$

$$\iff x_j \leq a'_{ij} \vee b_i \ (\forall i)(\forall j) \iff (1.2) .$$

$\square$

Corollary 1.1. *The system* (1.1) *has the greatest solution (with respect to the product order of B^n)*

(1.3)
$$x_j = \prod_{i=1}^{m}(a'_{ij} \vee b_i) \qquad (j = 1, \ldots, n) .$$

$\square$

Lemma 1.2. (Rudeanu [1977]). *The system of inequalities*

(1.4)
$$\bigvee_{j=1}^{n} a_{ij}x_j \geq b_i \qquad (i = 1, \ldots, m)$$

is consistent if and only if

(1.5)
$$b_i \leq \bigvee_{j=1}^{n} a_{ij} \qquad (i = 1, \ldots, m) ,$$

in which case the set of solutions is given by the system of recurrent inequalities

(1.6.1)
$$\bigvee_{i=1}^{m} b_i \prod_{j=2}^{n}(a'_{ij} \vee x'_j) \leq x_1 ,$$

(1.6.k)
$$\bigvee_{i=1}^{m} b_i(\prod_{j=1}^{k-1} a'_{ij}) \prod_{j=k+1}^{n}(a'_{ij} \vee x'_j) \leq x_k \qquad (k = 2, \ldots, n-1) ,$$

(1.6.n)
$$\bigvee_{i=1}^{m} b_i(\prod_{j=1}^{n-1} a'_{ij}) \leq x_n.$$

PROOF: By the method of eliminations of variables.

System (1.4) is equivalent to the equation

$$(1.7.1) \qquad \bigvee_{i=1}^{m} b_i(a'_{i1} \vee x'_1) \prod_{j=2}^{n}(a'_{ij} \vee x'_j) = 0 ,$$

which we split into the following two equations:

$$(1.8.1) \qquad x'_1 \bigvee_{i=1}^{m} b_i \prod_{j=2}^{n}(a'_{ij} \vee x'_j) = 0 ,$$

$$(1.7.2) \qquad \bigvee_{i=1}^{m} b_i a'_{i1} \prod_{j=2}^{n}(a'_{ij} \vee x'_j) = 0 .$$

At step k, $1 < k < n$, we have obtained the equation

$$(1.7.k) \qquad \bigvee_{i=1}^{m} b_i(\prod_{j=1}^{k-1} a'_{ij}) \prod_{j=k}^{n}(a'_{ij} \vee x'_j) = 0 ,$$

which we split into the equations

$$(1.8.k) \qquad x'_k \bigvee_{i=1}^{m} b_i(\prod_{j=1}^{k-1} a'_{ij}) \prod_{j=k+1}^{n}(a'_{ij} \vee x'_j) = 0 ,$$

$$(1.7.k+1) \qquad \bigvee_{i=1}^{m} b_i(\prod_{j=1}^{k} a'_{ij}) \prod_{j=k+1}^{n}(a'_{ij} \vee x'_j) = 0 .$$

At step n we have obtained the equation

$$(1.7.n) \qquad \bigvee_{i=1}^{m} b_i(\prod_{j=1}^{n-1} a'_{ij})(a'_{in} \vee x'_n) = 0 ,$$

which we split into the equations

$$(1.8.n) \qquad x'_n \bigvee_{i=1}^{m} b_i \prod_{j=1}^{n-1} a'_{ij} = 0 ,$$

$$(1.7.n+1) \qquad \bigvee_{i=1}^{m} b_i \prod_{j=1}^{n} a'_{ij} = 0 .$$

The consistency condition (1.7.n+1) is equivalent to (1.4), while the inequalities (1.8) are equivalent to (1.6). □

Corollary 1.2. *If the system* (1.4) *is consistent, then it has the greatest solution* $(1,\dots,1)$. □.

Theorem 1.1. (Löwenheim [1919], Rudeanu [1977]). *(i) The system of equations*

$$(1.9) \qquad \bigvee_{j=1}^{n} a_{ij}x_j = b_i \qquad (i=1,\dots,m)$$

is consistent if and only if

(1.10)
$$b_i \le \bigvee_{j=1}^{n} a_{ij} \prod_{h=1,h\neq i}^{m} (a'_{hj} \vee b_h) \qquad (i = 1, \ldots, m),$$

or equivalently,

(1.10′)
$$b_i \le \bigvee_{j=1}^{n} a_{ij} \prod_{h=1}^{m} (a'_{hj} \vee b_h) \qquad (i = 1, \ldots, m).$$

(ii) *When this is the case, the solutions are given by the system of recurrent inequalities*

(1.11.1)
$$\bigvee_{i=1}^{m} b_i \prod_{j=2}^{n} (a'_{ij} \vee x'_j) \le x_1 \le \prod_{i=1}^{m} (a'_{i1} \vee b_i),$$

(1.11.k)
$$\bigvee_{i=1}^{m} b_i \left(\prod_{j=1}^{k-1} a'_{ij}\right) \prod_{j=k+1}^{n} (a'_{ij} \vee x'_j) \le x_k \le$$
$$\le \prod_{i=1}^{m} (a'_{ik} \vee b_i) \qquad (k = 2, \ldots, n-1),$$

(1.11.n)
$$\bigvee_{i=1}^{m} b_i \prod_{j=1}^{n-1} a'_{ij} \le x_n \le \prod_{i=1}^{m} (a'_{in} \vee b_i).$$

PROOF: (i) For the consistency condition (1.10) see the paper by Löwenheim or BFE, Theorem 7.6. Clearly (1.10′) implies (1.10). Conversely, it follows from (1.10) that for each i,

$$b_i \le b_i \bigvee_{j=1}^{n} a_{ij} \prod_{h=1,h\neq i}^{m} (a'_{hj} \vee b_h) = \bigvee_{i=1}^{n} a_{ij} \prod_{h=1}^{m} (a'_{hj} \vee b_h).$$

(ii) Immediate from Lemmas 1.1 and 1.2.   □

Corollary 1.3. (Löwenheim [1919]). *The system* (1.9) *is consistent for every* $(b_1, \ldots, b_m) \in B^m$ *if and only if*

(1.12)
$$\bigvee_{j=1}^{n} a_{ij} \prod_{h=1,h\neq i}^{m} a'_{hj} = 1 \qquad (i = 1, \ldots, m).$$

PROOF: If conditions (1.12) hold, then

$$b_i \le \bigvee_{j=1}^{n} a_{ij} \prod_{h=1,h\neq i}^{m} a'_{hj} \le \bigvee_{j=1}^{n} a_{ij} \prod_{h=1,h\neq i}^{m} (a'_{hj} \vee b_h)$$

for all $b_1, \ldots, b_m$ and all i, therefore system (1.9) is consistent for every $b_1, \ldots, b_m \in B$. Conversely, if the last condition is fulfilled, then for each i take $b_i := 1$ and $b_h := 0$ for $h \neq i$, which transforms (1.10) into (1.12).   □

We are now going to study in some detail systems (1.9) with unique solutions and matrices $\|a_{ij}\|$ such that the corresponding systems (1.9) are consistent for any $b_1, \ldots, b_m$.

In the rest of this section $\|a_{ij}\|$ denotes an $m \times n$ matrix with entries in the Boolean algebra B.

Definition 1.1. The matrix $\|a_{ij}\|$ is said to be *row/column orthogonal (normal, orthonormal)* provided each row/each column is orthogonal (normal, orthonormal); cf. Comment to Lemma 5.1.1. □

Lemmas 1.3-1.5, Proposition 1.1 and Theorem 1.2 below are taken from Rudeanu [1983b].

Lemma 1.3. *The following implications hold:* $\|a_{ij}\|$ *is row normal and column orthogonal* $\implies$ (1.12) $\implies$ $\|a_{ij}\|$ *is row normal.*

PROOF: If $\|a_{ij}\|$ is row normal and column orthogonal, then for each i,

$$\bigvee_{j=1}^{n} \prod_{h=1,h\neq i}^{m} a'_{hj} = \bigvee_{j=1}^{n} \prod_{h=1,h\neq i}^{m} a_{ij}a'_{hj} = \bigvee_{j=1}^{n} a_{ij} = 1 \,,$$

while (1.2) implies that for each i,

$$\bigvee_{j=1}^{n} a_{ij} \geq \bigvee_{j=1}^{n} a_{ij} \prod_{h=1,h\neq i}^{m} a'_{hj} = 1 \,.$$

□

Lemma 1.4. *Suppose* $m \geq n$. *Then relations* (1.12) *hold if and only if* $\|a_{ij}\|$ *is row normal and column orthogonal.*

PROOF: Taking into account Lemma 1.3, it remains to prove that relations (1.12) imply the orthogonality of each column; without loss of generality it suffices to prove that $a_{11}a_{21} = 0$. Setting

$$a_i = \bigvee_{j=2}^{n} a_{ij} \prod_{h=1,h\neq i}^{m} a'_{hj} \qquad (i = 1,\ldots,m) \,,$$

it follows from (1.11) that

$$1 = \prod_{i=1}^{m} \bigvee_{j=1}^{n} a_{ij} \prod_{h=1,h\neq i}^{m} a'_{hj} = \prod_{i=1}^{m}(a_{i1} \prod_{h=1,h\neq i}^{m} (a'_{h1} \vee a_i))$$

$$\leq (a'_{21} \vee a_1)(a'_{11} \vee a_2)\prod_{i=3}^{m}(a'_{11} \vee a_i) = (a'_{21} \vee a_1)(a'_{11} \vee \prod_{i=2}^{m} a_i)$$

$$= a'_{21}(a'_{11} \vee \prod_{i=2}^{m} a_i) \vee a_1 a'_{11} \vee \prod_{i=1}^{m} a_i \leq a'_{21} \vee a'_{11} \vee \prod_{i=1}^{m} a_i \ ;$$

if we succeed in proving that $\prod_{i=1}^{m} a_i = 0$, it will follow that $a'_{21} \vee a'_{11} = 1$, as desired.

But $\prod_{i=1}^{m} a_i$ is a join of terms of the form

$$t_{j_1 \ldots j_m} = a_{1j_1} c_{1j_1} a_{2j_2} c_{2j_2} \ldots a_{mj_m} c_{mj_m} \ ,$$

where $j_1, \ldots, j_m \in \{2, \ldots, n\}$ and we have set

$$c_{ij} = \prod_{h=1, h \neq i}^{m} a'_{hj} \qquad (i = 1, \ldots, m; \ j = 2, \ldots, n) \ .$$

Since $m > n - 1$, for each $(j_1, \ldots, j_m) \in \{2, \ldots, n\}^m$, there exist two distinct indices $q, r \in \{1, \ldots, m\}$ such that $j_q = j_r$. Therefore

$$t_{j_1 \ldots j_m} \leq a_{qj_q} c_{rj_r} \leq a_{qj_q} a'_{qj_r} = a_{qj_q} a'_{qj_q} = 0 \ .$$

$\square$

Lemma 1.5. *If $\|a_{ij}\|$ is row normal and column orthogonal, then $m \leq n$.*

PROOF: Suppose, by way of contradiction, that $m > n$. Then the square matrix $\|a_{ij}\|_{i,j=1,\ldots,n}$ is also row normal and column orthogonal, therefore it is orthonormal (see e.g. BFE, Theorem 7.5). Hence for each $j \in \{1, \ldots, n\}$,

$$a_{n+1,j} = a_{n+1,j} \bigvee_{i=1}^{n} a_{ij} = \bigvee_{i=1}^{n} a_{n+1,j} a_{ij} = 0 \ ,$$

which contradicts the fact that row $n + 1$ is normal. $\square$

Proposition 1.1. *If system (1.9) is consistent for every $(b_1, \ldots, b_m) \in B^m$, then $m \leq n$.*

PROOF: Immediate from Corollary 1.3 of Theorem 1.1 and Lemmas 1.4 and 1.5. $\square$

The case $m = n$ is reminiscent of ordinary systems of linear equations.

Theorem 1.2. *Let $\|a_{ij}\|$ be a square Boolean matrix of order n. Then:*
(i) The system

(1.13)
$$\bigvee_{j=1}^{n} a_{ij} x_j = b_i \qquad (i = 1, \ldots, n)$$

is consistent for every $b_1, \ldots, b_n \in B$ if and only if the matrix $\|a_{ij}\|$ is orthonormal.

(ii) When this is the case, for each $(b_1, \ldots, b_n) \in B^n$, *system* (1.13) *has the unique solution*

(1.14)
$$x_j = \prod_{i=1}^{n} (a'_{ij} \vee b_i) \qquad (j = 1, \ldots, n) .$$

COMMENT: There are numerous conditions equivalent to the orthonormality of matrix $\|a_{ij}\|$; see e.g. BFE, Theorem 7.5.

PROOF: (i) By Corollary 1.3 of Theorem 1.1, Lemma 1.4 and the fact that for square matrices the condition in Lemma 1.4 is equivalent to the orthonormality of $\|a_{ij}\|$; cf. BFE, Theorem 7.5.

(ii) Suppose the matrix $\|a_{ij}\|$ is orthonormal. Let $(x_1, \ldots, x_n)$ be a solution of (1.13), which exists by (i). Take an index $k \in \{1, \ldots, n\}$, say $1 < k < n$. Using also (1.11), we obtain

$$\prod_{h=1}^{n} (a'_{hk} \vee b_h) = \left(\bigvee_{i=1}^{n} a_{ik}\right) \prod_{h=1}^{n} (a'_{hk} \vee b_h)$$

$$= \bigvee_{i=1}^{n} a_{ik} \prod_{h=1}^{n} (a'_{hk} \vee b_h) \leq \bigvee_{i=1}^{n} a_{ik} b_i$$

$$= \bigvee_{i=1}^{n} \left(\bigvee_{j=1, j \neq k}^{n} a_{ij} \right)' b_i$$

$$= \bigvee_{i=1}^{n} \left(\prod_{j=1}^{k-1} a'_{ij} \right) \left(\prod_{j=k+1}^{n} a'_{ij} \right) b_i$$

$$\leq \bigvee_{i=1}^{n} b_i \left(\prod_{j=1}^{k-1} a'_{ij} \right) \prod_{j=k+1}^{n} (a'_{ij} \vee x'_j)$$

$$\leq x_k \leq \prod_{h=1}^{n} (a'_{hk} \vee b_h) ,$$

which proves (1.14.k). Similar proofs establish (1.14.1) and (1.14.n).   □

We now return to matrices $\|a_{ij}\|$ of arbitrary dimensions $m \times n$. Theorem 1.3, Theorem 1.4 and its Corollaries 1.4-1.5 below are taken from Melter and Rudeanu [1984a]. The present theorems correct an error in the original formulations of Theorems 1 and 2, op. cit.

Theorem 1.3. *(i) The system*

(1.9)
$$\bigvee_{j=1}^{n} a_{ij} x_j = b_i \qquad (i = 1, \ldots, m)$$

has a unique solution if and only if the following conditions hold:

(1.15.1)
$$\bigvee_{i=1}^{m} b_i \prod_{j=2}^{n}(a'_{ij} \vee \bigvee_{h=1}^{m} a_{hj}b'_h) = \prod_{i=1}^{m}(a'_{i1} \vee b_i) \,,$$

(1.15.k)
$$\bigvee_{i=1}^{m} b_i(\prod_{j=1}^{k-1} a'_{ij}) \prod_{j=k+1}^{n}(a'_{ij} \vee \bigvee_{h=1}^{m} a_{hj}b'_h)$$
$$= \prod_{i=1}^{m}(a'_{ik} \vee b_i) \qquad (k = 2,\dots,n-1) \,,$$

(1.15.n)
$$\bigvee_{i=1}^{m} b_i(\prod_{j=1}^{n-1} a'_{ij}) = \prod_{i=1}^{m}(a'_{in} \vee b_i) \,,$$

(1.16)
$$b_i \leq \bigvee_{j=1}^{n} a_{ij} \qquad (i = 1,\dots,m).$$

(ii) When this is the case, the unique solution is (1.3).

PROOF: based on Theorem 1.1.

Suppose first that conditions (1.15) and (1.16) hold. For each $i \in \{1,\dots,m\}$, set

$$p_i = b_i \prod_{j=1}^{n}(a'_{ij} \vee \bigvee_{h=1}^{m} a_{hj}b'_h) \,.$$

It follows from (1.15.1) that

$$b_i(\prod_{j=2}^{n}(a'_{ij} \vee \bigvee_{h=1}^{m} a_{hj}b'_h)) \bigvee_{h=1}^{m} a_{h1}b'_h = 0 \,,$$

therefore

$$p_i = b_i a'_{i1} \prod_{j=2}^{n}(a'_{ij} \vee \bigvee_{h=1}^{m} a_{hj}b'_h) \,.$$

We continue by induction. Relations (1.15.1),\dots,(1.15.k-1) yield

$$p_i = b_i(\prod_{j=1}^{k-1} a'_{ij}) \prod_{j=k}^{n}(a'_{ij} \vee \bigvee_{h=1}^{m} a_{hj}b'_h) \,,$$

while (1.15.k) implies

$$b_i(\prod_{j=1}^{k-1} a'_{ij})(\prod_{j=k+1}^{n}(a'_{ij} \vee \bigvee_{h=1}^{m} a_{hj}b'_h)) \bigvee_{h=1}^{m} a_{hk}b'_h = 0 \,,$$

therefore

$$p_i = (\prod_{j=1}^{k} a'_{ij}) \prod_{j=k+1}^{n}(a'_{ij} \vee \bigvee_{h=1}^{m} a_{hj}b'_h) \,.$$

So (1.15.1),\dots,(1.15.n-1) yield

$$p_i = (\prod_{j=1}^{n-1} a'_{ij})(a'_{in} \vee \bigvee_{h=1}^{m} a_{hn}b'_h) \,,$$

while (1.15.n) implies

$$b_i(\prod_{j=1}^{n-1} a'_{ij}) \bigvee_{h=1}^{m} a_{hn}b'_h = 0 \,,$$

hence $p_i = \prod_{j=1}^{n} a'_{ij} = 0$ by (1.16).

Therefore $p_i = 0$ for all i, that is relations (1.10') hold, which shows that system (1.9) is consistent, by Theorem 1.1(i). On the other hand relations (1.15) show that the vector (1.3) satisfies conditions (1.11) as equalities, hence (1.3) is a solution of (1.9).

Further let $(x_1, \ldots, x_n)$ be an arbitrary solution of (1.9). Then it follows from (1.11.n) and (1.15.n) that $x_n = \prod_{i=1}^{m}(a'_{in} \vee b_i)$. As a matter of fact it follows immediately by reverse induction that $x_k = \prod_{i=1}^{m}(a'_{ik} \vee b_i)$ for $k = n, n-1, \ldots, 1$. □

Theorem 1.4. *The following conditions are equivalent for a Boolean matrix* $\|a_{ij}\|$ *of order* $m \times n$:

(i) *The system (1.9) has a unique solution for each* $(b_1, \ldots, b_m) \in B^m$;

(ii) *For every* $M \subseteq \{1, \ldots, m\}$,

(1.17.1)
$$\bigvee_{i \in CM} \prod_{j=2}^{n}(a'_{ij} \vee \prod_{h \in M} a_{hj}) = \prod_{i \in M} a'_{i1} \,,$$

(1.17.k)
$$\bigvee_{i \in CM}(\prod_{j=1}^{k-1} a'_{ij}) \prod_{j=k+1}^{n}(a'_{ij} \vee \bigvee_{h \in M} a_{hj})$$
$$= \prod_{i \in M} a'_{ik} \qquad (k = 2, \ldots, n-1) \,,$$

(1.17.n)
$$\bigvee_{i \in CM} \prod_{j=1}^{n-1} a'_{ij} = \prod_{i \in M} a'_{in} \,,$$

where $CM = \{1, \ldots, m\} \backslash M$, *while* $\bigvee \varnothing = 0$ *and* $\prod \varnothing = 1$,

(1.18)
$$\bigvee_{j=1}^{n} a_{ij} = 1 \qquad (i = 1, \ldots, m).$$

PROOF: It follows from Theorem 1.3 that condition (i) is equivalent to the fact that relations (1.15) and (1.16) hold for every $(b_1, \ldots, b_m) \in \{0, 1\}^m$; in view of the Müller-Löwenheim Verification Theorem (see e.g. BFE, Theorem 2.13), the latter property is equivalent to the fact that relations (1.15) and (1.16) hold for every $(b_1, \ldots, b_m) \in B^m$, and this is precisely condition (ii). □

Corollary 1.4. *Let* $\|a_{ij}\|$ *be an* $m \times n$ *Boolean matrix such that the system (1.9) has a unique solution for every* $(b_1, \ldots, b_m) \in B^m$. *Then:*

(i) *The matrix is row and column normal.*

(ii) If, moreover, the matrix is row orthogonal, then $m = n$ and the matrix is orthonormal.

PROOF: (i) By (1.18) and (1.17), respectively, for $M := \{1, \ldots, m\}$.

(ii) Lemma 1.5 applied to the transpose of $\|a_{ij}\|$ implies $n \leq m$, while $m \leq n$ by Proposition 1.1. Thus $m = n$, hence the matrix is orthonormal by BFE, Theorem 3.5. □

Corollary 1.5. *A square Boolean matrix $\|a_{ij}\|$ is orthonormal if and only if it fulfils conditions (1.17) and (1.18) with $m = n$.*

PROOF: By Theorems 1.2 and 1.4. □

Chistov [1994]* studied a system of equations similar to (1.9), namely

$$\bigvee_{j=1}^{n} a_{ij}x_j \vee c_i = x_i \qquad (i = 1, \ldots, n) ,$$

in view of applications to the problem of analyzing contact circuits.

2 Generalized minterms and interpolating systems

Let us recall the following fundamental facts about Boolean functions (see e.g. BFE, Chapter 1, §2 and the generalizations to Post functions in the present Ch.5, §2). A *Boolean function* $f : B^n \longrightarrow B$ is characterized by the fact that it has a unique representation of the form

$$(2.1) \qquad f(X) = \bigvee_{A} c_A X^A \qquad (\forall X \in B^n) ,$$

where $\bigvee_A$ means iterated join over all vectors $A \in \{0,1\}^n$, while with every $X = (x_1, \ldots, x_n) \in B^n$ and every $A = (\alpha_1, \ldots, \alpha_n) \in \{0,1\}^n$ is associated the *minterm*

$$(2.2) \qquad X^A = x_1^{\alpha_1} \ldots x_n^{\alpha_n} ,$$

where for every $x \in B$ and every $\alpha \in \{0,1\}$ we have set

$$(2.3) \qquad x^0 = x', x^1 = x .$$

If we introduce the *projections*

$$\pi_j : B^n \longrightarrow B, \; \pi_j(x_1 \ldots, x_n) = x_j \qquad (j = 1, \ldots, n) ,$$

then identity (2.1) expresses the equality

$$(2.4) \qquad f = \bigvee_{A} c_A \pi_1^{\alpha_1} \ldots \pi_n^{\alpha_n}$$

in the algebra of all Boolean functions from B^n to B and we refer to the 2^n

functions $\pi_1^{\alpha_1} \ldots \pi_n^{\alpha_n}$ as the *minterms in the variables* $x_1, \ldots x_n$ (see also 5.(2.5) in the proof of Theorem 5.2.1).

On the other hand, since in fact

$$(2.5) \qquad\qquad c_A = f(A) \qquad (\forall A \in \{0,1\}^n),$$

it follows that every function $c : \{0,1\}^n \longrightarrow B$ has a unique extension to a Boolean function $f : B^n \longrightarrow B$. Note that there is a bijection between the minterms $\pi_1^{\alpha_1} \ldots \pi_n^{\alpha_n}$ and the elements $(\alpha_1, \ldots, \alpha_n)$ of $\{0,1\}^n$.

In this section we introduce the concepts of *linear base* or *system of generalized minterms*, and *interpolating system*, which are the most general concepts which behave as the minterms and the set $\{0,1\}^n$. More exactly, there is a bijection between the set of all linear bases and the set of all interpolating systems, and each linear base and its associated interpolating system behave like the set of minterms and $\{0,1\}^n$, respectively. We follow Rudeanu [1983b] for generalized minterms and Melter and Rudeanu [1984a] for interpolating systems.

Definition 2.1. We say that a system $(f_1, \ldots, f_p)$ of Boolean functions $f_j : B^n \longrightarrow B$ $(j = 1, \ldots, n)$ is a *linear generator* of Boolean functions if every Boolean function $f : B^n \longrightarrow B$ can be written in the form

$$(2.6) \qquad\qquad f = c_1 f_1 \vee \ldots \vee c_p f_p$$

for some constants $c_1, \ldots, c_p \in B$ depending on f. If, moreover, for each f the representation (2.6) is unique, we say that $(f_1, \ldots, f_p)$ is a *linear base* or a *system of generalized minterms*. $\qquad\square$

Lemma 2.1. *The system* $(f_1, \ldots, f_p)$ *is a linear generator (a base) if and only if the system of equations*

$$(2.7) \qquad\qquad \bigvee_{j=1}^{p} f_j(A) x_j = f(A) \qquad (\forall A \in \{0,1\}^n)$$

has a solution (a unique solution) $(x_1, \ldots, x_n) \in B^n$ *for every choice of the elements* $f(A) \in B$, $A \in \{0,1\}^n$.

PROOF: In view of the Müller-Löwenheim Verification Theorem, the identity (2.6) holds if and only if

$$\bigvee_{j=1}^{p} c_j f_j(A) = f(A) \qquad (\forall A \in \{0,1\}^n).$$

$\qquad\square$

Proposition 2.1. *If* $(f_1, \ldots, f_p)$ *is a linear generator, then* $p \geq 2^n$.

PROOF: Immediate from Lemma 2.1 and Proposition 1.1. $\qquad\square$

Proposition 2.2. *The system* $(f_1, \ldots, f_p)$ *of Boolean functions* $f_j : B^n \longrightarrow B$ *is a linear generator if and only if*

$$(2.8) \qquad \bigvee_{j=1}^{p} f_j(A) \prod \{f_j'(C) \mid C \in \{0,1\}^n \ \& \ C \neq A\} = 1 \ (\forall A \in \{0,1\}^n) \ .$$

PROOF: Immediate from Lemma 2.1 and Corollary 1.3 of Theorem 1.1.   □

Thus linear generators are to be sought only among systems with at least 2^n functions, by Proposition 2.1, and they can be determined with the aid of Proposition 2.2. We are now going to determine all linear generators with exactly 2^n functions. We need a preliminary result which has also an intrinsic interest.

Proposition 2.3. *The following conditions are equivalent for a Boolean function $f : B^n \longrightarrow B$:*

   (i) the set $\{f(A) \mid A \in \{0,1\}^n\}$ is orthonormal ;
   (ii) the equation $f(x_1, \ldots, x_n) = 1$ has a unique solution ;
   (iii) f is of the form

$$(2.9) \qquad f(x_1, \ldots, x_n) = \prod_{j=1}^{n} (x_j + \xi_j) \ ,$$

where $\xi_1, \ldots \xi_n$ are constants from B and $+$ denotes the ring sum $x + y = xy' \vee x'y$.

PROOF: $(i) \Longleftrightarrow (ii)$: This is the equivalence $(i) \Longleftrightarrow (ii)$ in BFE, Theorem 6.6 (due to Whitehead), applied to f'.

   $(ii) \Longleftrightarrow (iii)$: The dual of the equivalence $(i) \Longleftrightarrow (ii)$ in BFE, Theorem 6.6, (due to Whitehead) states that equation $f(x_1, \ldots, x_n) = 1$ has the unique solution $(\eta_1, \ldots, \eta_n)$ if and only if f is of the form

$$f(x_1, \ldots, x_n) = \prod_{j=1}^{n} (x_j \vee \eta_j')(x_j' \vee \eta_j) \ ;$$

but $(x \vee \eta')(x' \vee \eta) = x\eta \vee x'\eta' = x + \eta'$.   □

Remark 2.1. Since $\xi + \alpha = \xi^{\alpha'}$ for $\alpha \in \{0,1\}$, identity (2.9) is equivalent to

$$(2.10) \qquad f(A) = \Xi^{A'} \qquad (\forall A \in \{0,1\}^n) \ ,$$

where we have set $\Xi = (\xi_1, \ldots, \xi_n)$ and $A' = (\alpha_1', \ldots, \alpha_n')$.   □

Lemma 2.2. *The following conditions are equivalent for a system*

$$(2.11) \qquad \Xi_i = (\xi_{i1}, \ldots, \xi_{in}) \qquad (i = 1, \ldots, m)$$

of vectors from B^n :

   (i) the vectors

$$(2.12) \qquad (\Xi_1^A, \ldots, \Xi_m^A) \qquad (A \in \{0,1\}^n)$$

are orthogonal ;
   (ii) the vectors (2.12) are normal ;
   (iii) the vectors (2.12) are orthonormal .

PROOF: Consider the $2^n \times 2^m$ matrix whose rows are the vectors (2.12). This matrix is column orthonormal because each column i consists of the 2^n minterms in the variables Ξ_i. Therefore the desired equivalence follows from BFE, Theorem 7.5, which states the equivalence of several properties of a square Boolean matrix, among which are the following: row orthogonal and column normal; row normal and column orthogonal; orthonormal.   □

Theorem 2.1. α) *The following conditions are equivalent for a system* $(f_0, f_1, \ldots, f_{2^n-1})$ *of Boolean functions* $f_i : B^n \longrightarrow B$ $(i = 0, 1, \ldots, 2^n - 1)$:

(i) *the system is a linear generator ;*

(ii) *the system is a linear base ;*

(iii) *the* $2^n \times 2^n$ *matrix* $\|f_i(A)\|$ *is orthonormal ;*

(iv) *the functions* f_i *are of the form*

$$(2.13) \qquad f_i(x_1, \ldots, x_n) = \prod_{j=1}^{n}(x_j + \xi_{ij}) \qquad (i = 0, 1, \ldots 2^n - 1)$$

and the vectors

$$(2.14) \qquad (\Xi_0^A, \Xi_1^A, \ldots, \Xi_{2^n-1}^A) \qquad (A \in \{0,1\}^n)$$

are orthogonal, where we have set

$$(2.15) \qquad \Xi_i = (\xi_{i1}, \ldots, \xi_{in}) \qquad (i = 0, 1, \ldots, 2^n - 1) \ ;$$

(v) *the functions* f_i *are of the form* (2.13) *and the vectors* (2.14) *are normal;*

(vi) *the functions* f_i *are of the form* (2.13) *and the vectors* (2.14) *are orthonormal .*

β) *When this is the case, the coefficients* c_i *in formula* (2.6) *are given by*

$$(2.16) \qquad c_i = \prod_{A}(f_i'(A) \vee f(A)) \qquad (i = 0, 1, \ldots, 2^n - 1) \ .$$

COMMENT: The minterms in the usual sense can be written in the form (2.13): we choose a bijection $i \mapsto (\beta_1, \ldots, \beta_n)$ between the sets $\{0, 1, \ldots, 2^n - 1\}$ and $\{0,1\}^n$, then we set

$$f_i(x_1, \ldots, x_n) = x_1^{\beta_1} \ldots x_n^{\beta_n} = \prod_{j=1}^{n}(x_j + \beta_j') \ .$$

So in this case we have $\xi_{ij} = \beta_j'$ for all j, and for every $A \in \{0,1\}^n$, $f_i(A)$ is 1 or 0 according as $A = (\beta_1, \ldots, \beta_n)$ or $A \neq (\beta_1, \ldots, \beta_n)$. Thus $\|f_i(A)\|$ is a permutation matrix; if the above bijection associates with each $i \in \{0, 1, \ldots, 2^n - 1\}$ its binary representation, then $\|f_i(A)\|$ is the identity matrix. In the case of generalized minterms (2.10) the elements ξ_{ij} need not be 0 or 1.

PROOF: $(i) \Longleftrightarrow (ii) \Longleftrightarrow (iii)$ and β : Immediate from Lemma 2.1 and Theorem 1.2.

$(iv) \Longleftrightarrow (v) \Longleftrightarrow (vi)$: By Lemma 2.2.

$(iii) \Longleftrightarrow (iv)$: The functions f_i are of the form (2.13) by Proposition 2.3. The vectors (2.14) are in fact

$$(2.17) \qquad (f_0(A'), f_1(A'), \ldots, f_{2^n-1}(A')) \qquad (A \in \{0,1\}^n)$$

by Remark 2.1. But when A runs over $\{0,1\}^n$, so does A', therefore the set of all the vectors (2.17) coincides with the set of all the vectors

$$(2.18) \qquad (f_0(A), f_1(A), \ldots, f_{2^n-1}(A)) \qquad (A \in \{0,1\}^n) ,$$

which is the set of rows of the matrix $\|f_i(A)\|$.

$(vi) \Longleftrightarrow (iii)$: As shown in the previous proof, the orthonormal vectors (2.14) are in fact the rows of the matrix $\|f_i(A)\|$. On the other hand, a new application of Remark 2.1 shows that the columns of this matrix are

$$(f_i(0, \ldots, 0), f_i(0, \ldots, 0, 1), \ldots, f_i(1, \ldots, 1))$$

$$= (\Xi_i^{(1,\ldots,1)}, \Xi_i^{(1,\ldots,1,0)}, \ldots, \Xi_i^{(0,\ldots,0)}) \qquad (i = 0, 1, \ldots, 2^n - 1) ,$$

that is, they are systems of minterms, hence they are orthonormal.   □

Corollary 2.1. *The following conditions are equivalent for a system (g, h) of Boolean functions $g, h : B \longrightarrow B$:*

(i) the system is a linear generator :

(ii) the system is a linear base ;

(iii) the functions g and h are of the form

$$(2.19) \qquad g(x) = x + a, \; h(x) = x + a' .$$

PROOF: Theorem 2.1 yields $g(x) = x+a$ and $h(x) = x+b$, then $\Xi_0 = a$, $\Xi_1 = b$, hence the vectors (a', b') and (a, b) are orthonormal.   □

Proposition 2.4. *Systems of linear generators $(f_1, \ldots, f_p)$ exist if and only if $p \geq 2^n$.*

PROOF: The condition $p \geq 2^n$ is necessary by Proposition 2.1. We have just determined all linear generators with 2^n functions and there is a coarse way of obtaining a linear generator with p functions, $p \geq 2^n$: take a linear generator $(f_0, f_1, \ldots, f_{2^n-1})$ and add $f_{2^n} = f_{2^n+1} = \ldots = f_{p-1} = 0$.   □

In view of Proposition 2.4 one might say that the systems of generalized minterms or linear bases, described in Theorem 2.1, are the best generalization of the classical system of minterms.

Definition 2.2. A system $(\Xi_i)_{i=1,\ldots,m}$ of m vectors $\Xi_i = (\xi_{i1}, \ldots, \xi_{in})$ from B^n will be called an *interpolating system* of points for Boolean functions of n variables provided every function $g : \Xi \longrightarrow B$ can be uniquely extended to a Boolean function $f : B^n \longrightarrow B$.   □

Lemma 2.3. *The following conditions are equivalent for a system* $(\Xi)_{i=1,\dots,m}$ *of vectors from* B^n:

(i) $(\Xi_i)_{i=1,\dots,m}$ *is an interpolating system*;

(ii) *for each* $(b_1,\dots,b_m) \in B^m$, *the system of equations*

(2.20)
$$\bigvee_A f(A)\Xi_i^A = b_i \qquad (i=1,\dots,m)$$

has a unique solution with respect to the unknowns $(f(A))_{A\in\{0,1\}^n}$;

(iii) $m = 2^n$ *and the vectors* (2.14) *are orthogonal*;

(iv) $m = 2^n$ *and the vectors* (2.14) *are normal*;

(v) $m = 2^n$ *and the vectors* (2.14) *are orthonormal*.

PROOF: $(iii) \Longleftrightarrow (iv) \Longleftrightarrow (v)$: By Lemma 2.2.

$(i) \Longleftrightarrow (ii)$: Condition (i) amounts to saying that for every $(b_1,\dots,b_m) \in B^m$ there is a unique Boolean function $f : B^n \longrightarrow B$ satisfying

$$f(\Xi_i) = b_i \qquad (i=1,\dots,m);$$

but the latter system is equivalent to (2.20) and we have obtained condition (ii). Note that system (2.20) is of the form (1.9), the $m \times 2^n$ matrix of coefficients being $\|\Xi_i^A\|$, which is row orthonormal.

$(ii) \Longrightarrow (v)$: In view of Corollary 1.4 of Theorem 1.4, we have $m = 2^n$ and the matrix $\|\Xi_i^A\|$ is orthonormal; but the vectors (2.14) are the columns of this matrix.

$(v) \Longrightarrow (ii)$: Define the functions f_i by (2.13). Then Theorem 2.1 shows that the matrix $\|f_i(A)\|$ is orthonormal, therefore, according to Theorem 1.2, the system (2.20) has a unique solution for each $(b_1,\dots,b_m) \in B^m$. $\qquad\square$

Theorem 2.2. α) *The following correspondences establish a bijection between interpolating systems and systems of generalized minterms: with each interpolating system* $(\Xi_i)_{i=0,\dots,2^n-1}$ *one associates the system of generalized minterms*

(2.13)
$$f_i(x_1,\dots,x_n) = \prod_{j=1}^n (x_j + \xi_{ij}) \qquad (i=0,1,\dots,2^n-1)$$

and with each system (2.13) *of generalized minterms one associates the interpolating system*

(2.15)
$$\Xi_i = (\xi_{i1},\dots,\xi_{in}) \qquad (i=0,1,\dots,2^n-1).$$

β) *The elements* ξ_{ij} *occurring in* (2.13) *are determined by*

(2.21)
$$\xi_{ij} = \bigvee\{f_i(A') \mid A \in \{0,1\}^n \ \& \ \alpha_j = 1\}$$
$$(j=1,\dots,n;\ i=0,1,\dots,2^n-1).$$

COMMENT: Beside the bijection established in this theorem, there is a bijection $f_i \leftrightarrow \Xi_i$ between the elements of each system (2.13) of generalized minterms

and the associated interpolating system (2.15); the latter bijection generalizes
the bijection between the set of minterms and the set $\{0,1\}^n$.

PROOF: β) Taking into account Remark 2.1 we obtain

$$\xi_{ij} = \bigvee \{\xi_{i1}^{\alpha_1} \cdots \xi_{i,j-1}^{\alpha_{j-1}} \xi_{ij} \xi_{i,j+1}^{\alpha_{j+1}} \cdots \xi_{in}^{\alpha_n} \mid$$

$$\mid \alpha_1, \ldots, \alpha_{j-1}, \alpha_{j+1}, \ldots, \alpha_n \in \{0,1\}\} = \bigvee \{\Xi_i^A \mid A \in \{0,1\}^n \ \& \ \alpha_j = 1\}$$

$$= \bigvee \{f_i(A') \mid A \in \{0,1\}^n \ \& \ \alpha_j = 1\} \, .$$

α) In view of Lemma 2.2, every interpolating system has 2^n elements and the
vectors (2.14) are orthonormal. Therefore, if $(\Xi_i)_{i=0,\ldots,2^n-1}$ is an interpolating
system, the functions f_i defined by (2.13) form a system of generalized minterms
by Theorem 2.1.

The same theorem shows that any system of generalized minterms is of the
form (2.13), where the 2^n vectors (2.14) constructed via (2.15) are orthonormal. Therefore Lemma 2.2 shows that $(\Xi_i)_{i=0,\ldots,2^n-1}$ is an interpolating system,
which is uniquely determined in view of β).

Finally it is plain that the correspondences α) establish a bijection.   □

Corollary 2.2. Let $(\Xi_i)_{i=0,\ldots,2^n-1}$ be an interpolating system. For every Boolean
function f and every point $X \in B^n$,

$$(2.22) \qquad\qquad f(X) = \bigvee_{i=0}^{2^n-1} f(\Xi_i) \bigvee_A \Xi_i^A X^A \, .$$

COMMENT: In the case of the classical interpolating system $(\Xi_i)_{i=0,\ldots,2^n-1} =$
$\{0,1\}^n$, each Ξ_i is of the form $\Xi_i = C \in \{0,1\}^n$, therefore $\bigvee_A \Xi_i^A X^A = X^C$
and formula (2.22) reduces to $f(X) = \bigvee_C f(C)X^C = \bigvee_A f(A)X^A$.
PROOF: Since we are given the Boolean function f, the parameters b_i in system
(2.20) are in fact $b_i = f(\Xi_i)$. Taking into account Theorem 1.2, the unique
solution of system (2.20) is

$$f(A) = \prod_{i=0}^{2^n-1} ((\Xi_i^A)' \vee f(\Xi_i))$$

$$= (\bigvee_{i=0}^{2^n-1} \Xi_i^A f'(\Xi_i))' = \bigvee_A \Xi_i^A f(\Xi_i) \quad (A \in \{0,1\}^n) \, ,$$

therefore, as $f(X) = \bigvee_A f(A)X^A$, formula (2.20) follows easily.   □

Corollary 2.3. *Another bijection between interpolating systems and systems
of generalized minterms is obtained by associating each system of generalized
minterms*

$$(2.13) \qquad f_i(x_1,\ldots,x_n) = \prod_{j=1}^{n}(x_j + \xi_{ij}) \qquad (i = 0,1,\ldots,2^n - 1)$$

with the interpolating system

$$(2.15') \qquad \Xi_i' = (\xi_{i1}',\ldots,\xi_{in}') \qquad (i = 0,\ldots,2^n - 1) .$$

COMMENT: In the classical case of the set of minterms and $\{0,1\}^n$, the above bijection is the classical one, which associates the minterm $x_1^{\alpha_1}\ldots x_n^{\alpha_n}$ with $(\alpha_1,\ldots,\alpha_n)$; cf. Comment to Theorem 2.1.

PROOF: Since clearly $\Xi'^A = \Xi^{A'}$, the vectors (2.14) coincide with the vectors

$$(2.14') \qquad ((\Xi_0)'^A, (\Xi_1')^A, \ldots, (\Xi_{2^n-1})'^A) \qquad (A \in \{0,1\}^n) ,$$

therefore we obtain a permutation of the set (2.15) of interpolating systems by mapping Ξ to Ξ'. Now the desired bijection is obtained by composing the bijection in Theorem 2.1 with the above permutation. □

3 Prime implicants and syllogistic forms

The resolution principle formulated by Robinson [1965] enables deduction in predicate calculus to be mechanized by means of a single rule of inference. The principle is of course applicable to Boolean problems, inasmuch as propositional logic is a subset of predicate logic. The basic approach to resolution applied in predicate calculus, theorem-proving by refutation, has nevertheless found little application in switching theory and related fields. An approach to propositional resolution given by Blake [1937] seems to us, however, to be applicable in a direct way to such applications.

Blake's dissertation, published 27 years before Robinson's paper, demonstrated that all of the consequents of a disjunctive normal form may be generated by repeated production of the consensus (propositional resolution) of pairs of terms, and that all of the prime implicants of the original function will be included in the resulting formula. Blake's dissertation is remarkable not only for presenting the essential idea of resolution, but also for anticipating many of the techniques later discovered by Quine [1952], [1955], [1959] and others for generating prime implicants. The theory of prime implicants has thus arisen independently to serve two quite different ends, viz., propositional inference via resolution (Blake) and formula minimization (Quine). While Quine's approach has been the basis for extended research, Blake's formulation remains virtually unknown.

Blake noted that the problem of finding consequents $g = 0$ of the equation $f = 0$ is essentially that of finding functions g such that $g \le f$. As a matter of fact, this had been shown by Löwenheim [1910] (see also BFE, Theorem 2.14), but in Blake's formulation the problem is solved by expressing the function f in a form he called "syllogistic"; this form enables all included disjunctive normal expressions to be read off by inspection.

Among the syllogistic forms of a truth function f there is one which Blake called the "simplified canonical form" and which we shall call the *Blake canonical form* for f and denote it by $BCF(f)$. This form turns out to be the disjunction of all of the prime implicants of f.

We begin this section by constructing an axiomatic theory of prime implicants within the general framework of finite join semilattices; the concepts of syllogistic representation and Blake canonical form are defined naturally within this framework. We next specialize the axiomatic theory to simple Boolean functions to obtain at once the classical theory of prime implicants, yet presented in terms of functions rather than expressions. Another specialization of the axiomatic theory will be a generalization of the classical theory of prime implicants to arbitrary Boolean functions. (In most cases the Boolean functions referred to in the literature are in fact truth functions, also called switching functions, and coincide essentially with simple Boolean functions in our sense.)

In this section we follow closely Brown and Rudeanu [1986], except Lemma 3.1 and Proposition 3.14. For a thorough study of Boolean reasoning in Blake's line and numerous applications see Brown [1990].

The axiomatic approach we begin with is due to Davio, Deschamps and Thayse [1978]. For another approach to minimization problems and prime implicants see Rudeanu [1964].

Definition 3.1. Let $(L; \vee)$ be a join semilattice. By a *generating system* we mean a subset G of L such that every element of L can be written as a join of elements from G. An *implicant* of an element $a \in L$ is defined as an element $g \in G$ such that $g \leq a$. The maximal elements (if any) of the set of implicants of an element a will be called the *prime implicants* of a.   □.

Remark 3.1. Generating systems do exist, for instance L itself. If L has a least element 0, then 0 belongs to every generating system.   □

Remark 3.2. A well-known argument shows that if L is a finite join semilattice then for every implicant g of a there is a prime implicant p of a such that $g \leq p$.□

In the sequel we work with an arbitrary but fixed generating system G.

Definition 3.2. By a *representation* of an element $a \in L$ we mean a subset $\{a_1, \ldots, a_s\}$ of L such that

$$(3.1) \qquad\qquad a = a_1 \vee \ldots \vee a_s \; ;$$

if $a_1, \ldots, a_s \in G$ the representation is called a *G-representation*.   □

For the sake of simplicity, however, we shall refer to (3.1) itself as a representation of a. We nevertheless keep in mind the exact definition given above for a representation, so that we will identify the representation (3.1) with any representation $a = b_1 \vee \ldots \vee b_t$ that differs from (3.1) only in the order of the elements and/or the number of occurrences of some elements.

Proposition 3.1. (Davio, Deschamps and Thayse [1978], Theorem 1.15). *In a finite join semilattice every element equals the disjunction of all of its prime implicants.*

PROOF: Given $a \in L$, let $\{p_1, \ldots, p_m\}$ be the set of prime implicants of a. Then $p_i \leq a$ ($i = 1, \ldots, m$) and therefore $p_1 \vee \ldots \vee p_m \leq a$. To prove the converse inequality, let $a = g_1 \vee \ldots \vee g_n$ be another G-representation of a. In view of Remark 3.2, for every integer $h \in \{1, \ldots, n\}$, there is a prime implicant p_{i_h} of a such that $g_h \leq p_{i_h}$. Then

$$a = g_1 \vee \ldots \vee g_n \leq p_{i_1} \vee \ldots \vee p_{i_m} \leq p_1 \vee \ldots \vee p_m \leq a.$$

□

Definition 3.3. The representation

(3.2) $$a = p_1 \vee \ldots \vee p_m$$

of an element $a \in L$ as the disjunction of all of its prime implicants will be denoted by $\mathrm{BCF}(a)$ and called the *Blake canonical form of a*. □

Clearly $\mathrm{BCF}(a)$ is unique. An important property of the Blake canonical form is related to the following

Definition 3.4. Let (3.1) and

(3.3) $$b = b_1 \vee \ldots \vee b_t$$

be two G-representations. We say that (3.3) is *formally included* in (3.1), written (3.3) $\ll$ (3.1) or $\{b_1, \ldots, b_t\} \ll \{a_1, \ldots, a_s\}$ or

(3.4) $$b_1 \vee \ldots \vee b_t \ll a_1 \vee \ldots \vee a_s,$$

if for every b_i there is some a_h such that $b_i \leq a_h$. □

Note that

(3.5) $$b_1 \vee \ldots \vee b_t \ll a_1 \vee \ldots \vee a_s \Longrightarrow b_1 \vee \ldots \vee b_t \leq a_1 \vee \ldots \vee a_s.$$

Definition 3.5. The G-representation (3.1) is *syllogistic* provided

(3.6) $$b_1 \vee \ldots \vee b_t \leq a_1 \vee \ldots \vee a_s \Longrightarrow b_1 \vee \ldots \vee b_t \ll a_1 \vee \ldots \vee a_s$$

for every subset $\{b_1, \ldots, b_t\} \subseteq G$. □

Remark 3.3. The G-representation (3.1) is syllogistic if and only if

(3.7) $$g \leq a_1 \vee \ldots \vee a_s \Longrightarrow g \ll a_1 \vee \ldots \vee a_s$$

for every element $g \in G$. □

Proposition 3.2. *In a finite join semilattice, a G-representation is syllogistic if and only if it contains all of the prime implicants of the represented element.*

PROOF: Consider the representation (3.1) and let $\{p_1, \ldots, p_m\} \subseteq \{a_1, \ldots, a_s\}$. If $g \leq a_1 \vee \ldots \vee a_s$ then g is an implicant of a, hence Remark 3.2 implies that

$g \leq p_i$ for some i, consequently $g \ll a_j$ for some j, that is, $g \ll a_1 \vee \ldots \vee a_s$. Therefore the G-representation (3.1) is syllogistic by Remark 3.3. Conversely, suppose the G-representation (3.1) is syllogistic and take a prime implicant p_i of a. Then $p_i \ll a_1 \vee \ldots \vee a_s$ again by Remark 3.3, that is, $p_i \leq a_j$ for some j. But a_j is an implicant of a, hence $a_j \leq p_k$ for some k by Remark 3.2. So $p_i \leq a_j \leq p_k$, therefore $p_i = a_j = p_k$ by the maximality of p_i. $\square$

Corollary 3.1. *The Blake canonical form is syllogistic.* $\square$

Proposition 3.3. (Davio, Deschamps and Thayse [1978], Theorem 1.16). *Suppose $(L; \wedge, \vee)$ is a finite lattice and let $a_1, \ldots, a_k \in L$. Then every prime implicant of $a_1 \wedge \ldots \wedge a_k$ is a prime implicant of $p_{h_1} \wedge \ldots \wedge p_{h_k}$ for some prime implicants p_{h_j} of a_j $(j = 1, \ldots, k)$.*

PROOF: Let p be a prime implicant of $a_1 \wedge \ldots \wedge a_k$. Then for every $j \in \{1, \ldots, k\}$ we have $p \leq a_j$, therefore $p \leq p_{h_j}$ for some prime implicant p_{h_j} of a_j. So $p \leq p_{h_1} \wedge \ldots \wedge p_{h_k}$ and in view of Remark 3.2 there is a prime implicant q of $p_{h_1} \wedge \ldots \wedge p_{h_k}$ such that $p \leq q$. It follows that $q \leq a_1 \wedge \ldots \wedge a_k$ by the maximality of p. $\square$

Proposition 3.4. *Suppose $(L; \wedge, \vee)$ is a finite lattice and G is a sub-meet-semilattice of L. Then for any elements $a_1, \ldots, a_k \in L$, every prime implicant of $a_1 \wedge \ldots \wedge a_k$ is of the form $p_{h_1} \wedge \ldots \wedge p_{h_k}$ for some prime implicants p_{h_j} of a_j $(j = 1, \ldots, k)$.*

PROOF: Proposition 3.3 yields

$$p \leq p_{h_1} \wedge \ldots \wedge p_{h_k} \leq a_1 \wedge \ldots \wedge a_k,$$

where the hypothesis implies $p_{h_1} \wedge \ldots \wedge p_{h_k} \in G$. Therefore $p = p_{h_1} \wedge \ldots \wedge p_{h_k}$ by the maximality of p. $\square$

Definition 3.6. Suppose further that L is a distributive lattice. For every $k \geq 2$ and for every system

(3.8.j) $a_j = a_{j1} \vee \ldots \vee a_{jn(j)}$ $(j = 1, \ldots, k)$

of k representations,

$$(3.8.1) \times \ldots \times (3.8.k)$$

is the representation of $a_1 \wedge \ldots \wedge a_k$ obtained by multiplying out the k representations $(3.8.1), \ldots, (3.8.k)$. $\square$

In other words, the representation $(3.8.1) \times \ldots \times (3.8.k)$ is

(3.9) $$a_1 \wedge \ldots \wedge a_k = \bigvee_{\varphi} a_{1\varphi(1)} \wedge \ldots \wedge a_{k\varphi(k)},$$

where φ runs over the set of all functions

(3.10) $$\varphi : \{1, \ldots, k\} \longrightarrow \bigcup_{j=1}^{k} \{1, \ldots, n(j)\}$$

having the property

(3.11) $\varphi(j) \in \{1, \ldots, n(j)\}$ $(j = 1, \ldots, k)$.

Proposition 3.5. *Suppose $(L; , \wedge, \vee)$ is a finite distributive lattice and G is a sub-meet-semilattice of L. For every $k \geq 2$ and every $a_1, \ldots, a_k \in L$, if (3.8.1), \ldots, (3.8.k) are syllogistic G-representations of $a_1, \ldots, a_k$, respectively, then (3.8.1) $\times \ldots \times$ (3.8.k) is a syllogistic representation of $a_1 \wedge \ldots \wedge a_k$.*

PROOF: In view of Proposition 3.2, it suffices to prove that every prime implicant p of $a_1 \wedge \ldots \wedge a_k$ occurs in the representation (3.8.1) $\times \ldots \times$ (3.8.k). But p is of the form described in Proposition 3.4, where for each $j \in \{1, \ldots, k\}$, p_{h_j} occurs in the representation (3.8.j), again by Proposition 3.2. Therefore p does occur in the representation (3.8.1) $\times \ldots \times$ (3.8.k).   □

Corollary 3.2. *$\mathrm{BCF}(a_1) \times \ldots \times \mathrm{BCF}(a_k)$ is a syllogistic representation of $a_1 \wedge \ldots \wedge a_k$.*   □

In order to specialize the above axiomatic setting to the classical theory of prime implicants, let us recall first that we have introduced (see e.g. BFE, Chapter 1, §2) the distinction between *Boolean functions* and *simple Boolean functions* (see also the present Definition 3.2.8 and Remark 3.2.6), which is in fact the specialization to Boolean algebras of the distinction between algebraic functions and polynomials, respectively (see e.g. Ch.2, §§ 2,3). To be specific, the simple Boolean functions are those Boolean functions $f : B^n \longrightarrow B$ (cf. (2.1)-(2.5)) for which $f(A) \in \{0,1\}$ for every $A \in \{0,1\}^n$, or equivalently, they are characterized by the representation

(3.12) $f(x_1, \ldots, x_n) = \bigvee \{X^A \mid A \in \{0,1\}^n \ \& \ f(A) = 1\}$;

cf. BFE, Theorem 1.7 (see also Proposition 5.2.2 and Corollary 5.2.7).

The functions referred to in the literature as "Boolean functions" are either simple Boolean functions in our sense or, even more particularly, *truth functions*, also called *switching functions*, i.e., functions $f : \{0,1\}^n \longrightarrow \{0,1\}$; (cf. BFE, Theorem 1.11 and Corollary; see also the present Remark 5.2.2 and Proposition 5.2.3). Since the algebra of all simple Boolean functions of n variables is isomorphic to the algebra of all truth functions of n variables by the Corollary of Theorem 1.21 in BFE, it is immaterial whether the classical theory of prime implicants applies to simple Boolean functions or to truth functions.

Now we specialize the above axiomatic theory as follows. In the rôle of L we take the Boolean algebra of all truth functions of n variables, while G is the set of all terms in the sense of Definition 3.7. below.

Definition 3.7. By a *term* we mean a truth function that can be represented in the form

(3.13)
$$x_{i_1}^{\alpha_1} \ldots x_{i_r}^{\alpha_r}, \{i_1, \ldots, i_r\} \subseteq \{1, \ldots, n\},$$

$$\{\alpha_1, \ldots, \alpha_r\} \subseteq \{0,1\}, 1 \leq k \leq n .$$ □

We refer to the above specialization of the axiomatic setting as the *functional theory* of prime implicants of truth functions; it matches the classical theory of prime implicants, which works with formal expressions instead of the truth functions they generate. Let us explain this. A *formal term* is an expression of the form (3.13) in which the indices $i_1, \ldots, i_r$ are pairwise distinct. The *expressions* are of the following sorts: formal terms, formal disjunctions $t_1 \vee \ldots \vee t_m$ of formal terms, and the symbol 0 (zero). Two expressions are regarded as equivalent provided they generate the same function. The set of formal terms is quasi-ordered by the following relation: the formal term (3.13) is included in the formal term $x_{j_1}^{\beta_1} \wedge \ldots \wedge x_{j_s}^{\beta_s}$ provided every *literal* $x_{j_k}^{\beta_k}$ of the latter term occurs in (3.13). This quasi-order is extended in a natural way to the set of all expressions (do not confuse with Definition 3.4); the maximality of prime implicants is defined with respect to the present quasi-order. Proposition 3.6 below implies that the quasi-order between expressions amounts to the usual pointwise order $\leq$ between the functions generated by the expressions. We have thus established a dictionary between the classical theory of prime implicants and the functional theory obtained by the specialization of the axiomatic setting. Therefore these theories are essentially the same. It is convenient to work within the functional framework, borrowing certain phrases from the classical approach; for instance, "literal of a term t" would mean a literal of the formal term that generates t; etc.

Proposition 3.6. *Let t and v be two terms, represented by the formal terms $x_{i_1}^{\alpha_1} \ldots x_{i_r}^{\alpha_r}$ and $x_{j_1}^{\beta_1} \ldots x_{j_s}^{\beta_s}$, respectively. Then $t \leq v$ if and only if*

(3.14)
$$\{j_1, \ldots, j_s\} \subseteq \{i_1, \ldots, i_s\} \ \& \ (j_k = i_h \implies \beta_k = \alpha_h)$$
$$\forall k = 1, \ldots, s; \ j = 1, \ldots, r \ .$$

PROOF: Well known.   □

Like Proposition 3.6, the next two propositions do not come from the axiomatic theory, but are specific to truth functions. We need two more definitions.

Definition 3.8. Two terms s and t are said to have an *opposition in the variable* x_i if either (i) $s \leq x_i$ and $t \leq x_i'$, or (ii) $s \leq x_i'$ and $t \leq x_i$.   □

Remark 3.4. A) It follows from Proposition 3.6 that the situations (i) and (ii) in the above definition can be characterized as follows: there exist two terms p, q independent of x_i and such that either (i) $s = px_i$ and $t = qx_i'$, or (ii) $s = px_i'$ and $t = qx_i$. B) Two terms may have several oppositions, a single opposition or none.   □

Proposition 3.7. (Blake [1937], Corollary of Theorem 10.3). *Suppose the terms s and t have exactly one opposition, say $s = s_1 x$ and $t = t_1 x'$, where the terms s_1 and t_1 are independent of x and have no opposition. Then*

(3.15)
$$s \vee t = s \vee t \vee s_1 t_1$$

is a syllogistic representation of $s \vee t$.

PROOF: Take $a := b := 1$ in Proposition 3.12 below.   □

Proposition 3.8. *Let the simple Boolean function f be expressed by*

(3.16) $$f = t_1 \vee \ldots \vee t_m \,,$$

where $t_1, \ldots, t_m$ are terms such that there is no opposition between any two of them. Then (3.16) is syllogistic.

PROOF: Let $p = x_{i_1}^{\alpha_1} \ldots x_{i_k}^{\alpha_k}$ be a prime implicant of f, where the indices $i_1, \ldots, i_k$ are pairwise distinct. Suppose first that every term t_j contains a literal $x_{h(j)}^{\beta_{h(j)}}$ not in p. Define $\gamma_{i_1} = \alpha_1, \ldots, \gamma_{i_k} = \alpha_k$, then for $j \in \{1, \ldots, m\}$ and $h(j) \notin \{i_1, \ldots, i_m\}$ define $\gamma_{h(j)} = \beta'_{h(j)}$; this is possible because any two distinct terms t_h and t_l have no opposition, therefore if $h(j) = h(l)$ then $\beta_{h(j)} = \beta_{h(l)}$. Finally take $\gamma_i \in \{0, 1\}$ arbitrarily for the other indices i. Then $p(\gamma_1, \ldots, \gamma_n) = 1$ and for every $j \in \{1, \ldots, m\}$ we have two possibilities: if $h(j) \in \{i_1, \ldots, i_m\}$, say $h(j) = i_l$, then $\beta_{h(j)} = \alpha'_l$, hence $\gamma_{h(j)}^{\beta_{h(j)}} = \gamma_{i_l}^{\alpha'_l} = \alpha_l^{\alpha'_l} = 0$, which implies $t_j(\gamma_1, \ldots, \gamma_n) = 0$, else $h(j) \notin \{i_1, \ldots, i_m\}$, in which case $\gamma_{h(j)}^{\beta_{h(j)}} = (\beta'_{h(j)})^{\beta_{h(j)}} = 0$, implying again $t(\gamma_1, \ldots, \gamma_n) = 0$. Therefore $f(\gamma_1, \ldots, \gamma_n) = 0$, in contradiciton with the fact that p is an implicant of f.

We have thus proved the existence of a term t_{j_0} such that all of its literals are in p, and therefore $p \leq t_{j_0}$ by Proposition 3.7. Since $t_{j_0} \leq f$, the maximality of p implies $p = t_{j_0}$. $\qquad\square$

The reader is referred to the monograph by Brown [1990] for a thorough study of Blake's theory, in both directions mentioned at the beginning of this section and with numerous applications.

We are now going to construct the announced theory of prime implicants for Boolean functions (not necessarily simple). The Boolean algebra B will henceforth be supposed to be finite. This hypothesis is not overly restrictive, because in practice we work in most cases with a finite number of Boolean functions, each of which is expressed using a finite number of constants. We can therefore replace the original Boolean algebra by the Boolean subalgebra generated by all these (finitely many) constants, and the latter subalgebra is finite.

Now we specialize the axiomatic theory as follows. In the rôle of L we take the Boolean algebra of all Boolean functions $f : B^n \longrightarrow B$, while G is the set of all genterms in the sense of Definition 3.9 below (do not confuse with the generalized minterms studied in §2).

Definition 3.9. By a *generalized term* or *genterm* we we mean any function of the form at, where $a \in B$ and $t : B^n \longrightarrow B$ is a function that can be represented in the form (3.13). By a slight extension of Definition 3.7, the function t will be called the *term* associated with the genterm. $\qquad\square$

In the remainder of this section we prove the uniqueness of the representation at for a non-null genterm (Proposition 3.9), we give a necessary condition for a genterm to be a prime implicant (Proposition 3.10) and prove that our concept of prime implicant reduces to the customary one in the case of a simple Boolean function (Proposition 3.11). We then prove directly Proposition 3.12, which generalizes Proposition 3.7, and show that such a generalization is not possible for

Proposition 3.8 (Remark 3.5). The final Proposition 3.14 will be used in the next section.

Proposition 3.9. *For all genterms* as, bt:

$$(3.17) \qquad\qquad as \leq bt \Longleftrightarrow (a \leq b \text{ and } s \leq t) \text{ or } a = 0 ;$$

$$(3.18) \qquad\qquad as = bt \Longleftrightarrow (a = b \text{ and } s = t) \text{ or } a = b = 0 .$$

PROOF: Suppose $as \leq bt$ and $a \neq 0$. Then $as \leq b$ and taking $X \in B^n$ such that $s(X) = 1$, it follows that $a \leq b$. Now suppose, by way of contradiction, that $s \not\leq t$. Then $s(Y) = 1$ and $t(Y) = 0$ for some $Y \in B^n$; but this contradicts $as \leq bt$. Thus $a \leq b$ and $s \leq t$. The converse implication from (3.17) is trivial. Finally (3.18) follows immediately from (3.17). □

Proposition 3.10. *If* cp *is a prime implicant of a Boolean function* f, *where* $c \in B$ *and* p *is a term, then*

$$(3.19) \qquad\qquad c = \prod_A (f(A) \vee p'(A)) \in f(B^n).$$

PROOF: According to a theorem of Whitehead (see e.g. BFE, Theorem 2.5),

$$\{f(X) \mid p(X) = 1\} = [a, b], \ a = \prod_A (f(A) \vee p'(A)), \ b = \bigvee_A f(A)p(A) .$$

Thus $[a, b] \subseteq f(B^n)$ and we will prove that $c = a$.

First we show that $ap \leq f$. This is trivial if $f(X) = 1$ identically; otherwise a theorem due to Löwenheim (see e.g. BFE, Theorem 2.14) reduces the proof to establishing that $ap(A) \leq f(A)$ for all $A \in \{0,1\}^n$. But $p(A) \in \{0,1\}$; if $p(A) = 0$ the inequality is trivial, otherwise $ap(A) = a \leq f(A) \vee p'(A) = f(A)$.

The Whitehead theorem mentioned above implies the existence of an element $X \in B^n$ such that $p(X) = 1$ and $f(X) = a$. Therefore $c = cp(X) \leq f(X) = a$. Thus $c \leq a$ and in fact $c = a$, otherwise $cp(X) = c < a = ap(X)$ and since $cp \leq ap$ we would have $cp < ap$, in contradiction with the maximality of the implicant cp. □

Proposition 3.11. *Let* f *be a simple non-zero Boolean function. Then the prime implicants of* f *obtained within the genterms are the same as the prime implicants of* f *in the customary sense.*

PROOF: Let p be a prime implicant in the customary sense. In view of Remark 3.2, $p \leq cq$ for some genterm cq that is a prime implicant of f. Then Proposition 3.9 implies that $c = 1$ and $p \leq q$, therefore $q(= cq)$ is a prime implicant, hence $p = q$ is a prime implicant in the sense of genterms. Conversely, suppose the genterm cp is a prime implicant. Then c fulfils (3.19) and since $f(A) \in \{0,1\}$ for all A because f is a simple Boolean function, it follows that $c \in \{0,1\}$, hence $c = 1$, therefore $cp = p$ is a prime implicant in the customary sense. □

Proposition 3.12. *Suppose the genterms s and t have exactly one opposition, say $s = as_1x$ and $t = bt_1x'$, where $a, b \in B$ and s_1, t_1 are terms independent of the variable x and have no opposition. Then*

(3.20) $$s \vee t = s \vee t \vee abs_1t_1$$

is a syllogistic representation of $s \vee t$.

PROOF: The identity (3.20) holds because

(3.21) $$abs_1t_1 = abs_1t_1(x \vee x') \leq as_1x \vee bt_1x' = s \vee t.$$

Further, let cp be a prime implicant of $s \vee t$; we must show that $cp \in \{s, t, abs_1t_1\}$.

If $p = p_1x$, where p_1 is a term independent of x, then $cp = cp_1x$ and from

(3.22) $$cp \leq s \vee t = as_1x \vee bt_1x',$$

we obtain by multiplication with x that $cp \leq as_1x = s \leq s \vee t$, therefore $cp = s$ by the maximality of cp.

If $p = p_1x'$ one proves similarly that $cp = t$.

If p is independent of x, then taking in turn $x := 1$ and $x := 0$ in (3.22), we obtain $cp \leq as_1$ and $cp \leq bt_1$, hence $cp \leq abs_1t_1$. Since abs_1t_1 is an implicant of $s \vee t$ by (3.21), the maximality of cp implies $cp = abs_1t_1$. □

Remark 3.5. While Proposition 3.12 generalizes Proposition 3.7, Proposition 3.8 cannot be extended to genterms. For example, take two elements $a, b \in B \backslash \{0, 1\}$ such that $a \vee b = 1$. Then xy is a prime implicant of the function $f(x, y) = ax \vee by$, therefore this representation is not syllogistic. □

The next proposition was proved by Brown and Rudeanu [1983] in the case of simple Boolean functions.

Proposition 3.13. *Consider the Blake canonical form (with respect to genterms) of a non-zero Boolean function $f : B^n \longrightarrow B$. Suppose x_n and x'_n to be factored from the genterms in which they appear, yielding the modified formula*

(3.23) $$f(x_1, \ldots, x_n) = r(x_1, \ldots, x_{n-1})x_n \vee s(x_1, \ldots, x_{n-1})x'_n \vee$$
$$\vee t(x_1, \ldots, x_{n-1}).$$

Then $rs \leq t$ and t is expressed in the Blake canonical form.

PROOF: Note first that $rs \leq f$, because $rsf' = rs(r'x_n \vee s'x'_n)t' = 0$.

Now let ap be a prime implicant of rs. Then $ap \leq f$, hence there is a prime implicant bq of f such that $ap \leq bq$. But bq occurs in (3.23). It cannot occur in rx_n, because this would imply $ap \leq rx_n$ and since p does not involve the variable x_n, taking $x_n := 0$ would yield $ap = 0$, which is impossible. Similarly bq cannot occur in sx'_n, therefore bq is a term of t. Thus $bq \leq t$ for every prime implicant of rs, hence $rs \leq t$.

Further, every genterm of t is a prime implicant of f, hence a prime implicant of t. Conversely, let ap be a prime implicant of t; we have to prove that ap occurs in t. But $ap \leq f$, hence $ap \leq bq$ for some prime implicant bq of f. It follows by Proposition 3.9 that $p \leq q$ and since p does not involve the variable x_n, nor does q. Therefore the prime implicant bq occurs in t, hence $ap \le bq \le t$, which implies $ap = bq$ by the maximality of ap.   □

We are now going to show that in a very particular case the determination of the prime implicants in the general sense of genterms reduces to the classical case of simple Boolean functions.

Lemma 3.1. *Let B be a free Boolean algebra and Γ its set of free generators, $card(\Gamma) = m$. Then the Boolean algebra of all simple Boolean functions $g : B^{n+m} \longrightarrow B$ is isomorphic to the Boolean algebra of all Boolean functions $f : B^n \longrightarrow B$, via the mapping $\varphi(g) = f$, $f(X) = g(X, \Gamma)$.*

PROOF: Recall first the well-known property that B is isomorphic to the Boolean algebra of all truth functions of m variables, which amounts to the fact that every $x \in B$ has a unique representation of the form $x = h(y)$, where $h : B^m \longrightarrow B$ is a simple Boolean function.

Now take a Boolean function $f : B^n \longrightarrow B$ and a simple Boolean function $g : B^{n+m} \longrightarrow B$, that is,

$$(3.24) \qquad g(X, Y) = \bigvee_A \bigvee_C g_{AC} X^A Y^C,$$

where A runs over $\{0,1\}^n$, C runs over $\{0,1\}^m$ and $g_{AC} \in \{0,1\}$. Then the condition $f = \varphi(g)$ amounts to $f(A) = g(A, \Gamma)$ for all A, that is,

$$(3.25) \qquad f(A) = \bigvee_C g_{AC} \Gamma^C \qquad (\forall A \in \{0,1\}^n).$$

In view of the property mentioned above, for every Boolean function f there is a unique system of elements $g_{AC} \in \{0,1\}$ such that relations (3.25) hold, i.e., $f = \varphi(g)$.

Thus φ is a bijection and it is also a homomorphism, in view of the behaviour of orthonormal systems (see e.g. BFE, Theorem 1.5, or the present Lemma 5.1.2).   □

Proposition 3.14. *Suppose B is a free Boolean algebra and Γ its set of free generators, $card(\Gamma) = m$, and let $f : B^n \longrightarrow B$ be a Boolean function, $0 \ne f \ne 1$. Denote by g the simple Boolean function associated with f in Lemma 3.1. Further, let*

$$(3.26) \qquad BCF(g) = \bigvee_{k \in A} p_{0k}(X) \vee \bigvee_{k \in C} q_{0k}(Y) \vee \bigvee_{i \in D} \bigvee_{k \in D(i)} p_i(X) q_{ik}(Y) ,$$

where A, C, D and $D(i), i \in D$, are finite sets and the terms corresponding to each of these sets are pairwise distinct; one or two of the sets A, C, D may be empty. Then

$$(3.27) \qquad BCF(f) = \bigvee_{k \in C} q_{0k}(\Gamma) \vee \bigvee_{k \in A} p_{0k}(X) \vee \bigvee_{i \in D} (\bigvee_{k \in D(i)} q_{ik}(\Gamma)) p_i(X) .$$

PROOF: We will use the isomorphism φ established in Lemma 3.1. First we note that the right side of (3.27) equals f. Then we prove that every prime implicant $\pi = ap(X)$, where a and $p(X)$ cannot be both equal to 1, is one of the genterms of (3.27), which are

$$\bigvee_{k \in D} q_{0k}(\Gamma); \ p_{0k}(X), \ k \in A; \ (\bigvee_k q_{ik}(\Gamma))p_i(X), i \in D .$$

We know that $a = h(\Gamma)$, where $h : B^m \longrightarrow B$ is a simple Boolean function. Suppose first $a \neq 1$, BCF(h) $= \bigvee_{j=1}^s r_j(Y)$. Then by applying φ^{-1} to the inequality

$$p(X)h(\Gamma) = \pi \leq f(X) ,$$

we deduce

$$p(X) \bigvee_{j=1}^s r_j(Y) \leq g(X,Y)$$

and since the representation (3.26) is syllogistic, we obtain s inequalities of the form

(3.28) $\qquad p(X)r_j(Y) \leq p_{i(j)}(X)q_{i(j)k(j)}(Y) \qquad (j=1,\ldots,s) ,$

hence by applying φ and taking into account Proposition 3.9 it follows that

$$p(X) \leq p_{i(j)}(X) \qquad (j=1,\ldots,s) ,$$

$$r_j(\Gamma) \leq q_{i(j)k(j)}(\Gamma) \qquad (j=1,\ldots,s) .$$

Since the given terms are pairwise disinct, it follows that the terms $p_{i(1)},\ldots,p_{i(s)}$ coincide, so that the previous inequalities reduce to $p(X) \leq p_i(X)$ for some $i \in D$ and

$$r_j(\Gamma) \leq q_{ik(j)}(\Gamma) \qquad (j=1,\ldots,s) ,$$

therefore

$$\pi \leq p_i(X) \bigvee_{j=1}^s q_{ik(j)}(\Gamma) \leq p_i(X) \bigvee_{k \in D(i)} q_{ik}(\Gamma) ,$$

which in fact is an equality because π is a prime implicant.

If $a = 1$ then from $p(X) = \pi \leq f(X)$ we get

$$p(X) = \varphi^{-1}(\pi) \leq g(X,Y) ,$$

hence

$$p(X) \leq \bigvee_{k \in A} p_{0k}(X) ,$$

therefore $p(X) \leq p_{0k}(X)$ for some $k \in A$, hence $\pi = p(X) = p_{0k}(X)$ by the maximality of π.

If $p(X) = 1$ then $a \neq 1$ is constructed as above and instead of (3.28) we obtain

$$r_j(Y) \le q_{0k(j)}(Y) \qquad (j = 1, \ldots, s),$$

where $k(1), \ldots, k(s) \in C$, hence

$$\pi = \bigvee_{j=1}^{s} r_j(\Gamma) \le \bigvee_{j=1}^{s} q_{0k(j)}(\Gamma) \le \bigvee_{k \in C} q_{0k}(\Gamma),$$

which is again an equality.

Conversely, let us prove that each genterm of (3.27) is a prime implicant of f. But every genterm is included in a prime implicant of f and since the latter is itself a genterm of (3.27), as we have just proved, it remains to show that there is no proper inclusion between the genterms of (3.27).

Proposition 3.9 immediately implies that $\bigvee_{j \in C} q_{0j}(\Gamma)$ and $p_{0k}(X)$ are incomparable for every $k \in A$, and that for every $i \in D$ we have

$$\bigvee_{j \in C} q_{0j}(\Gamma) \nleq (\bigvee_{k} q_{ik}(\Gamma)) p_i(X).$$

Then for every $k \in A$ and every $i \in D$,

$$p_{0k}(X) \ngeq (\bigvee_{k} q_{ik}(\Gamma)) p_i(X),$$

otherwise by applying φ^{-1} we would obtain

$$p_{0k}(X) \ge (\bigvee_{k} q_{ik}(Y)) p_i(X),$$

hence all $p_i(X) q_{ik}(Y) = p_{0k}(X)$, a contradiction.

Further we use the fact that Γ is a system of free generators. Thus for every $i \in D$ we have $\bigvee_{k} q_{ik}(\Gamma) \ne 1$, which implies that for every $j \in A$,

$$p_{0j}(X) \nleq (\bigvee_{k} q_{ik}(\Gamma)) p_i(X).$$

Similarly, for every $i \in D$,

$$\bigvee_{j \in C} q_{0j}(\Gamma) \ngeq (\bigvee_{k} q_{ik}(\Gamma)) p_i(X),$$

otherwise

$$\bigvee_{j \in C} q_{0j}(\Gamma) \ge \bigvee_{k} q_{ik}(\Gamma),$$

which is possible only if $C = D(i)$ and $\{q_{0j} \mid j \in C\} = \{q_{ik} \mid k \in D(i)\}$, which would imply $p_i(X) q_{ik}(Y) \le q_{0j}(Y)$, hence $p_i(X) q_{ik}(Y) = q_{0j}(Y)$, again a contradiction.

Finally if $i, j \in D$ and

$$(\bigvee_k q_{ik}(\Gamma))p_i(X) \le (\bigvee_k q_{jk}(\Gamma))p_j(X)\,,$$

then $p_i(X) \le p_j(X)$ and $\bigvee_k q_{ik}(\Gamma) \le \bigvee_k q_{jk}(\Gamma)$, hence $D(i) = D(j)$ and $\{q_{ik} \mid k \in D(i)\} = \{q_{jk} \mid k \in D(j)\}$; without loss of generality we may assume $q_{ik} = q_{jk}$ for all k. If $i \ne j$ then $p_i(X) < p_j(X)$, hence

$$p_i(X)q_{ik}(Y) < p_j(X)q_{ik}(Y) = p_j(X)q_{jk}(Y)\,,$$

a contradiction. Therefore $i = j$.   □

Example 3.1. Let B be the free Boolean algebra generated by $\Gamma = \{a_1, a_2\}$ and define $f : B^3 \longrightarrow B$ by

(3.29)
$$f = a_2 x_1 \vee a_1 x_3 \vee a_2 x_2' x_3 \vee a_2' x_1' x_2 \vee (a_1' \vee a_2')x_2' x_3' \vee$$
$$\vee\, a_1' a_2' x_1 x_2' \vee a_1' x_1' x_2 \vee a_1 a_2' x_2 x_3\,.$$

Notice that f has 8 genterms, while the associated function $g = \varphi^{-1}(f)$ has 9 terms. In view of Proposition 3.14, we regard f as a function g of 5 variables and determine BCF(g) by standard methods. After suitable factoring within the 14 terms of BCF(g) we find

(3.30)
$$\mathrm{BCF}(f) = a_2 x_1 \vee a_1 x_3 \vee a_2 x_2' x_3 \vee a_2' x_1' x_2 \vee (a_1' \vee a_2')x_2' x_3' \vee$$
$$\vee\, a_1' x_1 x_2' \vee a_1' x_1' x_2 \vee a_1 a_2' x_2 x_3 \vee a_1' a_2 \vee a_1 x_1 x_2 x_3 \vee (a_1' \vee$$
$$a_2')x_1' x_3' \vee x_1 x_2' x_3'\,.$$

                                                                           □

4 Reproductive solutions, recurrent inequalities and recurrent covers

A theorem due to Vaught [1954] states that a Horn sentence is valid in every Boolean algebra if and only if it holds in the two-element Boolean algebra $\{0, 1\}$; cf. Chang and Keisler [1973], Exercise 6.3.14. This generalizes the Müller-Löwenheim Verification Theorem, which refers only to identities and implications (see e.g. BFE, Theorems 2.13, 2.14) and on the other hand Banković [1997c] extended the Vaught theorem to Post algebras; cf. Theorem 5.3.6. We have mentioned in Ch.5 that the Vaught theorem has been used as an efficient tool for proving theorems on Boolean functions and equations. In fact,this line of research was initiated by Mijajlović [1977], [1980], who in the latter paper also gave an elementary proof of the Vaught theorem. Mijajlović was followed by Banković, who re-proved certain known results [1987a], [1993c] and established many new ones, namely various formulas defining the general solution or the reproductive solution of a Boolean equation [1984], [1987b], [1988], [1989b], the most general form of a general solution [1989a] and the most general form of a reproductive solution [1992c]. Then Banković [2000] generalized the last result to Post equations; cf. Proposition 5.3.11.

This section comprises three parts. In the first one we present two related results of Banković [1993a], [1995b]: a particular solution determined for any consistent Boolean equation and one more form of the reproductive solution. In the second part we prove that there exists a unique triangular reproductive solution satisfying a certain natural condition; cf. Brown and Rudeanu [1985]. The third part, following Brown and Rudeanu [1983], introduces the concept of recurrent cover, which provides all interval-based solutions of a Boolean equation.

The concept of reproductive solution is recalled in Definition 1.1.3; see also the comments before Corollary 3.8 of Theorem 3.3.3. Löwenheim's well-known reproductive solution of a Boolean equation requires the knowledge of a particular solution; see e.g. BFE, Theorem 2.11 and the discussion after that theorem, as well as Examples 2.33-2.36 for various possibilities of obtaining a particular solution (cf. Theorem 5.3.4, 5.(3.38) and Example 5.3.9). The next proposition provides a particular solution once and for all, as well as a reproductive solution.

Proposition 4.1. (Banković [1993a], [1995b]). *Suppose the Boolean equation in* n *unknowns*

$$(4.1) \qquad\qquad f(X) = 0$$

is consistent. Set $p = 2^n - 1$. *Then equation* (4.1) *has the particular solution*

$$(4.2) \quad \begin{aligned} \Xi &= f'(A)A \vee f(A)f'(A_1)A_1 \vee f(A)f(A_1)f'(A_2)A_2 \vee \ldots \\ &\quad \vee f(A)f(A_1)\ldots f(A_{p-1})f'(A_p)A_p \,, \end{aligned}$$

where $(A, A_1, \ldots, A_p)$ *is any permutation of* $\{0,1\}^n$, *and the reproductive solution*

$$(4.3) \quad \begin{aligned} \Phi(T) &= \bigvee_{i=0}^{p}(f'(A_i)A_i \quad \vee \quad f(A_i)f'(A_{i_1})A_{i_1} \quad \vee \\ &\quad f(A_i)f(A_{i_1})f'(A_{i_2})A_{i_2} \vee \ldots \\ &\quad \vee f(A_i)f(A_{i_1})\ldots f(A_{i_{p-1}})f'(A_{i_p})A_{i_p})T^{A_i} \,, \end{aligned}$$

where $(A_i, A_{i_1}, A_{i_2}, \ldots, A_{i_p})$ *is a permutation of* $\{0,1\}^n$ *for every* $i \in \{0,1,\ldots,p\}$.

PROOF: The orthogonal set

$$\{f'(A), f(A)f'(A_1), f(A)f(A_1)f'(A_2), \ldots, f(A)\ldots f(A_{p-1})f'(A_p)\}$$

is also orthonormal, because the consistency condition $\prod_A f(A) = 0$ of equation (4.1) and the identity $x' \vee xy = x' \vee y$ imply

$$f'(A) \vee f(A)f'(A_1) \vee f(A)f(A_1)f'(A_2) \vee \ldots \vee f(A)\ldots f(A_{p-1})f'(A_p)$$

$$= f'(A) \vee f'(A_1) \vee f'(A_2) \vee \ldots \vee f'(A_p) = 1 \,.$$

Taking into account the behaviour of Boolean functions with respect to orthonormal systems (see e.g. BFE, Theorem 4.6(iv); cf. the present Lemma 5.2.2), we obtain

$$f(\Xi) = f'(A)f(A) \vee f(A)f'(A_1)f(A_1) \vee \ldots \vee f(A)\ldots f(A_{p-1})f'(A_p)f(A_p) = 0 \,.$$

Now we must prove that $f(\Phi(T)) = 0$ for every $T \in B^n$, or equivalently, $f(\Phi(A)) = 0$ for every $A \in \{0,1\}^n$, say for A_i. But $\Phi(A_i)$ is the particular solution Ξ given by (4.2) for the permutation $(A_i, A_{i_1}, \ldots, A_{i_p})$.

Finally we have to prove the implication

(4.4) $$f(X) = 0 \Longrightarrow X = \Phi(X).$$

But setting $\Phi = (\varphi_1, \ldots, \varphi_n)$, the conclusion of (4.4) can be written in the form $\bigvee_{j=1}^{n}(x_j + \varphi_j(X)) = 0$, therefore the Verification Theorem (see e.g. BFE, Theorem 2.14) enables us to write (4.4) in the form

(4.5) $$\bigvee_{j=1}^{n}(\alpha_j + \varphi_j(\alpha)) \le f(A) \qquad (\forall A \in \{0,\}^n),$$

where α_j is the j-th component of A. Using again the notation A_i for an arbitrary vector of $\{0,1\}^n$ and $\alpha_j, \alpha_{j_1}, \ldots, \alpha_{j_p}$ for the j-th component of the vector $A, A_{i_1}, \ldots, A_{i_p}$, respectively, condition (4.5) can also be written in the form

(4.6) $$\alpha_j + \varphi_j(A_i) \le f(A_i) \ (j = 1, \ldots, n) \qquad (i = 0, \ldots, p).$$

But

$$\varphi_j(A_i) = f'(A_i)\alpha_j \vee f(A_i)f'(A_{i_1})\varphi_{j_1} \vee \ldots \vee f(A_i)\ldots f(A_{i_{p-1}})f'(A_{i_p})\alpha_{j_p}$$

is of the form

$$\varphi_j(A_i) = f'(A_i)\alpha_j \vee f(A_i)z,$$

hence if $\alpha_j = 0$ then

$$\alpha_j + \varphi_j(A_i) = \varphi_j(A_i) = f(A_i)z \le f(A_i),$$

while if $\alpha_j = 1$ then

$$\alpha_j + \varphi_j(A_i) = \varphi'_j(A_i) = f(A_i)(f'(A_i) \vee z') \le f(A_i).$$

$\square$

The following rather surprising result is somewhat in the same line of research.

Proposition 4.2. (Kečkić and Prešić [1984]). *If the Boolean equation in one unknown*

(4.7) $$f(x) := ax \vee bx' = 0$$

is consistent (i.e., $ab = 0$), then its unique reproductive solution is

(4.8) $$x = a't \vee bt',$$

or equivalently (using the ring operation $x + y = xy' \vee x'y$; see e.g. BFE, Chapter 1, §3),

(4.8') $$x = t + f(t).$$

PROOF: The reproductive solution (4.8) is well known (see e.g. BFE, Lemma 2.2) and the identity $t + f(t) = a't \vee bt'$ is easy to prove.

Conversely, let $x = \varphi(t)$ be a reproductive solution of equation (4.7). Then clearly $f(x) = 0 \Longleftrightarrow x = \varphi(x)$ and since $x = \varphi(x) \Longleftrightarrow x + \varphi(x) = 0$, the

Verification Theorem implies $f(x) = x + \varphi(x) = 0$, or equivalently $\varphi(x) = x + f(x)$.   □

We are now going to generalize this result, following Brown and Rudeanu [1985], to the effect that every consistent Boolean equation (not merely in one unknown) has a unique reproductive solution satisfying certain natural supplementary conditions.

Recall that the method of succesive elimination of variables (cf. BFE, Chapter 1, §4 and the present Ch.5, §3) provides a sequence of Boolean equations of the form

(4.9.1) $f_1(x_1, \ldots, x_n) := f(x_1, \ldots, x_n) = 0$,

(4.9.2) $f_2(x_2, \ldots, x_n) = 0$,

$$\cdots$$

(4.9.n-1) $f_{n-1}(x_{n-1}, x_n) = 0$,

(4.9.n) $f_n(x_n) = 0$,

(4.9.n+1) $f_{n+1} = 0$.

If condition (4.9.n+1) is fulfilled, we obtain a reproductive solution of the form

(4.10.1) $x_1 = \varphi_1(t_1, \ldots, t_n)$,

(4.10.2) $x_2 = \varphi_2(t_2, \ldots, t_n)$,

$$\cdots$$

(4.10.n-1) $x_{n-1} = \varphi_{n-1}(t_{n-1}, t_n)$,

(4.10.n) $x_n = \varphi_n(t_n)$,

where (4.10.n) is a reproductive solution of equation $(4.9'.n) = (4.9.n)$ in the unknown x_n, then (4.10.n-1) is a reproductive solution of the equation

(4.9'.n-1) $f_{n-1}(x_{n-1}, \varphi_n(t_n)) = 0$

in the unknown $x_{n-1}, \ldots$, (4.10.2) is a reproductive solution of the equation

(4.9'.2) $f_2(x_2, \varphi_3(t_3, \ldots, t_n), \ldots, \varphi_n(t_n)) = 0$

in the unknown x_2 and (4.10.1) is a reproductive solution of the equation

(4.9'.1) $f_1(x_1, \varphi_2(t_2, \ldots, t_n), \ldots, \varphi_n(t_n)) = 0$

in the unknown x_1.

We will study in some detail the functions $f_1, \ldots, f_n, f_{n+1}$ occurring in the method of succesive elimination of variables, so that the following definition is convenient.

Definition 4.1. The *eliminants*

(4.11) $$f_k : B^{n-k+1} \longrightarrow B \ (k = 1, \ldots, n), \ f_{n+1} \in B,$$

of a Boolean function $f : B^n \longrightarrow B$ are constructed recursively:

(4.12.1) $$f_1 = f \,,$$

(4.12.k) $$f_k(x_k, \ldots, x_n) = f_{k-1}(1, x_k, \ldots, x_n) f_{k-1}(0, x_k, \ldots, x_n)$$
$$(k = 2, \ldots, n) \,,$$

(4.12.n+1) $$f_{n+1} = f_n(1) f_n(0) \,.$$

□

Remark 4.1. It follows easily by induction that

(4.13.k) $$f_k(x_k, \ldots, x_n)$$
$$= \prod_{\alpha_1, \ldots, \alpha_{k-1} \in \{0,1\}} f(\alpha_1, \ldots, \alpha_{k-1}, x_k, \ldots, x_n)$$
$$(k = 2, \ldots, n) \,,$$

(4.13.n+1) $$f_{n+1} = \prod_A f(A) \,.$$

□

Definition 4.2. Every reproductive solution of the form (4.10) is said to be a *triangular reproductive solution* of the consistent Boolean equation $f(x_1, \ldots, x_n) = 0$. The triangular reproductive solution obtained by the method of succesive elimination of variables is called the *standard reproductive solution*. □

Note the following properties:

Lemma 4.1. *Let* $(\varphi_1, \ldots, \varphi_n)$ *be a triangular reproductive solution of the Boolean equation* (4.9.1) *and let* $f_1, \ldots, f_n, f_{n+1}$ *be the eliminants of the function* f. *Then:*

(i) For every $k \in \{1, \ldots, n\}$, $(\varphi_k, \varphi_{k+1}, \ldots, \varphi_n)$ *is a triangular reproductive solution of equation* (4.9.k).

(ii) Every solution $(x_1, \ldots, x_n) \in B^n$ *of equation* (4.9.1) *satisfies the relations*

(4.14.k) $$\varphi_k(x_k, \ldots, x_n) = \varphi_k(x_k, \varphi_{k+1}(x_{k+1}, \ldots, x_n), \ldots, \varphi_n(x_n))$$
$$(k = 1, \ldots, n - 1) \,.$$

PROOF: (i) Since for every k the eliminants of f_k are $f_k, f_{k+1}, \ldots, f_n, f_{n+1} = 0$, it suffices to prove (i) for $k := 2$.

Take $(t_2, \ldots, t_n) \in B^{n-1}$. Then

$$f_1(\varphi_1(t_1, \ldots, t_n), \varphi_2(t_2, \ldots, t_n), \ldots, \varphi_n(t_n)) = 0 \,,$$

hence the equation (4.9′.1) in the unknown x_1 is consistent, therefore

$$f_2(\varphi_2(t_2,\ldots,t_n),\ldots,\varphi_n(t_n))$$

$$= f_1(1,\varphi_2(t_2,\ldots,t_n),\ldots,\varphi_n(t_n))f_1(0,\varphi_2(t_2,\ldots,t_n),\ldots,\varphi_n(t_n)) = 0 \,.$$

Further let $(x_2,\ldots,x_n) \in B^{n-1}$ be a solution of equation (4.9.2). But (4.9.2) is the consistency condition for equation (4.9.1) viewed as an equation in the unknown x_1. Therefore (4.9.1) holds for the above $x_2,\ldots,x_n$ and some $x_1 \in B$. Since $(\varphi_1,\ldots,\varphi_n)$ is a reproductive solution of equation (4.9.1), we get

(4.15) $$\qquad\qquad x_k = \varphi_k(x_k,\ldots,x_n) \qquad (k=1,\ldots,n)$$

and the desired conclusion is provided by (4.15) for $k = 2,\ldots,n$.

(ii) is obtained from (4.15) by successive substitutions.   □

Example 4.1. Consider the equation $a'x_1' \vee ax_2' = 0$ and apply first the method of successive eliminations. The equation (4.9.2) is $f_2(x_2) := ax_2' = 0$ and has the reproductive solution $x_2 = \varphi_2(t_2) = a \vee t_2$. The equation (4.9′.1) is $a'x_1' \vee a(a \vee t_2)' = 0$, that is, $a'x_1' = 0$ and has the reproductive solution $x_1 = \varphi_1(t_1,t_2) = a' \vee t_1$, where we have introduced the fictitious variable t_2 in order to point out the triangular character of the solution. Note that the unique relation (4.14), namely $\varphi_1(x_1,x_2) = \varphi_1(x_1,\varphi_2(x_2))$, is identically satisfied.

However it is easy to check that $x_1 = \psi_1(t_1,t_2) := a' \vee t_1 \vee t_2'$, $x_2 = \psi_2(t_2) := a \vee t_2$, defines another triangular reproductive solution of the equation, for which (4.14) is not identically satisfied: $\psi_1(x_1,\psi_2(x_2)) = a' \vee x_1 \neq \psi_1(x_1,x_2)$.   □

Lemma 4.2. *The standard triangular reproductive solution of a consistent Boolean equation (4.1) satisfies the identities*

(4.16.k)
$$\varphi_k(x_k,\ldots,x_n)$$
$$= \varphi_k(\varphi_k(x_k,\ldots,x_n),\varphi_{k+1}(x_{k+1},\ldots,x_n),\ldots,\varphi_n(\varphi_n(x_n)))$$
$$(k=1,\ldots,n)\,.$$

COMMENT: This result was proved by M. Gotō in 1956 for the two-element Boolean algebra and in 1957 for any finite Boolean algebra, while the general case of an arbitrary Boolean algebra was settled by Rudeanu in 1965; cf. BFE, Theorem 2.8. Another proof was given by Lavit [1974]. The above compact form (4.16) is due to Brown and Rudeanu [1985].

PROOF: Using Proposition 4.2. Take an arbitrary vector $(t_1,\ldots,t_n)$. Since φ_n is a reproductive solution of the equation $f_n(x) = 0$, it follows that $f_n(\varphi_n(t_n)) = 0$, therefore $\varphi_n(t_n) = \varphi_n(\varphi_n(t_n))$.

Now suppose that relations (4.16.k+1),...,(4.16.n) are fulfilled. Then Proposition 4.2 implies the identity

$$\varphi_k(x_k,\ldots,x_n) = x_k + f_k(x_k,\varphi_{k+1}(x_{k+1},\ldots,x_n),\ldots,\varphi_n(x_n))\,,$$

therefore

$$\varphi_k(\varphi_k(t_k,\ldots,t_n),\varphi_{k+1}(t_{k+1},\ldots,t_n),\ldots,\varphi_n(\varphi_n(t_n))) = \varphi_k(t_k,\ldots,t_n)+$$

$$+f_k(\varphi_k(t_k,\ldots,t_n),\varphi_{k+1}(\varphi_{k+1}(t_{k+1},\ldots,t_n),\ldots,\varphi_n(t_n)),\ldots,\varphi_n(\varphi_n(t_n)))$$

$$= \varphi_k(t_k,\ldots,t_n) + f_k(\varphi_k(t_k,\ldots,t_n),\varphi_{k+1}(t_{k+1},\ldots,t_n),\ldots,\varphi_n(t_n))$$

$$= \varphi_k(t_k,\ldots,t_n)$$

by Lemma 4.1(i). $\qquad\qquad\qquad\qquad\qquad\qquad\qquad\qquad\qquad\qquad\qquad$ □

Example 4.1 points out the significance of the next result.

Proposition 4.3. *The standard triangular reproductive solution of a consistent Boolean equation is the unique triangular reproductive solution for which relations (4.14) hold identically.*

PROOF: Let $(\varphi_1,\ldots,\varphi_n)$ be the standard reproductive solution of equation (4.1). Then in view of Proposition 4.2 and Lemma 4.2 and using an abbreviated notation, we have

$$\varphi_k(x_k,\varphi_{k+1},\ldots,\varphi_n) = x_k + f_k(x_k,\varphi_{k+1}(\varphi_{k+1},\ldots,\varphi_n),\ldots,\varphi_n(\varphi_n))$$

$$= x_k + f_k(x_k,\varphi_{k+1},\ldots,\varphi_n) = \varphi_k(x_k,x_{k+1},\ldots,x_n)\,.$$

Now suppose that $(\psi_1,\ldots,\psi_n)$ is a triangular reproductive solution satisfying the identities (4.14). Then in view of Lemma 4.1, ψ_n is the unique reproductive solution of equation $f_n(x) = 0$, hence $\psi_n = \varphi_n$. Further assume by induction that $\psi_j = \varphi_j$ $(j = k+1,\ldots,n)$.

If $x = \psi_k(x,x_{k+1},\ldots,x_n)$, then taking into account the inductive hypothesis and the fact that $(\psi_k,\psi_{k+1},\ldots,\psi_n)$ is a reproductive solution of equation (4.9.k), we obtain

$$f_k(x,\varphi_{k+1},\ldots,\varphi_n) = f_k(x,\psi_{k+1},\ldots,\psi_n) = f_k(\psi_k,\psi_{k+1},,\ldots,\psi_n) = 0$$

and conversely, if $f_k(x,\varphi_{k+1},\ldots,\varphi_n) = 0$, then $f_k(x,\psi_{k+1},\ldots,\psi_n) = 0$, hence the reproductivity of $(\psi_k,\ldots,\psi_n)$ and (4.14) imply

$$x = \psi_k(x,\psi_{k+1},\ldots,\psi_n) = \psi_k(x,x_{k+1},\ldots,x_n)\,.$$

Thus

$$x = \psi_k(x,x_{k+1},\ldots,x_n) \Longleftrightarrow f_k(x,\varphi_{k+1},\ldots,\varphi_n) = 0$$

and since $x = \varphi_k$ is a reproductive solution of equation (4.9'.k), we have also

$$x = \varphi_k(x,x_{k+1},\ldots,x_n) \Longleftrightarrow f_k(x,\varphi_{k+1},\ldots,\varphi_n) = 0\,.$$

It follows that

$$x + \psi_k(x,x_{k+1},\ldots,x_n) = 0 \Longleftrightarrow x + \varphi_k(x,\varphi_{k+1},\ldots,\varphi_n) = 0\,,$$

whence $\psi_k = \varphi_k$ by the Verification Theorem. $\qquad\qquad\qquad\qquad\qquad$ □

We have seen in BFE, Chapter 3, §2 and in the above discussion the concern for obtaining the most general form of a reproductive solution, which means necessary and sufficient conditions in order that a family $\Phi = (\varphi_1, \ldots, \varphi_n)$ of Boolean functions define the reproductive/general solution of a given Boolean equation. On the other hand, the method of successive elimination of variables entails calculating the sequence

(4.12) $$f = f_1, f_2, \ldots, f_n, f_{n+1}$$

of *eliminants* of f; cf. Definition 4.1. These can be used to obtain either a reproductive solution of the equation, or a system of recurrent inequalities expressing the set of all the (particular) solutions. In the remaining of this section we introduce the concepts of *recurrent cover* and *subsumptive general solution*, which generalize the eliminants and the corresponding interval-based solution, respectively. A bijection is established between recurrent covers and subsumptive general solutions, which, in fact, provide all interval-based representations of the set of solutions. The simplification of subsumptive general solutions is also studied. We follow closely Brown and Rudeanu [1983].

Definition 4.3. The function $f : B^n \longrightarrow B$ is said to be *evanescible with respect to* $(b_k, \ldots, b_n) \in B^{n-k+1}$ if either $k > 1$ and there is an element $(x_1, \ldots, x_{k-1}) \in B^{k-1}$ such that $f(x_1, \ldots, x_{k-1}, b_k, \ldots, b_n) = 0$, or $k = 1$ and $f(b_1, \ldots, b_n) = 0$. The function f will called *evanescible* if equation $f(x_1, \ldots, x_n) = 0$ is consistent.   □

Remark 4.2, Theorem 4.1 and Corollary 4.1 paraphrase the method of succesive eliminations in terms of eliminants.

Remark 4.2. It follows from Remark 4.1 that for each $k \in \{1, \ldots, n\}$, f_k is evanescible if and only if $f_{n+1} = 0$.   □

Theorem 4.1. *The following conditions are equivalent for a Boolean function* $f : B^n \longrightarrow B$, *an integer* $k \in \{1, \ldots, n\}$ *and* $(x_k, \ldots, x_n) \in B^{n-k+1}$:

(i) f *is evanescible with respect to* $(x_k, \ldots, x_n)$;

(ii) $f_k(x_k, \ldots, x_n) = 0$;

(iii) $f_j(0, x_{j+1}, \ldots, x_n) \leq x_j \leq f_j'(1, x_{j+1}, \ldots, x_n)$ $(j = k, \ldots, n - 1)$,

$$f_n(0) \leq x_n \leq f_n'(1) .$$

                     □

Corollary 4.1. *The following conditions are equivalent for a Boolean function* $f : B^n \longrightarrow B$ *and* $(x_1, \ldots, x_n) \in B^n$:

(i) $f(x_1, \ldots, x_n) = 0$;

(ii) $f_j(0, x_{j+1}, \ldots, x_n) \leq x_j \leq f_j'(1, x_{j+1}, \ldots, x_n)$ $(j = 1, \ldots, n - 1)$,

$$f_n(0) \leq x_n \leq f_n'(1) .$$

                     □

Now our aim is to find all possible ways of describing the set of all the solutions by means of recurrent inequalities in the same way as in the above theorem. So we first "transform" the theorem into a definition.

Definition 4.4. Let $f : B^n \longrightarrow B$ be an evanescible Boolean function. Given a double sequence $(u_k, v_k)_{k=1,\dots,n}$ of Boolean functions

$$(4.17) \qquad u_k, v_k : B^{n-k} \longrightarrow B \quad (k = 1, \dots, n-1), \quad u_n, v_n \in B ,$$

we say that the system of inequalities

$$(4.18.j) \qquad u_j(x_{j+1}, \dots, x_n) \leq x_j \leq v_j(x_{j+1}, \dots, x_n) \qquad (j = 1, \dots, n-1) ,$$

$$(4.18.n) \qquad \qquad u_n \leq x_n \leq v_n$$

determines a *subsumptive general solution* of the equation $f(x_1, \dots, x_n) = 0$ provided for every $k \in \{1, \dots, n\}$ and every $(x_k, \dots, x_n) \in B^{n-k+1}$ the following conditions are equivalent:
  (i) f is evanescible with respect to $(x_k, \dots, x_n)$;
  (ii) relations (4.18.j) hold for $j = k, \dots, n$.   □

Definition 4.5. Let $f : B^n \longrightarrow B$ be a Boolean functions and $f_1, \dots, f_{n+1}$ its eliminants. A system $(g_k)_{k=1,\dots,n+1}$ of Boolean functions

$$(4.19) \qquad g_k : B^{n-k+1} \longrightarrow B \ (k = 1, \dots, n), \ g_{n+1} \in B$$

is called a *recurrent cover* of f provided

$$(4.20) \qquad g_k \vee g_{k+1} \vee \dots \vee g_n \vee g_{n+1} = f_k \qquad (k = 1, \dots, n+1) .$$

□

Remark 4.3. The eliminants form a recurrent cover and $g_{n+1} = f_{n+1}$ for any recurrent cover.   □

We are going to establish a bijection between subsumptive general solutions and recurrent covers.

Proposition 4.4. *Let* $f : B^n \longrightarrow B$ *and* (4.17) *be Boolean functions. Then the system* (4.18) *determines a subsumptive general solution of the equation* $f(x_1, \dots, x_n) = 0$ *if and only if the sequence of functions defined by*

$$(4.21.j) \qquad \begin{aligned} g_j(x_j, \dots, x_n) &= u_j(x_{j+1}, \dots, x_n)x_j' \vee v_j'(x_{j+1}, \dots, x_n)x_j \\ &(j = 1, \dots, n-1) , \end{aligned}$$

$$(4.21.n) \qquad g_n(x_n) = u_n x_n' \vee v_n' x_n ,$$

$$(4.21.n+1) \qquad g_{n+1} = 0 ,$$

is a recurrent cover of f.

PROOF: Suppose (4.18) is a subsumptive general solution of $f(x_1, \ldots, x_n) = 0$. Then the functions (4.17) are Boolean, hence so are the functions (4.21). Besides, the equation is consistent, therefore it follows from (4.13.n+1) and (4.21.n+1) that $f_{n+1} = 0 = g_{n+1}$, i.e., relation (4.20.n+1) holds. Now fix an index $k \in \{1, \ldots, n\}$. Taking into account Theorem 4.1 and relations (4.21) and using introduction of fictitious variables, the equivalence $(i) \Longleftrightarrow (ii)$ in Definition 4.4 becomes

$$f_k(x_k, \ldots, x_n) = 0 \Longleftrightarrow g_j(x_j, \ldots, x_n) = 0 \ (j = k, \ldots, n)$$

$$\Longleftrightarrow \bigvee_{j=k}^{n} g_j(x_j, \ldots, x_n) = 0 \Longleftrightarrow (\bigvee_{j=k}^{n+1} g_j)(x_k, \ldots, x_n) = 0 \ .$$

But f_k is evanescible by Remark 4.2, therefore the above equivalence implies $f_k = \bigvee_{j=k}^{n+1} g_j$ by the Müller-Löwenheim Verification Theorem (see e.g. BFE, Corollary of Theorem 2.14).

Conversely, suppose the sequence $(g_1, \ldots, g_{n+1})$ defined by (4.21) is a recurrent cover of f. Then $f_{n+1} = g_{n+1} = 0$ by Remark 4.3 and (4.21.n+1), hence f is evanescible by (4.13.n+1). Now fix an index $k \in \{1, \ldots, n\}$. Using in turn Theorem 4.1, (4.20), elimination of fictitious variables and (4.21), we obtain:

$$f \text{ is evanescible with respect to } (x_k, \ldots, x_n)$$

$$\Longleftrightarrow f_k(x_k, \ldots, x_n) = 0 \Longleftrightarrow (\bigvee_{j=k}^{n+1} g_j)(x_1, \ldots, x_n) = 0$$

$$\Longleftrightarrow \bigvee_{j=k}^{n} g_j(x_j, \ldots, x_n) = 0$$

$$\Longleftrightarrow u_j(x_{j+1}, \ldots, x_n)x_j' \vee v_j'(x_{j+1}, \ldots, x_n)x_j = 0 \ (j = k, \ldots, n)$$

$$\Longleftrightarrow u_j(x_{j+1}, \ldots, x_n) \leq x_j \leq v_j(x_{j+1}, \ldots, x_n) \qquad (j = k, \ldots, n) \ .$$

$\square$

Proposition 4.5. *Let $f : B^n \longrightarrow B$ and (4.19) be Boolean functions. Then the following conditions are equivalent:*

(i) the system (4.19) is a recurrent cover of f for which $g_{n+1} = 0$;

(ii) the system of inequalities

(4.22.j) $$g_j(0, x_{j+1}, \ldots, x_n) \leq x_j \leq g_j'(x_{j+1}, \ldots, x_n) \ (j = 1, \ldots, n-1) \ ,$$

(4.22.n) $$g_n(0) \leq x_n \leq g_n'(1) \ ,$$

determines a subsumptive general solution of the equation $f(x_1, \ldots, x_n) = 0$.

PROOF: Define a sequence of the form (4.17) by

(4.23.j)
$$u_j(x_{j+1},\ldots,x_n) = g_j(0,x_{j+1},\ldots,x_n) \,,$$

$$v_j(x_{j+1},\ldots,x_n) = g_j(1,x_{j+1},\ldots,x_n) \quad (j=1,\ldots,n-1) \,,$$

(4.23.n)
$$u_n = g_n(0), v_n = g_n(1) \,.$$

Apply Proposition 4.4 with the Boolean function (4.23) in the rôle of the functions (4.17). Then the system (4.18) becomes (4.22),while the functions (4.21.k) $(k=1,\ldots,n)$ become the given functions $g_1,\ldots,g_n$. Thus in view of Proposition 4.4, the system of inequalities (4.22) determines a subsumptive general solution of the equation $f=0$ if and only if the system of functions $g_1,\ldots,g_n,g_{n+1}=0$ is a recurrent cover of f. □

Theorem 4.2. *Let* $f : B^n \longrightarrow B$ *be an evanescible Boolean function. Then the mappings* (4.21) *and* (4.22) *establish a bijection between the subsumptive general solutions of the equation* $f(x_1,\ldots,x_n)=0$ *and the recurrent covers of the function* f *for which* $g_{n+1}=0$.

PROOF: A subsumptive general solution (4.18) is in fact determined by the system of functions u_k, v_k $(k=1,\ldots,n)$, so that as a matter of fact Proposition 4.5 takes the sequence $(g_1,\ldots,g_n,0)$ to the sequence determined by (4.23). But the maps (4.21) and (4.23) are clearly inverse to each other. □

We are now going to study in some detail the construction of eliminants and recurrent covers.

Proposition 4.6. *The eliminants* (4.12) *obey the following rule:* BCF (f_{k+1}) *is obtained from* BCF(f_k) *by deleting the genterms that involve* x_k $(k=1,\ldots,n)$.

PROOF: Let

$$f_k(x_k,\ldots,x_n) = r(x_{k+1},\ldots,x_n)x_k \vee s(x_{k+1},\ldots,x_n)x'_k \vee t(x_{k+1},\ldots,x_n)$$

be the expression obtained from BCF(x_k) by factoring x_k and x'_k from the genterms in which they appear. Then $f_{k+1} = (r \vee t)(s \vee t) = rs \vee t = t$ and t is expressed in the Blake canonical form by Proposition 3.13. □

Example 4.2. Consider the Boolean algebra B and the function f in Example 3.1. Then the Blake canonical forms of the eliminants of f are the following:

$$f_1 = f \text{ given by formula } (3.30) \,;$$

$$f_2 = a_1x_3 \vee a_2x'_2x_3 \vee (a'_1 \vee a'_2)x'_2x'_3 \vee a_1a'_2x_2x_3 \vee a'_1a_2 \,,$$

$$f_3 = a_1x_3 \vee a'_1a_2 \,,$$

$$f_4 = a'_1a_2 \,.$$

Notice that $f_4 \neq 0$ because a_1 and a_2 are free generators, therefore the equation $f=0$ is inconsistent. □

Proposition 4.7. *Let $f : B^n \longrightarrow B$ be an evanescible Boolean function and $f_1, \ldots, f_n, f_{n+1}$ its eliminants. Then the following conditions are equivalent for a system (4.19) of Boolean functions for which $g_{n+1} = 0$:*

(i) $\bigvee_{j=k}^{n} g_j = f_k$ $(k = 1, \ldots, n)$, i.e., (4.19) is a recurrent cover of f ;

(ii) $g_k \vee f_{k+1} = f_k$ $(k = 1, \ldots, n)$;

(iii) $f'_{k+1} f_k \leq g_k \leq f_k$ $(k = 1, \ldots, n)$;

(iv) $\bigvee_{j=1}^{n} g_j = h$ and

(4.24.k)
$$g_k(0, x_{k+1}, \ldots, x_n) g_k(1, x_{k+1}, \ldots, x_n)$$
$$\leq \bigvee_{k+1}^{n} g_j(x_j, \ldots, x_n) \qquad (k = 1, \ldots, n-1) ,$$

(4.24.n)
$$g_n(0) g_n(1) = 0 .$$

PROOF: $(i) \Longrightarrow (ii)$: By induction on k.

$(ii) \Longrightarrow (i)$: By reverse induction on k.

$(ii) \Longleftrightarrow (iii)$: Since $f_{k+1} \leq f_k$, it follows that

$$(ii) \Longleftrightarrow (g_k \vee f_{k+1}) f'_k \vee g'_k f'_{k+1} f_k = 0 \Longleftrightarrow g_k f'_k \vee g'_k f'_{k+1} f_k = 0 \Longleftrightarrow (iii).$$

$(ii) \Longrightarrow (iv)$: Since $(ii) \Longrightarrow (i)$, we have $\bigvee_{k=1}^{n} g_k = f_1 = f$. Then, using an abbreviated notation, we get $g_k(0) \vee f_{k+1} = f_k(0)$ and $g_k(1) \vee f_{k+1} = f_k(1)$, hence

$$g_k(0) g_n(1) \vee f_{k+1} = f_k(0) f_k(1) = f_{k+1} ,$$

that is, $g_k(0) g_k(1) \leq f_{k+1}$. Besides, $g_n(0) = f_n(0)$ and $g_n(1) = f_n(1)$, therefore

$$g_n(0) g_n(1) = f_n(0) f_n(1) = f_{n+1} = 0 .$$

$(iv) \Longrightarrow (i)$: By hypothesis $\bigvee_{j=1}^{n} g_j = f = f_1$. Suppose $\bigvee_{j=k}^{n} g_j = f_k$, where $k \leq n - 1$. Then

$$f_{k+1} = f_k(0) f_k(1) = (g_k(0) \vee \bigvee_{j=k+1}^{n} g_j)(g_k(1) \vee \bigvee_{j=k+1}^{n} g_j)$$

$$= g_k(0) g_k(1) \vee \bigvee_{j=k+1}^{n} g_j = \bigvee_{j=k+1}^{n} g_j .$$

$\square$

Proposition 4.8. *Let $f : B^n \longrightarrow B$ be an evanescible Boolean function, $f_1, \ldots, f_n, f_{n+1} = 0$ its eliminants and (4.17) a system of Boolean functions. Then the system of inequalities (4.18) determines a subsumptive general solution of the equation $f(x_1, \ldots, x_n) = 0$ if and only if for every $k = 1, \ldots, n$,*

(4.25.a)
$$f_k(0, x_{k+1}, \ldots, x_n) f'_k(1, x_{k+1}, \ldots, x_n)$$
$$\leq u_k(x_{k+1}, \ldots, x_n) \leq f_k(0, x_{k+1}, \ldots, x_n)$$

and

(4.25.b)
$$f'_k(1, x_{k+1}, \ldots, x_n) \leq v_k(x_{k+1}, \ldots, x_n)$$
$$\leq f_k(0, x_{k+1}, \ldots, x_n) \vee f'_k(1, x_{k+1}, \ldots, x_n) .$$

PROOF: Set (again in abbreviated notation) $g_k = u_k x_k' \lor v_k' x_k$ $(k = 1, \ldots, n)$, $g_{n+1} = 0$. Then, in view of Proposition 4.4, (4.17) determines a subsumptive general solution of the equation $f = 0$ if and only if $(g_1, \ldots, g_{n+1})$ is a recurrent cover of f. Taking into account Proposition 4.7, the latter condition is equivalent to

$$f_{k+1}' f_k \leq u_k x_k' \lor v_k' x_k \leq f_k \qquad (k = 1, \ldots, n) ,$$

or equivalently, for all $k = 1, \ldots, n$,

$$f_{k+1}' f_k(0) \leq u_k \leq f_k(0) \ \& \ f_{k+1}' f_k(1) \leq v_k' \leq f_k(1)$$

and since $f_{k+1} = f_k(0) f_k(1)$, the last inequalities become

$$f_k'(1) f_k(0) \leq u_k \leq f_k(0) \ \& \ f_k'(0) f_k(1) \leq v_k' \leq f_k(1) .$$

□

Given the eliminants $f_1, \ldots, f_n, f_{n+1}$ of a Boolean function f, Proposition 4.7.(iii) expresses the set of recurrent covers of f by a system of intervals, i.e., of "incompletely specified" functions. Thus, in the case of simple Boolean functions, well-known methods of minimization may be used to find a recurrent cover expressed by the simplest possible disjunctive form (joins of terms). The following proposition and its corollary show, however, that relatively simple recurrent covers may be found without recourse to formal processes of minimization.

Proposition 4.9. *Let the eliminants $f_1, \ldots, f_n$ of a Boolean function f of n variables be represented by expressions of the form*

(4.26.k)
$$f_k(x_k, \ldots, x_n) = r_k(x_{k+1}, \ldots, x_n) x_k \lor s_k(x_{k+1}, \ldots, x_n) x_k' \lor$$
$$\lor t_k(x_{k+1}, \ldots, x_n) \qquad (k = 1, \ldots, n) .$$

Then the Boolean functions defined by the formulas

(4.27.k)
$$g_k(x_k, \ldots, x_n) = r_k(x_{k+1}, \ldots, x_n) x_k \lor s_k(x_{k+1}, \ldots, x_n) x_k'$$
$$(k = 1, \ldots, n) ,$$

(4.27.n+1)
$$g_{n+1} = f_{n+1} ,$$

form a recurrent cover of f.

PROOF: Since

$$f_{k+1} = f_k(1, \ldots) f_k(0, \ldots) = (r_k \lor t_k)(s_k \lor t_k) = r_k s_k \lor t_k ,$$

we deduce that

$$f_{k+1}' f_k = (r_k' \lor s_k') t_k'(r_k s_k \lor s_k x_k' \lor t_k)$$
$$= s_k' t_k' r_k x_k \lor r_k' t_k' s_k x_k' \leq r_k x_k \lor s_k x_k' \leq f_k ,$$

whence the desired conclusion follows by Proposition 4.7(iii).   □

Corollary 4.2. *If the eliminants f_k are expressed in disjunctive forms (i.e., as joins of genterms), then a recurrent cover is obtained by deleting from each f_k the genterms that do not involve x_k.* □

Example 4.3. Consider the Boolean function (3.29) (with no condition on the Boolean algebra B), for which the eliminants of order > 1 were determined in Example 4.2. By applying Corollary 4.2 of Proposition 4.9 we obtain the recurrent cover

$$g_1 = a_2 x_1 \vee a_2' x_1' x_2 \vee a_1' a_2' x_1 x_2' \vee a_1' x_1' x_2 \,,$$

$$g_2 = a_2 x_2' x_3 \vee (a_1' \vee a_2') x_2' x_3' \vee a_1 a_2' x_2 x_3 \,,$$

$$g_3 = a_1 x_3 \,,$$

$$g_4 = a_1' a_2 \,,$$

which is simpler than the set of eliminants. □

Proposition 4.10. *Given a Boolean function f of n variables, define associated Boolean functions $g_1, \ldots, g_n, g_{n+1}$ by the following prescriptions:*

(4.28.1) $$g_1 = f \,,$$

(4.28.k) $$g_k = \bigvee \{p \mid p \le x_k \,\&\, p \not\le x_j \ (j = 1, \ldots, k-1)\} \quad (k = 2, \ldots, n) \,,$$

(4.28.n+1) $$g_{n+1} = \bigvee \{p \mid p \not\le x_i \ (i = 1, \ldots, n)\} \,,$$

where p denotes a prime implicant of f (with the convention $\bigvee \emptyset = 0$). Then $(g_1, \ldots, g_n, g_{n+1})$ is a recurrent cover of f.

PROOF: Each function g_k depends at most on the variables $x_k, \ldots, x_n$ and $g_k \vee \ldots \vee g_{n+1}$ is expressed as the disjunction of all the prime implicants that do not involve any of $x_1, \ldots, x_{k-1}$. But it follows from Proposition 4.6 that the last disjunction equals f_k, therefore relations (4.20) hold. □

Example 4.4. Consider again the Boolean algebra B and the function f in Examples 3.1 and 4.2. The prime implicants of f are given in formula (3.30), therefore Proposition 4.10 yields the following recurrent cover of f:

$$g_1 = f \ (\text{see } (3.29)) \,;$$

$$g_2 = a_2 x_2' x_3 \vee (a_1' \vee a_2') x_2' x_3' \vee a_1 a_2' x_2 x_3 \,,$$

$$g_3 = a_1 x_3 \,,$$

$$g_4 = a_1' a_2 \,.$$

So the recurrent covers in Examples 4.3 and 4.4 differ only in g_1, the latter system being simpler than the former. □

5 Generalized systems/solutions of Boolean equations

In the Introduction to Chapter 10 of BFE we wrote: "In this chapter we deal with the problem of solving Boolean inequalities $f(X) \neq 0$. Although we do not know very much about this problem (§1), this suffices to determine all possible syllogistic moods, thus completing the Aristotelian logic (§2). Equations and inequalities provide a systematic tool for solving the so-called 'intelligence problems' or 'logical problems', which are not merely simple puzzles, but have practical applications (§3).". In particular this led us to the following "Problem 10.1. Develop the theory of Boolean inequalities and that of alternative systems of Boolean equations"; see also BFE, Theorem 13.5 (McKinsey).

In this section we present, with certain slight improvements, part of the paper by Marriott and Odersky [1996][1], which may be viewed as the beginning of a solution for the above "Problem 10.1". As a matter of fact, the Marriott-Odersky paper was motivated by the fact that certain problems of query optimization in databases reduce to the solution of systems consisting of Boolean equations and Boolean inequalities, as shown by Helm, Marriott and Odersky [1991], [1995] (cf. Ch.14, §5), who also anticipated some of the results in the paper by Marriott and Odersky.

The section ends with a few words about certain concepts of generalized solutions of a Boolean equation that have been introduced in the literature.

Generally speaking, when we refer to a system of equations of any kind, we understand that those equations have to be fulfilled simultaneously; in other words, they are linked by logical conjunction. Thus the idea of a generalized system of equations is to link those equations by any logical function, not merely conjunction. In our case the generalized systems of Boolean equations will be the predicates over the given Boolean algebra, in the Boolean language without quantifiers. Let us state this more precisely.

Definition 5.1. The *generalized systems of Boolean equations* (GSBE's for short) over a Boolean algebra B are defined recursively as follows:

(i) every Boolean equation $f(X) = 0$ is a GSBE;

(ii) the negation, logical conjunction and logical disjunction of any GSBE's is a GSBE;

(iii) every GSBE is obtained by applying rules (i) and (ii) finitely many times.   □

It follows easily by induction that any GSBE involves finitely many variables, say $x_1, \ldots, x_n$. If each of these variables x_i is replaced by an element $a_i \in B$, one obtains a statement which is true or false. In other words, any GSBE is a predicate over the universe B, as we have anticipated, and in fact, each GSBE has infinitely many intepretations as a predicate, obtained by introduction of fictitious variables. Therefore the following definition makes sense.

---

[1] We have been unable to understand completely this paper.

Definition 5.2. Let $S(x_1, \ldots, x_n)$ denote a GSBE whose (free!) variables belong to the set $\{x_1, \ldots, x_n\}$. By a *solution* of $S(x_1, \ldots, x_n)$ is meant any vector $(a_1, \ldots, a_n) \in B^n$ such that the statement $S(a_1, \ldots, a_n)$ obtained by replacing each x_i by a_i is true. A GSBE which has solutions is said to be *consistent* or *satisfiable* . Two GSBE's $S(x_1, \ldots, x_n)$ and $T(x_1, \ldots, x_n)$ are said to be *equivalent* provided they have the same set of solutions.   □

Remark 5.1. In the above definition $S(x_1, \ldots, x_n)$ is not merely a GSBE, but a pair consisting of a GSBE and a set $\{x_1, \ldots, x_n\}$ which includes the set of variables actually involved in the GSBE; the inclusion may be strict. As a matter of fact, the same remark applies to any Boolean or lattice equation $f(x_1, \ldots, x_n) = 0$ and to any usual system of such equations.   □

We first prove that the problem of solving GSBE's reduces to a particular case of it.

Definition 5.3. An *elementary* GSBE is either a Boolean equation $f(X) = 0$, or a system of the form

$$(5.1) \qquad\qquad g_1(X) \neq 0 \,\&\, \ldots \,\&\, g_m(x) \neq 0 \,,$$

or of the form

$$(5.2) \qquad\qquad f(X) = 0 \,\&\, g_1(X) \neq 0 \,\&\, \ldots \,\&\, g_m(X) \neq 0 \,.$$

If $m = 1$ we will say that the GSBE is *atomic*. An atomic GSBE of the form $g(X) \neq 0$ will be called a *negated Boolean equation*.   □

Proposition 5.1. *Every GSBE is equivalent to a logical disjunction of elementary GSBE's, possibly a single elementary GSBE.*

PROOF: Same as the well-known proof that a truth function, or a simple Boolean function, can be written as a disjunction of terms. In our case the equality $=$ is replaced by logical equivalence $\Longleftrightarrow$ and the terms are of the form

$$f_1(X) = 0 \,\&\, \ldots \,\&\, f_p(X) = 0 \,\&\, g_1(X) \neq 0 \,\&\, \ldots \,\&\, g_m(X) \neq 0 \,,$$

where $p \geq 0$, $m \geq 0$ and $p + m > 0$. But, as we know, every system of Boolean equations

$$f_1(X) = 0 \,\&\, \ldots \,\&\, f_p(X) = 0$$

is equivalent to a single Boolean equation $f(X) = 0$.   □

Hence we obtain the following basic result:

Theorem 5.1. *The set of solutions of any GSBE is the union of the sets of solutions of several elementary GSBE's.*   □

Thus the problem of solving GSBE's is reducible to the problem of solving elementary GSBE's. Unfortunately, we have only partial results on the latter problem, even in the simplest case.

Proposition 5.2. *The following conditions are equivalent for an atomic GSBE*
$$(5.3) \hspace{3cm} f(X) = 0 \ \& \ g(X) \neq 0 \,,$$
where f is a simple Boolean function:

(i) (5.3) *is satisfiable ;*

(ii) *the negated Boolean equation $g(X) \not\leq f(X)$ is satisfiable ;*

(iii) $\bigvee_A f'(A)g(A) \neq 0$.

PROOF: $(i) \Longleftrightarrow (ii)$: If $f = 1$ (constant function) then both (5.3) and $g \not\leq f$ are not satisfiable. Otherwise f is evanescible and the Müller-Löwenheim Verification Theorem yields
$$\exists X(f(X) = 0 \ \& \ g(X) \neq 0)$$
$$\Longleftrightarrow \neg \forall X(f(X) = 0 \Rightarrow g(X) = 0) \Longleftrightarrow \neg \forall X(g(X) \leq f(X)) \,.$$

$(ii) \Longleftrightarrow (iii)$: The identities $g \leq f$ and $f'g = 0$ are equivalent. $\square$

Remark 5.2. The above equivalence $(ii) \Longleftrightarrow (iii)$ is valid even if the Boolean function f is not simple. $\square$

Proposition 5.3. *Suppose $\Xi \in B^n$ is a solution of the Boolean equation $f(X) = 0$. Then a vector $X \in B^n$ is a solution of system (5.3) if and only if it is of the form*
$$(5.4) \hspace{3cm} X = f(T)\Xi \vee f'(T)T \,,$$
for some $T \in B^n$ such that $g(X) \neq 0$; the latter condition is equivalent to
$$(5.5) \hspace{3cm} f(T)g(\Xi) \vee f'(T)g(T) \neq 0 \,.$$

PROOF: In view of Löwenheim's reproductive solution based on a particular solution (see e.g. BFE, Theorem 2.11; cf. the present Theorem 5.3.4), the solutions of the equation $f(X) = 0$ coincide with the vectors of the form (5.4). On the other hand, the behaviour of Boolean functions with respect to orthonormal systems (see e.g. BFE, Theorem 4.6(viii); cf. the present Lemma 5.2.2) shows that (5.4) implies
$$g(X) = f(T)g(\Xi) \vee f'(T)g(T) \,.$$
$\square$

Remark 5.3. If f is a simple Boolean function, then the consistency condition (iii) in Proposition 5.2 also yields a particular solution of (5.3) in $\{0,1\}^n$: there exists $A \in \{0,1\}^n$ such that $f'(A)g(A) \neq 0$, hence $g(A) \neq 0$ and $f'(A) = 1$, therefore $f(A) = 0$. See the discussion before Proposition 4.1 for various procedures of obtaining a particular solution Ξ in the general case. $\square$

The above two propositions imply that the problems of deciding the satisfiability of an atomic system and of solving such a system reduce to the correponsing problems for negated Boolean equations. The next proposition shows that atomic GSBE's are closed under elimination of variables.

Lemma 5.1. *For arbitrary elements* $a, b, c, d \in B$,

(5.6) $\qquad \exists x \, (a \leq x \leq b \,\&\, \neg(c \leq x \leq d)) \iff a \leq b \,\&\, \neg(b \leq d \,\&\, c \leq a)$.

PROOF: $\Longrightarrow$: Clearly $a \leq b$. If $b \leq d$ and $c \leq a$ then $x := a$ would satisfy $c \leq x \leq d$, a contradiction.

$\Longleftarrow$: If $b \not\leq d$ then the left member of (5.6) holds, as shown by taking $x := b$, while if $c \not\leq a$, the desired conclusion is obtained with $x := a$. $\qquad\square$

Proposition 5.4. *Let* $S(X)$ *denote the atomic system* (5.3), *where* f *and* g *are arbitrary Boolean functions, and let* $x \in X$. *For* $h \in \{f, g\}$ *and* $a \in B$, *let* $h(x = a)$ *denote what is obtained from* $h(X)$ *by taking* $x := a$. *Then*

(5.7)
$$\exists x S(X) \iff f(x = 0)f(x = 1) = 0 \,\&$$
$$\&f'(x = 1)g(x = 1) \lor f'(x = 0)g(x = 0) \neq 0 \,.$$

PROOF: Set $f(x = 0) = a$, $f(x = 1) = b$, $g(x = 0) = c$ and $g(x = 1) = d$. Then Lemma 4.1 yields

$$\exists x S(X) \iff \exists x(ax' \lor bx = 0 \,\&\, \neg(cx' \lor dx = 0))$$

$$\iff \exists x(a \leq x \leq b' \,\&\, \neg(c \leq x \leq d'))$$

$$\iff a \leq b' \,\&\, \neg(b' \leq d' \,\&\, c \leq a)$$

$$\iff ab = 0 \,\&\, (b'd \neq 0 \text{ or } a'c \neq 0) \iff ab = 0 \,\&\, b'd \lor a'c \neq 0 \,.$$

$\square$

Proposition 5.4 shows that the first stage of the method of successive elimination of variables can be applied to GSBE's too, but we do not know how to perform the second stage.

Example 5.1. The following GSBE was studied by Brown [1976] using a different approach:

(5.8) $\qquad a_1 \lor x_1 < a_2' \lor x_1 x_2 \text{ or } a_2 \lor a_1 x_1 \neq a_1 \lor x_2'$.

Setting $a_1 \lor x_1 = A$, $a_2' \lor x_1 x_2 = B$, $a_2 \lor a_1 x_1 = C$ and $a_1 \lor x_2' = D$, we see that system (5.8) is the disjunction of two elementary GSBE's, namely the atomic system

(5.9) $\qquad\qquad\qquad AB' = 0 \,\&\, A'B \neq 0$

and the negated Boolean equation

(5.10) $\qquad\qquad\qquad CD' \lor C'D \neq 0 \,.$

But

$$AB' = (a_1 \lor x_1)a_2(x_1' \lor x_2') = a_1 a_2(x_1' \lor x_2') \lor a_2 x_1 x_2' \,,$$
$$A'B = a_1' x_1'(a_2' \lor x_1 x_2) = a_1' a_2' x_1' \,,$$

hence the equation $AB' = 0$ has the particular solution $\xi_1 = \xi_2 = 1$, for which formulas (5.4) become

$$x_1 = f(T) \vee t_1 = a_1 a_2 (t_1' \vee t_2') \vee a_2 t_1 t_2' \vee t_1 = a_1 a_2 \vee t_1 \,,$$

$$x_2 = f(T) \vee t_2 = a_1 a_2 (t_1' \vee t_2') \vee a_2 t_1 t_2' \vee t_2 = a_1 a_2 \vee a_2 t_1 \vee t_2 \,,$$

which implies $A'B = a_1' a_2' t_1'$. Therefore Proposition 5.3 shows that the solutions of system (5.9) are given by

(5.11) $\qquad\qquad x_1 = a_1 a_2 \vee t_1, \; x_2 = a_1 a_2 \vee a_2 t_1 \vee t_2, \; a_1' a_2' \not\leq t_1 \,;$

this also implies that the consistency condition for (5.9) is

(5.12) $\qquad\qquad\qquad\qquad\qquad a_1' a_2' \neq 0 \,.$

Of course, the consistency condition can also be obtained by Proposition 5.2: for the negated Boolean equation $A'B \not\leq AB'$ is $A'B(A' \vee B) \neq 0$, that is, $A'B \neq 0$, i.e., $a_1' a_2' x_1' \neq 0$, whose consistency condition is (5.12).

Further, $CD' = a_1' a_2 x_2$ and $C'D = a_2'(a_1' \vee x_1')(a_1 \vee x_2')$, hence the negated equation (5.10) is in fact

(5.13) $\qquad\qquad a_1' a_2 x_2 \vee a_1' a_2' x_2' \vee a_1 a_2' x_1' \vee a_2' x_1' x_2' \neq 0 \,.$

It is easy to see that for the function g in the left member of (5.13) we have $\bigvee_A g(A) = 1 \neq 0$, therefore the negated Boolean equation (5.13) is consistent. We can write (5.13) in the equivalent form

(5.14) $\qquad x_2 \not\leq a_1 \vee a_2' \text{ or } a_1' a_2' \not\leq x_2 \text{ or } a_1 a_2' \not\leq x_1 \text{ or } a_2' \not\leq x_1 \vee x_2 \,.$

$\qquad\qquad\qquad\qquad\qquad\qquad\qquad\qquad\qquad\qquad\qquad\qquad\qquad$ $\square$

Now we deal with the problem of reducing the satisfiability of elementary GSBE's to that of atomic GSBE's. We will see that this is possible in sufficiently large Boolean algebras.

Recall that the *free Boolean algebra* FBn with $n (\geq 1)$ free generators $x_1, \ldots,$ x_n can be described as the Boolean algebra of all truth functions of n variables, the free generators being the variables $x_1, \ldots, x_n$ (see e.g. BFE, Theorem 1.21 and Corollary).

Lemma 5.2. *Let* $T \neq \emptyset$ *be a set of minterms in the variables* $x_1, \ldots, x_n$; $n \geq 1$. *Then there is a Boolean homomorphism* $\varphi : FBn \longrightarrow P(T)$ *such that for every minterm* t, $\varphi(t) \neq \emptyset \Longleftrightarrow t \in T$.

PROOF: Let φ be the homomorphism which extends the map

$$\varphi(x_i) = \{t \in T \mid t \leq x_i\} \qquad (i = 1, \ldots, n) \,.$$

Let t_0 be a minterm; without loss of generality we can take $t_0 = x_1 \ldots x_k x_{k+1}' \ldots x_n'$, as the proof in the general case will be the same up to a more sophisticated notation. Then

$$\varphi(t_0) = \{t \in T \mid t \leq x_1\} \cap \ldots \cap \{t \in T \mid t \leq x_k\} \cap$$

$$\cap \{t \in T \mid t \nleq x_{k+1}\} \cap \ldots \cap \{t \in T \mid t \nleq x_n\}$$

$$= \{t \in T \mid t \leq x_1 \ \& \ \ldots \ \& \ t \leq x_k \ \& \ t \leq x'_{k+1} \ \& \ \ldots \ \& \ t \leq x'_n\}$$

$$= \{t \in T \mid t = x_1 \ldots x_k x'_{k+1} \ldots x'_n\}$$

$= \{t_0\}$ or $\emptyset$ according as $t_0 \in T$ or $t_0 \notin T$. $\qquad\qquad\square$

Lemma 5.3. *Let B be a finite Boolean algebra, say $\operatorname{card}(B) = 2^m$. Let $n \geq 1$ and suppose $\varphi : FBn \longrightarrow B$ is a Boolean homomorphism. Then for any (possibly infinite) Boolean algebra B_1 with $\operatorname{card}(B_1) \geq 2^{m-1}$, there is a Boolean homomorphism $\varphi_1 : FBn \longrightarrow B_1$ such that for every minterm t, $\varphi_1(t) = 0 \Longleftrightarrow \varphi(t) = 0$.*

PROOF: Note first that B_1 contains a chain

(5.15) $$0 < c_1 < \ldots < c_{m-1} < 1 :$$

if B_1 is infinite this follows from Proposition 3.4 in Koppelberg [1989], while if B_1 is finite then it contains $m - 1$ atoms $a_1, \ldots, a_{m-1}$ and we take $c_k = a_1 \vee \ldots \vee a_k$ ($k = 1, \ldots, m - 1$).

The next step is to define

(5.16) $$b_1 = c_1, \ b_k = c'_{k-1} c_k \ (k = 2, \ldots, m - 1), \ b_m = c'_{m-1} ;$$

it follows readily from (5.15) that $b_k > 0$ for all k and $\{b_1, \ldots, b_m\}$ is an orthonormal set which has exactly m elements ($xy = 0$ and $x = y$ would imply $x = 0$).

Further let $\varphi_1 : FBn \longrightarrow B_1$ be the homomorphism which extends the map

(5.17) $$\varphi_1(x_i) = \bigvee \{b_k \mid a_k \leq \varphi(x_i)\} \qquad (i = 1, \ldots, n) .$$

Now take a minterm t; as in the previous proof, we can suppose without loss of generality that $t = x_1 \ldots x_h x'_{h+1} \ldots x'_n$. Taking into account that φ_1 is a homomorphism, the sequence (5.16) is orthonormal, $a_1, \ldots, a_{m-1}$ are atoms, φ is a homomorphism and using distributivity we obtain

$$\varphi_1(t) = (\bigvee \{b_k \mid a_k \leq \varphi(x_1)\}) \ldots (\bigvee \{b_k \mid a_k \leq \varphi(x_h)\}) \cdot$$

$$(\bigvee \{b_k \mid a_k \nleq \varphi(x_{h+1})\}) \ldots (\bigvee \{b_k \mid a_k \nleq \varphi(x_n)\})$$

$$= \bigvee \{b_{k_1} \ldots b_{k_n} \mid a_{k_1} \leq \varphi(x_1) \ \& \ \ldots \ \& \ a_{k_h} \leq \varphi(x_h) \ \&$$

$$\& \ a_{k_{h+1}} \leq (\varphi(x_{h+1}))' \ \& \ \ldots \ \& \ a_{k_n} \leq (\varphi(x_n))'\}$$

$$= \bigvee \{b_k \mid a_k \leq \varphi(x_1) \ldots \varphi(x_h) \varphi(x'_{h+1}) \ldots \varphi(x'_n)\} = \bigvee \{b_k \mid a_k \leq \varphi(t)\} .$$

Thus if $\varphi(t) = 0$ then $\varphi_1(t) = \bigvee \emptyset = 0$, whereas if $\varphi(t) > 0$ then $\varphi_1(t) > 0$ because all $b_k > 0$. $\qquad\qquad\square$

Proposition 5.5. *Suppose $f, g_1, \ldots, g_m : B^n \longrightarrow B$ are simple Boolean functions and $\operatorname{card}(B) \geq 2^{m-1}$. Then the following conditions are equivalent:*

(i) the elementary GSBE

$$(5.18) \qquad f(X) = 0 \ \& \ g_1(X) \neq 0 \ \& \ \ldots \ \& \ g_m(X) \neq 0$$

is satisfiable ;

(ii) each atomic GSBE

$$(5.19.\text{k}) \qquad f(X) = 0 \ \& \ g_k(X) \neq 0 \qquad (k = 1, \ldots, m)$$

is satisfiable ;

(iii) each negated Boolean equation

$$(5.20.\text{k}) \qquad g_k(X) \not\leq f(X) \qquad (k = 1, \ldots, m)$$

is satisfiable ;

(iv) $\bigvee_A f'(A) g_k(A) \neq 0 \qquad (k = 1, \ldots, m)$.

PROOF: $(i) \Longrightarrow (ii)$: Trivial.

$(ii) \Longrightarrow (i)$: Let T and T_k be the sets of minterms in the normal disjunctive forms of f and g_k, respectively $(k = 1, \ldots, m)$. For each $k = 1, \ldots, m$, the satisfiability of system (5.19.k) implies the existence of a minterm $t_k \in T_k \backslash T$. Now apply Lemma 5.2 with $T := \{t_1, \ldots, t_m\}$ and Lemma 5.3 with $B := \mathcal{P}(\{t_1, \ldots, t_m\})$, the homomorphism φ from Lemma 5.2 and $B_1 := B$. Consequently there is a homomorphism $\varphi : FBn \longrightarrow B$ such that for every minterm t,

$$\varphi_1(t) = 0 \Longleftrightarrow \varphi(t) = \emptyset \Longleftrightarrow t \notin \{t_1, \ldots, t_m\} .$$

This further implies

$$f(\varphi_1(x_1), \ldots, \varphi_1(x_n)) = \varphi_1(f(x_1, \ldots, x_n)) = \varphi_1\left(\bigvee_{t \in T} t\right) = \bigvee_{t \in T} \varphi_1(t) = 0$$

because $t \in T \Longrightarrow t \notin \{t_1, \ldots, t_m\}$ and similarly

$$g_k(\varphi_1(x_1), \ldots, \varphi_1(x_m)) = \bigvee_{t \in T_k} \varphi_1(t) \geq \varphi_1(t_k) \neq 0 .$$

$(ii) \Longleftrightarrow (iii) \Longleftrightarrow (iv)$: By Proposition 5.2.   □

Proposition 5.6. *The following property holds if and only if the Boolean algebra B is infinite: for every $m, n \geq 1$, every $m + 1$ simple Boolean functions $f, g_1, \ldots, g_m : B^n \longrightarrow B$, conditions (i) and (ii) in Proposition 5.5 are equivalent.*

PROOF: If B is infinite the equivalence holds by Proposition 5.5. If B is finite, say $\operatorname{card}(B) = 2^m$, then there is no chain $0 < x_1 < \ldots < x_m < 1$, although each of the atomic GSBE's $0 < x_1, \ x_i < x_{i+1} (i = 1, \ldots, m - 1), \ x_m \leq 1$ is satisfiable.   □

In the sequel we will study the transfer of the satisfiability of a GSBE expressed in terms of simple Boolean functions between various Boolean algebras. The problem makes sense in view of the following

Remark 5.4. A simple Boolean function $f : B^n \longrightarrow B$ is uniquely determined by the system of 2^n values $f(A) \in \{0,1\}$, $A \in \{0,1\}^n$ (cf. BFE, Theorem 1.7). Since $\{0,1\}$ is included in any Boolean algebra, the function f can be regarded as being defined over any Boolean algebra and we may refer to it as an *n-variable simple Boolean function*, regardless of any specific Boolean algebra. From a practical point of view, we interpret the operations and variables occurring in the expression of f as the operations and variables of the Boolean algebra we are working in. Therefore a GSBE expressed by simple Boolean functions can be regarded as being defined over any Boolean algebra.   □

Remark 5.5. Let B and B_1 be Boolean algebras. Since every GSBE is the logical disjunction of several elementary GSBE's, it follows that in order to prove that every GSBE expressed by simple Boolean functions and satisfiable in B is also satisfiable in B_1, it suffices to prove this property for elementary GSBE's.   □

Proposition 5.7. *If a GSBE expressed by simple Boolean functions is satisfiable in some infinite Boolean algebra then it is satisfiable in every infinite Boolean algebra.*

PROOF: Immediate from Remark 5.5 and Proposition 5.5.   □

Lemma 5.4. *If the sets M, N satisfy $\emptyset \neq M \subset N$, then there is an injective Boolean homomorphism $\varphi : \mathcal{P}(M) \longrightarrow \mathcal{P}(N)$.*

COMMENT: Since the inclusion mapping is not a Boolean homomorphism, this lemma is not trivial.
PROOF: Pick an element $a \in M$ and for every $Z \subseteq M$ set: if $a \in Z$ then $\varphi(Z) = Z \cup (N \backslash M)$, else $\varphi(Z) = Z$. Then it is plain that $\varphi(Z \cup Y) = \varphi(Z) \cup \varphi(Y)$. Further note that $\{Z, M \backslash Z, N \backslash M\}$ is an orthonormal system of $\mathcal{P}(N)$. Therefore if $a \in Z$ then

$$\varphi(M \backslash Z) = M \backslash Z = N \backslash (Z \cup (N \cup M)) = N \backslash \varphi(Z) ,$$

while if $a \notin Z$ then

$$\varphi(M \backslash Z) = (M \backslash Z) \cup (N \backslash M) = N \backslash Z = N \backslash \varphi(Z) .$$

Thus φ is a Boolean homomorphism. Besides, $\varphi(Z) = \emptyset \Longleftrightarrow Z = \emptyset$, which, in view of the ring structure of a Boolean algebra, proves that φ is an injection.   □

Corollary 5.1. *If system (5.18), where $f, g_1, \ldots, g_m$ are simple Boolean functions, is satisfiable in $\mathcal{P}(M)$, then it is satisfiable in $\mathcal{P}(N)$ as well.*

PROOF: Let X be a solution of (5.18) in $\mathcal{P}(M)$ and φ the homomorphism in Lemma 5.4. Then

$$f(\varphi^n(X)) = \varphi(f(X)) = \varphi(\emptyset) = \emptyset ,$$

$$g_k(\varphi^n(X)) = \varphi(g_k(X)) \neq \emptyset \text{ because } g_k(X) \neq \emptyset.   □$$

Proposition 5.8. *If a GSBE expressed by simple Boolean functions is satisfiable in some Boolean algebra B, then it is satisfiable in any Boolean algebra B_1 with* $\mathrm{card}(B_1) \geq \mathrm{card}(B)$.

PROOF: In view of Remark 5.5 it suffices to prove the assertion for an elementary GSBE, say (5.18), where $f, g_1, \ldots, g_m$ are simple Boolean functions. Taking into account the well-known representation of finite Boolean algebras as powersets (see e.g. BFE, Theorem 1.4), we may suppose, without loss of generality, that $B = \mathcal{P}(M)$, where M is a finite non-empty set.

It follows from Corollary 5.1 of Lemma 5.4 that (5.18) is satisfiable in the Boolean algebras $\mathcal{P}(N)$, where N is an infinite set, which implies that (5.18) is satisfiable in any infinite Boolean algebra, by Proposition 5.7.

Now suppose B_1 is a finite Boolean algebra with $\mathrm{card}(B_1) \geq \mathrm{card}(B)$. If $\mathrm{card}(B_1) = \mathrm{card}(B)$ then B_1 is isomorphic to B, therefore (5.18) is satisfiable in B_1, too. If $\mathrm{card}(B_1) > \mathrm{card}(B)$, then we may suppose, without loss of generality, that $B_1 = \mathcal{P}(N)$ where $M \subset N$, hence (5.18) is satisfiable in B_1 by the same corollary. $\square$

Corollary 5.2. *A GSBE expressed by simple Boolean functions is satisfiable in all Boolean algebras if and only if it is satisfiable in the two-element Boolean algebra.* $\square$

The paper by Cerny and Marin [1974] mentions the problem of solving generalized systems of Boolean equations within the context of decompositions of truth functions.

Another line of research consists in working with the conventional concepts of Boolean equation and system of Boolean equations but with a more general concept of solution. Thus Leont'ev and Tonoyan [1993] consider a system of equations of the form

$$(5.21) \qquad\qquad f_i(x_1, \ldots, x_n) = 0 \qquad (i = 1, \ldots, m)$$

and look for maximal consistent subsystems of (5.21); this may be viewed as an approximate solution in the case when system (5.21) is inconsistent. The same authors define a generalized solution of system (5.21) as a subset $M \subseteq B^n$ (as a matter of fact they work with $B := \{0, 1\}$) such that for each $i \in \{1, \ldots, m\}$ there is a solution $X \in M$ of the i-th equation (5.21). Another concept of generalized solution was suggested by Chajda [1973].

We also mention the paper by Balakin [1973] on random Boolean equations by quoting its review in Math. Rev. 48(1973),no.6,#13484: "For two n-dimensional Boolean vectors $X, Y \in \{0, 1\}^n$ and $I \subseteq \{1, 2, \ldots, n\}$, let $\delta_I(X, Y)$ be equal to 1 or 0 according to whether $x_i = y_i$ for all $i \in I$ or not. The author considers a system of random Boolean equations. Joins of expressions of the form $\delta_I(X, Y)$ are to be equal to 1, X being an unknown vector, Y fixed or the coordinates y_i mutually independent, and I being defined by sampling without replacement in each equation. If η denotes the number of missing coordinates x_i in the system, then the number of distinct solutions of the system converges to 2^η in probability.

Statistical interpretation: observations y_j are distorted outside the sets I_j, $j = 1, 2, \ldots, t$.".

7. Closure operators on Boolean functions

Several closure operators on the algebra of truth functions have been studied in the literature. This chapter begins with a preliminary section containing all necessary prerequisites on closure operators in general. Then we present, within the framework of arbitrary Boolean functions, some specific closures (increasing, decreasing, independent and simple) with an application to the problem of determining a partially defined Boolean function from a given class. Other applications will be included in Ch.8, §1, and in Ch.11, §3.

1 A general theory

We begin this section by recalling the prerequisites necessary to the understanding of the present chapter. Then we deal with arbitrary closure operators on Boolean functions.

For the prerequisites we follow Bourbaki [1963], Exercice §1.13. See also e.g. Balbes and Dwinger [1974], Chapter II, §4; Birkhoff [1967], Chapter V, §1; Davey and Priestley [1990], 2.20−2.23.

A *closure operator* or simply a *closure* on a poset L is an *extensive, idempotent* and *isotone* map $\varphi : L \to L$, which means that for all $x, y \in L$, 1) $x \leq \varphi(x)$, 2) $\varphi(\varphi(x)) = x$ and 3) $x \leq y \Longrightarrow \varphi(x) \leq \varphi(y)$. By a *Moore family* or a *closure system* of L is meant a subset $M \subseteq L$ such that for each element $x \in L$, the *principal filter* of M generated by x, i.e. the set

$$(1.1) \qquad M[x) = \{y \in M \mid x \leq y\} \,,$$

has least element.

It is easy to see that if φ is a closure on L then the set

$$(1.2) \qquad M_\varphi = \{x \in L \mid x = \varphi(x)\}$$

of *closed elements* is a Moore family of L. For if $x \in L$ then $M_\varphi[x)$ has the least element $\varphi(x)$, because $\varphi(x) \in M_\varphi$ by 2) above, and $x \leq \varphi(x)$ by 1), hence $\varphi(x) \in M_\varphi[x)$ and if $y \in M_\varphi[x)$, then from $x \leq y$ one infers $\varphi(x) \leq \varphi(y) = y$.

Conversely, if M is a Moore family of L then the map $\varphi_M : L \longrightarrow L$ defined by

$$(1.3) \qquad \varphi_M(x) \text{ is the first element of } M[x) \,,$$

is a closure operator. For $\varphi_M(x) \in M[x]$, therefore $x \leq \varphi_M(x)$, and if $x \leq y$ then $M[y] \subseteq M[x]$, hence $\varphi_M(x) \leq \varphi_M(y)$. Finally extensivity and isotony imply $\varphi_M(x) \leq \varphi_M(\varphi_M(x))$ and on the other hand $\varphi_M(x) \in M[x] \subseteq M$, hence $\varphi_M(x) \in M[\varphi_M(x)]$ and the last set has first element $\varphi_M(\varphi_M(x))$, therefore $\varphi_M(\varphi_M(x)) \leq \varphi_M(x)$.

Thus (1.2) establishes a map $\varphi \mapsto M_\varphi$ from the set of closure operators on L to the set of Moore families of L, while (1.3) defines a map $M \mapsto \varphi_M$ in the opposite sense. But

$$M_{\varphi_1} = M_{\varphi_2} \implies \forall x \, M_{\varphi_1}[x) = M_{\varphi_2}[x)$$

$$\implies \forall x \, \varphi_1(x) = \varphi_2(x) \iff \varphi_1 = \varphi_2 ,$$

therefore the map (1.2) is injective. Besides, for every x,

$$x \in M_{\varphi_M} \iff x = \varphi_M(x) \iff x \text{ is the first element of } M[x)$$

$$\implies x \in M[x) \implies x \in M ,$$

therefore $M_{\varphi_M} \subseteq M$. Conversely, if $x \in M$, then since $x \leq x$ it follows that $x \in M[x)$ and as $x \leq y$ for all $y \in M[x)$, it follows that x is the first element of $M[x)$, therefore $x \in M_{\varphi_M}$ as was seen before. So $M \subseteq M_{\varphi_M}$ and we have thus proved that $M = M_{\varphi_M}$, which shows that the map (1.2) is surjective. Therefore (1.2) is a *bijection between closure operators and Moore families* and (1.3) is its inverse. We shall simply write φ and M instead of φ_M and M_φ.

If L is a lattice then

$$(1.4) \qquad \varphi(x \wedge y) \leq \varphi)x) \wedge \varphi(y) \leq \varphi(x) \vee \varphi(y) \leq \varphi(x \vee y)$$

because $x \wedge y \leq x$ implies $\varphi(x \wedge y) \leq \varphi(x)$ and similarly $\varphi(x \wedge y) \leq \varphi(y)$, which establishes the first inequality (1.4), while the last inequality has a dual proof.

In this case a Moore family $M \subseteq L$ is a sub-meet-semilattice, i.e.,

$$(1.5) \qquad x \in M \,\&\, y \in M \implies x \wedge y \in M$$

because if $x, y \in M$ and φ is the closure associated with M, then $\varphi(x) \wedge \varphi(y) = x \wedge y \leq \varphi(x \wedge y)$ which, together with (1.4), imply $\varphi(x \wedge y) = x \wedge y$.

One proves similarly that if L is a complete lattice, then

$$(1.6) \qquad X \subseteq M \implies \inf X \in M \,;$$

this includes the case when $X = \emptyset$, as $\inf \emptyset = 1 \in M$ (because $1 \leq \varphi(1)$, hence $\varphi(1) = 1$). As is well known, (1.6) implies that *every Moore family is a complete lattice*, where the least upper bound in M of a subset $X \subseteq M$, denoted by $\sup_M X$, is $\inf \overline{X}$, where $\overline{X}$ denotes the set of upper bounds of X. One has in general $\sup X \leq \sup_M X$. In the specific case of a Moore family it is easily proved that

$$(1.7) \qquad \sup_M X = \varphi(\sup X) .$$

The concepts dual to the above ones are termed *dual closure operator* or *interior operator* and *dual Moore family*, respectively. The French school (see below) uses the terms *upper closure* (fermeture supérieure) and *upper net* (réseau

supérieur) for closure and Moore family, respectively, and *lower closure* (fermeture inférieure) and *lower net* (réseau inférieur) for the dual concepts. Of course,
the dual of the above results are valid.

Let us introduce the term *double Moore family* for a subset $M \subseteq L$ which is
both a Moore family and a dual Moore family of L. It follows from (1.5) and
its dual that a double Moore family of a lattice L is a sublattice of L. In the
case of a complete lattice L this result can be strengthened: a subset $M \subseteq L$ is
a double Moore family if and only if it is a sub-complete-lattice of L. Necessity
follows from (1.6) and its dual. Conversely, if M is a sub-complete-lattice of L,
then for every $x \in L$, we have $\inf M[x) \in M$ and since $x \leq y$ for every $y \in M[x)$,
it follows that $\inf M[x) \leq y$ for every $y \in M[x)$, therefore $\inf M[x) \in M[x)$,
showing that $\inf M[x)$ is in fact the first element of $M[x)$. Thus M is a Moore
family and also a dual Moore family, by the dual proof.

Proposition 1.1. (Lapscher, [1968], II.2, Théorème 1). *Suppose M is a double
Moore family and let $\overline{\varphi}$ and $\underline{\varphi}$ be the closure and the interior operators, respectively, associated with M. Then the following identities hold:*

(1.8)
$$\overline{\varphi}(x \vee y) = \overline{\varphi}(x) \vee \overline{\varphi}(y) \,,$$

(1.8′)
$$\underline{\varphi}(x \wedge y) = \underline{\varphi}(x) \wedge \underline{\varphi}(y) \,;$$

if the lattice is complete, then

(1.9)
$$\overline{\varphi}(\sup X) = \sup \overline{\varphi}(X) \,,$$

(1.9′)
$$\underline{\varphi}(\inf X) = \inf \underline{\varphi}(X) \,.$$

COMMENT: One says that $\overline{\varphi}$ is *additive*, while $\underline{\varphi}$ is *multiplicative*.
PROOF: Since M is a dual Moore family it follows that $0 \in M$, therefore
$\overline{\varphi}(\sup \varnothing) = \overline{\varphi}(0) = 0 = \sup \overline{\varphi}(\varnothing)$. Now suppose $X \neq \varnothing$. For every $x \in X$ we have

$$x \leq \overline{\varphi}(x) \leq \sup \overline{\varphi}(X) \text{ and } x \leq \sup X \,,$$

hence, using again the fact that M is a dual Moore family,

$$\sup X \leq \sup \overline{\varphi}(X) \in M \text{ and } \overline{\varphi}(X) \leq \overline{\varphi}(\sup X) \,,$$

therefore

$$\overline{\varphi}(\sup X) \leq \sup \overline{\varphi}(X) \text{ and } \sup \overline{\varphi}(X) \leq \overline{\varphi}(\sup X) \,,$$

which proves (1.9). The same proof with $X := \{x, y\}$ establishes (1.8), while
(1.8′) and (1.9′) follow by duality.   □

For a thorough study of the relationship between closure operators, Moore
families and their duals on semilattices, lattices and complete lattices, see Lapscher [1968].

We now turn to the case of Boolean algebras.

Proposition 1.2. (Lapscher [1968], II.2, Théorème 3). *Ley φ be a closure operator on a Boolean algebra $(B, \cdot, \vee, ', 0, 1)$ and M the Moore family associated with φ. Then the following conditions are equivalent:*

(i) M is closed with respect to $'$;

(ii) M is a Boolean subalgebra of B ;

(iii) $x \cdot \varphi(y) = 0 \Longrightarrow \varphi(x) \cdot \varphi(y) = 0$;

(iv) $\varphi(0) = 0$ & $\varphi(x \cdot \varphi(y)) = \varphi(x) \cdot \varphi(y)$.

COMMENT: According to P.R. Halmos, a closure φ satisfying $\varphi(x \cdot \varphi(y)) = \varphi(x) \cdot \varphi(y)$ is called a *hemimorphism*.

PROOF: (i)$\Longleftrightarrow$(ii): M is anyway closed with respect to meet, while $x \vee y = (x'y')'$.

(ii)$\Longrightarrow$(iv): $0 = 1'$ and $1 \in M$, hence $0 \in M$. Further, note that if $z \in M$ then $\varphi(xz) \leq \varphi(z) = z$, while the hypothesis implies $\varphi(a) \vee \varphi(b) = \sup_M \{\varphi(a), \varphi(b)\} = \varphi(a \vee b)$. Hence we infer in turn

$$\varphi(y)\varphi(x\varphi(y)) = \varphi(x\varphi(y)) \ \& \ \varphi(y)\varphi(x(\varphi(y))') = 0 ,$$

$$\varphi(y)\varphi(x) = \varphi(y)\varphi(x\varphi(y) \vee x(\varphi(y))') = \varphi(y)\varphi(x\varphi(y)) \vee \varphi(y)\varphi(x(\varphi(y))')$$
$$= \varphi(x\varphi(y)) .$$

(iv)$\Longrightarrow$(iii): Immediate.

(iii)$\Longrightarrow$(i): Take $x \in M$. Then $x'\varphi(x) = x'x = 0$, hence $\varphi(x')x = \varphi(x')\varphi(x) = 0$, therefore $\varphi(x') \leq x' \leq \varphi(x')$. □

Corollary 1.1. *Under the above conditions, M is a double Moore family. The associated interior operator φ° is related to φ by*

$$(1.10) \qquad\qquad \varphi^\circ x = (\varphi x')' \ \& \ \varphi x = (\varphi^\circ x')' ,$$

and φ satisfies (1.8), while φ° satisfies (1.8').

PROOF: It suffices to prove that $(\varphi x')'$ is the greatest element of M included in x. But $\varphi x' \in M$, hence $(\varphi x')' \in M$. Then $x' \leq \varphi x'$ implies $(\varphi x')' \leq x$. Finally if $y \in M$ and $y \leq x$ then $x' \leq y' \in M$, hence $\varphi x' \leq y'$, therefore $y \leq (\varphi x')'$. □

The study of closure operators on the Boolean algebra of switching functions (alias truth functions) of n variables in view of applications to logical design was initiated by Kuntzmann [1965] and continued by Lapscher [1968], Lavit [1974] and J. Deschamps [1975], [1977], [1990]. In this chapter and in Ch.11, §3, we present a few sample results in the general case of arbitrary (not necessarily simple) Boolean functions.

Notation. Given a Boolean algebra B, we denote by BFn the Boolean algebra of all Boolean functions $f : B^n \longrightarrow B$. □

A basic concept in switching theory is that of *uncompletey defined Boolean* (in fact, truth) *function*, which means a function with $0-1$ arguments and $0-1$ values, which may not be defined for certain values in $\{0, 1\}^n$. In universal algebra this is called a *partially defined function* $f : \{0, 1\}^n \xrightarrow{\ \circ\ } \{0, 1\}$, that is, a function

$f : \mathrm{dom} f \longrightarrow \{0,1\}$, where $\mathrm{dom} f \subseteq \{0,1\}^n$. Let $\underline{f}, \overline{f} : \{0,1\}^n \longrightarrow \{0,1\}$ be the extensions of f defined by $\underline{f}(A) = 0$ (by $\overline{f}(A) = 1$) for $A \in \{0,1\}^n \backslash \mathrm{dom} f$. Then $\underline{f}(A) \leq f(A) \leq \overline{f}(A)$ for every $A \in \mathrm{dom} f$. This suggests the following more general definition.

Definition 1.1. Let $\underline{f}, \overline{f} \in \mathrm{BFn}$ with $\underline{f} \leq \overline{f}$. Then the interval

(1.11) $$[\underline{f}, \overline{f}] = \{f \in \mathrm{BFn} \mid \underline{f} \leq f \leq \overline{f}\}$$

is called a *partially defined Boolean function* and every $f \in [\underline{f}, \overline{f}]$ is said to be a *representative* of $[\underline{f}, \overline{f}]$. □

Many problems in logical design reduce to problems of the following type: determine a convenient representative of a partially defined Boolean function, where "convenient" often means membership in a certain Moore family of Boolean functions; cf. the next sections. This explains the interest of the following result.

Proposition 1.3. (Lapscher [1968]). *Let* $[\underline{f}, \overline{f}]$ *be a partially defined Boolean function,* $\mathcal{M}$ *a double Moore family of* BFn *and* $\overline{\varphi}, \varphi$ *the closure and the interior operator, respectively, associated with* $\mathcal{M}$. *Then there exists a representative* $f \in [\underline{f}, \overline{f}] \cap \mathcal{M}$ *if and only if*

(1.12) $$\overline{\varphi} \underline{f} \leq \varphi \overline{f},$$

in which case these representatives form a sublattice of BFn *with first element* $\overline{\varphi} \underline{f}$ *and last element* $\varphi \overline{f}$.

PROOF: If $f \in [\underline{f}, \overline{f}] \cap \mathcal{M}$ then $\overline{\varphi} \underline{f} \leq \overline{\varphi} f = f = \varphi f \leq \varphi \overline{f}$, which proves necessity and the fact that the required representatives belong to $[\overline{\varphi} \underline{f}, \varphi \overline{f}]$. Conversely, if (1.12) holds, then both $\overline{\varphi} \underline{f}$ and $\varphi \overline{f}$ belong to $[\underline{f}, \overline{f}] \cap \mathcal{M}$. The last statement is obvious. □

Proposition 1.4. (J. Deschamps [1977], [1990]). *Let* $[\underline{f}, \overline{f}]$ *be a partially defined Boolean function,* $\mathcal{M}$ *a Moore family and* φ *the associated closure operator. Then there exists* $f \in [\underline{f}, \overline{f}] \cap \mathcal{M}$ *if and only if*

(1.13) $$\varphi \underline{f} \leq \overline{f}.$$

PROOF: Similar to (and simpler than) that of Proposition 1.2. □

In the rest of this chapter we study some specific closures.

2 Isotone and monotone closures

After the general presentation in §1, in this and the next section we study certain specific operators on the algebra of Boolean functions. Unless otherwise stated, we work with Boolean functions $f : B^p \longrightarrow B$ and use the following:

Notation. We denote a variable-vector and the set of its elements by the same capital letter. If X and Y are subvectors of the vector Z such that $X \cup Y = Z$ and $X \cap Y = \emptyset$, then we write $Z = (X, Y)$, regardless of the places of the components of X and Y in the vector Z. An equivalent notation is $Y = Z \backslash X$. We set $\dim X = n$ and $\dim Y = m$, where the *dimension* $\dim V$ of a vector V is the number of its components; so $n + m = p$. Unless otherwise stated, the letters A, C will stand for vectors in $\{0,1\}^n$.   □.

Definition 2.1. Let $k \in \{1, \ldots, p\}$. The Boolean function f is said to be:

(i) *isotone* or *increasing* in the variable z_i provided

$$\begin{aligned}(2.1)\qquad & a \leq b \Longrightarrow f(Z \backslash \{z_i\}, a) \leq f(Z \backslash \{z_i\}, b) \\ & (\forall a, b \in B ; \ (\forall Z \backslash \{z_i\} \in B^{p-1}) ;\end{aligned}$$

(ii) *antitone* or *decreasing* in the variable z_i provided

$$\begin{aligned}(2.2)\qquad & a \leq b \Longrightarrow f(Z \backslash \{z_i\}, a) \geq f(Z \backslash \{z_i\}, b) \\ & (\forall a, b \in B ; \ (\forall Z \backslash \{z_i\} \in B^{p-1}) ;\end{aligned}$$

(iii) *monotone* in the variable z_i provided it is isotone or antitone in z_i.   □

Definition 2.2. Let $Z = (X, Y)$ be the variable vector of the Boolean function f. Then f is called *isotone* or *increasing* (*antitone* or *decreasing* , *monotone*) in X provided it is isotone (antitone, monotone) in each variable of X; if $X = Z$ (hence $Y = \emptyset$), we simply say that f is *isotone* (*antitone*, *monotone*). Further, let $A \in \{0,1\}^n$. The function f will be called A-*monotone* provided it is isotone (antitone) in each variable x_i for which $\alpha_i = 1$ (for which $\alpha_i = 0$).   □

Remark 2.1. The function f is isotone (antitone) in X if and only if, for every $S, T \in B^n$ and every $Y \in B^m$,

$$S \leq T \Longrightarrow f(S, Y) \leq f(T, Y) \ (S \leq T \Longrightarrow f(S, Y) \geq f(T, Y)) .$$

□

The reader is referred to BFE, Chapter 11, §3, for several properties of isotone Boolean functions; one obtains by duality properties of antitone Boolean functions. For instance:

Lemma 2.1. *A Boolean function* $g : B^n \longrightarrow B$ *is isotone (antitone) if and only if its restriction to* $\{0,1\}^n$ *is isotone (antitone).*

FIRST PROOF: In view of Remark 2.1, the isotony condition can be written in the form

$$\bigvee_{i=1}^{n} x_i y_i' = 0 \Longrightarrow f(X) f'(Y) = 0 ,$$

which is equivalent to

$$f(A) f'(C) \leq \bigvee_{i=1}^{n} \alpha_i \gamma_i' \qquad (\forall A, C \in \{0,1\}^n)$$

by the Verification Theorem for functions of $2n$ variables. Since $\bigvee_{i=1}^{n} \alpha_i \gamma_i' \in \{0,1\}$, the last inequality is equivalent to

$$\bigvee_{i=1}^{n} \alpha_i \gamma_i' = 0 \implies f(A)f'(C) = 0 \, .$$

The proof for antitony is similar.

SECOND PROOF: From Proposition 5.3.13.   □

Theorem 2.1. *The set* Inc_X *of Boolean functions increasing in* X *and the set* Dec_X *of Boolean functions decreasing in* X *are double Moore families. The closure operators* inc_X, dec_X *and the interior operators* inc_X°, dec_X° *associated with them are given by the following formulas:*

$$(2.3) \qquad (\mathrm{inc}_X f)(X,Y) = \bigvee_A (\bigvee_{C \leq A} f(C,Y)) X^A \, ,$$

$$(2.4) \qquad (\mathrm{dec}_X f)(X,Y) = \bigvee_A (\bigvee_{C \geq A} f(C,Y)) X^A \, ,$$

$$(2.5) \qquad (\mathrm{inc}_X^{\circ} f)(X,Y) = \bigvee_A (\prod_{C \geq A} f(C,Y)) X^A \, ,$$

$$(2.6) \qquad (\mathrm{dec}_X^{\circ} f)(X,Y) = \bigvee_A (\prod_{C \leq A} f(C,Y)) X^A \, .$$

COMMENT: In the above formulas the index X is to be understood as a simplified notation for the set $\{i_1, \ldots, i_n\}$ of indices of the variables in the subvector X of Z, i.e., $X = (x_{i_1}, \ldots, x_{i_n})$.

PROOF: Denote the coefficients of X^A in formulas (2.3)–(2.6) by $i(A)$, $d(A)$, $i^{\circ}(A)$ and $d_{\circ}(A)$, respectively.

Then clearly

$$i^{\circ}(A), d^{\circ}(A) \leq f(A) \leq i(A), d(A) \, ,$$

thus proving that

$$\mathrm{inc}_X^{\circ} f, \mathrm{dec}_X^{\circ} f \leq f \leq \mathrm{inc}_X f, \mathrm{dec}_X f \, .$$

Now take A, C with $A \leq C$. Then $A_1 \leq A \implies A_1 \leq C$ and $A_1 \geq C \implies A_1 \geq A$, which implies that $i(A) \leq i(C)$, $d(C) \leq d(A)$, $i^{\circ}(C) \geq i^{\circ}(A)$ and $d^{\circ}(A) \geq d^{\circ}(C)$, hence it follows by Lemma 2.1 that $\mathrm{inc}_X f$ and $\mathrm{inc}_X^{\circ} f$ are isotone in X, while $\mathrm{dec}_X f$ and $\mathrm{dec}_X^{\circ} f$ are antitone in X.

Further, suppose $f \leq h$, where h is isotone in X. Then

$$i(A) \leq \bigvee_{C \leq A} h(C,Y) = h(A,Y)$$

for every A, therefore $\mathrm{inc}_X f \leq h$. We have thus proved that $\mathrm{inc}_X f$ is the least X−isotone upper bound of f and one concludes similarly that $\mathrm{dec}_X f$ is the least X−antitone upper bound of f, while $\mathrm{inc}_X^\circ f$ (while $\mathrm{dec}_X^\circ f$) is the greatest X−isotone (X−antitone) lower bound of f.   □

Most of Corollaries 2.1-2.7, Remark 2.2 and Proposition 2.1 are taken from Kuntzmann [1965], Chapitre VI, §2, and especially from Lapscher [1968], Chapitre III; however these authors confine to truth functions.

Corollary 2.1. *The operators* inc_X *and* dec_X *satisfy* (1.8), *while* inc_X° *and* dec_X° *satisfy* (1.8′).

PROOF: From Theorem 2.1 and Proposition 1.1.   □

Corollary 2.2. *Let* $\varphi \in \{\mathrm{inc}, \mathrm{dec}\}$. *If the Boolean function* g *is independent of* X, *then*

(2.7)
$$\varphi_X(fg) = (\varphi_X f)g \ \& \ \varphi_X^\circ(f \vee g) = \varphi_X^\circ f \vee g \,.$$

□

Corollary 2.3. *In the case of truth functions the following identities hold:*

(2.3′)
$$(\mathrm{inc}_X f)(X, Y) = \bigvee_{T \leq X} f(T, Y) \,,$$

(2.4′)
$$(\mathrm{dec}_X f)(X, Y) = \bigvee_{T \geq X} f(T, Y) \,,$$

(2.5′)
$$(\mathrm{inc}_X^\circ f)(X, Y) = \prod_{T \geq X} f(T, Y) \,,$$

(2.6′)
$$(\mathrm{dec}_X^\circ f)(X, Y) = \prod_{T \leq X} f(T, Y) \,.$$

PROOF: Formula (2.3) implies

(2.8)
$$(\mathrm{inc}_X f)(A, Y) = \bigvee_{C \leq A} f(C, Y) \qquad (\forall A \in \{0,1\}^n)$$

and in the case of truth functions (2.8) coincides with the identity (2.3′); similar proofs hold for (2.4′)−(2.6′).   □

Definition 2.3. For $C = (\gamma_1, \ldots, \gamma_n) \in \{0,1\}^n$ we set

(2.9)
$$\mathrm{Ker}\, C = \{i \mid \gamma_i = 0\} \ \& \ \mathrm{Coker}\, C = \{i \mid \gamma_i = 1\} \,.$$

□

Corollary 2.4. *Lat* $C \in \{0,1\}^n$ *and* f *a function of the form* $f(X, Y) = aX^C g(Y)$. *Then*

(2.10)
$$(\mathrm{inc}_X f)(X, Y) = a \prod \{x_i \mid i \in \mathrm{Coker}\, C\} g(Y) \,,$$

(2.11)
$$(\mathrm{dec}_X f)(X, Y) = a \prod \{x_k' \mid k \in \mathrm{Ker}\, C\} g(Y) \,.$$

PROOF: If $A \not\geq C$ then

$$A_1 \leq A \Longrightarrow A_1 \neq C \Longrightarrow f(A_1, Y) = 0 \,,$$

therefore (2.3) implies

$$(\mathrm{inc}_X f)(X, Y) = \bigvee_{A \geq C} (\bigvee_{A_1 \leq A} f(A_1, Y)) X^A = \bigvee_{A \geq C} f(C, Y) X^A$$

$$= f(C, Y) \bigvee_{A \geq C} X^A = ag(Y) \prod \{x_i \mid i \in \mathrm{Coker}\, C\}$$

and the proof of (2.11) is similar. □

Remark 2.2. It follows from Corollaries 2.1 and 2.4 that if a function f is expressed as a disjunction of terms of the form $aX^C g(Y)$, then $\mathrm{inc}_X f$ is obtained by deleting from each term all the factos x'_k , $k \in \mathrm{Ker}\, C$, while $\mathrm{dec}_X f$ is obtained by deleting from each term all the factors x_i , $i \in \mathrm{Coker}\, C$. □

To obtain $\mathrm{inc}^\circ_X f$ and $\mathrm{dec}^\circ_X f$ it is convenient to express f in dual form and apply the duals of the above properties. This task is left to the reader.

Example 2.1. Let us take $n := m := 1$, that is,

$$(2.12) \qquad f(x, y) = axy \vee bxy' \vee cx'y \vee dx'y' \,.$$

By applying Theorem 2.1 we obtain

$$(\mathrm{inc}_x f)(x, y) = (f(1, y) \vee f(0, y))x \vee f(0, y)x'$$

$$= (ay \vee by' \vee cy \vee dy')x \vee (cy \vee dy')x' = axy \vee bxy' \vee cy \vee dy' \,,$$

$$(\mathrm{dec}_x f)(x, y) = f(1, y)x \vee (f(1, y) \vee f(0, y))x'$$

$$= (ay \vee by')x \vee (ay \vee by' \vee cy \vee dy')x' = cx'y \vee dx'y' \vee ay \vee by' \,,$$

$$(\mathrm{inc}^\circ_x f)(x, y) = f(1, y)x \vee f(1, y)f(0, y)x'$$

$$= (ay \vee by')x \vee (ay \vee by')(cy \vee dy')x' = axy \vee bxy' \vee acy \vee bdy' \,,$$

$$(\mathrm{dec}^\circ_x f)(x, y) = f(1, y)f(0, y)x \vee f(0, y)x'$$

$$= (ay \vee by')(cy \vee dy')x \vee (cy \vee dy')x' = cx'y \vee dx'y' \vee acy \vee bdy' \,.$$

Note that the same result can be directly obtained by applying Remark 2.2. □

Proposition 2.1. *For every $W \subseteq X$, let $\varphi \in \{\mathrm{inc}_W, \mathrm{dec}_W, \mathrm{inc}^\circ_W, \mathrm{dec}^\circ_W\}$. Then for every two subvectors T, V of X,*

$$(2.13) \qquad \varphi_{T \cup V} = \varphi_T \circ \varphi_V \,.$$

PROOF: Let $\mathcal{M}_W$ denote any of the double Moore families Inc_W and Dec_W. It follows readily from Remark 2.1 that

$$(2.14) \qquad\qquad \mathcal{M}_{T \cup V} = \mathcal{M}_T \cap \mathcal{M}_V \ .$$

Now let φ_W be inc_W or dec_W; we will prove that for every f, $\varphi_T(\varphi_V f)$ is the least upper bound of f in $\mathcal{M}_{T \cup V}$.

But $f \leq \varphi_V f \leq \varphi_T \varphi_V f$. Further, note that in view of Theorem 2.1, if $g \in \mathcal{M}_V$ (i.e., g is V−isotone or V−antitone), then $\varphi_T g \in \mathcal{M}_V$ as well, therefore taking $g := \varphi_V f$ and using (2.14) we obtain $\varphi_T \varphi_V f \in \mathcal{M}_{T \cup V}$. Finally if $f \leq h \in \mathcal{M}_{T \cup V}$, then using again (2.14) we infer $\varphi_V f \leq \varphi_V h = h$, therefore $\varphi_T \varphi_V f \leq \varphi_T h = h$.

Property (2.13) for $\varphi_W \in \{\mathrm{inc}_W^\circ, \mathrm{dec}_W^\circ\}$ follows by duality.   □

Corollary 2.5. $\varphi_T \circ \varphi_V = \varphi_V \circ \varphi_T$.   □

Corollary 2.6. $\varphi_T, \varphi_V \leq \varphi_{T \cup V}$ for $\varphi_W \in \{\mathrm{inc}_W, \mathrm{dec}_W\}$ and $\varphi_T, \varphi_V \geq \varphi_{T \cup V}$ for $\varphi_W \in \{\mathrm{inc}_W^\circ, \mathrm{dec}_W^\circ\}$.   □

Corollary 2.7. If T is a subvector of V then $\varphi_T \leq \varphi_V$ for $\varphi_W \in \{\mathrm{inc}_W, \mathrm{dec}_W\}$ and $\varphi_V \leq \varphi_T$ for $\varphi_W \in \{\mathrm{inc}_W^\circ, \mathrm{dec}_W^\circ\}$.   □

Theorem 2.2. Let $A \in \{0,1\}^n$. The set Mon_A of A-monotone Boolean functions (cf. Definition 2.2) is a double Moore family and the associated closure operator mon_A and interior operator mon_A° satisfy

$$(2.15) \qquad \mathrm{mon}_A = \mathrm{inc}_{\mathrm{Coker}A} \circ \mathrm{dec}_{\mathrm{Ker}A} = \mathrm{dec}_{\mathrm{Ker}A} \circ \mathrm{inc}_{\mathrm{Coker}A} \ ,$$

$$(2.15') \qquad \mathrm{mon}_A^\circ = \mathrm{inc}_{\mathrm{Coker}A}^\circ \circ \mathrm{dec}_{\mathrm{Ker}A}^\circ = \mathrm{dec}_{\mathrm{Ker}A}^\circ \circ \mathrm{inc}_{\mathrm{Coker}A}^\circ \ .$$

COMMENT: $\mathrm{Ker}A$ and $\mathrm{Coker}A$ are sets of indices, therefore the notation used in formulas (2.15) and (2.15') is proper (unlike the notation in Theorem 2.1; cf. the comment to that theorem).

PROOF: Similar to the proof of Theorem 2.1, which is possible because

$$(2.16) \qquad\qquad \mathrm{Mon}_A = \mathrm{Inc}_{\mathrm{Coker}A} \cap \mathrm{Dec}_{\mathrm{Ker}A} \ .$$

□

Taking into account Corollaries 2.1−2.3 and Proposition 2.1, we see that Theorem 2.2 implies the following corollaries.

Corollary 2.8. The operator mon_A satisfies (1.8), while mon_A° satisfies (1.8').   □

Corollary 2.9. If the Boolean function g does not depend on the variables X, then

$$(2.17) \qquad \mathrm{mon}_A(fg) = (\mathrm{mon}_A f)g \ \& \ \mathrm{mon}_A^\circ(f \vee g) = \mathrm{mon}_A^\circ f \vee g \ .$$

□

Corollary 2.10. *Let T, V be subvectors of X, $T \cap V = \emptyset$, and $A \in \{0,1\}^{\dim T}$, $C \in \{0,1\}^{\dim V}$. Then*

(i) $\varphi_{A \cup C} = \varphi_A \circ \varphi_C = \varphi_C \circ \varphi_A$ $(\varphi \in \{\mathrm{mon}, \mathrm{mon}^\circ\})$;

(ii) mon_A, $\mathrm{mon}_C \leq \mathrm{mon}_{A \cup C}$ & mon_A°, $\mathrm{mon}_C^\circ \geq \mathrm{mon}_{A \cup C}$.   □

Corollary 2.11. *If T is a subvector of V, then $\mathrm{mon}_T \leq \mathrm{mon}_V$ and $\mathrm{mon}_T^\circ \geq \mathrm{mon}_V^\circ$.*   □

Remark 2.3. It follows from Theorem 2.2 and Remark 2.2 that if a function f is expressed as a disjunction of terms of the form $aX^C g(Y)$, then $\mathrm{mon}_A f$ is obtained by deleting from each term the factor x'_k, $k \in \mathrm{Ker}\,C$, and all the factors x_i, $i \in \mathrm{Coker}\,C$.   □

Proposition 2.2. *Take a vector $A \in \{0,1\}^n$. Denote $\mathrm{Ker}\,A = \{i_1, \dots, i_q\}$ and $\mathrm{Coker}\,A = \{k_1, \dots, k_r\}$ $(q + r = n)$. For any $X \in B^n$ set $X = (X^-, X^+)$, where $X^- = (x_{i_1}, \dots, x_{i_q})$, $X^+ = (x_{k_1}, \dots, x_{k_r})$. Then mon_A is given by formula*

$$(2.18) \qquad (\mathrm{mon}_A f)(X, Y) = \bigvee_C (\bigvee_{D^- \geq C^-} \bigvee_{D^+ \leq C^+} f(D, Y)) X^C .$$

PROOF: Denote by C_-, D_- vectors of $\{0,1\}^q$ and by C_+, D_+ vectors of $\{0,1\}^r$. It follows from Theorems 2.2 and 2.1 that

$$(\mathrm{mon}_A f)(X, Y) = (\mathrm{dec}_{\mathrm{Ker}\,A}(\mathrm{inc}_{\mathrm{Coker}\,A} f))(X^-, X^+, Y)$$

$$= \bigvee_{C_-} (\bigvee_{D_- \geq C_-} (\mathrm{inc}_{\mathrm{Coker}\,A} f)(C_-, X^+, Y))(X^-)^{C_-}$$

$$= \bigvee_{C_-} (\bigvee_{D_- \geq C_-} \bigvee_{C_+} (\bigvee_{D_+ \geq C_+} f(D_-, D_+, Y))(X^+)^{C_+})(X^-)^{C_-}$$

$$= \bigvee_{C_-} \bigvee_{D_- \geq C_-} \bigvee_{C_+} \bigvee_{D_+ \geq C_+} f(D_-, D_+, Y)(X^+)^{C_+}(X^-)^{C_-}$$

$$= \bigvee_{C_- C_+} \bigvee_{D_- \geq C_-} \bigvee_{D_+ \geq C_+} f(D_-, D_+, Y)X^{(C_-, C_+)} ,$$

which coincides with (2.18), because if we set $(C_-, C_+) = C$ and $(D_-, D_+) = D$, then the components of $C, D \in \{0,1\}^n$ are $C^- = C_-$, $C^+ = C_+$ and $D^- = D_-$, $D^+ = D_+$; besides, C runs over $\{0,1\}^n$.   □

Example 2.2. Let us take again the function (2.12) in Example 2.1 and $A = (1,0)$. Taking into account the results obtained in Example 2.1, we see that the function $g = \mathrm{inc}_x f$ is given by

$$g(x, y) = axy \vee bxy' \vee cy \vee dy' ,$$

hence

$$(\mathrm{mon}_{(1,0)}f)(x,y) = (\mathrm{dec}_y g)(x,y) = ax \vee bxy' \vee c \vee dy' \ ,$$

while the function $h = \mathrm{inc}_x^\circ f$ is given by

$$h(x,y) = axy \vee bxy' \vee acy \vee bdy' \ ,$$

therefore

$$(\mathrm{mon}_{(1,0)}^\circ f)(x,y) = (\mathrm{dec}_y^\circ h)(x,y) = h(x,1)h(x,0)y \vee h(x,0)y'$$

$$= (ax \vee ac)(bx \vee bd)y \vee (bx \vee bd)y' = ab(x \vee cd)y \vee b(x \vee d)y' \ .$$

We can also apply Proposition 2.3 and obtain

$$(\mathrm{mon}_{(1,0)}f)(x,y)$$

$$= (\bigvee_\varepsilon f(0,\varepsilon))x'y' \vee f(0,1)x'y \vee (\bigvee_\alpha \bigvee_\beta f(\alpha,\beta))xy' \vee (\bigvee_\alpha f(\alpha,1))xy$$

$$= (c \vee d)x'y' \vee cx'y \vee (a \vee b \vee c \vee d)xy' \vee (a \vee c)xy = ax \vee bxy' \vee c \vee dy' \ ,$$

as expected. The same result can be readily obtained by Remark 2.3.   □

Remark 2.4. The inequality $\varphi(fg) \leq (\varphi f)(\varphi g)$ holds for any closure φ (this follows immediately from $fg \leq f,g$). If $\varphi \in \{\mathrm{inc}_X, \mathrm{dec}_X, \mathrm{mon}_A\}$ and g is independent of X, then $\varphi g = g$ and Corollaries 2.2 and 2.9 yield the equality $\varphi(fg) = (\varphi f)g = (\varphi f)(\varphi g)$. This result cannot be sharpened by dropping the condition on g. Take e.g. $g := f'$. Then $\varphi(ff') = \varphi 0 = 0$, while e.g. $(\mathrm{inc}_x(ax \vee bx'))(\mathrm{inc}_x(a'x \vee b'x')) = (ax \vee b)(a'x \vee b') = (a+b)x.$ □

Lapscher [1968] has also studied some other Moore families of Boolean functions and the associated closure and/or interior operations, e.g., decomposable functions and antipermutable functions; the latter class includes the subclass of even functions and that of symmetric functions. The investigation was continued by Lavit [1974]. It may be considered that the research of the French school is directed towards the construction of partially defined Boolean functions having certain properties, in view of applications to switching theory. See also the next section and Ch.11, §3.

We conclude this section with a few results obtained in Rudeanu [1983a]. As a matter of fact, one of them is formula (2.3), but the focus of that paper is different.

Andreoli [1961] noted that for every Boolean function $f : B^n \longrightarrow B$, the function $f \circ f$ is isotone and, moreover, f is isotone if and only if $f \circ f = f$. For if $f(x) = ax \vee bx'$, then $(f \circ f)(x) = (a \vee b)x \vee abx'$, while f is isotone iff $a \geq b$, which is equivalent to $a \vee b = a$, $a \wedge b = b$. This property was generalized by Tošić [1978a] for functions of several variables, but only in the case of finite Boolean algebras.

This suggests the following

Definition 2.4. Let BFn denote the set of all Boolean functions $f : B^n \longrightarrow B$ and IBFn the subset of isotone Boolean functions. A mapping $\delta : \text{BFn} \longrightarrow \text{IBFn}$ is said to be an *isotony detector* provided $\text{IBFn} = \{f \in \text{BFn} \mid f = \delta f\}$. A mapping $\delta : \text{BFn} \longrightarrow \text{BFn}$ is called a *Boolean operator* provided each *discriminant* $(\delta f)(A)$, $A \in \{0,1\}^n$, of δf is a Boolean function of the discriminants $f(C)$, $C \in \{0,1\}^n$, of the function f. By a *Boolean isotony detector* we mean any isotony detector which is also a Boolean operator.   □

For instance, the Andreoli map $\delta f = f \circ f$ is a Boolean isotony detector, as well as its generalization by Tošić. Also, the isotone closure inc and the isotone interior inc°, obtained from inc_X and inc_X° in Theorem 2.1 for $m := 0$, are Boolean isotony detectors. We are now going to determine all Boolean isotony detectors.

Proposition 2.3. *A Boolean operator* $\delta : \text{BFn} \longrightarrow \text{BFn}$ *is a Boolean isotony detector if and only if, for every* $f \in \text{BFn}$ *and every* $A \in \{0,1\}^n$,

$$(2.19) \quad \begin{aligned} &\left(\textstyle\prod_{C \geq A} f(C)\right) \textstyle\prod_{E \not\geq A} \textstyle\prod_{D < E} \\ &\left(f'(D) \vee f(E) \vee \textstyle\bigvee_{E < A}(\delta f)(E) \leq (\delta f)(A) \leq \textstyle\bigvee_{C \leq A} f(C) \vee \right. \\ &\left. \textstyle\bigvee_{D \not\leq A} \textstyle\bigvee_{E > D} f(D)f'(E)\right). \end{aligned}$$

COMMENT: Conditions (2.19), written in the increasing order of the vectors A, yield a tree-like construction of all Boolean isotony detectors.

PROOF: The Boolean operator δ is an isotony detector if and only if for every f, (i) δf is isotone, and (ii) f is isotone $\Longrightarrow f = \delta f$ (as the converse implication follows from (i)). Taking into account Lemma 2.1, we see that (i) and (ii) are equivalent to

$$(2.20) \quad \bigvee_{E < A} (\delta f)(D) \leq (\delta f)(A) \qquad (\forall A \in \{0,1\}^n)$$

and

$$(2.21) \quad \begin{aligned} &f(A) = \textstyle\bigvee_{C \leq A} f(C) \ (\forall A \in \{0,1\}^n \Longrightarrow \\ &\Longrightarrow f(A) = (\delta f)(A) \qquad (\forall A \in \{0,1\}^n), \end{aligned}$$

respectively. Setting $f(A) = x_A$ and $(\delta f)(A) = y_A$, condition (2.21) can be successively written in the following equivalent forms:

$$x_A + \bigvee_{C \leq A} x_C = 0 \ (\forall A \in \{0,1\}^n) \Longrightarrow x_A = y_A \ (\forall A \in \{0,1\}^n),$$

$$\bigvee_A \left(x_A + \bigvee_{C \leq A} x_C\right) = 0 \Longrightarrow \bigvee_A (x_A + y_A) = 0,$$

$$\bigvee_A (x_A + y_A) \leq \bigvee_A \left(x_A + \bigvee_{C \leq A} x_C\right),$$

$$\left(\bigvee_A (x_A y_A' \vee x_A' y_A)\right) \prod_C \left(x_C \bigvee_{D \leq C} x_D \vee x_C' \prod_{D \leq C} x_D'\right) = 0,$$

$$\bigvee_A (x_A y'_A \vee x'_A y_A) \prod_C (x_C \vee \prod_{D \leq C} x'_D) = 0 \,,$$

$$x_A y'_A \prod_C (x_C \vee \prod_{D < C} x'_D) \vee x'_A y_A \prod_C (x_C \vee \prod_{D < C} x'_D) = 0 \; (\forall A \in \{0,1\}^n) \,,$$

$$x_A \prod_{C \neq A} (x_C \vee \prod_{D < C} x'_D) \leq y_A \leq x_A \vee \bigvee_C x'_C \bigvee_{D < C} x_D \; (\forall A \in \{0,1\}^n) \,,$$

$$x_A (\prod_{C > A} x_C) \prod_{C \nleq A} (x_C \vee \prod_{D < C} x'_D) \leq y_A$$

$$\leq x_A \vee \bigvee_{D < A} x_D \vee \bigvee_{C \neq A} x'_C \bigvee_{D < C} x_D \; (\forall A \in \{0,1\}^n) \,,$$

(2.22)
$$(\prod_{C \geq A} x_C) \prod_{C \nleq A} \prod_{D < C} (x_C \vee x'_D) \leq y_A$$

$$\leq \bigvee_{D \nleq A} x_D \vee \bigvee_{C \neq A} \bigvee_{D < C, D \nleq A} x'_C x_D \quad (\forall A \in \{0,1\}^n) \,,$$

and since $D < C \; \& \; D \nleq A \Longrightarrow C \neq A$, we see that (2.20) & (2.22) $\Longleftrightarrow$ (2.19).   □

Corollary 2.12. *A Boolean operator δ which satisfies*

(2.23) δf is isotone and $\mathrm{inc}^\circ f \leq \delta f \leq \mathrm{inc} f$ $(\forall f \in \mathrm{IBFn})$

is a Boolean isotony detector.

PROOF: Let $m(A) \leq (\delta f)(A) \leq M(A)$ denote the inequalities (2.19). Then (2.23) implies

$$m(A) \leq \prod_{C \geq A} f(C) \vee (\delta f)(A) = (\mathrm{inc}^\circ f)(A) \vee (\delta f)(A)$$

$$\leq (\delta f)(A) \leq (\mathrm{inc} f)(A) = \bigvee_{C \leq A} f(C) = M(A) \,.$$

□

Corollary 2.13. *A Boolean operator $\delta : \mathrm{BF1} \longrightarrow \mathrm{BF1}$ is a Boolean isotony detector if and only if it satisfies (2.23).*

PROOF: For $n := 1$ conditions (2.19) become (using $\prod \emptyset = 1$ and $\bigvee \emptyset = 0$)

$$f(0) f(1) \leq (\delta f)(0) \leq f(0) \,,$$

$$f(1) \vee (\delta f)(0) \leq (\delta f)(1) \leq f(0) \vee f(1) \,,$$

and they can be written in the form

$$(\mathrm{inc}^\circ f)(0) \leq (\delta f)(0) \leq \mathrm{inc} f)(0) \,,$$

$$(\mathrm{inc}^\circ f)(1) \vee (\delta f)(0) \leq (\delta f)(1) \leq (\mathrm{inc} f)(1) \,,$$

hence they imply (2.23)   □

Definition 2.5. For every $f \in$ BFn and every $i \in \{1, \ldots, n\}$, define the Boolean function f_i by

(2.24) $$f_i(x_1, \ldots, x_n) = f(x_1, \ldots, x_{i-1}, f(x_1, \ldots, x_n), x_{i+1}, \ldots, x_n)$$

and for every permutation $(i_1, \ldots, i_n)$ of $(1, \ldots, n)$, let

(2.25) $$f_{i_1 \ldots i_n} = f_{i_1}(f_{i_2}(\ldots (f_{i_n})) \ldots) .$$

$\square$

Lemma 2.2. *1) The operator $f \mapsto f_i$ is Boolean. 2) The function f_i is isotone in x_i and in each variable x_j such that f is isotone in x_j.*

PROOF: Set $Y = (x_1, \ldots, x_{i-1}, x_{i+1}, \ldots, x_n)$, or equivalently, $X = (Y, x_i)$. Also, let A denote vectors in $\{0, 1\}^n$ and C, D vectors in $\{0, 1\}^{n-1}$. Then

$$f(X) = \bigvee_A f(A) X^A = \bigvee_C f(C, 1) Y^C x_i \vee \bigvee_C f(C, 0) Y^C x_i' ,$$

$$f_i(X) = \bigvee_C f(C, 1) Y^C (\bigvee_D f(D, 1) Y^D x_i \vee \bigvee_D f(D, 0) Y^D x_i') \vee$$

$$\vee \bigvee_C f(C, 0) Y^C (\bigvee_D f'(D, 1) Y^D x_i \vee \bigvee_D f'(D, 0) Y^D x_i')$$

$$= \bigvee_C f(C, 1) Y^C x_i \vee \bigvee_C f(C, 1) f(C, 0) Y^C x_i' \vee \bigvee_C f(C, 0) f'(C, 1) Y^C x_i$$

$$= \bigvee_C (f(C, 1) \vee f(C, 0)) Y^C x_i \vee \bigvee_C f(C, 1) f(C, 0) .$$

Thus f_i is clearly isotone in x_i. Further, if f is isotone in x_j and $x_j \le y_j$ then $f(\ldots, x_j, \ldots) \le f(\ldots, y_j, \ldots)$, hence

$$f_i(\ldots, x_j, \ldots) = f(\ldots, x_j, \ldots, f(\ldots, x_j, \ldots), \ldots)$$

$$\le f(\ldots, x_j, \ldots, f(\ldots, y_j, \ldots), \ldots)$$

$$\le f(\ldots, y_j, \ldots, f(\ldots, y_j, \ldots), \ldots) = f_i(\ldots, y_j, \ldots) ,$$

where the remaining variables are fixed. Finally note that if $A = (C, \alpha_i)$, then

$$f(A) = (f(C, 1) \vee f(C, 0)) \alpha_i \vee \bigvee_D f(D, 1) f(D, 0) .$$

$\square$

The following result was first proved by Tošić [1978a] in the particular case of finite Boolean algebras.

Proposition 2.4. *For every permutation $(i_1, \ldots, i_n)$ of $(1, \ldots, n)$, the operator*

(2.26) $$\delta : \text{BFn} \longrightarrow \text{BFn}, \qquad \delta f = f_{i_1 \ldots i_n} ,$$

is a Boolean isotony detector.

COMMENT: See also square roots in Ch. 11, §2.

PROOF: A repeated application of Lemma 2.2 shows that δ is a Boolean operator and δf is isotone in each variable, hence it is an increasing Boolean function. Now we suppose f is an isotone Boolean function and must prove that $f = f_{i_1 \dots i_n}$. But we have seen that the Andreoli operator $f \mapsto f \circ f$ is an isotony detector for Boolean functions of one variable. It follows that $f = f_{i_1}$ and f_{i_1} is also isotone by Lemma 2.2. A repeated application of this argument yields $f_{i_1} = f_{i_1 i_2}, \dots, f_{i_1 \dots i_{n-1}} = f_{i_1 \dots i_{n-1} i_n}$, therefore $f = f_{i_1 \dots i_n}$.   □

Example 2.3. For $n := 2$ formulas (2.19) become

$$f(0,0)f(0,1)f(1,0)f(1,1) \le (\delta f)(0,0)$$

$$\le f(0,0) \vee f(0,1)f'(1,1) \vee f(1,0)f'(1,1) ,$$

$$f(0,1)f(1,1)(f'(0,0) \vee f(1,0)) \vee (\delta f)(0,0) \le (\delta f)(0,1)$$

$$\le f(0,0) \vee f(0,1) \vee f(1,0)f'(1,1) ,$$

$$f(1,0)f(1,1)(f'(0,0) \vee f(0,1)) \vee (\delta f)(0,0) \le (\delta f)(1,0)$$

$$\le f(0,0) \vee f(1,0) \vee f(0,1)f'(1,1) ,$$

$$f(1,1)(f'(0,0) \vee f(0,1))(f'(0,0) \vee f(1,0)) \vee (\delta f)(0,0) \vee (\delta f)(0,1) \vee (\delta f)(1,0)$$

$$\le (\delta f)(1,1) \le f(0,0) \vee f(0,1) \vee f(1,0) \vee f(1,1) .$$

□

The task of performing a similar study for *Boolean antitony detectors* is left to the reader.

We conclude by noting that an isotone truth function f has a unique representation of the form $f = t_1 \vee \dots \vee t_m$, where $t_1, \dots, t_m$ are pairwise uncomparable minterms in which all variables are uncomplemented. Tošić [1980] describes the function f by an $m \times n$ matrix whose rows are associated with $t_1, \dots, t_m$ in an obvious way and uses this representation to count the number of elements in various subclasses of isotone truth functions.

3 Independent and decomposition closures

After the presentation of the isotone and monotone operators in §2, in this section we first deal with independent closures and their relationship with the previous operators and with the concepts of consequence and consistency; a new independent closure is introduced for simple Boolean functions. We conclude with a closure operator related to the decomposition of Boolean functions.

Theorem 3.1 and its corollaries, Propositions 3.1–3.4 and Remark 3.1 below can be found in Lapscher [1968], especially in Chapitre III, §2. Yet Theorem 3.1 should be related to the well-known Schröder theorem asserting that the range of a Boolean function $f : B^n \longrightarrow B$ is the interval $[\prod f(A), \bigvee f(A)]$ (see e.g. BFE, Theorem 2.4).

Definition 3.1. A Boolean function $f : B^{n+m} \longrightarrow B$ is *independent* of the variables $X \in B^n$ provided $f(X, Y) = f(T, Y)$ for every $X, T \in B^n$ and every $Y \in B^m$. □

Theorem 3.1. *The set* Ind_X *of Boolean functions* $f : B^{n+m} \longrightarrow B$ *independent of X is a double Moore family. The closure operator* ind_X *and the interior operator* ind_X° *associated with* Ind_X *are given by formulas*

(3.1)
$$(\mathrm{ind}_X f)(Y) = \bigvee_A f(A, Y),$$

(3.2)
$$(\mathrm{ind}_X^\circ f)(Y) = \prod_A f(A, Y).$$

PROOF: The property $f(X, Y) \leq \bigvee_A f(A, Y)$ is well known and it implies that if a Boolean function $h : B^{n+m} \longrightarrow B$ independent of X satisfies

$$f(X, Y) \leq h(Y) \ (\forall X \in B^n) \qquad (\forall Y \in B^m),$$

then $\mathrm{ind}_X f \leq h$. The proof is completed by Corollary 1.1. □

Corollary 3.1. ind_X *satisfies* (1.8), *while* ind_X° *satisfies* (1.8'). □

Corollary 3.2. *If the Boolean function g is independent of X, then*

(3.3)
$$\mathrm{ind}_X(fg) = (\mathrm{ind}_X f)g \ \& \ \mathrm{ind}_X^\circ(f \vee g) = \mathrm{ind}_X^\circ f \vee g.$$

□

Remark 3.1. This result cannot be sharpened by dropping the condition on g. For instance, $\mathrm{ind}_X(ff') = \mathrm{ind}_X 0 = 0$, whereas e.g.

$$(\mathrm{ind}_x(ax \vee bx'))(\mathrm{ind}_x(a'x \vee b'x')) = (a \vee b)(a' \vee b') = a + b.$$

□

Proposition 3.1. *The following relations hold:*

(3.4)
$$\mathrm{ind}_X = \mathrm{dec}_X \circ \mathrm{inc}_X = \mathrm{inc}_X \circ \mathrm{dec}_X,$$

(3.5)
$$\mathrm{ind}_X^\circ = \mathrm{dec}_X^\circ \circ \mathrm{inc}_X^\circ = \mathrm{inc}_X^\circ \circ \mathrm{dec}_X^\circ.$$

PROOF: It suffices to prove (3.4). For every $A \in \{0, 1\}^n$,

$$(\mathrm{dec}_X(\mathrm{inc}_X f))(A, Y) = \bigvee_{C \geq A} (\mathrm{inc}_X f)(C, Y) = \bigvee_{C \geq A} \bigvee_{D \leq C} f(D, Y)$$

$$\leq \bigvee_D f(D, Y) \leq \bigvee_{C \geq A} \bigvee_{D \leq C} f(D, Y);$$

the last inequality holds because $(1,\ldots,1) \geq A$ and $D \leq (1,\ldots,1)$. Thus

$$(\mathrm{dec}_X(\mathrm{inx}_X f))(A,Y) = \bigvee_C f(C,Y) = (\mathrm{ind}_X f)(Y) \ (\forall A \in \{0,1\}^n) \,.$$

$\square$

Remark 3.2. It follows from Proposition 3.1 and Remark 2.2 that if a function f is expressed as a disjunction of terms of the form $aX^C g(Y)$, then $\mathrm{ind}_X f$ is obtained by deleting from each term all the factors x and x' with $x \in X$. $\square$

Proposition 3.2. Let $A \in \{0,1\}^n$ and define A' componentwise. Then

$$(3.6) \qquad\qquad \mathrm{ind}_X = \mathrm{mon}_{A'} \circ \mathrm{mon}_A = \mathrm{mon}_A \circ \mathrm{mon}_{A'} \,.$$

PROOF: By Remarks 2.3 and 3.1. $\square$

Proposition 3.3. Let T,V be subsets of X. Then:

(i) $\mathrm{ind}_{T \cup V} = \mathrm{ind}_T \circ \mathrm{ind}_V = \mathrm{ind}_V \circ \mathrm{ind}_T$;

(ii) $\mathrm{ind}_T, \mathrm{ind}_V \leq \mathrm{ind}_{T \cup V}$ and $\mathrm{ind}_T^\circ, \mathrm{ind}_V^\circ \geq \mathrm{ind}_{T \cup V}^\circ$;

(iii) if T is a subvector of V then $\mathrm{ind}_T \leq \mathrm{ind}_V$ and $\mathrm{ind}_T^\circ \geq \mathrm{ind}_V^\circ$.

PROOF: (i) follows by Remark 3.2; (ii) and (iii) are then obvious. $\square$

Proposition 3.4. For every Boolean function $f : B^{n+m} \longrightarrow B$,

$$(3.7) \qquad\qquad (\mathrm{ind}_X f)(Y) = (\mathrm{mon}_A f)(A,Y) \qquad (\forall A \in \{0,1\}^n)\,(\forall Y \in B^m)\,.$$

PROOF: We use Proposition 2.2 and note first that $A^- = (0,\ldots,0) \in \{0,1\}^r$ and $A^+ = (1,\ldots,1) \in \{0,1\}^r$. Then

$$(\mathrm{mon}_A f)(A,Y) = \bigvee_{C^- \geq A^-}\; \bigvee_{C^+ \leq A^+} f(C,Y)$$

$$= \bigvee_{C^-} \bigvee_{C^+} f(C,Y) = \bigvee_C f(C,Y) = (\mathrm{ind}_X f)(Y)\,.$$

$\square$

It is recommended that $\mathrm{ind}_X f$ be calculated by a conjoint application of (some of) the above properties. Yet the computation may be tedious.

Example 3.1. Take $n := 3, m := 4$ and $f = g_1 g_2 g_3 g_4$, where

$$g_1 = (ax_1 \vee x_2)y_1 \vee (a' \vee x_1')x_2'y_1' \,, \quad g_2 = ax_1 y_2 \vee (a' \vee x_1')y_2' \,,$$

$$g_3 = (x_1 \vee a'x_3)y_3 \vee x_1'(a \vee x_3')y_3' \,, \quad g_4 = x_2 y_4 \vee x_2'y_4' \,.$$

Then

$$\mathrm{ind}_{(x_3)}g_3 = \mathrm{ind}_{(x_3)}((x_1 \vee a'x_3)y_3 \vee x_1'(a \vee x_3')y_3') = (x_1 \vee a')y_3 \vee x_1'y_3' \, ,$$

$$g_1g_4 = ((ax_1 \vee x_2)y_1 \vee (a' \vee x_1')x_2'y_1')(x_2y_4 \vee x_2'y_4')$$

$$= x_2y_1y_4 \vee ax_1x_2'y_1y_4' \vee (a' \vee x_1')x_2'y_1'y_4' \, ,$$

$$\mathrm{ind}_{(x_2)}(g_1g_4) = y_1y_4 \vee ax_1y_1y_4' \vee (a' \vee x_1')y_1'y_4' \, ,$$

$$\mathrm{ind}_X f = \mathrm{ind}_{(x_1,x_2)}(\mathrm{ind}_{(x_3)}g_1g_2g_3g_4)$$

$$= \mathrm{ind}_{(x_1)}(\mathrm{ind}_{(x_2)}(g_1g_2g_4\mathrm{ind}_{(x_3)}g_3)) = \mathrm{ind}_{(x_1)}(g_2(\mathrm{ind}_{(x_3)}g_3)(\mathrm{ind}_{(x_2)}(g_1g_4))) \, ,$$

$$g_2(\mathrm{ind}_{(x_3)}g_3) = (ax_1y_2 \vee (a' \vee x_1')y_2')((x_1 \vee a')y_3 \vee x_1'y_3')$$

$$= ax_1y_2y_3 \vee a'y_2'y_3 \vee x_1'y_2'y_3' \, ,$$

$$g_2(\mathrm{ind}_{(x_3)}g_3)(\mathrm{ind}_{(x_2)}(g_1g_4))$$

$$= y_1y_4(ax_1y_2y_3 \vee a'y_2'y_3 \vee x_1'y_2'y_3') \vee ax_1y_1y_4'y_2y_3 \vee$$

$$\vee a'y_1'y_4'y_2'y_3 \vee x_1'y_1'y_4'y_2'y_3' \, ,$$

$$\mathrm{ind}_X f = y_1y_4(ay_2y_3 \vee a'y_2'y_3 \vee y_2'y_3') \vee ay_1y_2y_3y_4' \vee a'y_1'y_2'y_3y_4' \vee y_1'y_2'y_3'y_4' \, .$$

$\square$

J. Deschamps [1975], [1977], [1990] introduced a new technique in the study of closure operators on truth functions, with applications to disjoint decompositions of truth functions and to partially defined truth functions. The technique consists in duplicating the variables of each Boolean function and working with a new closure operator on the enlarged set of Boolean functions. Besides, the 1990 paper uses an axiomatic framework which includes in a compact form the closures inc_X, dec_X, mon_A and ind_X. We present below (Definitions 3.2–3.4 and Propositions 3.5–3.9) the Deschamps approach with certain modifications which seem to us indispensable.

Definition 3.2. In the set BF2p of Boolean functions $f : B^{2p} \longrightarrow B$ let the variable-vector be denoted by (Z, W), where $Z = (z_1, \ldots, z_p)$ and $W = (w_1, \ldots, w_p)$. We associate each subvector $X = (z_{i_1}, \ldots, z_{i_n})$ of Z with the subvector $T = (w_{i_1}, \ldots, w_{i_n})$ of W. For the convenience of notation, sometimes we re-denote $X = (x_1, \ldots, x_n), T = (t_1, \ldots, t_n)$. If $A = (\alpha_1, \ldots, \alpha_n) \in \{0,1\}^n$, we denote by $\widetilde{X}^A$ or simply by $\widetilde{X}$ the vector

$$\widetilde{X} = (x_1^{\alpha_1}, \ldots, x_n^{\alpha_n}) = (z_{i_1}^{\alpha_1}, \ldots, z_{i_n}^{\alpha_n}) \, .$$

$\square$

Definition 3.3. A family of closure operators $\varphi_{(X,A)}$ (also denoted by $\varphi_{\widetilde{X}_A}$ or simply by $\varphi_{\widetilde{X}}$) on BF2p, associated with the pairs (X, A), where X is a subvector of Z and $A \in \{0,1\}^{\dim X}$ (cf. Definition 3.2), is called a *system of relative closures* provided the following conditions are fulfilled:

(i) $\varphi_{\widetilde{X}}(f \vee g) = \varphi_{\widetilde{X}}f \vee \varphi_{\widetilde{X}}g$ (additivity) ;

(ii) if the Boolean function g is independent of X, then

$$\varphi_{\widetilde{X}}(gf) = g\varphi_{\widetilde{X}}f \; ;$$

(iii) if $X \cap Y = \emptyset$ then $\varphi_{(X,A)} \circ \varphi_{(Y,B)} = \varphi_{(X \cup Y, A \cup B)}$;

(iv) if the function f is independent of certain variables Y, then so is $\varphi_{\widetilde{X}}f$.   □

The notation inc_X, dec_X, ind_X used so far can be regarded as being of the form $\varphi_{(X,A)}$, where A is the vector $(1,\ldots,1)$, so that $\widetilde{X} = X$. Also, the notation mon_A can be regarded as $\mathrm{mon}_{(X,A)}$, where X is implicitly given by A. The above properties (i)–(iii) have been proved for all of the closures inc_X, dec_X, mon_A and ind_X. As a matter of fact, inc_X, dec_X and ind_X satisfy a stronger form of (iii), in which the hypothesis $X \cap Y = \emptyset$ is not assumed, while for mon_A this hypothesis is needed in the very definition of $\mathrm{mon}_{A \cup B}$. Property (iv), which is trivial for all of the above four specific closures, enables us to regard them as being defined on BF2p, not merely on BFp.

Definition 3.4. Consider the framework and notation established in Definitions 3.2 and 3.3. Associate with every $\widetilde{X} = (X, A)$ and every $C \in \{0,1\}^{\dim X}$, the subvectors

(3.8) $X|\mathrm{Ker}C = (x_{h_1}, \ldots, x_{h_r})$, $A|\mathrm{Ker}C = (\alpha_{h_1}, \ldots, \alpha_{h_r})$,

where $h_1, \ldots, h_r$ are the pairwise distinct elements of the set $\mathrm{Ker}C$ (cf. Definition 2.3). Set further

(3.8′) $\widetilde{X}|\mathrm{Ker}C = (X|\mathrm{Ker}C, A|\mathrm{Ker}C)$

and for any $f \in$ BF2p define

(3.9) $\Phi_{\widetilde{X}}f = \bigvee_C T^C \varphi_{\widetilde{X}|\mathrm{Ker}C}f$,

where C runs over $\{0,1\}^{\dim X}$ and T is associated with X in Definition 3.2. By convention we set $\varphi_{\widetilde{X}|\emptyset} = \varphi_{\emptyset} =$ the identity mapping on BF2p.   □

Example 3.2. Take $p := 3$, that is, $Z = (z_1, z_2, z_3)$ and $W = (w_1, w_2, w_3)$. The vector T associated with $X := (z_1, z_3)$ is $T = (w_1, w_3)$. Take $A := (1, 0)$. Then $\widetilde{X} = (w_1, w_3')$ and for every $f \in$ BF6,

$$\Phi_{(z_1, z_3')}f = w_1'w_3'\varphi_{(z_1, z_3')}f \vee w_1'w_3\varphi_{(z_1)}f \vee w_1w_3'\varphi_{(z_3')}f \vee w_1w_3f \;.$$

□

Proposition 3.5. *For every $\widetilde{X}$, $\Phi_{\widetilde{X}}$ is a closure operator which satisfies properties* (i)−(iii) *in Definition* 3.3.

PROOF: Each property of Φ follows from the corresponding property of φ. Property (iv) and additivity are clear; the latter implies isotony. Then

$$\Phi_{\widetilde{X}}f \geq \bigvee_{C} T^C f = f \bigvee_{C} T^C = f ,$$

$$\Phi_{\widetilde{X}}(\Phi_{\widetilde{X}}f) = \bigvee_{C} T^C \varphi_{\widetilde{X}|\mathrm{Ker}C}(\bigvee_{D} T^D \varphi_{\widetilde{X}|\mathrm{Ker}D}f)$$

$$= \bigvee_{C} T^C \bigvee_{D} T^D \varphi_{\widetilde{X}|\mathrm{Ker}C}(\varphi_{\widetilde{X}|\mathrm{Ker}D}f)$$

$$= \bigvee_{C} T^C \varphi_{\widetilde{X}|\mathrm{Ker}C}(\varphi_{\widetilde{X}|\mathrm{Ker}C}f) = \bigvee_{C} T^C \varphi_{\widetilde{X}|\mathrm{Ker}C}f = \Phi_{\widetilde{X}}f .$$

If g is independent of X then

$$\Phi_{\widetilde{X}}(gf) = \bigvee_{C} T^C \varphi_{\widetilde{X}|\mathrm{Ker}C}f = g \bigvee_{C} T^C \varphi_{\widetilde{X}|\mathrm{Ker}C}f = g\Phi_{\widetilde{X}}f .$$

Now take $\widetilde{X} = (X, A)$ and $\widetilde{Y} = (Y, D)$, where $X \cap Y = \emptyset$. Then the vectors T and V associated with X and Y, respectively, satisfy $T \cap V = \emptyset$. If E denotes a vector running over $\{0, 1\}^{\dim Y}$ then

$$\Phi_{\widetilde{X}}(\Phi_{\widetilde{Y}}f) = \bigvee_{C} T^C \varphi_{\widetilde{X}|\mathrm{Ker}C}(\bigvee_{E} V^E \varphi_{\widetilde{Y}|\mathrm{Ker}E}f)$$

$$= \bigvee_{C} T^C \bigvee_{E} V^E \varphi_{\widetilde{X}|\mathrm{Ker}C}(\varphi_{\widetilde{Y}|\mathrm{Ker}E}f)$$

$$= \bigvee_{C} \bigvee_{E} T^C V^E (\varphi_{\widetilde{X}|\mathrm{Ker}C} \circ \varphi_{\widetilde{Y}|\mathrm{Ker}E})(f)$$

$$= \bigvee_{C \cup E} (T \cup V)^{C \cup E} \varphi_{(\widetilde{X} \cup \widetilde{Y}, \mathrm{Ker}C \cup \mathrm{Ker}E)}f ,$$

which concludes the proof because $\mathrm{Ker}C \cup \mathrm{Ker}E = \mathrm{Ker}(C \cup E)$.   □

Proposition 3.6. *The closure $\Phi_{\widetilde{X}}$ is also given by formula*

(3.10)
$$\Phi_{\widetilde{X}}f = \bigvee_{C}(\prod\{t'_i \mid i \in \mathrm{Ker}C\})\varphi_{\widetilde{X}|\mathrm{Ker}C}f .$$

PROOF: The inequality $\leq$ clearly holds. To prove the converse, fix a vector C and compute

$$P = (\prod\{t'_i \mid i \in \mathrm{Ker}C\})(\varphi_{\widetilde{X}|\mathrm{Ker}C}f)(\Phi_{\widetilde{X}}f)$$

$$= \bigvee_{D} (\prod \{t'_i \mid i \in \mathrm{Ker}C\})(\varphi_{\widetilde{X}|\mathrm{Ker}C}f)\cdot$$

$$\cdot(\prod \{t'_j \mid j \in \mathrm{Ker}D\})(\prod \{t_k \mid k \in \mathrm{Coker}D\})\varphi_{\widetilde{X}|\mathrm{Ker}D}f\ .$$

The non-null D−terms are characterized by $\mathrm{Ker}C \cap \mathrm{Coker}D = \emptyset$, or equivalently $\mathrm{Ker}C \subseteq \mathrm{Ker}D$, in other words by the fact that $\overline{X}|\mathrm{Ker}C$ is a subvector of $\overline{X}|\mathrm{Ker}D$; denote this situation by $D \succ C$. Then

$$P = \bigvee_{D \succ C} (\prod \{t'_i \mid i \in \mathrm{Ker}C\})(\varphi_{\widetilde{X}|\mathrm{Ker}C}f)\cdot$$

$$\cdot(\prod \{t'_j \mid j \in \mathrm{Ker}D\backslash\mathrm{Ker}C\}) \prod \{t_k \mid k \in \mathrm{Coker}D\}$$

$$= (\prod \{t'_i \mid i \in \mathrm{Ker}C\})(\varphi_{\widetilde{X}|\mathrm{Ker}C}f)\cdot$$

$$\cdot \bigvee_{D \succ C} (\prod \{t'_j \mid j \in \mathrm{Ker}D\backslash\mathrm{Ker}C\})(\prod \{t_k \mid k \in \mathrm{Coker}D\})$$

$$= (\prod \{t'_i \mid i \in \mathrm{Ker}C\})\varphi_{\widetilde{X}|\mathrm{Ker}C}f$$

because

$$D \succ C \Longrightarrow \mathrm{Coker}C = (\mathrm{Ker}D\backslash\mathrm{Ker}C) \cup \mathrm{Coker}D\ ,$$

so that the terms of the disjunction $\bigvee_{D \succ C}$ are exactly the minterms in $T|\mathrm{Coker}C$. We have thus proved that

$$(\prod \{t'_i \mid i \in \mathrm{Ker}C\})\varphi_{\widetilde{X}|\mathrm{Ker}C}f \le \varPhi_{\widetilde{X}}f$$

for every C, which establishes the inequality $\ge$ in (3.10).   □

Corollary 3.3. *If the function f is independent of W, then $\varPhi_{\widetilde{X}}f$ has an expression in which the variables in W appear only in complemented form w'_i, hence $\varPhi_{\widetilde{X}}f$ is decreasing in T.*

PROOF: From Proposition 3.6 and property (iv) in Definition 3.3.   □

Example 3.3. The formula established in Example 3.2 can be written in the simpler form

$$\varPhi_{(z_1,z'_3)}f = w'_1 w'_3 \varphi_{(z_1,z'_3)}f \vee w'_1\varphi_{(z_1)}f \vee w'_2\varphi_{(z'_3)}f \vee f\ .$$

□

Proposition 3.7 below is a generalization of the corresponding properties in J. Deschamps [1975], [1990], necessary in order to include the operator mon_A.

Proposition 3.7. *For every subvector* $X = (z_{i_1}, \ldots, z_{i_n})$ *of* Z, *every* $C = (\gamma_1, \ldots, \gamma_n) \in \{0,1\}^n$, *every* $A, D, \in \{0,1\}^p$ *and every* $f \in BFp$, *the following conditions are equivalent:*

$$(\varphi_{(X,C)}f)(Z) = (\Phi_{(Z,A)}f)(Z,D) ; \tag{3.11}$$

$$X = Z|\mathrm{Ker}D \ \& \ C = A|\mathrm{Ker}D ; \tag{3.12}$$

$$\alpha_{i_j} = \gamma_j \ (j = 1, \ldots, n) \ \& \ \mathrm{Ker}D = \{i_1, \ldots, i_n\} . \tag{3.13}$$

PROOF: Formula (3.9) and property (iv) in Definition 3.3 imply

$$(\Phi_{(Z,A)}f)(Z,D) = (\varphi_{(Z,A)|\mathrm{Ker}D}f)(Z) ,$$

which proves the equivalence (3.11)$\Longleftrightarrow$(3.12). Further, condition (3.12) is equivalent to

$$(X,C) = (Z,A)|\mathrm{Ker}D ; \tag{3.14}$$

but since $(X,C) = (z_{i_1}^{\gamma_1}, \ldots, z_{i_n}^{\gamma_n})$ and $(Z,A) = (z_1^{\alpha_1}, \ldots, z_p^{\alpha_p})$, condition (3.14) is equivalent to (3.13). $\qquad\square$

Remark 3.3. Proposition 3.7 can be applied by taking instead of Z, any subvector Y of Z such that X is a subvector of Y. This amounts to regard $Z\backslash Y$ as a set of parameters rather than variables. $\qquad\square$

Example 3.4. Take $p := 3, Z := (x_1, x_2, x_3), X := (z_1, z_3), C := (1,0)$. Then $\widetilde{X} = (X,C) = (z_1, z_3')$. In order to apply Proposition 3.7 we must take A of the form $A = (1, \alpha_2, 0)$, while D is determined by the condition $\mathrm{Ker}D = \{1,3\}$, that is $D = (0,1,0)$. We thus obtain

$$(\varphi_{(z_1, z_3')}f)(Z) = (\Phi_{(z_1, z_2^{\alpha_2}, z_3')}f)((Z, (0,1,0)) .$$

Alternatively, in view of Remark 3.3, we can regard z_2 as a parameter rather than as a variable, i.e., we can take $p := 2, Z := (z_1, z_2) = X$ and again $C = (1,0)$. In this case $A = (1,0)$ and $\mathrm{Ker}D = \{1,2\}$, i.e., $D = (0,0)$. We thus obtain (see also Example 3.2):

$$(\varphi_{(z_1, z_2')}f)(z_1, z_2) = (\Phi_{(z_1, z_2')}f)(z_1, z_2; 0, 0) .$$

$\qquad\square$

Proposition 3.8. *The function* $\Phi_{(X,A)}f$ *can be constructed by the following prescriptions. Express* $f(Z,W)$ *as a disjunction of terms of the form* $\mu(X)g(Z\backslash X, W)$, *where* g *is a Boolean function and* $\mu(X)$ *is a monomial (i.e., a product of factors of the form* x^γ, *with* $x \in X$ *and* $\gamma \in \{0,1\}$*), and in each term replace each factor* x^γ *by*

$$w'\varphi_{(x^\alpha)}(x^\gamma) \vee x^\gamma , \tag{3.15}$$

where w *is the* W*−variable associated with* x.

PROOF: Since Φ is additive, it suffices to prove the proposition for a single term μg, and since $\Phi_{\widetilde{X}}(\mu g) = (\Phi_{\widetilde{X}}\mu)g$ by property (ii), it remains to prove the result for monomials.

Re-denote $X = (z_{i_1}, \ldots, z_{i_n}) = (x_1, \ldots, x_n)$, hence $(X, A) = (x_1^{\alpha_1}, \ldots, x_n^{\alpha_n})$. Now μ is of the form

$$\mu(X) = x_{j_1}^{\gamma_1} \ldots x_{j_q}^{\gamma_q} \;;$$

set $Y = (x_{j_1}, \ldots, x_{j_q}), C = (\alpha_{j_1}, \ldots, \alpha_{j_q}), (Y, C) = (\widetilde{y}_1, \ldots, \widetilde{y}_q)$ and $x_{j_k}^{\gamma_k} = v_k$ $(k = 1, \ldots, q)$. Then properties (i) and (ii) yield

$$\Phi_{(X,A)}\mu = \Phi_{(Y,C)}(\Phi_{(X \backslash Y, A \backslash C)}\mu) = \Phi_{(Y,C)}\mu$$

$$= (\Phi_{(\widetilde{y}_1)} \circ \ldots \circ \Phi_{(\widetilde{y}_q)})\mu = (\Phi_{(\widetilde{y}_1)} \circ \ldots \circ \Phi_{\widehat{(\widetilde{y}_{q-1})}})(\Phi_{(\widetilde{y}_q)}\mu)$$

$$= (\Phi_{(\widetilde{y}_1)} \circ \ldots \circ \Phi_{\widehat{(\widetilde{y}_{q-1})}})(v_1 \ldots v_{q-1}(\Phi_{(\widetilde{y}_q)}v_q))$$

$$= ((\Phi_{(\widetilde{y}_1)} \circ \ldots \circ \Phi_{\widehat{(\widetilde{y}_{q-1})}})(v_1 \ldots v_{q-1}))\Phi_{(\widetilde{y}_q)}v_q = \ldots$$

$$= (\Phi_{(\widetilde{y}_1)}v_1)(\Phi_{(\widetilde{y}_2)}v_2) \ldots (\Phi_{(\widetilde{y}_q)}v_q)$$

and it remains to compute an arbitrary factor of this product. Taking into account extensivity, we get

$$(\Phi_{(x^\alpha)})(x^\gamma) = w^0 \varphi_{(x^\alpha)}(x^\gamma) \vee w^1 x^\gamma = w' \varphi_{(x^\alpha)}(x^\gamma) \vee x^\gamma \;.$$

$\square$

Corollary 3.4. *Let* $\mathrm{INC}_X, \mathrm{DEC}_X$ *and* MON_A *denote the operators (3.9) corresponding to* $\mathrm{inc}_X, \mathrm{dec}_X$ *and* mon_A, *respectively. In order to compute* $\mathrm{INC}_X f$ *one replaces each* x_i' *by* $x_i' \vee w_i'$, *while* $\mathrm{DEC}_X f$ *is obtained by changing each* x_i *to* $x_i \vee w_i'$. *The function* $\mathrm{MON}_A f$ *is constructed by applying for each* i *the former or the latter transformation, according as* $\alpha_i = 1$ *or* 0.

PROOF: In the case $\varphi = \mathrm{inc}$ we have $A = (1, \ldots, 1)$, so that taking in turn $\gamma := 1$ and $\gamma := 0$ in (3.15) we obtain $w' \mathrm{inc}_{(x)}(x) \vee x = w' x \vee x = x$ and $w' \mathrm{inc}_{(x)}(x') \vee x' = w' \cdot 1 \vee x'$, respectively. The proof for dec_X is similar. $\square$

Example 3.5. Consider a function f of the form

$$f(X, Y) = a(Y)xy' \vee b(Y)xz'v \vee c(Y)xzt' \vee d(Y)x'v' \;,$$

where $X = (x, y, z, t)$. Then

$$(\mathrm{MON}_{(x,y,z',t,v')}g)(X, Y) = a(Y)x(y' \vee w_y') \vee b(Y)xz'\vee$$

$$\vee c(Y)x(z \vee w_z')(t' \vee w_t') \vee d(Y)(x' \vee w_x')v' \;.$$

$\square$

There is also a dual theory of interior operators $\Phi^\circ_{\widetilde{X}}$. For instance, in order to compute $\mathrm{INC}^\circ_X f$ and $\mathrm{DEC}^\circ_X f$ one performs the transformations $x'_i \mapsto x'_i w'_i$ and $x_i \mapsto x_i w'_i$, respectively.

Now let $\mathcal{M}_{\widetilde{X}}$ be the Moore family associated with a relative closure $\varphi_{\widetilde{X}}$ and let $[\underline{f}, \overline{f}]$ be a partially defined Boolean function. According to Proposition 1.3, there exists $f \in [\underline{f}, \overline{f}] \cap \mathcal{M}_{\widetilde{X}}$ if and only if

$$(3.16) \qquad\qquad \varphi_{\widetilde{X}}\underline{f} \leq \overline{f} .$$

The technique developed so far enables us to find all the vectors $\widetilde{X}$ for which such a function f does exist.

Proposition 3.9. *Let $[\underline{f}, \overline{f}]$ be a partially defined Boolean function of* BFp *and $A \in \{0,1\}^p$. The identity (3.16) is fulfilled for a subector $\widetilde{X} = (X, C)$ of (Z, A) if and only if $\widetilde{X}$ is of the form (3.12), where D is a $0-1$ solution of the equation in W (cf. Definition 3.2)*

$$(3.17) \qquad\qquad \mathrm{ind}_Z((\Phi_{(Z,A)}\underline{f}) \cdot \overline{f}') = 0 .$$

PROOF: The identity (3.16) is equivalent to the identity $(\varphi_{\widetilde{X}}\underline{f}) \cdot \overline{f}' = 0$, which, in view of Proposition 3.7, can be written in the form

$$(3.18) \qquad\qquad ((\Phi_{(Z,A)}\underline{f})(Z, D)) \cdot \overline{f}'(Z) = 0 ,$$

where $\widetilde{X}$ and (Z, A) are related by (3.12). In other words, (3.16) $\Longleftrightarrow$ (3.12)&(3.18). Since

$$g(Z) = 0 \ (\forall Z \in B^p) \Longleftrightarrow g(A) = 0 \ (\forall A \in \{0,1\}^p) \Longleftrightarrow \bigvee_A g(A) = 0$$

for every $g \in \mathrm{BFp}$, the identity (3.18) is equivalent to

$$\mathrm{ind}_Z(((\Phi_{(Z,A)}\underline{f})(Z, D)) \cdot \overline{f}'(Z) = 0 ,$$

which means precisely that D is a solution of equation (3.17). $\qquad\square$

Remark 3.4. Corollary 3.3 of Proposition 3.6 and Remark 3.2 ensure that equation (3.17) has a form in which the unknowns appear only in complemented form w'_i. The simplest way of constructing (3.18) is to apply the procedure described in Proposition 3.8 (which, in the case of the operators inc, dec and mon, is specified in Corollary 3.4); then (3.17) can be obtained by applying Remark 3.2.

$\qquad\square$

Example 3.6. Let $p := 5$, $Z := (x, y, z, t, v)$, $A := (1, 1, 0, 1, 0)$ and the closure mon_A. Take

$$\underline{f} = axy' \vee bxz'v \vee cxzt' \vee dx'v' ,$$

$$\overline{f} = bx \vee cz \vee y' \vee v' ,$$

where $a, b, c, d \in B$. Taking into account Example 3.5, we get

$$\mathrm{MON}_{(x,y,z',t,v')}\underline{f} = ax(y' \vee w_y') \vee bxz' \vee cx(z \vee w_z')(t' \vee w_t') \vee d(x' \vee w_x')v' \ ,$$

$$\overline{f}' = (b' \vee x')(c' \vee z')yv \ ,$$

$$(\mathrm{MON}_{(x,y,z',t,v')}\underline{f}) \cdot \overline{f}' = ab'xyv(c' \vee z')w_y' \vee b'cxyvz'w_z'(t' \vee w_t') \ ,$$

hence equation (3.17) becomes

$$ab'w_y' \vee b'cw_z'w_t' = 0 \ .$$

If $ab', b'c \notin \{0,1\}$, the 0–1 solutions of this equation are given by $w_y' = w_z'w_t' = 0$. Therefore there are $3\times4=12$ solutions for D, namely

$$(\delta_1, 1, 0, 1, \delta_5), (\delta_1, 1, 1, 0, \delta_5), (\delta_1, 1, 1, 1, \delta_5) \quad (\delta_1, \delta_5 \in \{0,1\}) \ ,$$

and since $(Z, A) = (x, y, z', t, v')$, this yields the following solutions for $\widetilde{X}$:

$$(x, z', v'), (x, z'), (z', v'), (x, t, v'), (x, t), (t, v'), (x, v'), (x), (v') \ .$$

□

In the next part of this section, following Brown and Rudeanu [1981], we point out the relationship of the independent closure with the concepts of consistency and consequence, and we introduce a new independent closure operator for simple Boolean functions.

There are three major rôles played by the operator ind_X, namely in connection with consistency, consequences and functional dependence. We confine below to the first two aspects, while functional dependence will be studied in Ch. 8, §1.

We already know that $(\mathrm{ind}_X f)(Y) = 1$ is the *consistency* condition of the equation $f(X, Y) = 1$ with respect to X; this is the well-known Boole-Schröder theorem (see e.g. BFE, dual of Theorem 2.3, or the present Proposition 5.3.5).

Further let us introduce

Definition 3.5. 1) An *Y-upper-bound* of a function $f \in \mathrm{BFp}$ is a function $h \in \mathrm{BFp}$ independent of X and such that

(3.19) $$f(X, Y) \leq h(Y) \qquad (\forall X \in B^n) (\forall Y \in B^m) \ .$$

2) An *Y-consequence* of a Boolean equation $f(X, Y) = 1$ is a Boolean equation $h(Y) = 1$ such that

(3.20) $$f(X, Y) = 1 \Longrightarrow h(Y) = 1 \qquad (\forall X \in B^n) (\forall Y \in B^m) \ ;$$

it is understood that the various forms of the function h define the same equation $h(Y) = 1$.

□

The next remark is essentially due to Poretski.

Remark 3.5. In other words, using Theorem 3.1, the Y-upper-bounds of the function f are the upper bounds of f in the set Ind_X and these upper bounds form the principal filter $\text{Ind}_X[\text{ind}_X f)$ generated by $\text{ind}_X f$ in Ind_X; cf. 7.(1.1). Therefore a function h independent of X is an Y-upper-bound of f iff $\text{ind}_X f \leq h$.   □

Remark 3.6. Suppose the equation $f(Z) = 1$ is consistent. Then the Y-upper-bounds of f coincide with the functions occurring in the Y-consequences of the equation, by the Verification Theorem (see e.g. BFE, dual of Theorem 2.14). In particular not only $\text{ind}_X f$ is the *least Y-upper-bound* of f, but also $(\text{ind}_X f)(Y) = 1$ is the *least Y-consequence* of the equation $f(Z) = 1$, to the effect that $h(Y) = 1$ is an Y-consequence of $f(Z) = 1$ iff $\text{ind}_X f \leq h$ (by the first part of this remark and Remark 3.5). So the least Y-consequence determines all of the Y-consequences.   □

A more particular case is that in which equation $f(X,Y) = 1$ is consistent for any Y:

Remark 3.7. The following conditions are equivalent:
  (i) $\forall Y \in B^m \; \exists X \in B^n f(X,Y) = 1$;
  (ii) $\text{ind}_X f = 1$ holds identically ;
  (iii) the unique Y-consequence of the equation $f(X,Y) = 1$ is the identity $1=1$.   □

Remark 3.8. The conclusions of Remark 3.6 fail if the equation $f(Z) = 1$ is inconsistent. For $\text{ind}_X f$ is still the least Y-upper-bound of f, whereas any equation of the form $h(Y) = 1$ is an Y-consequence of $f(Z) = 1$, therefore the least Y-consequence is the (inconsistent) equation $0=1$. Yet in the case of simple Boolean functions the conclusions of Remark 3.6 are recaptured for inconsistent equations as well. For if $f(Z) = 1$ is an inconsistent simple Boolean equation, then $f = 0$ identically, hence $\text{ind}_X f = 0$.   □

We continue the above remark by a deeper insight of simple Boolean functions.

Theorem 3.2. *The set* SBFp *of simple Boolean functions* $f : B^p \longrightarrow B$ *is a double Moore family. The closure operator* sim *and the interior operator* sim° *associated with* SBFp *are given by formulas*

$$(3.21) \qquad (\text{sim} f)(Z) = \bigvee \{ Z^C \mid C \in \{0,1\}^p \ \& \ f(C) \neq 0 \} ,$$

$$(3.22) \qquad \text{sim}° f = (\text{sim} f')' .$$

PROOF: Clearly $\text{sim} f$ is a simple Boolean function and

$$f(Z) = \bigvee \{ f(C) Z^C \mid C \in \{0,1\}^p \ \& \ f(C) \neq 0 \} \leq (\text{sim} f)(Z) .$$

If $g \in$ SBFp and $f \leq g$, then whenever $f(C) \neq 0$ it follows from $f(C) \leq g(C)$ that $g(C) = 1$, hence $Z^C = g(C) Z^C \leq g(Z)$; this proves that $\text{sim} f \leq g$.
The proof is concluded by Corollary 1.1.   □

Corollary 3.5. sim *satisfies* (1.8), *while* sim° *satisfies* (1.8′).

PROOF: From Corollary 1.1.   □

Corollary 3.6. *If the Boolean function g is simple, then*

$$(3.23) \qquad \text{sim}(fg) = (\text{sim}f)g \ \& \ \text{sim}°(f \vee g) = \text{sim}°f \vee g .$$

PROOF: For every $C \in \{0,1\}^p$, it follows from $g(C) \in \{0,1\}$ that

$$f(C)g(C) \neq 0 \iff (f(C) \neq 0 \ \& \ g(C) = 1) ,$$

hence

$$(\text{sim}f)g = (\bigvee\{Z^C \mid f(C) \neq 0\})(\bigvee\{Z^D \mid g(D) = 1\})$$

$$= \bigvee\{Z^C \mid f(C) \neq 0 \ \& \ g(C) = 1\} = \text{sim}(fg) .$$

□

The duals of the properties established below are left to the reader.

Remark 3.9. If a is a constant $\neq 0$ and g is a simple Boolean function, then sim$a$=1 by (3.21), hence $\text{sim}(ag) = g$ by Corollary 3.6. Now Corollary 3.5 implies that if the function f is expressed as a disjunction of terms of the form ag, then sim$f$ is obtained by deleting a from each term.   □

Remark 3.10. The above Corollary 3.6 cannot be sharpened by dropping the condition on g. For instance, $\text{sim}(ff') = \text{sim}0 = 0$, whereas e.g. if $a \in B\backslash\{0,1\}$ then $(\text{sim}(ax))(\text{sim}(a' \vee x')) = x \cdot 1 = x$.   □

Definition 3.6. A *simple Y-upper-bound* of a function $f \in$ BFp is an Y-upper-bound h which is also a simple Boolean function. A *simple Y-consequence* of an equation $f(X,Y) = 1$ is a simple Boolean equation $h(Y) = 1$ which is also an Y-consequence of the former equation.   □

Theorem 3.3. *The set* Ind_XSBFp *of simple Boolean functions independent of* X *is a double Moore family. The associated closure operator and interior operator are provided by formulas*

$$(3.24) \qquad (\text{ind}_X\text{sim}f)(Y) = (\text{sim ind}_Xf)(Y) = \bigvee\{Y^C \mid \bigvee_A f(A,C) \neq 0\} ,$$

$$(3.25) \qquad (\text{ind}_X\text{sim})°f = (\text{ind}_X\text{sim}f')' .$$

PROOF: Since the Boolean function sim$f$ is simple, it follows that $\text{ind}_X\text{sim}f$ is also a simple Boolean function; besides, it does not depend on X. Then $f \leq$ sim$f \leq \text{ind}_X\text{sim}f$. Further if g is a simple Y-upper-bound of f, it follows that sim$f \leq$ sim$g = g$, therefore $\text{ind}_X\text{sim}f \leq \text{ind}_Xg = g$.

We have thus proved that $\mathrm{ind}_X\mathrm{sim}f$ is the least simple Y-upper-bound of f and one proves similarly that $\mathrm{sim}\,\mathrm{ind}_X f$ has the same property; this establishes the first equality (3.24). But

$$\mathrm{ind}_X f = \bigvee_A f(A,Y) = \bigvee_A \bigvee_C f(A,C)Y^C$$

$$= \bigvee_C \bigvee_A f(A,C)Y^C = \bigvee_C Y^C \bigvee_A f(A,C) \,,$$

hence Remark 3.9 implies

$$\mathrm{sim}\,\mathrm{ind}_X f = \bigvee \{ Y^C \mid \bigvee_A f(A,C) \neq 0 \} \,.$$

The proof is concluded by Corollary 1.1. $\qquad\square$

Corollary 3.7. *The operator* $\mathrm{ind}_X\mathrm{sim}$ *satisfies* (1.8), *while* $(\mathrm{ind}_X\mathrm{sim})^\circ$ *satisfies* (3.8′). $\qquad\square$

Corollary 3.8. *If g is a simple Boolean function independent of X, then* $(\mathrm{ind}_X\mathrm{sim})(fg) = (\mathrm{ind}_X\mathrm{sim}f)g$. $\qquad\square$

Remark 3.11. It follows easily from Remark 3.10 that the above corollary cannot be sharpened by dropping the condition on g. $\qquad\square$

Remark 3.12. It is also easily shown, using Remark 3.6, that if the equation $f(Z) = 1$ is consistent, then the simple Y-upper-bounds of the function f coincide with the functions occurring in the simple Y-consequences of the equation. In particular $(\mathrm{ind}_X\mathrm{sim}f)(Y) = 1$ is the least simple Y-consequence of the equation $f(Z) = 1$. $\qquad\square$

Remark 3.13. It follows from Remark 3.7 that the following conditions are equivalent:
(i) $\mathrm{ind}_X\mathrm{sim}f=1$ holds identically ;
(ii) the unique simple Y-consequence of the equation $f(X,Y) = 1$ is the identity $1 = 1$. $\qquad\square$

Example 3.7. Consider again the function $\underline{f}$ in Example 3.6 and take $X := (x)$, hence $Y = (y,z,t,v)$. Then

$$\mathrm{sim}\underline{f} = xy' \vee xz'v \vee xzt' \vee x'v' \,,$$

$$\mathrm{ind}_X\mathrm{sim}\underline{f} = y' \vee z' \vee zt' \vee v' = y' \vee z' \vee t' \vee v' \,,$$

$$\mathrm{ind}_X\underline{f} = ay' \vee bz'v \vee czt' \vee dv' \,,$$

$$\mathrm{sim}\,\mathrm{ind}_X\underline{f} = y' \vee z'v \vee zt' \vee v' = y' \vee z' \vee t' \vee v' \,.$$

$\qquad\square$

This study will be continued in the next chapter.

We conclude this section with one more closure operator, related to the problem of decomposing a Boolean function, to be dealt with in Ch. 11. Thus Definition 3.7 and Theorem 3.4 below are pertinent both to this chapter and to Ch. 11.

Definition 3.7. Let $f : B^{n+m} \longrightarrow B$ and $h_k : B^n \longrightarrow B$ $(k = 1, \ldots, r)$ be Boolean functions. Set $H = (h_1, \ldots, h_r)$. Then f is said to be H-decomposable if there is a Boolean function $g : B^{r+m} \longrightarrow B$ such that

(3.26) $f(X, Y) = g(h_1(X), \ldots, h_r(X), Y) \; (\forall X \in B^n) \; (\forall Y \in B^m)$.

$\square$

Theorem 3.4. (Lapscher [1968]). Let $h_k : B^n \longrightarrow B$ $(k = 1, \ldots, r)$ be Boolean functions and $H = (h_1, \ldots, h_r)$. The set Dcp_H of H-decomposable Boolean functions $f : B^{n+m} \longrightarrow B$ is a double Moore family and the associated closure operator and interior operator are provided by formulas

(3.27) $(\mathrm{dcp}_H f)(X, Y) = \bigvee_{C \in \{0,1\}^r} H^C(X) \mathrm{ind}_X(f(X, Y) H^C(X))$,

(3.28) $\mathrm{dcp}_H^\circ f = (\mathrm{dcp}_H f')'$.

PROOF: For every $C \in \{0, 1\}^r$, the function $g_C : B^m \longrightarrow B$ defined by

$$g_C(Y) = \mathrm{ind}_X(f(X, Y) H^C(X))$$

is Boolean, therefore the function (3.27) satisfies Definition 3.7 with the function g defined by

$$g(Z, Y) = \bigvee_{C \in \{0,1\}^r} Z^C g_C(Y) .$$

Then $f \leq \mathrm{dcp}_H f$ because

$$f(X, Y) \cdot (\mathrm{dcp}_H f)(X, Y) = \bigvee_C f(X, Y) H^C(X) \mathrm{ind}_X(f(X, Y) H^C(X))$$

$$= \bigvee_C f(X, Y) H^C(X) = f(X, Y) \bigvee_C H^C(X) = f(X, Y) .$$

Finally if $f \leq u \in \mathrm{Dcp}_H$, then setting $u(X, Y) = \bigvee_C H^C u_C(Y)$ (cf . (3.26)), it follows that for every $C \in \{0, 1\}^r$,

$$f(X, Y) H^C(X) \leq u(X, Y) H^C(X) = H^C(X) u_C(Y) \leq u_C(Y) ,$$

hence

$$\mathrm{ind}_X(f(X, Y) H^C(X)) \leq u_C(Y) ,$$

therefore (3.27) implies

$$(\mathrm{dcp}_H f)(X,Y) \le \bigvee_C H^C(X) u_C(Y) = u(X,Y) .$$

The proof is again concluded by Corollary 1.1.   □

Corollary 3.9. *The operator* dcp_H *satisfies* (1.8), *while* dcp_H° *satisfies* (1.8′).
                                                                   □

Corollary 3.10. *If the function* g *is* H-*decomposable, then* $\mathrm{dcp}_H(fg) = (\mathrm{dcp}_H f)g$
and $\mathrm{dcp}_H^\circ(f \vee g) = \mathrm{dcp}_H^\circ f \vee g.$ □

From a practical pont of view, note that f is H-decomposable iff $f = \mathrm{dcp}_H f$
iff $f = \mathrm{dcp}_H^\circ f.$

J. Deschamps [1977] has used the operators ind_X, ind_X° and $\Phi_{\widetilde{X}}$ in the study
of decompositions of truth functions.

8. Boolean transformations

We have studied in Ch.7, §3 the relationship of the closure operator ind_X with consistency and consequences of Boolean equations. In §1 of this chapter we use the same tool in order to study functional dependence of Boolean functions. We point out a natural concept of dependence of a family of Boolean functions and a stronger concept which turns out to be the Moore-Marczewski concept of dependence. Then we investigate Boolean transformations $F : B^n \longrightarrow B^m$: in §2 we study the range of a Boolean transformation, while §3 deals with injectivity. The last section is devoted to fixed points of lattice and Boolean transformations.

1 Functional dependence of Boolean functions

In this section, following Brown and Rudeanu [1981], we deal with systems of Boolean equations of the form

(1.1) $$f_i(x_1, \ldots, x_n) = y_i \qquad (i = 1, \ldots, m) ,$$

which we write in the compact form

(1.2) $$F(X) = Y ,$$

where $F = (f_1, \ldots, f_m)$, $X \in B^n$, $Y \in B^m$. We will use the letter A for vectors in $\{0, 1\}^n$ and the letters C, D for vectors in $\{0, 1\}^m$. A useful abbreviation will be $F^C(A)$ for $(F(A))^C$.

Definition 1.1. The *resolvent* of system (1.2) is the function $r \in \mathrm{BF}(n + m)$ such that (1.2) is equivalent to

(1.3) $$r(X, Y) = 1 ,$$

while the *eliminant* of system (1.2) is $\mathrm{ind}_X r$. $\qquad\square$

Proposition 1.1. *The following identities hold:*

(1.4) $$r(X, Y) = \prod_{i=1}^{m} (y_i f_i(X) \vee y_i' f_i'(X)) ,$$

$$(1.5) \qquad r(X,Y) = \bigvee_A \bigvee_C F^C(A) X^A Y^C \,,$$

$$(1.6) \qquad (\mathrm{ind}_X r)(Y) = \bigvee_A \prod_{i=1}^{m} (y_i f_i(A) \vee y_i' f_i'(A)) \,,$$

$$(1.7) \qquad (\mathrm{ind}_X r)(Y) = \bigvee_C (\bigvee_A F^C(A)) Y^C \,.$$

PROOF: Relation (1.4) is obvious and it implies (1.6). Besides, setting $C = (\gamma_1, \ldots, \gamma_m)$ and $D = (\delta_1, \ldots, \delta_m)$, property (1.4) also implies

$$r(A,B) = \prod_{i=1}^{m} (\gamma_i f_i(A) \vee \gamma_i' f_i'(A))$$

$$= \bigvee_D \gamma_1^{\delta_1} f_1^{\delta_1}(A) \ldots \gamma_m^{\delta_m} f_m^{\delta_m}(A) = \bigvee_D C^D F^D(A) = F^C(A) \,,$$

which proves (1.5). From (1.5) we obtain $r(A,Y) = \bigvee_C F^C(A) Y^C$, hence (1.7) holds because

$$(\mathrm{ind}_X r)(Y) = \bigvee_A \bigvee_C F^C(A) Y^C = \bigvee_C \bigvee_A F^C(A) Y^C \,.$$

$\square$

We now resume the last part of Ch.7, §3, starting with Definition 7.3.5, in the particular case of the reseolvent r.

Definition 1.2. A *relation of Boolean functional dependence* or simply a *relation of functional dependence* connecting $f_1, \ldots, f_m \in \mathrm{BF}n$ is an identity of the form

$$(1.8) \qquad h(f_1(X), \ldots, f_m(X)) = 1 \,,$$

also written as

$$(1.9) \qquad h \circ F = 1 \,,$$

where $h : B^m \longrightarrow B$ is a Boolean function. The relation is said to be *trivial* if $h = 1$ identically. A vector $F = (f_1, \ldots, f_m) \in (\mathrm{BF}n)^m$, also called a *family* of Boolean functions, is said to be *functionally dependent* or simply *dependent* if there is a non-trivial relation of functional dependence connecting $f_1, \ldots, f_m$; otherwise one says that F is *functionally independent*. $\square$

Remark 1.1. A necessary condition for the independence of F is that $f_1, \ldots, f_m$ be pairwise distinct. The property of functional independence is *hereditary*, i.e., every subfamily of a functionally independent vector is also functionally independent. The property of functional dependence is *co-hereditary*, i.e., if F is functionally dependent and F is a subfamily of G, then G is also functionally dependent. $\square$

Remark 1.2. The Y-consequences of system (1.2), which by definition are the Y-consequences of equation (1.3) (cf. Definition 7.3.5), can be identified with the relations of functional dependence connecting $f_1, \ldots, f_m$. For the implication $Y = F(X) \Longrightarrow h(Y) = 1$ can be written in the equivalent form $h(F(X)) = 1$. $\square$

Remark 1.3. Since the equation $r(Z) = 1$ has the solutions $Z = (X, F(X))$, we can apply Remarks 7.3.6 and 1.2, which imply that the Y-upper-bounds h of r coincide with the functions h occurring in the relations of functional dependence $h \circ F = 1$. In particular the identity $\mathrm{ind}_X r \circ F = 1$ is *the least relation of functional dependence* connecting F, to the effect that $h \circ F = 1$ is an identity iff $\mathrm{ind}_X r \leq h$. So the least relation of functional dependence determines all of them. $\square$

To continue the study the following definition is in order.

Definition 1.3. A *Boolean transformation* is a map of the form

(1.10) $$F = (f_1, \ldots, f_m) : B^n \longrightarrow B^m ,$$

where $f_1, \ldots, f_m \in$ BFn. Thus

(1.11) $$F(X) = (f_1(X), \ldots, f_m(X)) \qquad (\forall X \in B^n) .$$

$\square$

Theorem 1.1. *The following conditions are equivalent for a Boolean transformation* $F : B^n \longrightarrow B^m$:

(i) F is surjective ;

(ii) $\forall C \in \{0,1\}^m \; \exists X \in B^n \; F(X) = C$;

(iii) $\prod_C \bigvee_A F^C(A) = 1$;

(iv) $\mathrm{ind}_X r = 1$ *identically* ;

(v) The family F is functionally independent .

COMMENT: The equivalences (i)$\Longleftrightarrow$(iii) $\Longleftrightarrow$(v) are due to Whitehead [1901] and Löwenheim [1919]; see e.g. BFE, Theorem 8.3. The equivalence (i)$\Longleftrightarrow$(ii) was remarked by Kuntzmann [1965] and taken by him as a definition of independence in the class of truth functions.

PROOF: (i)$\Longleftrightarrow$(iv)$\Longleftrightarrow$(v): By Remarks 7.3.7 and 1.2.

(iii)$\Longleftrightarrow$(iv): By Proposition 1.1 (1.7).

(ii)$\Longleftrightarrow \forall C \; \exists X \; r(X, C) = 1 \Longleftrightarrow \forall C \; (\mathrm{ind}_X r)(C) = 1 \Longleftrightarrow$(iv). $\square$

Corollary 1.1. (Kuntzmann [1965]). *If* $m > n$ *then* $(f_1, \ldots, f_n)$ *is functionally dependent.*

PROOF: Let μ denote an arbitrary mapping $\mu : \{0,1\}^m \longrightarrow \{0,1\}^n$ and $C = (\gamma_1, \ldots, \gamma_m)$. Then

$$\prod_C \bigvee_A F^C(A) = \prod_C \bigvee_A f_1^{\gamma_1}(A) \ldots f_m^{\gamma_m}(A)$$

$$= \bigvee_\mu \prod_C f_1^{\gamma_1}(\mu(C)) \dots f_m^{\gamma_m}(\mu(C)) = \bigvee_\mu \prod_C F^C(\mu(C))$$

and since $m > n$, for every μ there exist $D, E \in \{0,1\}^m$ with $D \neq E$ and $\mu(D) = \mu(E)$, hence

$$\prod_C F^C(\mu(C)) \leq F^D(\mu(D)) F^E(\mu(E)) = 0 ,$$

therefore $\prod_C \bigvee_A F^C(A) = 0$. □

Corollary 1.2. *If the transformation F is surjective, then $n \geq m$.*

COMMENT: This is not trivial, since the Boolean algebra B may be infinite.
PROOF: From Theorem 1.1 and Corollary 1.1. □

Corollary 1.3. *Let $f \in BF1$. Then $\{f\}$ is functionally independent if and only if f is of the form $f(x) = x + a$.*

PROOF: Using the previous calculation we can write independency in the form

$$1 = \prod_\gamma \bigvee_\alpha f^\gamma(\alpha) = \bigvee_\mu \prod_\gamma f^\gamma(\mu(\gamma)) = \bigvee_\mu f(\mu(1)) f'(\mu(0)) = f(0) f'(1) \vee f(1) f'(0) ,$$

which is equivalent to $f(1) = f'(0)$, i.e., $f(x) = x + f(0)$. □

Remark 1.4. $\mathrm{ind}_Y r = 1$ identically, because the system of equations $F(X) = Y$ is trivially consistent with respect to Y. □

Remark 1.5. The above theory of functional dependence cannot be generalized by taking elements of an arbitrary Boolean algebra insted of Boolean functions. For 0 fulfils $0' = 1$, while every element $a \neq 0$ fulfils $h(a) = 1$, where $h \neq 1$ is the Boolean function $h(x) = x + a'$. Thus every nonempty set would be dependent. □

Example 1.1. Take $n := 3$, $m := 4$ and

$$f_1(X) := ax_1 \vee x_2 , \quad f_2(X) := ax_1 , \quad f_3(X) := x_1 \vee a'x_3 , \quad f_4(X) := x_2 .$$

Then Corollary 1.3 of Theorem 1.1 and Remark 1.1 show that the singleton $\{f_4\}$ is independent, while all the other subfamilies of $F = \{f_1, f_2, f_2, f_3, f_4\}$ are dependent. In view of Remark 1.3 and Example 7.3.1, the least relation of functional dependence connecting F is

$$(af_2 f_3 \vee a' f_2' f_3 \vee f_2' f_3') f_1 f_4 \vee a f_1 f_2 f_3 f_4' \vee a' f_1' f_2' f_3 f_4' \vee f_1' f_2' f_3' f_4' = 1 .$$

□

Another technique for determining $\mathrm{ind}_X r$ and hence the functional dependence or independence of a family F consists in using the Blake canonical form; cf. Brown and Rudeanu [1988]; see also Ch.6, §3.

We now specialize the previous results to simple upper bounds and consequences.

Proposition 1.2. *The operator* sim ind_X *(= ind_X sim) satisfies the following identities:*

(1.12)
$$(\mathrm{sim\,ind}_X f)(Y) = \bigvee \{Y^C \mid \bigvee_A F^C(A) \neq 0\},$$

(1.13)
$$(\mathrm{sim\,ind}_X f)(Y) = \prod \{y_1^{\delta_1} \vee \ldots \vee y_m^{\delta_m} \mid D \in \{0,1\}^m \,\&\, \bigvee_{i=1}^m f_i^{\delta_i} = 1\}.$$

PROOF: Note first that $r(A,C) = F^C(A)$ by Proposition 1.1.

Identity (1.12) follows by Theorem 7.3.3. Further, denote by $g(Y)$ the right side of (1.13). Then for any $C \in \{0,1\}^m$,

$$g(C) = 0 \iff \exists D \in \{0,1\}^m \bigvee_{i=1}^m f_i^{\delta_i} = 1 \,\&\, \gamma_1^{\delta_1} \vee \ldots \vee \gamma_m^{\delta_m} = 0$$

$$\iff \exists D \in \{0,1\}^m \bigvee_{i=1}^m f_i^{\delta_i} = 1 \,\&\, \delta_i = \gamma_i' \ (i = 1, \ldots, m)$$

$$\iff \bigvee_{i=1}^m f_i^{\gamma_i'} = 1 \iff \prod_{i=1}^m f_i^{\gamma_i} = 0$$

$$\iff \forall X \in B^n \ 0 = \prod_{i=1}^m f_i^{\gamma_i}(X) = \prod_{i=1}^m (f_i(X) + \gamma_i')$$

$$= \prod_{i=1}^m (f_i(X)\gamma_i \vee f_i'(X)\gamma_i') = r(X,C) \iff \forall A \in B^n \ r(A,C) = 0$$

$$\iff \bigvee_A r(A,C) = 0 \iff \bigvee_A F^C(A) = 0 \iff (\mathrm{sim\,ind}_X f)(C) = 0$$

by (1.12). Thus the simple Boolean functions g and $\mathrm{sim\,ind}_X f$ coincide on $\{0,1\}^m$, hence they are identical. □

Example 1.2. Consider the system $F(X) = Y$ in Example 1.1. Its resolvent r was constructed in Example 7.3.1, where it was shown that

$$\mathrm{ind}_X r = (ay_2y_3 \vee a'y_2'y_3 \vee y_2'y_3')y_1y_4 \vee ay_1y_2y_3y_4' \vee ay_1'y_2'y_3y_4' \vee y_1'y_2'y_3'y_4',$$

therefore

$$\mathrm{sim\,ind}_X r = (y_2y_3 \vee y_2'y_3 \vee y_2'y_3')y_1y_4 \vee y_1y_2y_3y_4' \vee y_1'y_2'y_3y_4' \vee y_1'y_2'y_3'y_4'$$

$$= (y_2' \vee y_3)y_1y_4 \vee y_1y_2y_3y_4' \vee y_1'y_2'y_4'.$$

The application of Proposition 1.2 to this example seems tedious. □

Example 1.3. Take $n := 2$, $m := 2$, $f_1(x_1, x_2) := ax_1$, $f_2(x_1, x_2) := bx_2$. Then

$$f_1^{\alpha_1} \vee f_2^{\alpha_2} = (ax_1)^{\alpha_1} \vee (bx_2)^{\alpha_2} .$$

If $ab \neq 0$, then none of the elements a', b' and $a' \vee b'$ is 1, hence $f_1^{\alpha_1} \vee f_2^{\alpha_2}$ is not the constant 1, for any α_1, α_2. Therefore (1.13) yields $\mathrm{sim}\,\mathrm{ind}_X r = \prod \emptyset = 1$. If $ab = 0$ but $a \neq 0 \neq b$, then the identity $f_1^{\alpha_1} \vee f_2^{\alpha_2} = 1$ holds only for $\alpha_1 = \alpha_2 = 0$, hence (1.13) implies $\mathrm{sim}\,\mathrm{ind}_X r = y_1' \vee y_2'$.

The same results are of course obtained by applying in turn ind_X and sim, or by applying in turn sim and ind_X.   □

The understanding of the next definition requires some preparation. Moore [1910] has introduced the concept of *completely independent* system of propositions: this means a system of propositions $(p_1, \ldots, p_m)$ such that none of the minterms

$$p_1^{\gamma_1} \& \ldots \& p_m^{\gamma_m} \qquad (\gamma_1, \ldots, \gamma_m \in \{0, 1\})$$

(where p^γ is p or $\neg p$ according as $\gamma = 1$ or 0) is an identically false proposition (that is, the negation of a tautology). This property is much stronger than the independence of the system, for the latter property is equivalent to the fact that none of the propositions

$$p_1 \& \ldots \& p_{i-1} \& \neg p_i \& p_{i+1} \& \ldots \& p_m \qquad (i = 1, \ldots, m)$$

is identically false. On the other hand, Marczewski [1958] has used the framework of universal algebra (see e.g. Ch.2) to obtain a common generalization of many concepts called "independence" in various fields of mathematics. A system $(a_1, \ldots, a_m)$ of elements of an algebra A is said to be *independent* if every map from $\{a_1, \ldots, a_m\}$ to A can be extended to a homomorphism from the subalgebra generated by $a_1, \ldots, a_m$ to A. It turns out (cf. Marczewski [1960]) that in the case of Boolean algebras this concept reduces to Moore's complete independence.

Definition 1.4. A *relation of simple Boolean dependence* connecting a family $F = \{f_1, \ldots, f_m\}$ of Boolean functions is a relation of functional dependence $h \circ F = 1$ for which h is a simple Boolean function. Note that $f_1, \ldots, f_m$ need not be simple. The family F is said to be *Moore-Marczewski dependent* if there is a non-trivial relation of simple Boolean dependence connecting $f_1, \ldots, f_m$; otherwise one says that F is *Moore-Marczewski independent*.   □

The fact that Definition 1.4 is actually a proper specialization of Moore's original concept to the algebra of Boolean function will become clear in Theorem 1.2 (iv).

Remark 1.6. A necessary condition for the Moore-Marczewski independence of F is that $f_1, \ldots, f_m$ be pairwise distinct. The Moore-Marczewski independence is a hereditary property, while the property of being Moore-Marczewski dependent is a co-hereditary property (cf. Remark 1.1).   □

Remark 1.7. The simple Y-consequences of system (1.12) can be identified with the relations of simple Boolean dependence connecting F (cf. Remark 1.2). $\qquad\square$

Remark 1.8. The simple Y-upper-bounds h of r coincide with the functions h occurring in the relations of simple Boolean dependence of $h \circ F = 1$. In particular $(\operatorname{sim ind}_X r) \circ F = 1$ is the *least relation of simple Boolean dependence* connecting F (cf. Remark 1.3). $\qquad\square$

Theorem 1.2. *The following conditions are equivalent for a family* $F = \{f_1, \ldots, f_m\} \subset \mathrm{BF}n$:
   (i) F is Moore-Marczewski independent ;
   (ii) $\operatorname{sim ind}_X r = 1$;
   (iii) $\bigvee_A F^C(A) \neq 0 \qquad (\forall C \in \{0,1\}^m)$;
   (iv) $f_1^{\gamma_1} \ldots f_m^{\gamma_m} \neq 0 \qquad (\forall C \in \{0,1\}^m)$;
   (v) $f_1^{\gamma_1} \vee \ldots \vee f_m^{\gamma_m} \neq 1 \qquad (\forall C \in \{0,1\}^m)$.

COMMENT: The symbol $\neq$ in (iv) and (v) means "not identical to".
PROOF: (i)$\Longleftrightarrow$(ii): According to Definition 1.4, F is Moore-Marczewski independent iff the only relation of simple Boolean dependence on F is the trivial one and in view of Remark 1.7 this is further equivalent to the fact that the only simple Y-consequence of $r(X,Y) = 1$ is $1 = 1$; the last condition is equivalent to (ii) by Remark 7.3.13.
   (ii)$\Longleftrightarrow$(iii) by (1.12).
   (ii)$\Longleftrightarrow$(v): Formula (1.13) shows that $\operatorname{sim ind}_X = 1$ iff the set in the right-hand side is empty, that is, iff $\bigvee_{i=1}^m f_i^{\gamma_i} \neq 1$.
   (iv)$\Longleftrightarrow$(v): The negation of $\bigvee_i f_i^{\gamma_i} = 1$ can be written $\prod_i f_i^{\delta_i} \neq 0$, where $\delta_i = \gamma_i'$. On the other hand, when $(\gamma_1, \ldots, \gamma_m)$ runs over $\{0,1\}^m$, so does $(\gamma_1', \ldots, \gamma_m')$. $\qquad\square$

Corollary 1.4. *Let $f \in \mathrm{BF}1$. Then $\{f\}$ is Moore-Marczewski independent if and only if $0 \neq f \neq 1$.* $\qquad\square$

Corollary 1.5. *The concept of functional independence is actually stronger than that of Moore-Marczewski independence.*

PROOF: Trivially a functionally independent family is also Moore-Marczewski independent. The converse does not hold, as shown e.g. by Corollaries 1.4 and 1.3. $\qquad\square$

Example 1.4. Let us study the Moore-Marczewski independence of the family $\{f_1, f_2\} \subset \mathrm{BF}1$, where $f_1 := ax$ and $f_2 := bx$. Condition (iii) of Theorem 1.2 becomes

$$(a' \vee x')(b' \vee x') \neq 0 \,\&\, (a' \vee x')bx \neq 0 \,\&\, ax(b' \vee x') \neq 0 \,\&\, axbx \neq 0 ,$$

that is,

$$a'b' \vee x' \neq 0 \ \& \ a'bx \neq 0 \ \& \ ab'x \neq 0 \ \& \ abx \neq 0 \,,$$

which is clearly equivalent to

$$a'b \neq 0 \ \& \ ab' \neq 0 \ \& \ ab \neq 0 \,.$$

$\square$

Example 1.4 shows in particular that Corollary 1.1 of Theorem 1.1 is no longer valid for Moore-Marczewski dependence.

Finally notice that, in contrast to Remark 1.5, the definition of Moore-Marczewski independence makes sense without any change for elements of an arbitrary Boolean algebra instead of Boolean functions.

2 The range of a Boolean transformation

After the study of functional dependence and Moore-Marczewski dependence, in the rest of this chapter we focus on Boolean transformations, already introduced in §1. First we study the range of a Boolean transformation, following Rudeanu [1975c].

In this and the next section we work with a concept of Boolean transformation which is more general than the one introduced in Definition 1.3.

Definition 2.1. By a *generalized Boolean transformation* or simply a *Boolean transformation* we mean a map

(2.1) $F = (f_1, \ldots, f_m) : D \longrightarrow B^m \,,$

where $(f_1, \ldots, f_m) \in$ BFn and D is a *Boolean domain*, that is,

(2.2) $D = \{X \in B^n \mid d(X) = 1\} \neq \emptyset \,,$

where $d \in$ BFn. Thus

(2.3) $F(X) = (f_1(X), \ldots, f_m(X))$ $(\forall X \in D) \,.$

$\square$

In particular if $d = 1$ identically then $D = B^n$ and

(2.4) $F : B^n \longrightarrow B^m$

is a Boolean transformation in the sense of Definition 1.3.

Notation. Let again A and C denote vectors of $\{0,1\}^n$ and $\{0,1\}^m$, respectively.

$\square$

Proposition 2.1. *The range of a Boolean transformation* (2.1) *is characterized by the Boolean equation*

(2.5) $$\bigvee_C (\bigvee_A d(A) F^C(A)) Y^C = 1 \,.$$

PROOF: As in §1, let r denote the resolvent of the system $F(X) = Y$. Then the equation $d(X)r(X,Y) = 1$ expresses the condition: $X \in D$ and $F(X) = Y$. Therefore Y belongs to the range of F if and only if the above equation is consistent with respect to X. Taking into account Proposition 1.1, the latter condition can be written

$$1 = (\text{ind}_X(dr))(Y) = \bigvee_A d(A)r(A.Y)$$

$$= \bigvee_A d(A) \bigvee_C F^C(A)Y^C = \bigvee_C \bigvee_A d(A)F^C(A)Y^C .$$

$\square$

Corollary 2.1. *The range of a Boolean transformation (2.4) is characterized by the Boolean equation*

(2.6)
$$\bigvee_C (\bigvee_A F^C(A))Y^C = 1 .$$

$\square$

Corollary 2.2. *The transformation (2.1) has a given range*

(2.7)
$$R = \{Y \in B^m \mid h(Y) = 1\} ,$$

where $h \in \text{BF}m$, if and only if

(2.8)
$$\bigvee_A d(A)F^C(A) = h(C) \qquad (\forall C \in \{0,1\}^m) .$$

$\square$

Corollary 2.3. *The transformation (2.4) has a given range (2.7) if and only if*

(2.9)
$$\bigvee_A F^C(A) = h(C) \qquad (\forall C \in \{0,1\}^m) .$$

$\square$

Corollary 2.4. *The transformation (2.1) takes a constant value $\Xi \in B^m$ if and only if*

(2.10)
$$\bigvee_A d(A)F^C(A) = \Xi^C \qquad (\forall C \in \{0,1\}^m) .$$

PROOF: Apply Corollary 2.2 to the case when the equation $g(Y) = 1$ has the unique solution $Y = \Xi = (\xi_1, \ldots, \xi_m)$. This means that $h(Y) = \prod_{i=1}^m (y_i + \xi_i)'$, which is equivalent to the fact that for every $C \in \{0,1\}^m$,

$$h(C) = \prod_{i=1}^m (\xi_i + \gamma_i)' = \prod_{i=1}^m \xi_i^{\gamma_i} .$$

$\square$

Corollary 2.5. *The transformation (2.4) takes a constant value $\Xi \in B^m$ if and only if*

(2.11)
$$\bigvee_A F^C(A) = \Xi^C \qquad (\forall C \in \{0,1\}^m).$$

$\square$

Corollary 2.6. (BFE, Theorem 8.2). *The transformation (2.1) is surjective if and only if*

(2.12)
$$\prod_C \bigvee_A d(A) F^C(A) = 1.$$

PROOF: Surjectivity means that equation (2.6) in Corollary 2.1 is identically satisfied, that is, all the coefficients equal 1. $\square$

Corollary 2.7. (cf. Theorem 1.1). *The transformation (2.4) is surjective if and only if*

(2.13)
$$\prod_C \bigvee_A F^C(A) = 1.$$

COMMENT: This result is due to Whitehead and Löwenheim; see e.g. BFE, Theorem 8.2, and the present Theorem 1.1. $\square$

Corollary 2.8. *The range of a Boolean transformation $f : D \longrightarrow B$, where D is given by (2.2), is the interval*

(2.14)
$$[\prod_A (d'(A) \vee f(A)), \bigvee_A d(A) f(A)].$$

COMMENT: This theorem is due to Whitehead; see e.g. BFE, Theorem 2.5.
PROOF: For $m := 1$ equation (2.5) reduces to

$$(\bigvee_A d(A) f(A)) y \vee (\bigvee_A d(A) f'(A)) y' = 1,$$

or equivalently,

$$(\prod_A (d'(A) \vee f'(A)) y \vee (\prod_A (d'(A) \vee f(A)) y' = 0,$$

whose set of solution is the interval (2.14). $\square$

Corollary 2.9. *The range of a Boolean function $f : B^n \longrightarrow B$ is the interval*

(2.15)
$$[\prod_A f(A), \bigvee_A f(A)].$$

COMMENT: This fundamental theorem goes back to Schröder; see e.g. BFE, Theorem 2.4. $\square$

Proposition 2.2. *The transformation (2.1) takes a constant value if and only if*

(2.16)
$$\bigvee_C (\prod_A (d'(A) \vee F^C(A))) \bigvee_A d(A) F^C(A) = 1 \ .$$

PROOF: In view of the dual of Theorem 6.7 (iii) in BFE, the condition that equation (2.5) in Proposition 2.1 have a unique solution is $\bigvee_C \varphi(C) \prod_{E \neq C} \varphi'(E) = 1$, where $\varphi(C)$ stands for the coefficient of Y^C in (2.15). So $\varphi(C) = \bigvee_A d(A) F^C(A)$, while

$$\prod_{E \neq C} \varphi'(E) = \prod_{E \neq C} \prod_A (d'(A) \vee (F^E(A))') = \prod_A \prod_{E \neq C} (d'(A) \vee (F^E(A))')$$

$$= \prod_A (d'(A) \vee \prod_{E \neq C} (F^E(A))') = \prod_A (d'(A) \vee (\bigvee_{E \neq C} F^E(A))') = \prod_A (d'(A) \vee F^C(A)) \ .$$

□

Corollary 2.10. *The transformation (2.4) takes a constant value if and only if*

(2.17)
$$\bigvee_C \prod_A F^C(A) = 1 \ .$$

□

Example 2.1. Here are some of the formulas of this section in the case $m := 1$:

(2.5′)
$$(\bigvee_A d(A) f(A)) y \vee (\bigvee_A d(A) f'(A)) y' = 1 \ ,$$

(2.8′)
$$\bigvee_A d(A) f(A) = g(1) \ \& \ \bigvee_A d(A) f'(A) = g(0) \ ,$$

(2.10′)
$$\bigvee_A d(A) f(A) = \xi \ \& \ \bigvee_A d(A) f'(A) = \xi' \ ,$$

(2.12′)
$$\bigvee_A d(A) f(A) = \bigvee_A d(A) f'(A) = 1 \ ,$$

(2.16′)
$$(\prod_A (d'(A) \vee f(A))) \prod_A d(A) f(A) \vee$$
$$\vee (\prod_A (d'(A) \vee f'(A))) \prod_A d(A) f'(A) = 1 \ .$$

□

3 Injectivity domains of Boolean transformations

After the study of the range of a Boolean transformation and in particular of surjectivity, we now carry out a somewhat similar study for injectivity. The main idea is to determine the injectivity domains of a Boolean transformation, that is, all the injective restrictions of that transformation and in particular the bijectivity domains, that is, the injective restrictions which preserve the range. Most results were given in Rudeanu [1978a].

We keep the framework of Definition 2.1 and use formulas (2.1)-(2.4).

Definition 3.1. An *injectivity domain* of a generalized Boolean transformation (2.1) is a Boolean domain

$$(3.1) \qquad G = \{X \in B^n \mid g(X) = 1\} \subseteq D$$

such that the restriction $F|_G$ is injective. If, moreover, $F(G) = F(D)$, then G is called a *bijectivity domain* of the transformation F. □

Remark 3.1. In other words, a bijectivity domain is a complete set of representatives for the cosets of D modulo $\ker F$. □

Notation. We choose an arbitrary but fixed ordering

$$(3.2) \qquad A_0, A_1, \ldots, A_N \text{ , where } N = 2^n - 1 \text{ ,}$$

of the elements of $\{0,1\}^n$ and for any function $\varphi \in \mathrm{BF}n$ we set

$$(3.3) \qquad \varphi(k) = \varphi(A_k) \qquad (k = 0, 1, \ldots, N) \text{ ,}$$

$$(3.4) \qquad \delta(k, h) = \bigvee_{i=1}^{m} (f_i(k) + f_i(h)) \qquad (k, h = 0, 1, \ldots, N) \text{ .}$$

□

Proposition 3.1. *Any transformation (2.1) has non-empty injectivity domains G; they are characterized by the system of conditions*

$$(3.5) \qquad \bigvee_{k=0}^{N} g(k) = 1 \text{ ,}$$

$$(3.6) \qquad g(k) \le d(k) \qquad (k = 0, 1, \ldots, N) \text{ ,}$$

$$(3.7) \qquad g(k)g(h) \le \delta(k, h) \qquad (k, h = 0, 1, \ldots, N \text{ ; } k \ne h) \text{ ,}$$

or equivalently, by the system

$$(3.8.0) \qquad \prod_{h=1}^{N} g'(h) \le g(0) \le d(0) \prod_{h=1}^{N} (g'(h) \vee \delta(0, h)) \text{ ,}$$

$$(3.8.k) \qquad g(k) \le d(k) \prod_{h=k+1}^{N} (g'(h) \vee \delta(k, h)) \qquad (k = 1, \ldots, N-1) \text{ ,}$$

(3.8.N) $$g(N) \leq d(N) .$$

COMMENTS: 1) Relations (3.8.N), (3.8.N-1), ..., (3.8.0), taken in this order, yield a recursive parametric representation of the set of injectivity domains.

2) For a generalization to injective pseudo-Boolean transformations see Ghilezan [1979b].

PROOF: Clearly relations (3.5) and (3.6) express the conditions $G \neq \emptyset$ and $G \subseteq D$, respectively.

The injectivity condition

$$(\forall X)\,(\forall \Xi)\,(g(X) = 1 \,\&\, g(\Xi) = 1 \,\&\, F(X) = F(\Xi) \Longrightarrow X = \Xi)$$

can be written in the form

$$(\forall X)\,(\forall \Xi)\,(g(X)g(\Xi) \prod_{i=1}^{m}(f_i(X) + f_i(\Xi))' = 1 \Longrightarrow \prod_{i=1}^{m}(x_i + \xi_i)' = 1) ,$$

which, in view of the Verification Theorem, is also equivalent to

(3.9)
$$g(k)g(h) \prod_{k=1}^{m}(f_i(k) + f_i(h))' \leq \prod_{i=1}^{m}((A_k)_i + (A_h)_i)'$$
$$(k, h = 0, 1, \ldots, N) .$$

But the right-hand side of (3.9) is 1 or 0 according as $k = h$ of $k \neq h$, therefore conditions (3.9) reduce to

(3.10) $$g(k)g(h)\delta'(k, h) = 0 \qquad (k, h = 0, 1, \ldots, N \,;\, k \neq h) .$$

Thus the non-empty injectivity domains are characterized by (3.5)-(3.7). Besides, such domains do exist, for instance the singletons $\{X\}$ with $X \in D$ (since a singleton is characterized by a Boolean equation with unique solution). Therefore it remains to prove that the system (3.5)-(3.7) is equivalent to (3.8).

Note that the first inequality (3.8.0) is a paraphrase of (3.5), while conditions (3.6) can be written in the form

(3.6'.k) $$g(k)d'(k) = 0 \qquad (k = 0, 1, \ldots, N) .$$

As to conditions (3.7), we have obtained them in the form (3.10), which we now rewrite as

(3.11.k) $$g(k) \bigvee_{h=k+1}^{N} g(h)\delta'(k, h) = 0 \qquad (k = 0, 1, \ldots, N - 1) .$$

But (3.6'.N) is a translation of (3.8.N), while the remaining equations (3.6'.k) and equations (3.11.k) can be put together in the form

(3.12.k) $$g(k)(d'(k)\vee \bigvee_{h=k+1}^{N} g(h)\delta'(k, h)) = 0 \qquad (k = 0, 1, \ldots, N-1) .$$

Equation (3.12.0) is equivalent to the second inequality (3.8.0), while (3.12.k) is equivalent to (3.8.k) for $k = 1, \ldots, N - 1$.   □

Corollary 3.1. *Any transformation* (2.4) *has non-empty injectivity domains G; they are characterized by the system of relations* (3.5),(3.7) *or equivalently, by the system*

$$(3.13.0) \qquad \prod_{h=1}^{N} g'(h) \leq g(0) \leq \prod_{h=1}^{N} (g'(h) \vee \delta(0, h)) \,,$$

$$(3.13.k) \qquad g(k) \leq \prod_{h=k+1}^{N} (g'(h) \vee \delta(k, h)) \qquad (k = 1, \ldots, N - 1) \,.$$

□

Corollary 3.2. (BFE, Theorem 8.1). *The Boolean transformation* (2.1) *is injective if and only if*

$$(3.14) \qquad d(k)d(h) \leq \delta(k, h) \qquad (h, k = 0, 1, \ldots, N \, ; \, h \neq k) \,.$$

□

Corollary 3.3. (BFE, Corollary 1 of Theorem 8.1). *The Boolean transformation* (2.4) *is injective if and only if*

$$(3.15) \qquad \delta(k, h) = 1 \qquad (h, k = 0, 1, \ldots, N \, ; \, h \neq k) \,.$$

□

Corollary 3.4. (BFE, dual of Corollary 2 of Theorem 8.1). *Let $f \in \mathrm{BF}n$. Then the Boolean transformation*

$$(3.16) \qquad f|_D : D \longrightarrow B$$

is injective if and only if

$$(3.17) \qquad d(k)d(h) \leq f(k) + f(h) \qquad (h, k = 0, 1, \ldots, N \, ; \, h \neq k) \,.$$

□

For the case $m := 1$, $D := B^n$, see BFE, Proposition 8.1. We are now going to take $n := 1$.

Remark 3.2. In the case $n := 1$, the Boolean domains of a Boolean algebra B are its intervals, because a Boolean domain is defined by a Boolean equation in one unknown $d(x) = 1$, whose solution set is the interval $[d'(0), d(1)]$.   □

Corollary 3.5. *The interval defined by a Boolean equation $g(x) = 1$ is an injectivity domain of the Boolean function $f : B \longrightarrow B$ if and only if*

$$(3.18) \qquad g'(\alpha_1) \leq g(\alpha_0) \leq g'(\alpha_1) \vee (f(1) + f(0)) \,,$$

where (α_0, α_1) is the sequence (3.2).

COMMENT: Relations (3.18) with $(1,0)$ in the rôle of the sequence (α_0, α_1), were first found by McKinsey [1936b]; see also BFE, Theorem 11.1.

PROOF: Take $n := 1$ in Corollary 3.1. This implies $N = 1$, hence system (3.13) reduces to (3.13.0), which coincides with (3.18). $\square$

Proposition 3.2. *The following conditions are equivalent for a Boolean domain G:*

(i) G is a bijectivity domain of the transformation (2.1) ;

(ii) G fulfils (3.5), (3.6), (3.7) and

$$(3.19) \qquad d(k) \le \bigvee_{h=0}^{N} g(h)\delta'(k,h) \qquad (k = 0,1,\dots,N) \, ;$$

(iii) G fulfils

$$(3.20) \qquad d(k) \le \bigvee_{h=0}^{N} g(h)\delta'(k,h) \prod_{j=0,j\neq h}^{N} (g'(j)\vee\delta(k,j)) \ (k=0,1,\dots,N) \, ;$$

(iv) G fulfils

$$(3.21.0) \qquad \begin{aligned} &\prod_{h=1}^{N} g'(h) \vee \bigvee_{h=0}^{N} d(h) \prod_{j=1}^{N}(g'(j) \vee \delta(h,j)) \\ &\le g(0) \le d(0) \prod_{h=1}^{N}(g'(h) \vee \delta(0,h)) \, , \end{aligned}$$

$$(3.21.k) \qquad \begin{aligned} &\bigvee_{h=0}^{N} d(h)(\prod_{j=0}^{k-1}\delta(h,j)) \prod_{j=k+1}^{N}(g'(j) \vee \delta(h,j)) \le g(k) \\ &\le d(k) \prod_{h=k+1}^{N}(g'(h) \vee \delta(k,h)) \qquad (k=1,\dots,N-1) \, , \end{aligned}$$

$$(3.21.N) \qquad \bigvee_{h=0}^{N} d(h) \prod_{j=0}^{N-1} \delta(h,j) \le g(N) \le d(N) \, .$$

COMMENT: As for Proposition 3.1.

PROOF: (i)$\Longleftrightarrow$(ii): We have to introduce surjectivity, which we express in the following equivalent forms:

$$(\forall X \in D)\,(\exists \Xi \in G)\ F(X) = F(\Xi) \, ,$$

$$d(X) = 1 \Longrightarrow (\exists \Xi)\ g(\Xi) \prod_{i=1}^{m}(f_i(X) + f_i(\Xi))' = 1 \, ,$$

$$d(X) = 1 \Longrightarrow \bigvee_{h=0}^{N} g(h)(\bigvee_{i=1}^{m}(f_i(X) + f_i(h)))' = 1 \, ,$$

and this implication is equivalent to (3.19) by the Verification Theorem.

(ii)$\Longleftrightarrow$(iv): We have already seen in Proposition 3.1 the equivalence of systems (3.5)-(3.7) and (3.8). In view of Lemma 6.1.2, system (3.19) is equivalent to

$$(3.22.0) \qquad \bigvee_{h=0}^{N} d(h) \prod_{j=1}^{N} (\delta(h,j) \vee g'(j)) \leq g(0) \, ,$$

$$(3.22.k) \qquad \bigvee_{h=0}^{N} d(h)(\prod_{j=0}^{k-1} \delta(h,j)) \prod_{j=k+1}^{N} (\delta(h,j) \vee g'(j)) \leq g(k) \ (k=1,\ldots,N-1),$$

$$(3.22.N) \qquad \bigvee_{h=0}^{N} d(h) \prod_{j=0}^{N-1} \delta(h,j) \leq g(N).$$

The logical conjunction of systems (3.8) and (3.22) is precisely (3.21).

(i)$\Longleftrightarrow$(iii): G is a bijectivity domain if and only if, for every $X \in D$ there is a unique $\Xi \in G$ for which $F(X) = F(\Xi)$, that is, a unique Ξ for which

$$(3.23) \qquad g(\Xi) \prod_{i=1}^{m} (f_i(X) + f_i(\Xi))' = 1 \, ,$$

and a theorem due to Parker and Bernstein [1955] (see also BFE, Theorem 6.7) shows that equation (3.23) has a unique solution if and only if

$$(3.24) \qquad \begin{aligned} &\bigvee_{h=0}^{N} g(h)(\prod_{i=1}^{m}(f_i(X) + f_i(h))') \cdot \\ &\cdot \prod_{j=0, j \neq h}^{N}(g'(j) \vee \bigvee_{i=1}^{m}(f_i(X) + f_i(j))) = 1 \, . \end{aligned}$$

Therefore condition (i) is equivalent to $d(X) = 1 \Longrightarrow (3.24)$ and the latter implication is equivalent to (3.20) by the Verification Theorem.   □

Corollary 3.6. *The following conditions are equivalent for a Boolean domain* G:

(i) G *is a bijectivity domain of the transformation* (2.4) ;

(ii) G *fulfils* (3.5), (3.7) *and*

$$(3.25) \qquad \bigvee_{h=0}^{N} g(h)\delta'(k,h) = 1 \qquad (k=0,1,\ldots,N) ;$$

(iii) G *fulfils*

$$(3.26) \qquad \bigvee_{h=0}^{N} g(h)\delta'(k,h) \prod_{j=0, j \neq h}^{N} (g'(j) \vee \delta(k,j)) = 1 \qquad (k=0,1,\ldots,N) ;$$

(iv) G fulfils

(3.27.0)
$$\prod_{h=1}^{N} g'(h) \vee \bigvee_{h=0}^{N} \prod_{j=1}^{N}(g'(j)\vee\delta(h,j)) \leq g(0) \leq \prod_{h=1}^{N}(g'(h)\vee\delta(0,h)) \,,$$

(3.27.k)
$$\bigvee_{h=0}^{N}(\prod_{j=0}^{k-1}\delta(h,j))\prod_{j=k+1}^{N}(g'(j)\vee\delta(h,j)) \leq g(k)$$
$$\leq \prod_{h=k+1}^{N}(g'(h)\vee\delta(k,h)) \qquad (k=1,\ldots,N-1)\,,$$

(3.27.N)
$$\prod_{j=0}^{N-1}\delta(h,j) \leq g(N)\,.$$

$\square$

Corollary 3.7. *The interval defined by a Boolean equation $g(X) = 1$ is a bijectivity domain of the Boolean function $f : B \longrightarrow B$ if and only if*

(3.28) $$g(0) \vee g(1) = 1 \ \& \ g(0)g(1) = f(0) + f(1)\,.$$

COMMENTS: 1) Here it is immaterial whether 0 and 1 have their usual meaning or the meaning (3.2)!

2) An equivalent form of this corollary, which generalizes previous results of Schmidt [1922], was given in BFE, Proposition 11.5.

PROOF: Set $\delta = \delta(0,1)$ and apply Corollary 3.6 (iii), which yields $g(0)g(1) \leq \delta$, $g(0) \vee g(1)\delta' = 1$ and $g(0)\delta' \vee g(1) = 1$. By applying distributivity, the last two equations split into $g(0) \vee g(1) = 1$, $g(0) \vee \delta' = 1$ and $\delta' \vee g(1) = 1$. Now the last two equations read $\delta \leq g(0)$ and $\delta \leq g(1)$, hence they are equivalent to $\delta \leq g(0)g(1)$. Finally the first and the last inequality yield $g(0)g(1) = \delta$. $\square$

Remark 3.3. Corollaries 3.2 and 3.3 of Proposition 3.1 have no analogues for bijectivity domains, because D (because B^n) is a bijectivity domain of the transformation (2.1) (transformation (2.4)) if and only if this transformation is injective. $\square$

For related results the reader is referred to BFE, Chapters 8 and 11.

4 Fixed points of lattice and Boolean transformations

A classical theorem due to Tarski [1955] states that evry isotone map $f : L \longrightarrow L$ from a complete lattice into itself has *fixed points*, i.e., elements $x \in L$ such that $f(x) = x$. As a matter of fact, the converse is also true: if the lattice L has the property that every isotone map $f : L \longrightarrow L$ has a fixed point, then L is a complete lattice; cf. Davis [1955]. A related result is the Kleene fixed point theorem. Replace the completeness of L by the weaker condition of

$\omega-completeness$, which requires that every *increasing sequence* of elements of L (i.e., $(x_n)_{n \in \mathbb{N}}$ where $x_n \leq x_{n+1}$ for all n) have sup, and replace the isotony of f by the stronger condition of $\omega-continuity$, which requires that f commute with sup of every increasing sequence. Under these conditions the map $f : L \longrightarrow L$ has fixed points; cf. Kleene [1952].

Most of this section follows Rudeanu [1980]; the main result is that every isotone Boolean transformation has fixed points. We conclude with a generalization of the concept of fixed point, due to Tošić [1978b].

The starting point is the following generalization of Definition 1.3.

Definition 4.1. Suppose L is a lattice and m, n are positive integers. A *lattice transformation* is a map of the form

$$(4.1) \qquad F = (f_1, \ldots, f_m) : L^n \longrightarrow L^m ,$$

where $f_1, \ldots, f_m : L^n \longrightarrow L$ are lattice functions. Thus

$$(4.2) \qquad F(X) = (f_1(X), \ldots, f_m(X)) \qquad (\forall X \in L^n) .$$

□

Definition 4.2. The lattice transformation (4.1) is said to be *isotone* ($\omega-$ *continuous*) if the functions $f_1, \ldots, f_m$ are isotone ($\omega-$continuous). □

So far we have denoted the meet operation in a lattice by $\wedge$ and the meet in a Boolean algebra as a product; both notations will be used in this section.

Lemma 4.1. Let $(L; \wedge, \vee, 0)$ be an $\omega-$complete distributive lattice with zero. Suppose further that for every $a \in L$, the function

$$(4.3) \qquad f_a : L \longrightarrow L , \quad f_a(x) = a \wedge x \ (\forall x \in L)$$

is $\omega-$continuous. Then every lattice transformation $F : L^n \longrightarrow L^m$ is $\omega-$continuous.

PROOF: According to the hypothesis,

$$(4.4) \qquad a \wedge \bigvee_{n \in \mathbb{N}} x_n = \bigvee_{n \in \mathbb{N}} (a \wedge x_n)$$

for every increasing sequence $(x_n)_{n \in \mathbb{N}}$. This implies that

$$(4.5) \qquad (\bigvee_{n \in \mathbb{N}} x_n) \wedge (\bigvee_{n \in \mathbb{N}} y_n) = \bigvee_{n \in \mathbb{N}} (x_n \wedge y_n)$$

provided the sequences $(x_n)_{n \in \mathbb{N}}$ and $(y_n)_{n \in \mathbb{N}}$ are increasing. For (4.4) implies

$$(\bigvee_{n \in \mathbb{N}} x_n) \wedge (\bigvee_{n \in \mathbb{N}} y_n) = \bigvee_{n \in \mathbb{N}} ((\bigvee_{m \in \mathbb{N}} x_m) \wedge y_n)$$

$$= \bigvee_{n \in \mathbb{N}} \bigvee_{m \in \mathbb{N}} (x_m \wedge y_n) = \bigvee_{p \in \mathbb{N}} (x_p \wedge y_p) ,$$

because $x_m \wedge y_n \leq x_p \wedge y_p$, where $p = \max(m, n)$.

Further, (4.5) implies that if $(x_{1n})_{n \in \mathbb{N}}, \ldots, (x_{pn})_{n \in \mathbb{N}}$ are increasing sequences, then

$$(4.6) \qquad (\bigvee_{q \in \mathbf{N}} x_{1q}) \wedge \ldots \wedge (\bigvee_{q \in \mathbf{N}} x_{pq}) = \bigvee_{q \in \mathbf{N}} (x_{1q} \wedge \ldots \wedge x_{pq}) .$$

On the other hand, every simple lattice function $f : L^n \longrightarrow L$ is a disjunction of *lattice monomials* $x_{i_1} \wedge \ldots \wedge x_{i_p}$, $\{i_1, \ldots, i_p\} \subseteq \{1, \ldots, n\}$; cf. Grätzer [1978], Theorem II.2.1(the proof is in fact similar to that of the present Lemma 3.3.1). Therefore every lattice function is a disjunction between a constant b (possibly 0) and a disjunction of lattice monomials and/or of terms of the form $a \wedge \mu$, where $a \in L$ and μ is a lattice monomial (this variant of Goodstein's Theorem 3.3.1 is due to the fact that our lattice L is not supposed to have 1) .

It follows from the above representation of a lattice function and formula (4.6) that every lattice function $f : L \longrightarrow L$ satisfies

$$(4.7) \qquad f(\bigvee_{q \in \mathbf{N}} x_{1q}, \ldots, \bigvee_{q \in \mathbf{N}} x_{nq}) = \bigvee_{q \in \mathbf{N}} f(x_{1q}, \ldots, x_{nq}) ,$$

provided the sequences occurring in (4.7) are increasing. In other words, every lattice function is ω−continuous. In view of Definition 4.2, this is the desired result.   □

Recall that a σ−*Boolean algebra* is a Boolean algebra in which every countable subset has inf and sup.

Corollary 4.1. *If B is a σ−Boolean algebra, then every isotone Boolean transformation $F : B^n \longrightarrow B^m$ is ω−continuous.*

PROOF: It is well known (see e.g. BFE, Theorem 11.3) that every isotone Boolean function has the representation given in Theorem 3.3.1, which is a representation of the more general form used in the proof of Lemma 4.1. On the other hand, a σ−Boolean algebra is known to fulfil (4.4) for any sequence (this follows e.g. from Proposition 3.2.4).   □

The above corollary implies further that every isotone Boolean transformation has fixed points, but in Theorem 4.2 below we will prove this for an arbitrary Boolean algebra.

Proposition 4.1. *Under the conditions of Lemma 4.1, every lattice transformation $F : L^n \longrightarrow L^n$ has fixed points.*

PROOF: From Lemma 4.1 and the Kleene fixed point theorem.   □

Theorem 4.1. *Let B be an arbitrary Boolean algebra. An element $X \in B^n$ is a fixed point of a Boolean transformation $F : B^n \longrightarrow B^n$ if and only if*

$$(4.8) \qquad \bigvee_A (F(A))^A X^A = 1 .$$

PROOF: The fixed point condition $f(X) = X$ can be written in the form $g(X) = 0$, where

$$g(X) = \bigvee_{i=1}^{n} (f_i(X) + x_i) .$$

The equivalent equation $g'(X) = 1$ is (4.8), because, setting $A = (\alpha_1, \ldots, \alpha_n)$,

$$g'(A) = \prod_{i=1}^{n}(f_i(A) + \alpha_i + 1) = \prod_{i=1}^{n} f_i^{\alpha_i} = (F(A))^A .$$

□

Corollary 4.2. *A Boolean transformation* $F : B^n \longrightarrow B^n$ *has fixed points if and only if*

(4.9)
$$\bigvee_A (F(A))^A = 1 .$$

□

Theorem 4.2. *Every isotone Boolean transformation* $F : B^n \longrightarrow B^n$ *has fixed points.*

PROOF: We have to prove (4.9). Setting $f_i' = g_i$ $(i = 1, \ldots, n)$, condition (4.9) can be written in the form

$$\prod_A (g_1^{\alpha_1}(A) \vee \ldots \vee g_n^{\alpha_n}(A)) = 0 ,$$

which, in view of distributivity, is equivalent to

(4.10)
$$\bigvee_\varphi \prod_A (g_{\varphi(A)})^{\alpha_{\varphi(A)}}(A) = 0 ,$$

where φ runs over all the mappings $\varphi : \{0,1\}^n \longrightarrow \{1, \ldots, n\}$.

Now suppose that for a given φ there exist $C, D \in \{0,1\}^n$ such that

(4.11) $C < D$, $\varphi(C) = \varphi(D)$ (say $= i$), $\gamma_i = 0$, $\delta_i = 1$.

Then

$$\prod_A (g_{\varphi(A)})^{\alpha_{\varphi(A)}}(A) \leq g_i'(C)g_i(D) = f_i(C)f_i'(D) = 0$$

by the isotony of F. Therefore, in order to prove (4.10) it suffices to show that for every φ there exist C, D satisfying (4.11).

It is convenient to use the isomorphism

$$A \mapsto \mathrm{Coker} A = \{i \in \{1, \ldots, n\} \mid \alpha_i = 1\}$$

(cf. Definition 7.2.3) between the posets $(\{0,1\}^n, \leq)$ and $(\mathcal{P}(\{1, \ldots, n\}), \subseteq)$. Then (4.11) becomes

(4.12)
$$\mathrm{Coker} C \subset \mathrm{Coker} D ,$$
$$\varphi(\mathrm{Coker} C) = \varphi(\mathrm{Coker} D) = i \in \mathrm{Coker} D \backslash \mathrm{Coker} C$$

and C, D are determined as follows. Define $i_0 = \varphi(\emptyset)$ and

$$i_h = \varphi(\{i_0, \ldots, i_{h-1}\}) \qquad (h = 2, \ldots, n) \, .$$

Since $i_0, i_1, \ldots, i_n \in \{1, \ldots, n\}$, it follows that $i_j = i_k$ for some $j \neq k$. Let j be the least index for which there exists $k > j$ with $i_j = i_k$. Take

$$\mathrm{Coker} C = \{i_0, \ldots, i_{j-1}\} \text{ or } \mathrm{Coker} C = \emptyset$$

according as $j > 0$ or $j = 0$; take also $\mathrm{Coker} D = \{i_0, \ldots, i_{k-1}\}$ and $i = i_j = i_k$. Then (4.12) holds.   □

Corollary 4.3. (Scognamiglio [1960]). *A Boolean function* $f : B \longrightarrow B$ *has fixed points if and only if it is isotone, in which case the set of fixed points is the range* $[f(0), f(1)]$.

COMMENT: For other properties equivalent to the isotony of the function f see BFE, Theorem 11.2.

PROOF: For $n := 1$ condition (4.9) becomes $f'(0) \vee f(1) = 1$, or equivalently, $f(0) \leq f(1)$, which is equivalent to isotony (cf. McKinsey [1936b]; see also BFE, Theorem 2.13). Equation (4.8) becomes $f'(0)x' \vee f(1)x = 1$ and its solutions are $f(0) \leq x \leq f(1)$.   □

Proposition 4.2. *Let* $(L; \wedge, \vee, 0)$ *be a distributive lattice with zero. Then every lattice function* $f : L \longrightarrow L$ *has the least fixed point* $f(0)$.

PROOF: The function f is of the form $f(x) = (a \wedge x) \vee b$ or $f(x) = x \vee b$ (cf. the proof of Lemma 4.1). In both cases $f(0) = b = f(b)$. The fixed point $f(0)$ is the least one because $f(0) \leq f(x)$ for any x.   □

Example 4.1. Consider the Boolean transformation

$$F : B^2 \longrightarrow B^2 \, , \quad F(x, y) = (y', xy \vee x'y') \, ,$$

which is not isotone. Further,

$$f_1'(0,0)f_2'(0,0) \vee f_1'(0,1)f_2(0,1) \vee f_1(1,0)f_2'(1,0)$$

$$\vee f_1(1,1)f_2(1,1) = 0 \vee 0 \vee 1 \vee 0 = 1 \, ,$$

showing that condition (4.9) is fulfilled, therefore fixed points do exist. Equation (4.8) reduces to $1 \cdot xy' = 1$, hence the unique fixed point is (1,0).
    Note that

$$F(0,0) = (1,1) \, , \quad F(1,1) = (0,1) \, , \quad F(0,1) = (0,0) \, ,$$

therefore in this example the unique fixed point cannot be reached by the Kleene procedure, which constructs the least fixed point as $\sup f^n(0)$.   □

    See Ghilezan [1980a] for an attempt to study the fixed points of a Boolean transformation $F : D \longrightarrow B^n$, $D \subseteq B^n$, and for the Boolean matrix equation $AX = X$.

    An interesting generalization of the concept of fixed point, which is motivated by switching theory, was given by Tošić [1978b].

Definition 4.3. A function $f : B^n \longrightarrow B$ *preserves the constant* $c \in B$ *if* $f(c,\ldots,c) = c$. $\qquad\qquad\qquad\qquad\qquad\qquad\qquad\qquad\qquad\qquad\qquad\qquad\square$

The following result was proved by Tošić under the unnecessary assumption that B is finite.

Proposition 4.3. *A Boolean function* $f : B^n \longrightarrow B$ *preserves constants if and only if* $f(0,\ldots,0) \le f(1,\ldots,1)$, *in which case the set of preserved constants is the interval* $[f(0,\ldots,0), f(1\ldots,1)]$.

COMMENT: As mentioned in the proof of Corollary 4.3, in the case $n := 1$ the property $f(0) \le f(1)$ is equivalent to isotony, therefore Proposition 4.3 is a generalization of Corollary 4.3.

PROOF: Since, setting $A = (\alpha_1,\ldots,\alpha_n)$,

$$f(c,\ldots,c) = \bigvee_A f(A)c^{\alpha_1}\ldots c^{\alpha_n} = \bigvee_{\alpha \in \{0,1\}^n} f(\alpha,\ldots,\alpha)c^\alpha$$

$$= f(0,\ldots,0)c' \vee f(1,\ldots,1)c\,,$$

it follows that

$$f(c,\ldots,c) = c \Longleftrightarrow f(0,\ldots,0)c' \vee f(1,\ldots,1)c = c$$

$$\Longleftrightarrow f(0,\ldots,0)c' \vee f'(1,\ldots,1)c = 0 \Longleftrightarrow f(0,\ldots,0) \le c \le f(1,\ldots,1)\,.$$

$$\square$$

Corollary 4.5 below was noticed by Tošić.

Corollary 4.4. *An isotone Boolean function preserves all the constants.* $\qquad\square$

Corollary 4.5. *Suppose* $f : B^n \longrightarrow B$ *is a simple Boolean function. Then there are four possibilities:*

(i) *if* $f(0,\ldots,0) = f(1,\ldots,1) = 0$, *then* f *preserves only the constant 0 ;*

(ii) *if* $f(0,\ldots,0) = f(1,\ldots,1) = 1$, *then* f *preserves only the constant 1 ;*

(iii) *if* $f(0,\ldots,0) = 0$ *and* $f(1,\ldots,1) = 1$, *then* f *preserves any constant ;*

(iv) *if* $f(0,\ldots,0) = 1$ *and* $f(1,\ldots,1) = 0$, *then* f *preserves no constant .*

$$\square$$

The classification devised in Corollary 4.5 was simultaneously discovered by Krnić [1978], who used it as a new approach in studying bases of truth functions and in particular Sheffer functions.

9. More on solving Boolean equations

This chapter is devoted to several important additions to the theory of Boolean equations.

In §1 we present various methods that have been proposed in the literature for solving Boolean equations and, in particular, truth equations or special classes of such equations. The study of Boolean equations with unique solution was initiated by Bernstein [1932] and Parker and Bernstein [1955] (see e.g. BFE, Chapter 6, §2). The particular case of systems of linear Boolean equations with unique solution is included in Ch.6, §1 of the present book, while in §2 of this chapter we resume the study of uniqueness in the line of Parker and Bernstein. Whereas most Boolean equations studied in this book are defined over an arbitrary Boolean algebra, in §3 of this chapter we deal with quadratic truth equations[1], which have several applications. This includes in particular the study of quadratic truth equations with unique solution. The last section deals with the use of computers for solving Boolean equations and in particular quadratic truth equations.

1 Special methods for solving Boolean equations

As was shown in BFE, Chapter 2, §5, a way of obtaining an efficient solution of Boolean equations is irredundancy. Two kinds of irredundancies have been studied: a recursive construction of the set of all the solutions, such that each solution is generated only once (the Davio-Deschamps method; cf. BFE, Theorem 2.10), and the construction of a set of values of the parameters occurring in a general/reproductive solution, for which there is a bijection between that set and the set of all the solutions (cf. BFE, Theorem 2.9). The former idea has been continued by J.-P. Deschamps [1971] and Brown [1982], while Lavit [1976] follows the latter idea.

We present below the Brown and Lavit papers and a related theorem of Brown and Rudeanu [2001]. The next topic is the minimization of the number of parameters in a general solution as approached by Banković [1989/90], [1992a]. Then we briefly survey several papers devoted to the same problem of

---

[1] Usually called quadratic Boolean equations in the literature; see the Preface.

solving efficiently Boolean equations and in particular truth equations. Further, we explain an algorithm due to Zakrevskiĭ [1975a] for testing the satisfiability of a truth function expressed as a ring sum of terms. Finally we list a few papers dealing with special types of truth equations and conclude by mentioning a quite special approach to arbitrary truth equations.

Definitions 1.1, 1.2 and Proposition 1.1 are in the spirit of Brown [1982], while Definitions 1.3, 1.4 and Propositions 1.2-1.4 are taken almost verbatim from that paper. Yet in Propositions 1.2 and 1.3 we have succeeded to drop the finiteness assumption from the original theorems.

Definition 1.1. Let $\Gamma = \{g_1, \ldots, g_m\}$ be an n-element subset of a Boolean algebra B. The Γ-*minterms* are the elements of the form $\Gamma^C = g_1^{\gamma_1} \ldots g_m^{\gamma_m}$, where $(\gamma_1, \ldots, \gamma_m) = C \in \{0,1\}^m$. □

Remark 1.1. The Γ-minterms are not necessarily distrinct; in particular some of them may be 0. However $\{\Gamma^C \mid C \in \{0,1\}^m\}$ is an orthonormal system. □

Definition 1.2. Let Γ be a finite subset of the Boolean algebra B. By an *orthogonal (orthonormal) Γ-representation* of a Boolean function $f : B^n \longrightarrow B$ we mean an identity of the form

$$(1.1) \qquad f(X) = \bigvee_{j=1}^{p} c_j f_j(X),$$

where $f_1, \ldots, f_p$ are simple non-zero Boolean functions, $\{c_1, \ldots, c_p\}$ is an orthogonal (orthonormal) system and c_j is a Γ-minterm whenever f_j is not a term. □

Proposition 1.1. *Every Boolean function has an orthogonal Γ-representation for some $\Gamma \subset B$.*

PROOF: It follows from Proposition 2.3.5 that any Boolean function $f : B^n \longrightarrow B$ can be written in the form $f(X) = g(X, \Gamma)$, where $g : B^{n+m} \longrightarrow B$ is a simple Boolean function and $\Gamma \subseteq B^m$, $m \geq 0$; for $m = 0$ it is understood that there is no Γ. So $f(X) = \bigvee_C g(X, C)\Gamma^C$, where C runs over $\{0,1\}^m$. Since each $g(X, C) \neq 0$ is a term or a disjunction of terms, we can write

$$(1.2.0) \qquad f(X) = f_1(X) \vee f_2(X),$$

$$(1.2.1) \qquad f_1(X) = \bigvee\{g(X, C)\Gamma^C \mid \Gamma^C \neq 0 \ \& \ g(X, C) \text{ is a term}\},$$

$$(1.2.2) \qquad f_2(X) = \bigvee\{g(X, C)\Gamma^C \mid \Gamma^C \neq 0 \ \& \ g(X, C) \text{ has } \geq 2 \text{ terms}\}.$$

But (1.2.1) can be further written in the form

$$(1.2.1') \qquad\qquad f_1(X) = \bigvee_{j=1}^{k} c_j f_j(X) \, ,$$

where $f_1, \ldots, f_k$ are the distinct terms $g(X, C)$ occurring in (1.2.1) and each c_j is a Γ−minterm or a disjunction of Γ−minterms, while a mere change of notation $\Gamma^C = c_j$ transforms (1.2.2) into

$$(1.2.2') \qquad\qquad f_2(X) = \bigvee_{j=k+1}^{p} c_j f_j(X) \, .$$

Now the representation (1.1) is just a compact form of the system (1.2.0), (1.2.1'), (1.2.2'). Besides, since the Γ−minterms occurring in (1.2.1) and (1.2.2) are pairwise distinct, it follows that $\{c_1, \ldots, c_p\}$ is an orthogonal system $\qquad\square$

Corollary 1.1. *If B is a finite Boolean algebra, then every Boolean function $f : B^n \longrightarrow B$ can be represented in the form (1.1), where $f_1, \ldots, f_p$ are simple non-zero Boolean functions, $\{c_1, \ldots, c_p\} \subseteq B$ is an orthogonal system and c_j is an atom whenever f_j is not a term.*

PROOF: Since B is generated by the atoms $a_1, \ldots, a_m$ (see e.g. BFE, Chapter 1, §1), we can take $\Gamma := \{a_1, \ldots, a_m\}$. Then the non-zero Γ−minterms are exactly the m atoms $a_i = a_i \prod_{j=1, j \neq i}^{m} a_j'$. $\qquad\square$

Definitions 1.3, 1.4 and Propositions 1.2-1.4 below are due to Brown [1982], yet in Propositions 1.2 and 1.3 we have succeeded to drop the finiteness assumption from the original theorems.

Proposition 1.2. *A Boolean equation $f(X) = 1$ is consistent if and only if the function f has an orthonormal Γ−representation for some $\Gamma \subset B$.*

PROOF: If the equation $f(X) = 1$ is consistent then, taking into account that all $g(A, C) \in \{0, 1\}$, the construction in the proof of Proposition 1.1 yields

$$\bigvee_{j=1}^{p} c_j = \bigvee \{\Gamma^C \mid g(X, C) \text{ is a term}\} \vee$$

$$\vee \bigvee \{\Gamma^C \mid g(X, C) \text{ has } \geq 2 \text{ terms}\} = \bigvee \Gamma^C \mid g(X, C) \neq 0\}$$

$$= \bigvee \{\Gamma^C \mid \exists A \in \{0, 1\}^n \; g(A, C) = 1\} = \bigvee_{C} \bigvee_{A} g(A, C) \Gamma^C$$

$$= \bigvee_{A} \bigvee_{C} g(A, C) \Gamma^C = \bigvee_{A} g(A, \Gamma) = \bigvee_{A} f(A) = 1 \, .$$

Conversely, suppose the function f has a representation of the form described above. For each $j \in \{1, \ldots, p\}$, pick a term t_j of f_j and set $h(X) = \bigvee_{j=1}^{p} c_j t_j(X)$. Then

$$\bigvee_{A} h(A) = \bigvee_{A} \bigvee_{j} c_j t_j(A) = \bigvee_{j} c_j \bigvee_{A} t_j(A) = \bigvee_{j} c_j \cdot 1 = 1 \, ,$$

therefore the equation $h(X) = 1$ is consistent; but $h(X) \leq f(X)$, hence the equation $f(X) = 1$ is consistent as well. $\qquad\square$

As we have seen, in most cases the structure of the set of solutions of a Boolean equation is rather complicated: for any given order of the variables, each variable ranges within an interval which depends on the previous variables. A favourable particular case is that when the set of solutions is a block in the sense of the following definition.

Definition 1.3. The *block* determined by n intervals $[a_i, b_i] \subseteq B$ $(i = 1, \dots, n)$ is their Cartesian product, i.e.,

(1.3)
$$[a_1, b_1] \times \dots \times [a_n, b_n]$$
$$= \{(x_1, \dots, x_n) \mid a_i \leq x_i \leq b_i \ (i = 1, \dots, n)\} .$$

$\qquad\square$

Proposition 1.3. α) *The set of solutions of a consistent Boolean equation $f(X) = 1$ is a block if and only if the function f can be represented in the form*

(1.4)
$$f(X) = \bigvee_{j=1}^{p} c_j t_j(X) ,$$

where $\{c_1, \dots, c_p\}$ is an orthonormal system and $t_1, \dots, t_p$ are terms.

β) *When this is the case, the block (1.3) of solutions is determined by the intervals*

(1.5)
$$a_i = \bigvee\{c_j \mid t_j(X) \leq x_i\}; \ b_i = \bigvee\{c_j \mid t_j(X) \not\leq x_i'\}$$
$$(i = 1, \dots, n) .$$

COMMENT: The constant function 1 is included in the definition of terms.
PROOF: Suppose f has the form (1.4). For each $i \in \{1, \dots, n\}$ and each $j \in \{1, \dots, p\}$, condition $t_j(X) \leq x_i$ means that $t_j(X)$ is a conjunction of literals including x_i, or equivalently, $t_j'(X)$ is a disjunction of literals including x_i'; a similar remark holds for $t_j(X) \leq x_i'$. It follows that $f'(X) = \bigvee_{j=1}^{p} c_j t_j'(X)$ can be written in the form

$$f'(X) = \bigvee_{i=1}^{n} (x_i' \bigvee\{c_j \mid t_j(X) \leq x_i\} \vee x_i \bigvee\{c_j \mid t_j(X) \leq x_i'\}) ,$$

therefore the equation $f(X) = 1$, or equivalently, $f'(X) = 0$, is equivalent to the system

$$x_i' a_i \vee x_i \bigvee\{c_j \mid t_j(X) \leq x_i'\} = 0 \qquad (i = 1, \dots, n) ,$$

whose solutions are, in view of orthonormality,

$$a_i \leq x_i \leq (\bigvee\{c_j \mid t_j(X) \leq x_i'\})' = b_i \qquad (i = 1, \dots, n) .$$

Conversely, suppose the Boolean equation has the solution set (1.3). Setting

(1.6)
$$g(X) = \bigvee_{i=1}^{n} (a_i x_i' \vee b_i' x_i) ,$$

we have
$$g(X) = 0 \iff a_i \leq x_i \leq b_i \ (i = 1, \ldots, n) \,,$$

therefore $f = g'$ by the Verification Theorem. On the other hand, setting $\Gamma = \{a_1, \ldots, a_n, b_1, \ldots, b_n\}$, we have $g(X) = h(X, \Gamma)$, where $h : B^{3n} \longrightarrow B$ is defined by

$$h(X, Y) = \bigvee_{i=1}^{n} (y_i x_i \vee y'_{n+1} x_i) \,.$$

Let $C = (\gamma_1, \ldots, \gamma_{2n}) \in \{0, 1\}^{2n}$ and notice that if $\gamma_i = 1$ and $\gamma_{n+i} = 0$ for some i, then $\Gamma^C = 0$ because $a_i b'_i = 0$. It follows that

(1.7)
$$g(X) = \bigvee_C h(X, C) \Gamma^C = \bigvee \{ h(X, C) \Gamma^C \mid \gamma_i \leq \gamma_{n+i}$$
$$(i = 1, \ldots, n) \}$$

and in this expansion each $h(X, C)$ is either 0 or a (disjunction of) literal(s) (it cannot be 1). Therefore, using orthonormality, we obtain

(1.8)
$$f(X) = g'(X) = \bigvee \{ h'(X, C) \Gamma^C \mid \gamma_i \leq \gamma_{n+i} \ (i = 1, \ldots, n) \} \,,$$

where each $h'(X, C)$ is a term. $\qquad\qquad\qquad\qquad\qquad\qquad\qquad \Box$

Example 1.1. Take $n := 2$. Consider two intervals $[a_1, b_1]$ and $[a_2, b_2]$. Then (1.6) reduces to
$$g(x_1, x_2) = a_1 x'_1 \vee b'_1 x_1 \vee a_2 x'_2 \vee b'_2 x_2$$

and the expansion (1.7) becomes

$$g(x_1, x_2) = a'_1 b'_1 a'_2 b'_2 (x_1 \vee x_2) \vee a'_1 b'_1 a'_2 b_2 x_1 \vee a'_1 b'_1 a_2 b_2 (x_1 \vee x'_2) \vee$$

$$\vee a'_1 b_1 a'_2 b'_2 x_2 \vee a'_1 b_1 a'_2 b_2 \cdot 0 \vee a'_1 b_1 a_2 b_2 x'_2 \vee$$

$$\vee a_1 b_1 a'_2 b'_2 (x'_1 \vee x_2) \vee a_1 b_1 a'_2 b_2 x'_1 \vee a_1 b_1 a_2 b_2 (x'_1 \vee x'_2) \,,$$

the relations $a_i \leq b_i \ (i = 1, 2)$ showing that the other 7 minterms in a_1, a_2, b_1, b_2 are zero and that the above expansion can be simplified as follows:

$$g(x_1, x_2) = b'_1 b'_2 (x_1 \vee x_2) \vee b'_1 a'_2 b_2 x_1 \vee b'_1 a_2 (x_1 \vee x'_2) \vee a'_1 b_1 b'_2 x_2 \vee$$

$$\vee a'_1 b_1 a'_2 b_2 \cdot 0 \vee a'_1 b_1 a_2 x'_2 \vee a_1 b'_2 (x'_1 \vee x_2) \vee a_1 a'_2 b_2 x'_1 \vee a_1 a_2 (x'_1 \vee x'_2) \,,$$

so that the corresponding expression (1.8) (or (1.4)) is

$$f(x_1, x_2) = b'_1 b'_2 x'_1 x'_2 \vee b'_1 a'_2 b_2 x'_1 \vee b'_1 a_2 x'_1 x_2 \vee$$

$$\vee a'_1 b_1 b'_2 x'_2 \vee a'_1 b_1 a'_2 b_2 \vee a'_1 b_1 a_2 x_2 \vee$$

$$\vee a_1 b'_2 x_1 x'_2 \vee a_1 b_1 a'_2 b_2 x_1 \vee a_1 a_2 x_1 x_2 \,.$$

$$\Box$$

The next proposition may be viewed as an attempt to reduce the solution of an arbitrary consistent Boolean equation $f(X) = 1$ to the favourable situation described in Proposition 1.3.

Definition 1.4. The *constituents* of a Boolean function f with respect to a Γ–representation (1.1) of it are the Boolean functions of the form

$$(1.9) \qquad f_\varphi(X) = \bigvee_{j=1}^{p} c_j t_{\varphi(j)}(X),$$

where for each $j \in \{1, \ldots, p\}$, $t_{\varphi(j)}$ is a term of f_j.   □

Proposition 1.4. *Let S be the solution set of a consistent Boolean equation $f(X) = 1$ over a finite Boolean algebra B. For each constituent f_φ with respect to the orthonormal Γ–representation of f constructed in Corollary 1.1, let S_φ be the solution set of the equation $f_\varphi(X) = 1$. Then the sets S_φ are blocks and $S = \bigcup S_\varphi$.*

PROOF: The sets S_φ are blocks by Proposition 1.3. Since $f \leq f_\varphi$ for every constituent f_φ, it follows that $\bigcup S_\varphi \subseteq S$. To prove the converse, take $\Xi \in S$. It follows from $\bigvee_{j=1}^{p} c_j f_j(\Xi) = 1$ by multiplication with c_j that $c_j f_j(\Xi) = c_j$, or equivalently, $c_j \leq f_j(\Xi)$ for all j. If f_j is a term, set $t_{\varphi(j)} = f_j$. Otherwise c_j is an atom and since every atom is join-irreducible (see e.g. Balbes and Dwinger [1974], Theorem III.2), it follows that $c_j \leq t_{\varphi(j)}(\Xi)$ for some term $t_{\varphi(j)}$ of f_j. We have thus constructed a constituent f_φ of the function f and this implies $f_\varphi(\Xi) \geq f(\Xi) = 1$, that is, $\Xi \in S_\varphi$.   □

The paper by Brown [1982] includes a discussion of prior related work as well. Thus Davio and J.-P. Deschamps [1969] have obtained a partition into blocks of the set of solutions (see also BFE, Theorem 2.10). The latter author [1971] has characterized the maximal blocks of solutions in terms of prime implicants.

We are now going to prove a result which is related to Proposition 1.2 and also to the case when a Boolean equation $f(X) = 1$ has a unique solution; cf. Theorem 2.1.

First we need some notation. Consider again $f \in BFn$ represented in the form

$$(1.10) \qquad f(X) = g(X, \Gamma),$$

where $g : B^{n+m} \longrightarrow B$ is a simple Boolean function and $\Gamma \subseteq B^m$, $m \geq 0$. Further, set

$$(1.11) \qquad I = \{0,1\}^n, \quad J = \{C \in \{0,1\}^m \mid \Gamma^C \neq 0\},$$

$$(1.12) \qquad F = \{\varphi : J \longrightarrow I \mid X^{\varphi(C)} \leq g(X,C) \; (\forall X \in B^n) \; (\forall C \in J)\},$$

$$(1.13) \qquad g_C : B^n \longrightarrow B, \quad g_C(X) = g(X,C) \; (\forall C \in J),$$

and by a slight abuse of notation let $< \Gamma >$ denote the subalgebra of B generated by the components of the vector Γ.

Remark 1.2. The set $\{\Gamma^C \mid C \in J\}$ is clearly orthonormal and none of its proper subsets is normal. For if $K \subset J$ and $\bigvee_{C \in K} \Gamma^C = 1$ then, multiplying the equality by Γ^D, where $D \in J \backslash K$, we would obtain $0 = \Gamma^D$, a contradiction. $\square$

Lemma 1.1. *For every* $A, C \in I$,

$$X^A \leq g(X, C) \ (\forall X \in B^n) \Longleftrightarrow g(A, C) = 1 .$$

PROOF: $\Longrightarrow$: Take $X := A$.

$\Longleftarrow$: We have $A^A = 1 = g(A, C)$, while if $D \in I \backslash \{A\}$ then $D^A = 0 \leq g(D, C)$. Thus $D^A \leq g(D, C) \ (\forall D \in I)$. $\square$

Lemma 1.2. *The following conditions are equivalent:*

(i) $\Xi \in < \Gamma >^N$ *and* $g(\Xi, \Gamma) = 1$;

(ii) $\Xi = (\xi_1, \ldots, \xi_n)$ *is of the form*

$$\text{(1.14)} \qquad \xi_h = \bigvee_{C \in J} p_{hC} , \ p_{hC} \in \{0, 1\} \ (\forall C \in J) \ (h = 1, \ldots, n) ,$$

where

$$\text{(1.15)} \qquad g(p_{1C}, \ldots, p_{nC}, C) = 1 \qquad (\forall C \in J) .$$

PROOF: It is well known that the elements of $< \Gamma >$ are of the form (1.14) (see e.g. BFE, Theorem 1.20, or the present Proposition 2.3.2). It follows from (1.14) that for every $A \in I$,

$$\Xi^A = \xi_1^{\alpha_1} \ldots \xi_n^{\alpha_n}$$

$$= (\bigvee_C p_{1C}^{\alpha_1} \Gamma^C) \ldots (\bigvee_{C \in J} p_{nC}^{\alpha_n} \Gamma^C) = \bigvee_{C \in J} P_C^A \Gamma^C ,$$

where we have set $P_C = (p_{1C}, \ldots, p_{nC})$, hence

$$g(\Xi, \Gamma) = \bigvee_{A \in I} g(A, \Gamma) \Xi^A = \bigvee_{A \in I} (\bigvee_{C \in J} g(A, C) \Gamma^C)(\bigvee_{C \in J} P_C^A \Gamma^C)$$

$$= \bigvee_{A \in I} \bigvee_{C \in J} g(A, C) P_C^A \Gamma^C = \bigvee_{C \in J} (\bigvee_{A \in I} g(A, C) P_C^A) \Gamma^C ,$$

therefore Remark 1.2 implies

$$g(\Xi, \Gamma) = 1 \Longleftrightarrow (\forall C \in J) \bigvee_{A \in I} g(A, C) P_C^A = 1$$

$$\Longleftrightarrow (\forall C \in J) \ (\exists A \in I) \ g(A, C) = 1 \ \& \ P_C = A$$

$$\Longleftrightarrow (\forall C \in J) \ g(P_C, C) = 1 .$$

$\square$

Lemma 1.3. *If the elements p_{hC} satisfy (1.15), then one obtains a map $\varphi \in F$
(cf. (1.12)) by setting $\varphi(C) = (p_{1C}, \ldots, p_{nC})$ ($\forall C \in J$). Conversely, if $\varphi \in F$
then the elements p_{hC} defined by $(p_{1C}, \ldots, p_{nC}) = \varphi(C)$ ($\forall C \in J$) satisfy (1.15).*

PROOF: The map φ associated with (1.15) in the above way satisfies $g(\varphi(C), C) =
1$ ($\forall C \in J$), hence $\varphi \in F$ by Lemma 1.1. Conversely, if $\varphi \in F$, define $\Xi \in < \Gamma >^n$
by (1.14), where

$$p_{hC} = \varphi(C)_h \ (h = 1, \ldots, n) \ (C \in J) \ ;$$

then $g(\varphi(C), C) = 1$ by Lemma 1.1, therefore $f(\Xi) = g(\Xi, \Gamma) = 1$ by Lemma
1.2.   □

Theorem 1.1. (Brown and Rudeanu [2001]). α) *The following conditions are
equivalent* (cf. (1.10)-(1.13))*:*

   (i) the equation

(1.16) $$f(X) = 1$$

is consistent ;

   (ii) each equation

(1.17.C) $$g(X, C) = 1 \, , \qquad C \in J$$

is consistent ;

   (iii) equation (1.16) has at least a solution in $< \Gamma >^n$;

   (iv) each equation (1.17.C), $C \in J$, has at least a solution in $\{0, 1\}^n$;

   (v) the $2^n \times p$ matrix $\| g(A, C) \|$ is column normal ;

   (vi) $F \neq \emptyset$;

   (vii) $g_C \neq 0$ ($\forall C \in J$) .

   β) *When the foregoing hold, there is a bijection $\varphi \mapsto \Xi_\varphi = (\xi_{\varphi 1}, \ldots, \xi_{\varphi n})$
between F and the solutions in $< \Gamma^n >$ of equation (1.16), where*

(1.18) $$\xi_{\varphi h} = \bigvee \{ \Gamma^C \mid C \in J \ \& \ X^{\varphi(C)} \leq x_h \} \qquad (h = 1, \ldots, n) \ .$$

COMMENTS: 1) The paper by Brown and Rudeanu [2001] also studies the representation (1.10) in some detail.

   2) Theorem 1.1.α) is a refinement of Theorem 9.3 (i) in BFE, which states
the equivalence (i)$\Longleftrightarrow$(ii).

PROOF: It follows from (1.10) that condition (i) becomes

(1.19) $$1 = \bigvee_{A \in I} g(A, \Gamma) = \bigvee_{A \in I} \bigvee_{C \in J} g(A, C) \Gamma^C = \bigvee_{C \in J} (\bigvee_{A \in I} g(A, C)) \Gamma^C \ .$$

But since $\bigvee_{A \in I} g(A, C) \in \{0, 1\}$ for every $C \in J$, Remark 1.2 shows that (1.19)
is further equivalent to

(1.20) $$\bigvee_{A \in I} g(A, C) = 1 \qquad (\forall C \in J) \ .$$

Thus (i)$\Longleftrightarrow$(1.20); but (1.20)$\Longleftrightarrow$(ii) and (1.20)$\Longleftrightarrow$(v). Moreover, since all $g(A,C) \in \{0,1\}$, condition (1.20) is also equivalent to

$$(1.21) \qquad\qquad (\forall C \in J) \; (\exists A \in I) \; g(A,C) = 1$$

and clearly (1.21)$\Longleftrightarrow$(iv), while the representation $g_C(X) = \bigvee_{A \in I} g(A,C) X^A$ (cf. (1.13)) shows that (1.21)$\Longleftrightarrow$ (vii). Besides, using Lemma 1.1, we obtain

$$(1.21) \Longleftrightarrow \exists \varphi : J \longrightarrow I \, , \; g(\varphi(C),C) = 1 \; (\forall C \in J) \Longleftrightarrow \exists \varphi \in F \Longleftrightarrow (vi) \, .$$

Finally Lemmas 1.2 and 1.3 show that (iii)$\Longleftrightarrow$(vi) and also the existence of a bijection between F and the solutions in $< \Gamma >^n$ of (1.16) , the solution Ξ_φ associated with $\varphi \in F$ being

$$\xi_{\varphi h} = \bigvee_{C \in J} \varphi(C)_h \Gamma^C = \bigvee \{\Gamma^C \mid C \in J \; \& \; \varphi(C)_h = 1\} \; (h = 1,\ldots,n) \, .$$

This formula coincides with (1.18) because clearly $\varphi(C)_h = 1 \Longleftrightarrow X^{\varphi(C)} \leq x_h$.

$\square$

Example 1.2. Suppose $\Gamma = \{a_1, a_2\}$, where $a_1 \neq 0 \neq a_2$, $a_1 a_2 = 0$ and $a_1 \vee a_2 \neq 1$. Using the well-known bijection between $\{0,1\}^p$ and $[0, 2^p - 1]$, we have $\Gamma^0 = a_1' a_2'$, $\Gamma^1 = a_1' a_2 = a_2$, $\Gamma^2 = a_1 a_2' = a_1$, $\Gamma^3 = 0$ and $J = \{0,1,2\}$. Let further $n := 3$ and

$$f(x_1, x_2, x_3) = (a_2 \vee a_1' x_1') x_2 x_3 \vee a_1 x_1' x_3' \, ,$$

hence

$$g(x_1, x_2, x_3, y_1 . y_2) = x_2 x_3 (y_2 \vee x_1' y_1') \vee x_1' x_3' y_1 \, ,$$

therefore

$$g_0(X) = g(X, 0, 0) = x_1' x_2 x_3 = X^3 \, ,$$

$$g_1(X) = g(X, 0, 1) = x_2 x_3 = x_1 x_2 x_3 \vee x_1' x_2 x_3 = X^7 \vee X^3 \, ,$$

$$g_2(X) = g(X, 1, 0) = x_1' x_3' = x_1' x_2 x_3' \vee x_1' x_2' x_3' = X^2 \vee X^0 \, .$$

Now the condition $X^{\varphi(C)} \leq g_C(X)$, which characterizes the functions $\varphi \in F$, yields $\varphi(0) = 3$, $\varphi(1) \in \{3,7\}$, $\varphi(2) \in \{0,2\}$. Therefore the set F consists of the following functions φ, represented below in the form $(\varphi(0), \varphi(1), \varphi(2))$:

$$(3,3,0) \, , \; (3,3,2) \, , \; (3,7,0) \, , \; (3,7,2) \, .$$

On the other hand, since $0 = (0,0,0)$, $1 = (0,1,0)$, $3 = (0,1,1)$ and $7 = (1,1,1)$, formulas (1.18) become

$$\xi_{\varphi 1} = \bigvee \{\Gamma^C \mid \varphi(C) = 7\} \, ,$$

$$\xi_{\varphi 2} = \bigvee \{\Gamma^C \mid \varphi(C) \in \{2,3,7\}\} \, ,$$

$$\xi_{\varphi 3} = \bigvee \{\Gamma^C \mid \varphi(C) \in \{3,7\}\} \, ,$$

hence we obtain the solutions

$$(0, \Gamma^0 \vee \Gamma^1, \Gamma^0 \vee \Gamma^1), (0, 1, \Gamma^0 \vee \Gamma^1), (\Gamma^1, \Gamma^0 \vee \Gamma^1, \Gamma^0 \vee \Gamma^1), (\Gamma^1, 1, \Gamma^0 \vee \Gamma^1),$$

that is, the solutions in $< \Gamma >^3$ are

$$(0, a_1', a_1'), \ (0, 1, a_1'), \ (a_2, a_1', a_1'), \ (a_2, 1, a_1').$$

$\square$

The paper by Martin and Nipkow [1988] rediscovers Theorem 9.3 (i) in BFE (cf. Comment to Theorem 1.1) and constructs the functions g_C, $C \in J$, in the case when g is expressed in ring form. In the notation of Theorem 1.1, the construction runs as follows. Let

$$f(X) = a + \sum_{h=1}^{r} a_h t_h(X),$$

where a and the a_hs are elements of B, while the t_hs are ring terms, that is, products of variables. Set

$$x_i = \sum_{C \in J} x_{iC} \Gamma^C,$$

where the x_is are 0-1-valued functions of x_i; define also $X_C = (x_{1C}, \ldots, x_{nC})$. Then for any ring term

$$t(X) = \prod_{i \in M} x_i, \ M \subseteq \{1, \ldots, n\},$$

it follows, using orthogonality, that

$$t(X) = \prod_{i \in M} \sum_{C \in J} x_{iC} \Gamma^C = \sum_{C \in J} \prod_{i \in M} x_{iC} \Gamma^C = \sum_{C \in J} t(X_C) \Gamma^C,$$

therefore, since the non-zero Γ^Cs are the atoms of the Boolean algebra $< \Gamma >$,

$$f(X) = a + \sum_{h=1}^{r} a_h \sum_{C \in J} t_h(X_C) \Gamma^C$$

$$= a + \sum_{C \in J} \sum_{h=1}^{r} a_h t_h(X_C) \Gamma^C = a + \sum_{C \in J} \sum \{t_h(X_C) \Gamma^C \mid a_h \Gamma^C \neq 0\}$$

$$= a + \sum_{C \in J} \Gamma^C \sum \{t_h(X_C) \mid a_h \geq \Gamma^C\}.$$

Consequently, setting $a = \sum_{C \in J} a_C \Gamma^C$, where the a_Cs are in $\{0, 1\}$, we obtain

$$g(X, C) = a_C + \sum \{t_h(X_C) \mid a_h \geq \Gamma^C\} \qquad (C \in J).$$

So if for some $C \in J$ we have $a_C = 0$ and $a_h \Gamma^C = 0$ for all h, then equation $g(X, C) = 1$ is inconsistent and hence so is the original equation $f(X) = 1$. Otherwise for each $C \in J$ a particular solution of the truth equation $g(X, C) = 1$ can be easily found and since the variables X_C, $C \in J$, are pairwise distinct, this immediately yields a particular solution of $f(X) = 1$.

Example 1.3. Consider again the Boolean algebra in Example 1.2 and the function

$$f = a_1 xyz + a_2 yz + a_2' xz + a_1' xy + a_1' a_2' x + (a_1 \vee a_2)y + a_1' z + a_2 .$$

Setting $x = a_1' a_2' x_0 + a_2 x_1 + a_1 x_2$ and similarly for y and z, we obtain

$$f = a_1(a_1' a_2' x_0 y_0 z_0 + a_2 x_1 y_1 z_1 + a_1 x_2 y_2 z_2) +$$

$$+ a_2(a_1' a_2' y_0 z_0 + a_2 y_1 z_1 + a_1 y_2 z_2) + a_2'(a_1' a_2' x_0 z_0 + a_2 x_1 z_1 + a_1 x_2 z_2) +$$

$$+ a_1'(a_1' a_2' x_0 y_0 + a_2 x_1 y_1 + a_1 x_2 y_2) + a_1' a_2'(a_1' a_2' x_0 + a_2 x_1 + a_1 x_2) +$$

$$+ (a_1 \vee a_2)(a_1' a_2' y_0 + a_2 y_1 + a_1 y_2) + a_1'(a_1' a_2' z_0 + a_2 z_1 + a_1 z_2) + a_2$$

$$= a_1 x_2 y_2 z_2 + a_2 y_1 z_1 + a_1' a_2' x_0 z_0 + a_1 x_2 z_2 + a_1' a_2' x_0 z_0 + a_2 x_1 y_1 +$$

$$+ a_1' a_2' x_0 + a_2 y_1 + a_1 y_2 + a_1' a_2' z_0 + a_2 z_1 + a_2$$

$$= a_1' a_2'(x_0 z_0 + x_0 y_0 + x_0 + z_0) + a_2(y_1 z_1 + x_1 y_1 + y_1 + z_1 + 1) +$$

$$+ a_1(x_2 y_2 z_2 + x_2 z_2 + y_2) .$$

$\square$

Another approach to solving efficiently Boolean equations is via an irredundant system of values given to the parameters of a general solution, as suggested in BFE, Theorem 2.9. This result was generalized by Lavit [1976], to the effect that in a Boolean algebra the equivalence classes modulo a principal ideal are isomorphic and each equivalence class provides an irredundant system of values for the parameters of a general solution; cf. Proposition 1.5, Corollary 1.2 and Remarks 1.3 and 1.4 below.

Proposition 1.5. *For every three elements* a, b, c *of a Boolean algebra* B, *the intervals* $[ba', b \vee a]$ *and* $[ca', c \vee a]$ *are isomorphic via the mapping*

$$(1.22) \qquad \varphi : [ba', b \vee a] \longrightarrow [ca', c \vee a] , \quad \varphi(x) = ca' \vee xa .$$

COMMENT: It is well known that an order isomorphism between two lattices is in fact a lattice isomorphism; cf. Birkhoff [1967], Lemma II.3.2. Since an order isomorphism clearly preserves 0 and 1, it follows that an order isomorphism between two Boolean algebras is in fact a bounded-lattice isomorphism, hence a Boolean isomorphism.

PROOF: In view of the above comment it suffices to prove that φ is an order isomorphism. But $ca' \leq ca' \vee xa \leq ca' \vee a = c \vee a$, therefore the map φ is well defined; besides, it is clearly isotone. Now set

(1.23) $$\psi : [ca', c \vee a] \longrightarrow [ba', b \vee a] , \ \psi(y) = ba' \vee ya ;$$

then ψ is also well defined and isotone. Moreover, if $x \in [ba', b \vee a]$ then

$$\psi(\varphi(x)) = ba' \vee (ca' \vee xa)a = ba' \vee xa = (ba' \vee x)(ba' \vee a) = x(b \vee a) = x ,$$

therefore $\psi \circ \varphi$ is the identity of $[ba', b \vee a]$ and similarly $\varphi \circ \psi$ is the identity of $[ca', c \vee a]$.   □

Remark 1.3. It is easy to see that the interval $[ba', b \vee a]$ is in fact the equivalence class of b modulo the congruence generated by the principal ideal $[0, a]$ (see e.g. Balbes and Dwinger [1974], Theorem II.9.17, or Birkhoff [1967], Lemma II.4.3, or Grätzer [1978], Theorem III.3.4).   □

Corollary 1.2. *Suppose the Boolean equation*

(1.24) $$ax \vee bx' = 0$$

is consistent. Then for every $c \in B$, formula

(1.25) $$\varphi(t) = b \vee a't , \ t \in [c(a \vee b), c \vee a'b']$$

provides an irredundant representation of the set $[b, a']$ of solutions of equation (1.24).

PROOF: Taking $a := a'b'$, $c := b$ and $x := t$ in Proposition 1.5, formula (1.21) becomes

$$\varphi : [c(a \vee b), c \vee a'b'] \longrightarrow [b(a \vee b), b \vee a'b'] = [b, b \vee a'] = [b, a'] ,$$

$$\varphi(t) = b(a \vee b) \vee ta'b' = b \vee ta' .$$

   □

Remark 1.4. In particular taking in turn $c := 1$ and $c := 0$ in formula (1.25), we obtain the intervals $[a \vee b, 1]$ and $[0, a'b']$. The latter interval reduces (1.25) to $x = b \vee a't$, $t \leq a'b'$, while the irredundant solution provided in BFE, Theorem 2.9, is $x = b \vee p$, $p \leq a'b'$.   □

   Another way towards an efficient solving of Boolean equation is the minimization of the number of parameters in the general solution, whenever possible. This approach goes back to Löwenheim [1910], who called it "economic coding"; see BFE, Chapter 9, §4, where the economic codings obtained by Löwenheim, Carvallo and Brown for simple Boolean equations and parametric Boolean equations are reported. One more result in this direction is

Proposition 1.6. (Banković [1989/90], [1992a]). *The general solution of a consistent equation in n unknowns $f(X) = 0$ can be written in the form*

(1.26) $$x_i = h_i(t_1, \ldots, t_p)   (i = 1, \ldots, n) ,$$

where $p \leq n$, if and only if

(1.27)
$$\bigvee_S \prod \{f(A) \mid A \in \{0,1\}^n \backslash S\} = 1 \, ,$$

where S runs over the subsets of $\{0,1\}^n$ of cardinality 2^p.

PROOF: Using the Vaught theorem, we first show that this proposition is a Horn sentence, then we prove it for the two-element Boolean algebra.

A slight modification in Lemma 5.3.2 (the number of parameters is p instead of n) ensures that the sentence "(1.26) is the general solution of equation $f(X) = 0$" is an atomic formula $\rho(X, H, T)$, where $X = \{x_1, \ldots, x_n\}$, $T = \{t_1, \ldots, t_p\}$ and $H = \{h_k(A) \mid k \in \{1, \ldots, n\} \,\& A \in \{0,1\}^n\}$. For fixed n and p, property (1.27) is of the form $g(F) = 0$, where g is a simple Boolean function in the 2^n variables $F = \{f(A) \mid A \in \{0,1\}^n\}$. Therefore our proposition can be successively written in the equivalent forms

$$\exists H \rho(X, H, T) \iff g(F) = 0 \, ,$$

$$(\exists H \rho(X, H, T) \implies g(F) = 0) \,\& \, (g(F) = 0 \implies \exists H \rho(X, H, T)) \, ,$$

$$\exists H(\rho(X, H, T) \implies g(F) = 0) \,\& \, \exists H(g(F) = 0 \implies \rho(X, H, T)) \, .$$

An obvious adaptation of the discussion before Theorem 5.3.6 to the case of Boolean algebras shows that the above formula is of the form $\exists H \alpha \,\& \, \exists H \beta$, where α and β are elementary Horn formulas, hence it is a Horn formula.

Now suppose the Boolean algebra is $\{0,1\}$. Then the range of the Boolean transformation (1.26) has at most 2^p elements. Conversely, every subset $S \subseteq \{0,1\}^n$ with $\text{card}(S) \leq 2^p$ can be represented in the parametric form (1.26) (take arbitrary values $h_k(A)$ $(k = 1, \ldots, n)$ for $A \in \{0,1\}^n \backslash S$). Therefore the general solution of the consistent truth equation can be written in the form (1.26) if and only if there are at most 2^p solutions. So it remains to prove that the latter condition is equivalent to (1.27).

But (1.27) says that there is a set $S \subseteq \{0,1\}^n$ such that

(1.28) $\text{card}(S) = 2^p \,\& \, f(A) = 1 \, \forall A \in \{0,1\}^n \backslash S \, .$

Let S_0 be the set solutions to $f(X) = 0$. If S_0 has cardinality $\leq 2^p$ then every subset $S \subseteq \{0,1\}^n$ such that $S_0 \subseteq S$ and $\text{card}(S) = 2^p$ satisfies (1.28), therefore (1.27) holds. Otherwise S_0 has cardinality $> 2^p$, hence for any subset $S \subseteq \{0,1\}^n$ of cardinality 2^p we have $S_0 \not\subseteq S$, therefore $f(A) = 0$ for some $A \in \{0,1\}^n \backslash S$. Consequently there is no set satisfying (1.28), so that (1.27) fails.   □

Another favourable situation consists in the fact that certain variables remain arbitrary in the general or reproductive solution. This situation is characterized in

Remark 1.5. (Banković [1989/90], [1992a]). Suppose $f(x_1, \ldots, x_n) = 0$ is a consistent Boolean equation and

(1.29) $\{1, \ldots, n\} = \{i_1, \ldots, i_p\} \cup \{j_1, \ldots, j_q\}$
        $\& \, \{i_1, \ldots, i_p\} \cap \{j_1, \ldots, j_q\} = \emptyset \, .$

Then the equation has a general (reproductive) solution in which the variables $x_{i_1}, \ldots, x_{i_p}$ are arbitrary if and only if

(1.30) $\forall x_{i_1} \ldots \forall x_{i_p} \exists j_1 \ldots \exists j_q \, f(x_1, \ldots, x_n) = 0$,

or equivalently,

(1.31) $$\bigvee_{\alpha_{i_1}, \ldots, \alpha_{i_p}} \prod_{\alpha_{j_1}, \ldots, \alpha_{j_q}} f(\alpha_1, \ldots, \alpha_n) = 0 .$$

□

The solution of systems of Boolean equations of the form $f_k(X) = 1$ involves multiplication of several functions expressed in disjunctive form; the complexity of this process depends on the order in which multiplications are performed. Zhuravlev and Platonenko [1984] determined the optimal order of multiplications by solving an integer programming problem, more exactly, by determining a maximal matching in a bipartite graph. The problem of an efficient multiplication was also studied by Zakrevskiĭ [1971], Chapter 3, §6, pp. 18-24, Platonenko [1983]* and Dyukova [1987]*.

Posherstnik [1979] suggested the solution of truth equations with many unknowns by decomposition of the original problem into several problems containing a smaller amount of information to be processed. The method used by Karepov and Lipskiĭ [1974]*, Teslenko [1974] and Matrosova [1975a]* is solving by enumeration along a tree. Thus e.g. the backtracking algorithm devised by Teslenko for solving a truth equation $f(x_1, \ldots, x_n) = 1$ entails calculating

(1.32) $f(x_1, \ldots, x_{m-1}, \alpha_m, \alpha_{m+1}, \ldots, \alpha_n)$,

where $\alpha_m \ldots, \alpha_n$ are fixed in $\{0, 1\}$. If the result is 1, a family of solutions has been found and the algorithm stops; if the result is 0, the scanning continues by omitting the 2^{m-1} vectors $(x_1, \ldots, x_{m-1}, \alpha_m, \ldots, \alpha_n)$; if (1.32) is not a constant, in the next step one computes $f(x_1, 0, \ldots, 0, 1, \alpha_{m+1}, \ldots, \alpha_n)$.

Cerny and Marin [1974] use matrix methods for solving truth equations. Tapia and Tucker [1980] suggest a $2^n \times 2^m$−matrix method for solving a truth equation $f(x_1, \ldots, x_n, y_1, \ldots, y_m) = 0$ with respect to the unknowns $x_1, \ldots, x_n$; the method also gives the number of solutions. Counting the number of solutions of certain equations is also the concern of Ustinov [1980], Igamberdyev [1989]* and Gorshkov [1996]*. For instance, Gorshkov points out certain cases in which the counting problem is polynomial, e.g. the case of systems of truth equations expressed in linear ring form:

$$\sum_{j=1}^{n} a_{ij} x_j = 1 \qquad (i = 1, \ldots, m) .$$

At this point we recall that a truth function f is said to be *satisfiable* provided the equation $f(x_1, \ldots, x_n) = 1$ is consistent. The *satisfiability problem* consists in deciding whether or not a truth function expressed in conjunctive form is satisfiable. An equivalent formulation of this problem consists in deciding whether or not a truth equation $g(x_1, \ldots, x_n) = 0$ is consistent, where g is expressed in disjunctive form (the negative answer meaning that $g = 1$ identically, i.e., g is a *tautology*). As a matter of principle, the problem can be solved by a backtracking algorithm (see e.g. Rudeanu [1973]), but the computations may be tedious and even prohibitive in certain cases. In fact the theory of computational complexity establishes that this problem is NP-complete; see e.g. Cook [1971]* or Garey and Johnson [1979]. Quite recently, Grozea [2000] encoded in polynomial time a conjunctive normal form as a sequence of fixed-length binary representation symbols. He then derived a primitive recursive function which solves the satisfiability problem. This has important consequences on the lower part of Grzegorczyk's hierarchy of primitive recursive classes.

To be sure, the satisfiability problem for smaller classes of truth functions may be polynomial. This possibility was studied by Schaefer [1978]*: "Let $S = \{R_1, \ldots, R_m\}$ [...] be any finite set of logical relations. Define an S–formula to be any conjunction of clauses of the form $R_j(x_1, \ldots, x_k)$ [...]. Let SAT(S) denote the set of all satisfiable S–formulas. The main theorem of the paper establishes six conditions such that if S satisfies one of them then SAT(S) is polynomial-time decidable, and otherwise SAT(S) is NP-complete. This result is then applied to obtain new NP-complete problems. One of them is the following: given a graph G, can the nodes of G be colored such that each node has exactly one neighbor the same color as itself?" (cf. Math. Rev. # 80d:68058).

Zakrevskiĭ [1975a], pp. 66-74, studied a variant of the satisfiability problem, namely for a truth function expressed as a ring sum of terms (possibly including the degenerate term 1). His solution is based on the following

Proposition 1.7. *Consider a truth function*

$$(1.33) \qquad\qquad f(X) = xg(Y) + x'h(Y) + k(Y) \,,$$

where $x \in X$ and $Y = X \backslash \{x\}$. Then either $g + h$ is satisfiable, in which case f is satisfiable, or the variable x is fictitious and $f = g + k = h + k$.

PROOF: If $g + h$ is not satisfiable, then $g + h = 0$ (identically), hence $g = h$, therefore (1.33) implies $f = g + k$. Now suppose $g + k$ is satisfiable, say $g(A) + h(A) = 1$. Then there exist four possibilities:

   (i) $g(A) = 1$, $h(A) = 0$, $k(A) = 0$;
   (ii) $g(A) = 0$, $h(A) = 1$, $k(A) = 1$;
   (iii) $g(A) = 1$, $h(A) = 0$, $k(A) = 1$;
   (iv) $g(A) = 0$, $h(A) = 1$, $k(A) = 0$.

In cases (i) and (ii) we have $f(1, A) = 1$, while in cases (iii) and (iv) we obtain $f(0, A) = 1$. $\qquad\square$

Corollary 1.3. *If the function f is satisfiable and the variable x is not fictitious, then $g + h$ is satisfiable.* $\qquad\square$

Clearly Proposition 1.7 and Corollary 1.3 yield an algorithm for testing the satisfiability of the function f and for providing a particular solution of the equation $f(X) = 1$ if the function is satisfiable. The algorithm has two stages. At each step j of the first stage the dimension $n - j + 1$ of the current problem is reduced to $n - j$. If the function obtained at the end of the first stage is not satisfiable, then all the intermediate functions are not satifiable, too; hence the variable which has been eliminated is fictitious and the algorithm is resumed with the function written in the form $f = g + k$. If the function f is not satisfiable, then all of its variables are fictitious and the algorithm detects this situation. Otherwise one passes to the second stage of the algorithm, which consists in a recursive construction of a solution as indicated in the proof of the proposition.

Example 1.4. Consider the truth function

$$(1.34.0) \qquad\qquad f = \varphi_1 + \varphi_2 + \varphi_3 ,$$

$$(1.34.1) \qquad \varphi_1 = x_1 x_3 x_5' x_6' + x_2' x_3' x_4' x_5 + x_2' x_3 x_4 + x_1' x_3 x_5 x_6 + x_2 x_4' ,$$

$$(1.34.2) \qquad \varphi_2 = x_2' x_4' x_5' x_6 + x_1' x_3' x_5 + x_1 x_4' x_5 + x_2' x_3 + x_2 x_4 x_5' ,$$

$$(1.34.3) \qquad \varphi_3 = x_1' x_3 x_6' + x_1' x_4 x_5 + x_2 x_3' x_5' x_6' + x_2' x_3' x_4' x_5' x_6 + x_2 x_5' x_6 .$$

Using the notation g_i, h_i, k_i for the functions g, h, k relative to a variable x_i, we have

$g_1 = x_3 x_5' x_6' + x_4' x_5$, $h_1 = x_3 x_5 x_6 + x_3' x_5 + x_3 x_6' + x_4 x_5$, $k_1 = $ the other terms.

For the new function

$$g_1 + h_1 = x_3 x_5' x_6' + x_4' x_5 + x_3 x_5 x_6 + x_3' x_5 + x_3 x_6' + x_4 x_5$$

we have

$$g_3 = x_5' x_6' + x_5 x_6 + x_6' , \ h_3 = x_5 , \ k_3 = x_4' x_5 + x_4 x_5 .$$

For the new function

$$g_3 + h_3 = x_5' x_6' + x_5 x_6 + x_6' + x_5$$

we have

$$g_5 = x_6 + 1 , \ h_5 = x_6' , \ k_5 = x_6'$$

and the new function $g_5 + h_5 = 0$ is not satisfiable. It follows that the previous sums $g_3 + h_3$ and $g_1 + h_1$ are not satisfiable, too. Therefore $g_1 = h_1$ and the variable x_1 is fictitious in the function f, which can be written

$$(1.35.0) \qquad\qquad f = g_1 + k_1 = \psi_1 + \psi_2 ,$$

$$(1.35.1) \qquad \psi_1 = x_3 x_5' x_6' + x_4' x_5 + x_2' x_3' x_4' x_5 + x_2' x_3 x_4 + x_2 x_4' ,$$

$$(1.35.2) \qquad \psi_2 = x_2' x_4' x_5' x_6 + x_2' x_3 + x_2 x_4 x_5' + x_2 x_3' x_5' x_6' + x_2' x_3' x_4' x_5' x_6 + x_2 x_5' x_6$$

(this is the example given by Zakrevskiǐ). We have

$$g_2 = x_4' + x_4 x_5' + x_3' x_5' x_6' + x_5' x_6 ,$$

$$h_2 = x_3' x_4' x_5' + x_3 x_4 + x_4' x_5' x_6 + x_3 + x_3' x_4' x_5' x_6 \ ,$$

$$k_2 = x_3 x_5' x_6' + x_4' x_5 \ .$$

For the function $g_2 + h_2$ we have

$$g_3 = x_4 + 1 \ ,$$

$$h_3 = x_4' x_5 + x_4' x_5' x_6 + x_5' x_6' \ ,$$

$$k_3 = x_4' x_5' x_6 + x_4' + x_4 x_5' + x_5' x_6 \ .$$

For the function $g_3 + h_3$ we have

$$g_4 = 1 \ , \ h_4 = x_5 + x_5' x_6 \ , \ k_4 = 1 + x_5' x_6' \ .$$

For the function $g_4 + h_4$ we have

$$g_5 = 1 \ , \ h_5 = x_6 \ , \ k_5 = 1$$

and the function $g_5 + h_5$ is satisfiable: the unique solution is $x_6 := 0$.

We have $g_5(0) = 1$, $h_5(0) = 0$, $k_5(0) = 1$, that is, we are in the case (iii) pointed out in the proof of Proposition 1.7, therefore we take $x_5 := 0$. Now

$$g_4(0,0) = 1 \ , \ h_4(0,0) = 0 \ , \ k_4(0,0) = 0 \ ,$$

hence we are in case (i) and we take $x_4 := 1$. Further,

$$g_3(1,0,0) = 0 \ , \ h_3(1,0,0) = 1 \ , \ k_3(1,0,0) = 1 \ ,$$

which means case (ii) and this yields $x_3 := 1$. Finally

$$g_2(1,1,0,0) = 1 \ , \ h_2(1,1,0,0) = 0 \ , \ k_2(1,1,0,0) = 1 \ ,$$

hence $x_2 := 0$ according to case (iii). We have thus found the particular solution $(0,1,1,0,0)$ of the equation $f = 1$, where the function f is given by (1.35); taking x_1 arbitrarily we obtain a solution corresponding to the original function defined by (1.34). □

The paper by Zakrevskiĭ and Utkin [1975] provides more particular solutions, but still not all of them. A combinatorial approach to the same problem is due to A. G. Levin [1978]. By using the so-called Kronecker matrix product, Kucharev, Shmerko and Yanushkevich [1991]* bring the two sides of a ring-form truth equation $f = g$ to a certain 2^n−term normal form from which the solutions are found by a simple inspection; cf. Yanushkevich [1998].

Serikov [1972] considers a truth equation $f(X,Y) = 1$ to be solved with respect to Y in the Boolean algebra of truth functions. The consistency condition is $c(X) = 1$, where we have set $c(X) = \bigvee_C f(X,C)$. Then, according to an idea which goes back to Voigt [1890], the original equation can be replaced by the equivalent unconstrained equation $f(X,Y) \geq c(X)$. If the latter equation is written in the form $c'(X) \vee f(X, Y) = 1$ and $F(X)$ is a particular solution of it, then the Löwenheim formula yields the reproductive solution

$$X = F(X)c(X)f'(X, Q(X)) \vee Q(X)(c'(X) \vee f(X, Q(X))),$$

where the function vector $Q(X)$ is the arbitrary parameter; cf. Rudeanu [1974b]. A slightly different form of the reproductive solution had previously been obtained by Serikov (op. cit.).

Several papers suggested algorithms for solving certain special types of truth equations and studied the complexities of these algorithms: Bak Khyng Khang [1975]*, Leont'ev and Nurlybaev [1975], Matrosova [1975a]*, [1975b]*, Zakrevskiĭ [1975b]*, Egiazaryan [1976], Mikhaĭlovskiĭ [1977]*, Kossovskiĭ [1978]*, Kabulov and Baĭzhumanov [1986]*, and Dyukova [1989]*. Thus e.g. Egiazaryan reduced the solution of certain systems of truth equations in ring form to an optimization problem in 0-1-variables, while Kossovskiĭ proved that "the decidability problem for functional Boolean equations by nondeterministic Turing machines requires exponential time" (cf. Math. Rev. # 80d:68062). See Katerinochka, Koroleva, Madatyan and Platonenko [1988] for a short survey of Soviet papers in the field.

Like some of the previous authors and independently of them, Aleksanyan [1989] suggests that systems of truth equations could be conveniently solved by using the ring form. More precisely, Aleksanyan is interested in representing a truth function in the form $f = k_1 \vee \ldots \vee k_s$, where k_i are products of linear functions, in such a way that the length s of the representation be minimized and, if possible, $k_i k_j = 0$ for $i \neq j$. Unfortunately, certain points of this paper have remained obscure to us.

Two quite particular systems of Boolean equations were solved in Berţi [1973] and Perić [1978]. The latter paper was generalized in Rudeanu [1995b] .

In contrast, Reischer, Simovici, Stojmenović and Tošić [1997] solved the general Boolean equation $f(x_1, \ldots, x_n) = 0$ written in ring form by the method of elimination of variables. They also established the following result:

Proposition 1.8. *The number of distinct consistent Boolean equations in n unknowns over the 2^r-element Boolean algebra is $(2^{2^n} - 1)^r$.*

PROOF: There are as many such equations as the number of possible choices for the 2^n coefficients $f(A)$, $A \in \{0,1\}^n$ in such a way that $\prod_A f(A) = 0$. Therefore it suffices to prove the following more general lemma: the number of all the sequences $(a_1, \ldots, a_m) \in B^m$ that satisfy $a_1 a_2 \ldots a_m = 0$ is $(2^m - 1)^r$. Since every 2^r-element Boolean algebra is isomorphic to the Boolean algebra $\mathcal{P}(\mathbf{r})$ of all the subsets of the set $\mathbf{r} = \{1, \ldots, r\}$, it suffices to prove the lemma in the latter algebra. In this case the condition becomes $A_1 \cap A_2 \cap \ldots \cap A_m = \emptyset$. For each element $i \in \mathbf{r}$ there are 2^m possibilities as concerns its membership in $A_1, A_2, \ldots, A_m$, among which the case $i \in A_1 \cap A_2 \cap \ldots \cap A_m$ is not acceptable, hence there are $2^m - 1$ possibilities left. Since the choices for the r elements $1, 2, \ldots, r$ are independent of each other, we obtain the total number $(2^m - 1)^r$.

$\square$

We conclude this section by mentioning quite a special approach to truth equations. It was remarked long ago that the operations of the two-element Boolean algebra can be expressed in terms of the usual arithmetic operations $+, -, \cdot$ as follows: $x \vee y = x + y - x \cdot y$, $xy = x \cdot y$, $x' = 1 - x$. Thus every truth equation $f(X) = 1$ is equivalent to a *pseudo-Boolean* inequality $g(X) \geq 1$, that is, an inequality expressed in terms of ordinary arithmetic operations but in which the variables take only the values 0,1. Conversely, for every pseudo-Boolean inequality there exists a truth equation having the same set of solutions. This equivalence is one of the basic ideas of *pseudo-Boolean programming*, which is a collection of Boolean and combinatorial techniques for solving $\{0, 1\}$-optimization problems; cf. Hammer and Rudeanu [1968]. The equivalence between truth equations, pseudo-Boolean inequalities and pseudo-Boolean equations was rediscovered by Bär and Rohleder [1967]*, Bär [1972]* and Meiler and Bär [1974]. The last paper discusses conditions for a truth equation to be equivalent to a linear inequality.

Papaioannou and Barrett [1975] apply the above transformations in order to express every truth function as a $\{0, 1\}$-valued real polynomial (*pseudo-Boolean function* for short) $T(f)$, which they call the *real transform* of the function f and suggest the use of this technique for transferring certain Boolean problems to the real domain. For instance, a system of truth equations $f_j(X) = 0$ $(j = 1, \ldots, m)$ is consistent if and only if the pseudo-Boolean function $\sum_{j=1}^{m} T(f_j)$ (real sum) is 0, in which case the solutions of the system coincide with the minimizing points of the pseudo-Boolean function and the latter points are determined by pseudo-Boolean programming. Del Picchia [1976] also devises a *numerical transform* of a truth function and uses it in order to solve truth equations. Nisan and Szegedy [1994] study in some detail the degree of the polynomial $T(f)$. For this line of research we refer the reader to the above three papers and to the literature quoted there.

2 Boolean equations with unique solution

The study of Boolean equations with unique solution was begun by Whitehead [1901] and continued by Bernstein [1932] and Parker and Bernstein [1955]; cf. BFE, Chapter 6, §2. This line of research was resumed 30 years after the publication of the Parker-Bernstein paper, as we are going to report in this section, while the uniqueness of the solution in the special case of quadratic truth equations will be studied in the next section.

Theorem 2.1 matches Theorem 1.1 and in particular it uses the notation $(1.10)-(1.13)$, (1.16) and (1.17).

Theorem 2.1. (Brown and Rudeanu [2001]). α) *The following conditions are equivalent:*

(i) the equation

(1.16)
$$f(X) = 1$$

has a unique solution ;

(ii) each equation

(1.17.C)
$$g(X,C) = 1, \qquad C \in J,$$

has a unique solution ;

(iii) equation $g(X,\Gamma) = 1$ has a unique solution in $< \Gamma >^n$;

(iv) each equation (1.17.C) has a unique solution in $\{0,1\}^n$;

(v) the $2^n \times p$ matrix $\| g(A,C) \|$ is column orthonormal ;

(vi) F is a singleton ;

(vii) each g_C , $C \in J$, is a minterm .

β) When the foregoing hold, the unique solution $\Xi = (\xi_1,\ldots,\xi_n)$ belongs to $< \Gamma >^n$ and is given by

(2.1)
$$\xi_h = \bigvee \{\Gamma^C \mid C \in J \ \& \ g(X,C) \leq x_h\} \qquad (h = 1,\ldots,n) .$$

PROOF: α) (iv)⟺(vii): Obvious.

(vi)⟺(vii): Both conditions are equivalent to the following one: for each $C \in J$ there is a unique $A \in I$ such that $X^A \leq g(X,C)$ ($\forall C \in J$).

(ii)⟺(vii): Clearly the simple Boolean equation (1.17.C) has a unique solution iff $g(X,C)$ is a minterm, in which case the unique solution is in $\{0,1\}$.

(ii)⟺(v): This was proved by Whitehead [1901]: see e.g. BFE, Theorem 6.7, (i)⟺(ii), applied to the equation $g'(X,C) = 0$.

(ii)⟹(i): Equation (1.16) is consistent by Theorem 1.1. Therefore, according to BFE, Proposition 9.4, (i)⟺(ii), applied to equation $f'(X) = 0$, the single solution of it is $\Xi = \bigvee_{C \in J} \Xi_C$, where for each $C \in J$, Ξ_C is the unique solution of equation (1.17.C).

(i)⟹(iv): Each equation (1.17.C) is consistent by Theorem 1.1. For each $C \in J$, let $\Xi_C = (\xi_{C1},\ldots,\xi_{Cn})$ and $\Upsilon_C = (\upsilon_{C1},\ldots,\upsilon_{Cn})$ be solutions of (1.17.C) in $\{0,1\}^n$. Applying once again Proposition 9.4 in BFE we get

$$\bigvee_{C \in J} \Gamma^C \xi_{Ch} = \bigvee_{C \in J} \Gamma^C \upsilon_{Ch} \qquad (h = 1,\ldots,n) .$$

For each $C \in J$ we take the meet of both sides with Γ^C and obtain $\Gamma^C \xi_{Ch} = \Gamma^C \upsilon_{Ch}$ for all C and all h, that is, $\Gamma^C(\xi_{Ch} + \upsilon_{Ch}) = 0$, hence $\xi_{Ch} + \upsilon_{Ch} = 0$. This proves $\Xi_C = \Upsilon_C$.

(i)⟺(iii): According to a result of Parker and Bernstein [1955] (see also BFE, Theorem 6.6, (i)⟺(iv)), a vector Ξ is the unique solution of equation (1.16) if and only if

(2.2)
$$\xi_h = \bigvee_{A \in I, \alpha_h = 1} f(A) = \bigvee_{A \in I, \alpha_h = 1} \bigvee_{C \in J} g(A,C)\Gamma^C \qquad (h = 1,\ldots,n) .$$

If (i) holds, relations (2.2) show that $\varXi \in < \varGamma >^n$. If (iii) holds, then the Parker-Bernstein theorem implies (2.2), where $\varXi$ is the unique solution in $< \varGamma >^n$ and f stands for the restriction of f to $< \varGamma >^n$. But the values $f(A)$ are the same for the function f itself and its restriction to $< \varGamma >^n$, so that relations (2.2) show that $\varXi$ is the unique solution of equation (1.16) within the Boolean algebra B.

β) The identity $X^{\varphi(C)} \leq g(X,C)$ holds for the unique function $\varphi \in F$ by the very definition (1.12) of F. But $g(X,C)$ is a minterm by condition (vii) in α), therefore $g(X,C) = X^{\varphi(C)}$, hence formula (1.18) in Theorem 1.1 reduces to (2.1).   □

Example 2.1. Take

$$f(X) = a_1' x_1' x_2 (a_2 x_3' \vee a_2' x_3) \vee a_1 x_1 x_2' x_3 ,$$

where a_1 and a_2 are as in Example 1.2. Then $m = 3$, $n = 2$ and

$$g_0(X) = g(X,0,0) = x_1' x_2 x_3 = X^3 ,$$

$$g_1(X) = g(X,0,1) = x_1' x_2 x_3' = X^2 ,$$

$$g_2(X) = g(X,1,0) = x_1 x_2' x_3 = X^5 ,$$

hence the unique function $\varphi \in F$, defined by $X^{\varphi(C)} = g_C(X)$, is given by $\varphi(0) = 3$, $\varphi(1) = 2$, $\varphi(2) = 5$, therefore the unique solution (2.1) is determined as follows:

$$\xi_1 = \bigvee \{ \varGamma^C \mid g_C(X) \leq x_1 \} = \varGamma^2 = a_1 ,$$

$$\xi_2 = \bigvee \{ \varGamma^C \mid g_C(X) \leq x_2 \} = \varGamma^0 \vee \varGamma^1 = a_1' a_2' \vee a_2 = a_1' \vee a_2 = a_1' ,$$

$$\xi_3 = \bigvee \{ \varGamma^C \mid g_C(X) \leq x_3 \} = \varGamma^0 \vee \varGamma^2 = a_1' a_2' \vee a_1 = a_2' \vee a_1 = a_2' .$$

□

The problem of characterizing equations with unique solutions has also been studied in the case of Boolean equations expressed in ring form. Consider a Boolean equation

(2.3)
$$\sum_{S \subseteq N} \prod_{i \in S} x_i = 1 ,$$

where $N = \{1, \ldots, n\}$ (and $a_\emptyset$ is the constant term of the polynomial). Then the following result holds:

Proposition 2.1. (Parker and Bernstein [1955]; see also BFE, Theorem 6.9). *The following conditions are equivalent for the Boolean equation (2.3):*

(i) *equation (2.3) has a unique solution ;*

(ii) $\sum_{S \subseteq N} (\prod_{S \subseteq T \subseteq N}) \prod_{S \subsetneq V \subseteq N} a_V' = 1$;

(iii) $a_N = 1$ *and* $a_{S \cap T} = a_S a_T$ *for every* $S, T \subseteq N$.   □

However the unique solution of equation (2.3) is not actually found in Proposition 2.1. We fill this gap in the following

Proposition 2.2. (Rudeanu [1993]). *If the equivalent conditions in Proposition 2.1 are fulfilled, then the unique solution is*

$$(2.4) \qquad\qquad x_i = a'_{N\setminus\{i\}} \qquad\quad (i = 1, \ldots, n) \,.$$

PROOF: It follows from (iii) that if $S \subseteq T \subseteq N$, then $a_S = a_S a_T$, i.e., $a_S \leq a_T$. Therefore $\prod_{S\subseteq T\subseteq N} a_T = a_S$ and (ii) reduces to

$$(2.5) \qquad\qquad \sum_{S\subseteq N} \prod_{S\not\subseteq V\subseteq N} a'_V = 1 \,.$$

Besides, for every $V \subseteq N$ such that $S \not\subseteq V$ there exists $i \in S\setminus V$, hence $V \subseteq N\setminus\{i\}$, therefore $a'_{N\setminus\{i\}} \leq a'_V$. By multiplying all these inequalities for a fixed S, all V such that $S \not\subseteq V \subseteq N$ and all $i \in S\setminus V$, we obtain

$$(2.6) \qquad\qquad \prod_{i\in S} a'_{N\setminus\{i\}} \leq \prod_{S\not\subseteq V\subseteq N} a'_V \,,$$

because each $i \in S$ occurs whenever $V \subseteq N\setminus\{i\}$. On the other hand

$$(2.7) \qquad\qquad \prod_{S\not\subseteq V\subseteq N} a'_V \leq \prod_{i\in S} a'_{N\setminus\{i\}}$$

because the sets $N\setminus\{i\}$ are some of the sets V.

Now (2.6) and (2.7) imply

$$\prod_{S\not\subseteq V\subseteq N} a'_V = \prod_{i\in S} a'_{N\setminus\{i\}} \,,$$

so that (2.5) can be written in the form

$$\sum_{S\subseteq N} \prod_{i\in S} a'_{n\setminus\{i\}} = 1 \,.$$

$\square$

Example 2.2. We write down explicitly some of the above formulas for $n := 1, 2, 3$, simplifying a bit the subscripts and taking into account the following remark: the equality $a_{S\cap T} = a_S a_T$ holds trivially for $S = T$ and follows from $a_N = 1$ for $S = N$ or $T = N$.

For $n := 1$ equation (2.3) becomes

$$a_1 x + a_0 = 1$$

and has a unique solution iff $a_1 = 1$; the unique solution is $x = a'_0$.

For $n := 2$ equation (2.3) becomes

$$a_{12} xy + a_1 x + a_2 y + a_0 = 1$$

and conditions (iii) reduce to $a_{12} = 1$, $a_1a_2 = a_0$, $a_1a_0 = a_0$ and $a_2a_0 = a_0$. The last two conditions express the inequalities $a_0 \leq a_1$ and $a_0 \leq a_2$, so that the second condition is equivalent to $a_1a_2 \leq a)$. The unique solution is $x = a_2'$, $y = a_1'$.

For $n := 3$ equation (2.3) becomes

$$a_{123}xyz + a_{23}yz + a_{13}xz + a_{12}xy + a_1x + a_2y + a_3z + a_0 = 1,$$

the conditions (iii) are $a_{123} = 1$ and ... (left to the reader). The unique solution is $x = a_{23}'$, $y = a_{13}'$, $z = a_{12}'$. $\square$

Remark 2.1. The solution (2.4) depends only on n out of the 2^n coefficients so long as the conditions are met. It would be interesting to investigate those classes of Boolean equations whose solutions depend on only a small subset of the coefficients. $\square$

We mention here the paper by Banković [1987a], where the Whitehead-Parker- Bernstein Theorem 6.7 in BFE is proved using the Vaught-theorem technique; cf. introduction to Ch.6 §4.

For quadratic truth equations with unique solution see the next section.

See also Proposition 10.1.4.

3 Quadratic truth equations

In this section we study truth equations which can be expressed in a disjunctive form involving only quadratic and linear terms. In particular we deal with quadratic truth equations having a unique solution. For applications see Ch.14 §1.

Stricto sensu, a *quadratic truth equation* is a truth equation of the form

(3.1) $f(x_1, \ldots, x_n) := t_1 \vee \ldots \vee t_m = 0$,

where each t_k is a term of the form $x_i^\alpha x_j^\beta$, where $i, j \in \{1, \ldots, n\}$ and $\alpha, \beta \in \{0, 1\}$.

However if $i = j$ then $xi^\alpha x_j^\beta$ vanishes if $\beta \neq \alpha$ and reduces to x_i^α if $\beta = \alpha$. Moreover, if equation (3.1) contains a *linear term* x_i^α, then $x_i = \alpha'$ in every solution of the equation, if any; in particular if equation (3.1) contains two terms of the form x_i^α and $x_i^{\alpha'}$, then it is inconsistent. Therefore we may suppose without loss of generality that equation (3.1) contains no linear terms.

Definition 3.1. The Greek letters $\xi, \eta, \ldots$ will stand for *literals*, i.e., elements of the set

(3.2) $\Lambda = \{x_1, \ldots, x_n, x_1', \ldots, x_n'\}$;

(the word "literal" will be used in the same spirit as the word "variable", i.e., both as a name for an unspecified element (of $\{0, 1\}$) and as a letter of an alphabet (viz., Λ)). The *variable* $|\xi|$ of a *literal* ξ is defined by

(3.3) $| x_i | = | x_i' | = x_i$ $(i = 1, \ldots, n)$.

□

Further, note that if equation (3.1) contains two terms of the form $\xi\eta$ and $\xi\eta'$ (of the form $\xi\eta'$ and $\xi'\eta$, of the form $\xi\eta$ and $\xi'\eta'$), then in every solution of the equation, if any, we have $\xi = 0$ ($\xi = \eta$, $\xi = \eta'$) and hence we can drop one of the unknowns $|\xi|, |\eta|$. Therefore we may suppose, without loss of generality, that none of the above situations holds.

We can summarize the above discussion as follows:

Definition 3.2. An *irreducible* quadratic truth equation is a truth equation of the form (3.1) such that I) each term t_k is of the form $x_i^\alpha x_j^\beta$ with $i, j \in \{1, \ldots, n\}$, $i \neq j$, while $\alpha, \beta \in \{0, 1\}$, and II) for each $i, j \in \{1, \ldots, n\}$, $i \neq j$, the equation contains at most one term of the form $x_i^\alpha x_j^\beta$.   □

Definition 3.3. (Minoux [1992]). Two equations (of any kind! cf. Ch.1) of the form $E_1(x_1, \ldots, x_n)$ and $E_2(x_1, \ldots, x_n, x_{n+1}, \ldots, x_{n+p})$ are *equivalent* provided there is a bijection between their sets of solutions.   □

Remark 3.1. Every quadratic truth equation is equivalent to an irreducible one.   □

Definition 3.4. A *pure* quadratic truth equation is an equation of the form (3.1) in which the terms are of the form $x_i x_j$ $(i \neq j)$ and/or of the form $x_i x_j'$ $(i \neq j)$.□

Remark 3.2. Every pure quadratic truth equation has the solution $(0, \ldots, 0)$.   □

Definition 3.5. A *switch* on a *variable* x_i is a transformation of the form $x_i' \mapsto y_i$, $x_i \mapsto y_i'$, where $y_i \notin \Lambda$. A *switch* on a *set* S of variables is a switch on each variable $x_i \in S$ such that the new variables y_i are pairwise distinct.   □

Proposition 3.1. (Simeone [1979]). *Let (3.1) be a quadratic truth equation with no linear terms. Then the equation is consistent if and only if it can be transformed into a pure one by a switch on some set S of variables.*

PROOF: The condition is sufficient by Remark 3.2. Conversely, we are going to prove that if equation (3.1) has the solution $A = (\alpha_1, \ldots, \alpha_n) \in \{0, 1\}^n$, then the switch on the set $S = \{x_i \in A \mid \alpha_i = 1\}$ transforms (3.1) into a pure equation. Consider a term $\xi\eta$ of equation (3.1), hence $|\xi| \neq |\eta|$. If $|\xi|, |\eta| \notin S$ then the term $\xi\eta$ is not altered and it cannot be of the form $x_i' y_j'$ (because $\alpha_i = \alpha_j = 0$), therefore it suffices to study the case $|\xi| = x_i \in S$. Set $|\eta| = x_j$. If $\xi\eta = x_i x_j$ then $x_j \notin S$, hence the term is transformed into $y_i' x_j$. If $\xi\eta = x_i x_j'$ then $x_j \in S$, hence the term is transformed into $y_i' y_j'$. If $\xi = x_i'$ then the transformed term contains the factor y_i.   □

Remark 3.3. If equation (3.1) is irreducible then the equation obtained from it by the switch constructed in Proposition 3.1 is also irreducible. This is easily shown by studying the three cases considered in the proof of Proposition 3.1. For instance, if $\xi\eta = x_i x_j$, then the transformed equation cannot include any of the terms $y_i x_j$ or $y_i' x_j'$ or $y_i x_j'$, otherwise equation (3.1) would contain one of the terms $x_i' x_j$ or $x_i x_j'$ or $x_i' x_j'$, respectively, contradicting irreducibility. $\qquad\square$

Corollary 3.1. *Every quadratic truth equation is equivalent to a pure irreducible equation.* $\qquad\square$

So far we have seen that a simple inspection of a quadratic truth equation allows its reduction to an irreducible one and, if a particular solution is known *a priori*, to a pure irreducible equation. However there may also exist hidden fixations of variables, $x_i = \alpha$, and/or identifications of variables, $x_j = x_k^\beta$. It will be seen below that the methods devised for solving quadratic truth equations include in particular the detection of all fixations and identifications of variables.

In the sequel we present several theoretical results and the solution methods based on them.

We begin with a very general lemma.

Lemma 3.1. (Rudeanu [1995a]). *The following conditions are equivalent for every two distinct literals ξ, η of a truth function g (not necessarily quadratic):*

(i) $\xi \leq \eta$ in every solution of the equation $g(X) = 0$, if any ;

(ii) $\xi\eta'$ is an implicant of the function g .

COMMENT: Properly speaking, condition (i) states that the values ξ_0, η_0 given to the literals ξ, η in any solution... satisfy $\xi_0 \leq \eta_0$. Similar comments hold for Propositions 3.2 and 3.2.

PROOF: Condition (i) can successively be written in the following equivalent forms: $g(X) = 0 \implies \xi \leq \eta$, then $g(X) = 0 \implies \xi\eta' = 0$ and finally $\xi\eta' \leq g(X)$. $\qquad\square$

Hypothesis. In the sequel we assume that all of the variables $x_1, \ldots, x_n$ do occur in the quadratic truth equation (3.1); unless otherwise stated, no other condition on equation (3.1) will be required. $\qquad\square$

Definition 3.6. (Aspvall, Plass and Tarjan [1979]). We associate with equation (3.1) the binary relation I ("implication") on Λ defined as follows:

$$(3.4) \qquad\qquad \xi I \eta \iff \xi\eta' \text{ is a term of (3.1) .}$$

$\qquad\square$

Remark 3.4. The following properties hold:

(i) $\xi\eta$ is a term of (3.1)$\iff \xi I \eta'$;

(ii) $\xi I \eta \iff \eta' I \xi'$;

(iii) $\xi I^+ \eta \iff \eta' I^+ \xi'$.

(Recall that the *transitive closure* of a binary relation R is $R^+ = \bigcup_{n \geq 1} R^n$, where $R^1 = R$ and $R^{n+1} = R^n \circ R$). $\qquad\square$

Lemma 3.2. (Rudeanu [1995a]). *For every two distinct literals ξ, η, the implications (i)$\Longrightarrow$(ii)$\Longrightarrow$(iii) hold, where we have set*

(i) *$\xi\eta'$ is a prime implicant of the function f (cf. (3.1)) ,*

(ii) *$\xi I^{+}\eta$,*

(iii) *$\xi\eta'$ is an implicant of the function f .*

PROOF: (i)$\Longrightarrow$(ii): It is well known that every prime implicant can be obtained by a repeated application of the *consensus* operation $xy \vee x'z \mapsto yz$, therefore we will prove that for every $k \geq 0$, if $\xi\eta'$ is obtained by k applications of consensus, then $\xi I^{+}\eta$. If $k = 0$ then $\xi\eta'$ is a term of the function f, hence $\xi I\eta$, therefore $\xi I^{+}\eta$. For the inductive step suppose the last consensus was $\xi\zeta \vee \eta'\zeta' \mapsto \xi\eta'$. Then $\xi I^{+}\zeta'$ and $\zeta' I^{+}\eta$ by the inductive hypothesis, hence $\xi I^{+}\eta$.

(ii)$\Longrightarrow$(iii): We are going to prove that for every $n \geq 1$, if $\xi I^{n}\eta$ then $\xi\eta'$ is an implicant of the function f. If $n = 1$ then $\xi I\eta$, hence $\xi\eta'$ is a term of the function f, therefore it is an implicant. For the inductive step suppose $\xi I^{n}\zeta$ and $\zeta I\eta$ for some $\zeta \in \Lambda$. Then $\xi\zeta'$ is an implicant of f by the inductive hypothesis, while $\zeta\eta'$ is a term of f, therefore $\xi\eta' \leq \xi\zeta' \vee \zeta\eta' \leq f$.   □

Remark 3.5. It was noted in Rudeanu [1995a] that the implications in Lemma 3.1 cannot be strengthened. Consider, for instance, the equation

$$(3.5) \qquad\qquad xy' \vee yz \vee x'z \vee zt = 0 .$$

Then the prime implicants are xy' and z. So zt is not a prime implicant, but zIt', hence $zI^{+}t'$. On the other hand, zt' is an implicant, but clearly $\neg(zI^{+}t)$.   □

Proposition 3.2. *The following conditions are equivalent for every two distinct literals ξ, η:*

(i) *$\xi\eta'$ and $\xi'\eta$ are implicants of the function f ;*

(ii) *$\xi I^{+}\eta$ and $\eta I^{+}\xi$;*

(iii) *$\xi = \eta$ in every solution of equation (3.1), if any .*

COMMENT: The equivalence (ii)$\Longleftrightarrow$(iii) is due to Aspvall, Plass and Tarjan [1979].

PROOF: Immediate from Lemmas 3.1 and 3.1.   □

Proposition 3.3. *The following conditions are equivalent for a literal ξ:*

(i) *ξ is an implicant of f ;*

(ii) *ξ is a prime implicant of f ;*

(iii) *$\xi I^{+}\xi'$;*

(iv) *$\xi = 0$ in every solution of equation (3.1), if any .*

COMMENT: The equivalence (iii)$\Longleftrightarrow$(iv) is due to Hansen and Jaumard [1985], while the equivalence with conditions (i) and (ii) was remarked by Rudeanu [1995a].

PROOF: Taking $\eta := \xi'$ in Lemma 3.2 we obtain (ii)$\Longrightarrow$(iii) $\Longrightarrow$(i), while the implication (i)$\Longrightarrow$(ii) is obvious. On the other hand (iv)$\Longleftrightarrow$(i) by Lemma 3.1, again with $\eta := \xi'$. $\square$

Example 3.1. Let us resume equation (3.5) in Remark 3.5. We have

$$I = \{(x,y),(y',x'),(y,z'),(z,y'),(x',z'),(z,x),(z,t'),(t,z')\} = I^1 ,$$

$$I^2 = I^1 \cup \{(x,z'),(z,y),(z,x'),(y',z')\} ,$$

$$I^3 = I^2 \cup \{(z,z')\} ,$$

$$I^4 = I^3 = I^+ ,$$

therefore there is exactly one variable fixed in every solution, namely $z := 0$. This reduces equation (3.5) to $xy' = 0$, which is immediately solved. $\square$

Theorem 3.1. (Aspvall, Plass and Tarjan [1979]). *The quadratic truth equation* (3.1) *is consistent if and only if there is no variable* x_k *such that* $x_k I^+ x'_k$ *and* $x'_k I^+ x_k$.

COMMENT: Since we are not satisfied with the original proof of sufficiency, we give below a proof based on Proposition 3.3.
PROOF: If equation (3.1) is consistent, then $x_k I^+ x'_k$ and $x'_k I^+ x_k$ for some k would imply that $x_k = 0$ and $x'_k = 0$ in every solution, which is impossible. Conversely, if the function f is a tautology, then x_k and x'_k are prime implicants, hence $x_k I^+ x'_k$ and $x'_k I^+ x_k$. $\square$

Corollary 3.2. (Hansen and Jaumard [1985]). *The quadratic truth equation* (3.1) *has a unique solution if and only if, for each* $k \in \{1,\ldots,n\}$, $x_k I^+ x'_k$ *or* $x'_k I^+ x_k$ *but not both.*

PROOF: From Proposition 3.3 and Theorem 3.1. $\square$

At this point we find it convenient to describe informally other practical consequences of Propositions 3.2 and 3.3.

Several algorithms are known which determine the transitive closure of a relation; their complexity is $O(p^3)$, where p is the number of elements in the support set. Therefore Propositions 3.3 and 3.2 yield algorithms for establishing whether or not a quadratic truth equation is consistent and, in the affirmative case, for determining all fixations of variables $x_j = x_k^{\alpha_{jk}}$ that hold in every solution.

As a matter of fact there is a more convenient approach to the identifications of variables. Aspvall, Plass and Tarjan [1979] describe relation I as a graph Γ whose set of vertices is Λ and there is an arc (ξ,η) whenever $\xi I \eta$. Then the condition $\xi I^+ \eta$ means the existence of a path from ξ to η, so that condition (ii) in Proposition 3.2 expresses the fact that vertices ξ and η belong to the same *strongly connected component* of the graph. Therefore one can benefit from the existence of linear-time algorithms for the determination of the strongly connected components of a graph.

The next step consists in replacing all fixed variables by their values and in replacing, for each strongly connected component, all of its literals by just one of them; then one applies idempotency and absorption to delete all redundant terms. The equation obtained in this way, called the *reduced equation associated* with the original equation (3.1), is clearly equivalent to (3.1) and irreducible.

In order to solve the equation after all possible simplifications have been made, the concept of solution has to be conveniently worked out. Aspvall, Plass and Tarjan [1979] noted that a solution of equation (3.1) is in fact α) a solution of (3.1) viewed as an equation in the $2n$ unknowns $x_1, \ldots, x_n, x'_1, \ldots, x'_n$, such that β) x_i and x'_i have complementary values, for every $i = 1, \ldots, n$. Further, since the terms $\xi \eta'$ associated with the pairs $(\xi, \eta) \in I$ - or equivalently, with the arcs (ξ, η) of the graph Γ - exhaust all the terms of equation (3.1), it follows that property α) amounts to saying that condition (*) $\xi = 1 \Longrightarrow \eta = 1$ holds for every $(\xi, \eta) \in I$. But this happens if and only if (*) holds for every $(\xi, \eta) \in I^+$. In graph-theoretical terms this means that (*) holds for every arc of Γ if and only if it holds for every path of Γ, which is quite clear. Thus α) means that (*) holds for every path of Γ, which is further equivalent to the following condition α'): if a vertex ξ has value 1 then all its successors have value 1, and if ξ has value 0 then all its predecessors have value 0.

One more idea was introduced by Hansen and Jaumard [1985]. Suppose the vertices ξ of the graph Γ are labelled with positive integers $\ell(\xi)$ such that $\gamma 1$) $\ell(\xi) < \ell(\eta)$ for any arc (ξ, η). Condition $\gamma 1$) is clearly equivalent to $\gamma 1'$) $\ell(\xi) < \ell(\eta)$ whenever $\xi \neq \eta$ and there is a path from ξ to η. It follows that for every vertex ξ, the successors of ξ have labels $> \ell(\xi)$, while the labels of the predecessors of ξ are $< \ell(\xi)$. Suppose further that $\gamma 2$) $\ell(\xi) + \ell(\xi') = 2n + 1$ for every $\xi \in \Lambda$. If one takes $\xi := 0 \Longleftrightarrow \ell(\xi) \leq n$, or equivalently, $\xi := 1 \Longleftrightarrow \ell(\xi) \geq n + 1$, one obtains a solution of equation (3.1), i.e., the above properties α) and β) hold. For clearly $\gamma 1$) implies the equivalent variant α') of α), while $\gamma 2$) shows that the cases 1) $\ell(\xi), \ell(xi') \leq n$ and 2) $\ell(\xi), \ell(\xi') \geq n + 1$ are impossible, therefore either $\ell(\xi) \leq n < \ell(\xi')$, or $\ell(\xi') \leq n < \ell(\xi)$, which implies that β) holds.

Hansen and Jaumard [1985] have actually devised a labelling algorithm in the case when the graph Γ associated to the equation is circuit-free. Recall that the *in-degree* of a vertex ξ is the number of arcs having the target ξ. Since the finite graph Γ is circuit-free, there exists at least one vertex having the in-degree 0. The algorithm scans the vertices in the increasing order of their in-degrees. Let ℓ be the current value of the label; the first value is $\ell := 0$. The general step is the following: $\ell := \ell + 1$, choose the new vertex ξ, $\ell(\xi) := \ell$, $\ell(\xi') = 2n + 1 - \ell$, drop the vertices ξ and ξ' from the graph.

It follows from the previous discussion that the algorithm stops and yields a particular solution of the equation. Besides, a possible switch on certain variables transforms the equation into a pure one. For let $y_1, \ldots, y_n$ be the literals ξ such that $\ell(\xi) \leq n$. Since $y_k := 0$ $(k = 1, \ldots, n)$ is a particular solution, the equation cannot contain any term of the form $y'_i y'_j$. Moreover, if certain variables y_i occur only in the complemented form y'_i, (that is, in terms of the form $y'_i y_j$), then the switches $z_i = y_i'$ transform the equation into a pure equation in which all the uncomplemented variables do appear.

Last but not least, note that the above technique is applicable to any quadratic truth equation, to the effect that one applies the above algorithm to the *reduced* equation associated with the given equation (3.1). For the graph associated with the latter equation is circuit free: the existence of a circuit would contradict the fact that its vertices belong to different strongly connected components.

See also the survey by Hammer and Simeone [1987], including in particular a survey of quadratic pseudo-Boolean functions and applications.

Example 3.2. Consider the truth equation

$$x_1' x_2' \vee x_1 x_3 \vee x_1' x_4 \vee x_3' x_4' \vee x_4 x_5' = 0 .$$

The strongly connected components are $\{x_1, x_3', x_4\}$ $(x_1 I x_3' I x_4 I x_1)$, $\{x_2\}$ and $\{x_5\}$. Therefore $x_1 = x_3' = x_4$ in any solution, and the reduced equation is

$$x_1' x_2' \vee x_1 x_5' = 0 .$$

Now $n = 3$ and the vertices with in-degree equal to 0 are x_2' and x_5'. We set e.g. $\ell(x_2') = 1$, hence $\ell(x_2) = 6$. Then the algorithm yields $\ell(x_5') = 2$, hence $\ell(x_5) = 5$. For the labels 3,4 we are faced with the graph having the vertices x_1, x_1' and no arc. We set e.g. $\ell(x_1) = 3$, hence $\ell(x_1') = 4$. The literals with $\ell(\xi) \leq 3$ are x_2', x_5' and x_1. This corresponds to the particular solution $x_2' := x_5' := x_1 := 0$, $x_1' := x_5 := x_2 := 1$. The switches $y_2 = x_2'$, $y_5 = x_5'$, transform the reduced equation into the pure equation $x_1' y_2 \vee x_1 y_5 = 0$, in which all of the uncomplemented variables x_1, y_2, y_5 do appear. Finally note the particular solution $(0, 1, 1, 0, 1)$ of the original equation.   □

Another algorithm which either detects the inconsistency of a given quadratic truth equation or transforms it into an equivalent pure equation, thus providing a particular solution, was devised by Simeone [1979]; see also Petreschi and Simeone [1980]*.

Proposition 3.4 below, which prepares the next theorem, may also be viewed as an explanation of the rôle played by the transitive closure I^+ in solving equation (3.1). Note that equation (3.1) can also be written in the form

(3.1') $$\bigvee \{\xi \eta' \mid \xi I \eta\} = 0$$

and this suggests introducing the equation

(3.6) $$\bigvee \{\xi \eta' \mid \xi I^+ \eta\} = 0 .$$

Proposition 3.4. *Equations* (3.1) *and* (3.6) *are equivalent.*

PROOF: Every term of (3.1) is a term of (3.6), therefore every solution of (3.6), if any, satisfies (3.1) as well. Conversely, let $\xi \eta'$ be a term of (3.6). Then $\xi \eta'$ is an implicant of the function f by Lemma 3.2. This implies that $\xi \eta' = 0$ in every solution of equation $f = 0$, if any (cf. Comment to Lemma 3.1).   □

We will need the following

Lemma 3.3. *If equation* (3.1) *is pure, so is* (3.6).

PROOF: We show that if two terms $\xi\eta$ and $\zeta\theta$ produce a new term by transitivity, this term is pure as well. From the hypothesis we have $\xi I \eta'$ and $\zeta I \theta'$, hence $\zeta = \eta'$ and the new term corresponds to $\xi I^+\theta'$, hence it is $\xi\theta$. If $\xi = x'$ and $\theta = y'$, then one of the two terms $\xi\eta = x'\eta$ and $\zeta\theta = \eta'y'$ is not pure, which is a contradiction. $\square$

Definition 3.7. Let us set $N = \{1, \ldots, n\}$ and

(3.7) $\qquad D_k = \{i \in N \mid x_k I x_i'\}\,, \ E_k = \{j \in N \mid x_k I x_j\} \qquad (k \in N)\,.$

$\square$

Lemma 3.4. (Crama, Hammer, Jaumard and Simeone [1986], [1987]). *If* (3.1) *is a pure quadratic truth equation and* $(\alpha_1, \ldots, \alpha_n) \in \{0,1\}^n$ *is a particular solution of it, then*

(3.8) $\qquad \alpha_k = \alpha_k (\prod_{i \in D_k} \alpha_i')(\prod_{j \in E_k} \alpha_j) \qquad (k = 1, \ldots, n)\,.$

PROOF: Equation (3.1) can be written in the form

$$\bigvee_{k=1}^{n} x_k ((\bigvee_{i \in D_k} x_i) \vee (\bigvee_{j \in E_k} x_j')) = 0$$

(where it may happen that $D_k = \emptyset$ and/or $E_k = \emptyset$), hence it is equivalent to the system

$$x_k \leq (\prod_{i \in D_k} x_i')(\prod_{j \in E_k} x_j) \qquad (k = 1, \ldots, n)\,.$$

$\square$

Theorem 3.2. (Crama, Hammer, Jaumard and Simeone [1986], [1987]). *Suppose equation* (3.1) *is irreducible, pure and consistent. If* I *is transitive, then* (3.1) *has the reproductive solution*

(3.9) $\qquad x_k = p_k (\prod_{i \in D_k} p_i')(\prod_{j \in E_k} p_j) \qquad (k = 1, \ldots, n)\,.$

Conversely, if equation (3.1) *has the reproductive solution* (3.9) *and all the variables* $x_1, \ldots, x_n$ *do occur in* (3.1), *then* I *is transitive.*

PROOF: Suppose first that I is transitive and take $(p_1, \ldots, p_n) \in \{0,1\}^n$. In view of Lemma 3.4, we only have to prove that the vector $(\alpha_1, \ldots, \alpha_n)$ defined by

(3.10) $\qquad \alpha_k = p_k (\prod_{i \in D_k} p_i')(\prod_{j \in E_k} p_j) \qquad (k = 1, \ldots, n)$

satisfies (3.1). Fix an integer $k \in \{1, \ldots, n\}$. If $i \in D_k$ then $\alpha_k \alpha_i \leq p'_i p_i = 0$. Now take $j \in E_k$; to prove that $\alpha_k \alpha'_j = 0$, suppose $\alpha_j = 0$ and prove $\alpha_k = 0$. In view of (3.8.j), three cases may occur: 1) $p_j = 0$, hence $\alpha_k = 0$; 2) $p'_h = 0$ for some $h \in D_j$, in which case $x_j I x'_h$ and since $x_k I x_j$ because $j \in E_k$, it follows by transitivity that $x_k I x'_h$, that is, $h \in D_k$, therefore $\alpha_k \leq p'_h = 0$; 3) $p_h = 0$ for some $h \in E_j$, in which case one proves similarly that $\alpha_k \leq p_h = 0$.

Conversely, suppose I is not transitive. Then there exist $\xi, \eta, \zeta \in \Lambda$ such that $\xi I \eta$ and $\eta I \zeta$ but $\neg(\xi I \zeta)$. The irreducibility of equation (3.1) implies that the variables $\mid \xi \mid, \mid \eta \mid$ and $\mid \zeta \mid$ are pairwise distinct, therefore, using possibly some preliminary switches, we may suppose without loss of generality that $\xi = x_k$, $\eta = x_j$ and $\zeta = x_i$, where k, j, i are pairwise distinct. So $x_k I x_j$, $x_j I x_i$ but $\neg(x_k I x_i)$, that is, we have $j \in E_k$, $i \in E_j$ but $i \notin E_k$. Since the sets $\{k\}$, D_k and E_k are pairwise distinct (again by irreducibility), we can take $p_k := 1$, $p_h := 0$ if $h \in D_k$ and $p_h := 1$ if $h \in E_k$. Then (3.10) yields $\alpha_k = 1$. Since the variable x_i does occur in the irreducible equation (3.1) and $i \notin E_k$, it follows that $i \in D_k$, hence $p_i = 0$. As $i \in E_j$, formula (3.10.j) yields $\alpha_j \leq p_i = 0$, therefore $\alpha_k \alpha'_j = 1$. On the other hand $x_k x'_j$ is a term of (3.1) because $j \in E_k$. Thus the vector (3.9) obtained for the above choice of $p_1, \ldots, p_n$ is not a solution of (3.1). $\square$

Corollary 3.3. *If equation (3.1) is consistent, then the reduced equation associatd with it has a reproductive solution of the form (3.9).*

PROOF: Let (3.1R) and (3.6R) denote the reduced equation associated with (3.1) and the equation of the form (3.6) associated with (3.1R), respectively. All of these equations are equivalent (cf. Proposition 3.4), hence they are consistent. We have noted that (3.1R) is a pure equation, hence equation (3.6R) is also pure by Lemma 3.3. Now Theorem 3.2 implies that equation (3.6R) has a reproductive solution of the form (3.9) and this is a reproductive solution of equation (3.1R), too. $\square$

Corollary 3.4. *Every consistent quadratic truth equation (3.1) has a reproductive solution which is obtained from the reproductive solution (3.9) of the reduced equation (3.1R) associated with it by using the bijection between the solutions of equation (3.1) and those of the equivalent equation (3.1R) (in the sense of Definition 3.3).* $\square$

Example 3.3. Let us resume the equation in Example 3.2. We have seen that the corresponding equation (3.1R) is $x'_1 y_2 \vee x_1 y_5 = 0$. The equation (3.6R) is $x'_1 y_2 \vee x_1 y_5 \vee y_2 y_5 = 0$. So

$$D_1 = \{5\}, \; E_1 = \emptyset, \; D_2 = \{5\}, \; E_2 = \{1\}, \; D_5 = \{1, 2\}, \; E_5 = \emptyset.$$

Formulas (3.9) become $x_1 = p_1 p'_5$, $y_2 = p_2 p'_5 p_1$, $y_5 = p_5 p'_1 p'_2$. Finally we come back to the other variables: $x_1 = x'_3 = x_4$, $x_2 = y'_2$, $x_5 = y'_5$. Therefore we have obtained the reproductive solution

$$x_1 = p_1 p'_5, \; x_2 = p'_2 \vee p_5 \vee p_1, \; x_3 = p'_1 \vee p_5, \; x_4 = p_1 p'_5, \; x_5 = p'_5 \vee p_1 \vee p_2.$$

$\square$

Example 3.4. Let us resume equation (3.5) in Remark 3.5. We have seen that the corresponding equation (3.1R) is $xy' = 0$, which coincides with (3.6R). Setting $x_1 = x$, $x_2 = y'$, we obtain the equation $x_1 x_2 = 0$, for which $D_1 = \{2\}$, $E_1 = \emptyset$, $D_2 = \{1\}$, $E_2 = \emptyset$, therefore the reproductive solution is $x_1 = p_1 p_2'$, $x_2 = p_2 p_1'$. The lost variables are z, which is fixed at 0, and t, which has remained arbitrary. Therefore we have obtained the reproductive solution

$$x = p_1 p_2' \,,\; y = p_2' \vee p_1 \,,\; z = 0 \,,\; t = p_4 \,.$$

□

Some of the above techniques can be extended to a class of truth equations which generalize pure quadratic truth equations. Proposition 3.5, Corollary 3.5 and Example 3.5 below are due to Minoux [1992] (in dual form).

Proposition 3.5. *Suppose* (3.1) *is a truth equation without linear terms in which every term* t_k *contains at most one complemented variable* x_i'.[2] *Then:*

α) *Equation* (3.1) *is consistent and the following condition is necessary in order that the solution be unique:*

$$(3.11) \qquad \forall i \in \{1, \ldots, n\} \; \exists k \in \{1, \ldots, m\} \; \exists j \in \{1, \ldots, n\} \backslash \{i\} \; t_k = x_i x_j' \,.$$

β) *When condition* (3.11) *is fulfilled, there exist* p *variables* $x_{i_1}, \ldots, x_{i_p}$, $p \geq 2$, *such that* $x_{i_1} = x_{i_2} = \ldots = x_{i_p}$ *in every solution of equation* (3.1).

PROOF: α) Clearly $x_i := 0$ $(i = 1, \ldots, n)$ is a solution of equation (3.1).

Now suppose condition (3.11) is not fulfilled. Let i be an index which invalidates (3.11). Then we claim that $x_i := 1$, $x_j := 0$ $(j \neq i)$ is a solution of equation (3.1). It suffices to show that every term t_k which contains x_i is taken to 0 by the above values. But either t_k contains more than two factors, in which case at least one of them is uncomplemented and $\neq x_i$, according to the hypothesis on the function f, or t_k contains exactly two factors, in which case it is of the form $t_k = x_i x_h$, $h \neq i$, according to the hypothesis on i.

β) Take an arbitrary variable x_{j_1}. In view of (3.11) there exists an index $j_2 \neq j_1$ such that (3.1) contains the term $x_{j_1} x_{j_2}'$; this implies that $x_{j_1} \leq x_{j_2}$ in every solution of equation (3.1). We continue in this way and obtain a sequence of pairwise distinct indices $j_1, j_2, \ldots$ such that $x_{j_1} \leq x_{j_2} \leq \ldots$ in every solution of (3.1). Clearly after a finite number of steps, say r, we reach again a previously obtained index j_q. Thus

$$x_{j_1} \leq x_{j_2} \leq \ldots \leq x_{j_r} = x_{j_q}$$

in every solution of equation (3.1), where $1 \leq q < r - 1$ because there are no linear terms. Thus

---

[2] In some of the papers mentioned in the present section, formulas with this property are called Horn formulas; they should not be confused, however, with Horn fomulas in the sense of Ch.5, §3.

$$x_{j_q} = x_{j_q+1} = \ldots = x_{j_{r-1}} = x_{j_q}$$

in every solution of (3.1) and the number of these variables is $r - q \geq 2$.   □

Corollary 3.5. *Suppose every term of the truth equation (3.1) contains at most one complemented variable. Then the equation is consistent and there is an algorithm which decides whether the solution of the equation is unique.*

PROOF: Consistency follows from the fact that after the variables corresponding to linear terms are stuck at 0 or 1, we are faced with an equation satisfying the hypotheses of Proposition 3.5.α).

The algorithm runs as follows. At each step of the algorithm we are faced with an equation of the form (3.1) in which every term contains at most one complemented variable. We stuck at 0 or 1 the variables corresponding to the linear terms. If the remaining equation does not fulfil condition (3.11) then the solution is not unique and the algorithm stops. Otherwise we identify certain variables as shown in the proof of Proposition 3.5. The resulting equation is the input for the next step.

Since the number of variables decreases at each step (cf. β) in Proposition 3.5), the algorithm stops either when condition (3.11) is not fulfilled and hence the solution is not unique, or when the transformed equation is the identity $0 = 0$, in which case the solution is unique.   □

Example 3.5. Consider the truth equation

$$x_7' \vee x_9 \vee x_6' x_7 \vee x_3' x_1 x_7 \vee x_2' x_3 x_6 \vee x_2' x_4 x_6 \vee x_1' x_2 x_7$$

$$\vee x_4' x_5 x_7 \vee x_9' x_7 x_8 \vee x_5' x_4 x_6 \vee x_7' x_1 x_2 \vee x_2' x_5 x_6 x_8$$

$$\vee x_4' x_1 x_3 x_6 \vee x_6' x_4 x_5 x_7 \vee x_4' x_1 x_3 x_7 x_8 \vee x_9' x_2 x_3 x_4 x_6 x_7 = 0 \,.$$

The linear terms yield $x_7 = 1$, $x_9 = 0$, hence the equation reduces to

$$x_6' \vee x_3' x_1 \vee x_2' x_3 x_6 \vee x_2' x_4 x_6 \vee x_1' x_2 \vee x_4' x_5 \vee x_8 \vee x_5' x_4 x_6$$

$$\vee x_2' x_5 x_6 x_8 \vee x_4' x_1 x_3 x_6 \vee x_6' x_4 x_5 \vee x_4' x_1 x_3 x_8 \vee x_2 x_3 x_4 x_6 = 0 \,.$$

Two new linear terms have appeared, which imply $x_6 = 1$, $x_8 = 0$, hence the equation is further reduced to

$$x_3' x_1 \vee x_2' x_3 \vee x_2' x_4 \vee x_1' x_2 \vee x_4' x_5 \vee x_5' x_4 \vee x_4' x_1 x_3 \vee x_2 x_3 x_4 = 0 \,.$$

Since there are no more linear term and condition (3.11) is fulfilled, we apply twice Proposition 3.5.β). From $x_3' x_1 \vee x_2' x_3 \vee x_1' x_2 = 0$ we obtain $x_1 \leq x_3 \leq x_2 \leq x_1$, hence $x_1 = x_2 = x_3$, while from $x_4' x_5 \vee x_5' x_4 = 0$ we obtain $x_5 \leq x_4 \leq x_5$, hence $x_4 = x_5$. Now the equation reduces to

$$x_1' x_4 \vee x_4' x_1 \vee x_1 x_4 = 0 \,.$$

The hypotheses of Proposition 3.5.β) are again satisfied. We have $x_4 \leq x_1 \leq x_4$, hence $x_1 = x_4$ and the equation reduces to

$$x_1 = 0 \, .$$

This equation coincides with its solution. The algorithm returns the unique solution $(0,0,0,0,0,1,1,0,0)$. $\quad\square$

For the computer implementation of the solution of quadratic truth equations see the next section.

4 Boolean equations on computers

In BFE, Chapter 15, §4, we have briefly reported several attempts to implement the solution of Boolean equations on computers. We should add here the program devised by Bordat [1975], which is in the same line. As we are going to show in this section, subsequent research on this problem refers mainly to quadratic truth equations, although a more general approach is not missing. The field is open for much future research.

We have already mentioned in §2 that Simeone [1979] has devised an algorithm which transforms a quadratic truth equation into an equivalent pure equation or detects the inconsistency of the given equation, if this is the case. The general step of the algorithm runs as follows. Choose one of the terms to be eliminated, say $x'y'$, and one of its two variables, say x. Construct a tree $T(x)$ of upper bounds of x with respect to the partial order I (cf. Definition 3.5) until one of the following two situations occurs: α) two upper bounds $u, v \in \{x_1, \ldots, x_n\}$ are found such that uv is a term of the given equation (3.1), or β) $T(x)$ exhausts all of the uncomplemented upper bounds of x and case α) has not been met.

In case α) let z be the greatest lower bound of u and v in the tree $T(x)$. It follows that $xIzIu$ and zIv; since $uv = 0$ in any solution, all the variables located on the path in $T(x)$ from x to z are forced to 0. After these variables are replaced by 0 in the equation, it may happen that some other variables are stuck to 0. Anyway, at least x is fixed at 0.

In case β) one performs a switch on all the variables in $T(x)$. This transforms the "negative" term $x'y'$ either into a "positive" term x_1y_1 or into a "mixed" term x_1y' and on the other hand no new "negative" term $u'v'$ is produced. For otherwise $u'v'$ would be the result of applying the switch either on the "positive" term uv with $u, v \in T(x)$, or on a "mixed" term, say uv', with $u \in T(x)$ and $v \notin T(x)$. But the former variant contradicts the fact that situation α) has not occurred, while in the latter variant the term uv' shows that uIv and hence $v \in T(x)$, again a contradiction.

Summarizing, the general step of the algorithm produces an equation which is equivalent to the input equation and in which either the number of variables or the number of "negative" terms has decreased. Therefore, after finitely many steps the algorithm stops and returns one of the following three answers: 1) a pure quadratic equation equivalent to the given equation (if all "negative" terms heve been eliminated and some (possibly all) variables have remained); 2) the unique solution of the equation (if each variable has been fixed at one and only one of the values 0,1); 3) the equation is inconsistent (if some variable has been stuck both at 0 and at 1).

The complexity of the algorithm is $O(mn)$, where n is the number of variables and m is the number of terms. The paper by Petreschi and Simeone [1980]* reports computational experiments on 50 randomly generated test problems ranging from 50 to 500 variables.

The algorithm given by Rudeanu [1995a] solves a quadratic truth equation by making use both of some specific theoretical results (cf. §3) and of techniques of pseudo-Boolean programming (cf. Hammer and Rudeanu [1968]). In connection with these techniques we only recall here that the set of solutions to a problem involving 0−1 variables - a particular case is precisely the problem of solving a truth equation - may often be conveniently partitioned into *families* or *hyperplanes of solutions*. Such a family is a set of solutions characterized by the fact that certain variables have fixed 0−1 values, while the other variables are arbitrary. It turns out that in the case of quadratic truth equations it is convenient to work with *generalized families of solutions*, which may include one more kind of variables, namely variables $x_j = x_i$ (or $x_j = x_i'$) for all the elements of the generalized family. The algorithm is aimed at producing the (possibly empty!) list of generalized families of solutions and it consists of two stages.

In the first stage, for each variable x_i $(i = 1, \ldots, n)$ one determines the set of all its upper bounds with respect to I^+ (i.e., $\xi \in \Lambda$ with $x_i I^+ \xi$) and the set of all its lower bounds with respect to I^+ (i.e., $\xi \in \Lambda$ with $\xi I^+ x_i$), unless either one finds that x_i' is an upper bound, in which case one fixes x_i and all the already found lower bounds of x_i at 0, or one finds that x_i' is a lower bound, in which case one fixes x_i and all the already found upper bounds of x_i at 1 (cf. Proposition 3.3). If neither case occurs one checks whether there is a variable x_j, $j \neq i$, such that either x_j or x_j' is simultaneously an upper bound and a lower bound of x_i, in which case every solution satisfies $x_i = x_j$ or $x_i = x_j'$, respectively. Should $x_i = 0$ or $x_i = 1$ or $x_i = x_j$ or $x_i = x_j'$ occur, this substitution is carried out in the equation and this operation may result in new fixed variables, which the algorithm determines.

In view of Propositions 3.2 and 3.3, all the equalities of the above type are determined. Therefore if the equation is inconsistent (cf. Theorem 3.1) or if it has a unique generalized family of solutions (in particular if it has a unique family of solutions or in the more particular case when it has a unique solution), then this result is obtained in the first stage and the algorithm stops. Otherwise the equation which remains after all the substitutions detected in the first stage have been performed, is solved in the second stage by a bifurcation process (i.e., splitting according to the 0−1 values of certain variables; cf. Hammer and Rudeanu [1968]) which provides the set of all the solutions expressed as a list of generalized families of solutions. The paper by Rudeanu [1995a] also reports computational experience.

The paper by A.G. Levin [1978], mentioned in §1, reports computational experience as well.

In BFE,Chapter 15, §5, we have suggested the computer implementation of the general algebraic (formula-handling) methods for solving Boolean equations, presented in BFE in great detail (and resumed in the present book). The first step in this direction was taken by Diallo [1983], who wrote computer programs for the method of successive elimination of variables and for the Löwenheim formula of the reproductive solution, both of them in the quite particular case of quadratic truth equations. A third approach to solving this type of equations was the pseudo-Boolean technique of bifurcations. The Diallo thesis reports computational experience with the three algorithms, run on the same set of examples. This includes a discussion of the compared merits of the three methods.

A further step in the above direction was taken by Sofronie [1989], who wrote a program for obtaining the reproductive solution of an arbitrary Boolean equation $f(x_1, \ldots, x_n) = 0$ written in ring form, by the method of successive elimination of variables. Computational experience is reported for a set of problems which includes several examples previously given in the literature on Boolean equations.

The next steps in the algebraic direction are still expected.

Another approach to handling truth functions on computers is via graph theory. Thus e.g. Bryant [1986] develops a graph-based data structure for representing truth functions and an associated set of manipulation algorithms. Here are some of the conclusions of this paper: "We have shown that by taking a well-known graphical representation of Boolean functions[3] and imposing a restriction on the vertex labels, the minimum size graph representing a function becomes a canonical form. Furthermore, given any graph representing a function, we can reduce it to a canonical form graph in nearly linear time. Thus our reduction algorithm not only minimizes the amount of storage required to represent a function and the time required to perform symbolic operations on the function, it also makes such tasks as testing for equivalence, satisfiability or tautology very simple. We have found this property valuable in many applications. [...] By combining concepts from Boolean algebra with techniques from graph algorithms, we achieve a high degree of efficiency. That is, the performance is limited more by the sizes of the data structures rather than by the algorithms that operate on them.". The computational experience refers to "the problem of verifying that the implementation of a logic function (in terms of combinational logic gate network) satisfies its specification (in terms of Boolean expressions)".

For the status of the art in graph representation of truth functions, see Drechsler and Becker [1998]. This important field of research does not completely meet, however, the requirement of the problem raised in BFE and mentioned above. For there still remain restrictions which are not pertinent to the very nature of the problem, but are imposed by the graph-based structure of the data.

---

[3] Truth functions in our terminology.

10. Boolean differential calculus

Boolean differential calculus is a field which was initiated in the fifties under the impetus of applications to switching theory, such as fault diagnosis, hazard detection, decomposition of functions and analysis and synthesis of switching circuits. From a mathematical viewpoint, Boolean differential calculus establishes Boolean analogues of certain basic concepts and results of differential calculus of real functions.

A short introduction to this field can be found in BFE, Chapter 14, §2. A more detailed survey, together with a generalization to multiple-valued logic, is provided in Davio, Deschamps and Thayse [1978], Chapter 7, while a very detailed monograph of the field is Thayse [1981b]; see also Thayse [1981a] and Bochmann and Posthoff [1981]*. For the status of logic differential calculus in multiple-valued logic design see Yanushkevich [1998].

This chapter is an attempt to point out results taken from papers outside the references quoted in the works mentioned above. The presentation is self-contained.

1 An informal discussion

Several Boolean analogues of the concepts of partial derivative have been suggested in the literature. Most of them fall under the following general scheme: with every Boolean function f is associated the Boolean function Df defined by

(1.1) $$(Df)(X) = (\varphi f)(X) \bullet (\psi f)(X),$$

where the functions $\varphi, \psi : \mathrm{BF}n \longrightarrow \mathrm{BF}n$ and the Boolean function $\bullet : B^2 \longrightarrow B$ are fixed. This is a slight generalization of a remark made by Thayse [1981b].

The best known such concept is probably that of (*Boolean*) *partial derivative*, defined as follows. Suppose $f : B^n \longrightarrow B$ is a Boolean function. Then for each $i \in \{1, \ldots, n\}$, the partial derivative $\partial f / \partial x_i : B^{n-1} \longrightarrow B$ is defined as follows:

(1.2)
$$\frac{\partial f}{\partial x_i}(x_1, \ldots, x_{i-1}, x_{i+1}, \ldots, x_n)$$
$$= f(x_1, \ldots, x_n) + f(x_1, \ldots, x_{i-1}, x'_i, x_{i+1}, \ldots, x_n)$$

and the partial derivatives of higher order are defined according to the scheme

(1.3)
$$\frac{\partial^2 f}{\partial x_i \partial x_j} = \frac{\partial}{\partial x_j}\left(\frac{\partial f}{\partial x_i}\right).$$

The following properties are well known and easy to prove:

(1.4)
$$(\partial f/\partial x_i)(x_1,\ldots,x_{i-1},x_{i+1},\ldots,x_n) = f(x_1,\ldots,x_{i-1},0,x_{i+1},$$
$$\ldots,x_n) + f(x_1,\ldots,x_{i-1},1,x_{i+1},\ldots,x_n),$$

(1.5)
$$\frac{\partial f}{\partial x_i} = 0 \iff f \text{ does not depend on } x_i,$$

(1.6)
$$\frac{\partial(cf)}{\partial x_i} = c\frac{\partial f}{\partial x_i} \text{ if } c \text{ does not depend on } x_i,$$

(1.7)
$$\frac{\partial f'}{\partial x_i} = \frac{\partial f}{\partial x_i},$$

(1.8)
$$\frac{\partial(f+g)}{\partial x_i} = \frac{\partial f}{\partial x_i} + \frac{\partial g}{\partial x_i},$$

(1.9)
$$\frac{\partial(fg)}{\partial x_i} = f\frac{\partial g}{\partial x_i} + g\frac{\partial f}{\partial x_i} + \frac{\partial f}{\partial x_i}\frac{\partial g}{\partial x_i},$$

(1.10)
$$\frac{\partial^2 f}{\partial x_i \partial x_j} = \frac{\partial^2 f}{\partial x_j \partial x_i},$$

(1.11)
$$\frac{\partial^2 f}{\partial x_i \partial x_i} = 0.$$

Several conditions on Boolean functions can be expressed in terms of Boolean derivatives.

Thus, a function of the form

(1.12)
$$f(x_1,\ldots,x_n) = a + \sum_{i=1}^{n} a_i x_i$$

is called *linear*. For every $f \in \mathrm{BFn}$, $i \in \{1,\ldots,n\}$, and $\alpha \in \{0,1\}$, let $f(x_i = \alpha)$ denote the function of $n-1$ variables obtained from f by taking $x_i := \alpha$. Consider the following condition depending on f, i and α:

(1.13) $f(x_i = 1)$ and $f(x_i = 0)$ are linear and $\dfrac{\partial f}{\partial x_i}$ is a constant .

Then the following proposition states more accurately a result obtained by Tikhonenko [1974]:

Proposition 1.1. *The following conditions are equivalent for a Boolean function* $f : B^n \longrightarrow B$:

(i) f *is linear* ;

(ii) *condition* (1.13) *holds for some i and some α* ;

(iii) condition (1.13) holds for all i and all α.

PROOF: Since (i)$\Longrightarrow$(iii) is immediate and (iii)$\Longrightarrow$ (ii) is trivial, it remains to prove (ii)$\Longrightarrow$(i). Suppose (ii), say for $\alpha = 0$. Then

$$f(X) = x_i f(x_i = 1) + x_i' f(x_i = 0) = (x_i' + 1)f(x_i = 1) + x_i' f(x_i = 0)$$

$$= x_i' \frac{\partial f}{\partial x_i} + f(x_i = 1) = \frac{\partial f}{\partial x_i} + \frac{\partial f}{\partial x_i} \cdot x_i + f(x_i = 1) .$$

The proof for $\alpha = 1$ is quite similar.   □

Consider also the following definitions. We say that a Boolean function f is:

(i) $a-$*linear in* x_i, if $f = x_i + a$,

(ii) $(a, b)-$*meet decomposable in* x_i, if $f = (x_i + a)b$,

(iii) $(a, b)-$*join decomposable in* x_i, if $f = (x_i + a) \vee b$,

where a and b are Boolean functions that do not depend on x_i. To translate the above properties into the language of Boolean derivatives, we use the representation of an arbitrary Boolean function f in the form

(1.14) $$f = c_1 x_i + c_0 x_i' ,$$

where $c_\alpha = f(x_i = \alpha)$ ($\alpha \in \{0, 1\}$) do not depend on x_i. Note that (1.4) becomes

(1.15) $$\frac{\partial f}{\partial x_i} = c_1 + c_0 .$$

Proposition 1.2. (Posthoff [1978]; see also Bochmann and Posthoff [1979]). *For ever y $f \in$ BFn and every $i \in \{1, \ldots, n\}$,*

(i) *f is $a-$linear in x_i for some a $\Longleftrightarrow \partial f / \partial x_i = 1$;*

(ii) *f is $(a, b)-$meet decomposable in x_i for some b*
     $\Longleftrightarrow \partial f / \partial x_i = f(x_1, \ldots, x_{i-1}, a', x_{i+1}, \ldots, x_n)$;

(iii) *f is $(a, b)-$join decomposable in x_i for some b*
     $\Longleftrightarrow \partial f / \partial x_i = f'(x_1, \ldots, x_{i-1}, a, x_{i+1}, \ldots, x_n)$.

PROOF: If $\partial f / \partial x_i = 1$ then $c_1 + c_0 = 1$ by (1.15), hence $c_1 = c_0'$, therefore (1.14) becomes $f = x_i + c_0$.

If $\partial f / \partial x_i = c_1 a' + c_0 a$, then (1.14) implies $c_1 a + c_0 a' = 0$, hence

$$(x_i + a)(c_1 a' + c_0 a) = x_i(c_1 a' + c_0 a) + c_0 a$$

$$= x_i(c_1 a' + c_0 a + c_1 a + c_0 a') + c_0 a + a'(c_1 a + c_0 a')$$

$$= x_i(c_1 + c_0) + c_0 = c_1 x_i + c_0(x_i + 1) = f(X) .$$

If $\partial f / \partial x_i = (c_1 a + c_0 a')'$, then $c_1 + c_0 + 1 = c_1 a + c_0 a'$, hence $c_1 a' + c_0 a = 1$, therefore

$$(x_i + a) \vee (c_1 a + c_0 a') = x_i + a + c_1 a + c_0 a' + (x_i + a)(c_1 a + c_0 a')$$

$$= x_i(c_1 a + c_0 a' + 1) + a + c_0 a' = x_i(c_1 + c_0) + a(c_1 a' + c_0 a) + c_0 a'$$

$$= x_i(c_1 + c_0) + c_0 = f(X) .$$

The converse implications are obvious.   □

In Ch.7 we have referred to the problem of determining partially defined Boolean functions satisfying various properties. As was seen in the comment to Definition 7.1.1, a partially defined Boolean function can be identified with the interval

(1.16) $[p, q] = \{ f \in \text{BFn} \mid p \leq f \leq q \} ,$

where $p, q \in \text{BFn}$ and $p \leq q$. The parametric representation $f = p \vee qt$ of the interval (1.16) can be given the form

(1.17) $f = p + rt ,$ where $r = p'q ,$

because $p \vee qt = p \vee p'qt = p + p'qt.$

Proposition 1.3. (Bochmann [1977]; see also Bochmann and Posthoff [1979]). *For every $i \in \{1, \ldots, n\}$, the interval (1.16) contains a function independent of x_i if and only if*

(1.18) $\dfrac{\partial p}{\partial x_i} (\dfrac{\partial r}{\partial x_i})' r' = 0 .$

PROOF: In view of (1.5), (1.8) and (1.9), the function (1.17) is independent of x_i if and only if

$$0 = \frac{\partial f}{\partial x_i} = \frac{\partial p}{\partial x_i} + r\frac{\partial t}{\partial x_i} + t\frac{\partial r}{\partial x_i} + \frac{\partial r}{\partial x_i}\frac{\partial t}{\partial x_i} .$$

Setting $t = c_1 x_i + c_0 x_i'$, the above identity becomes

$$\frac{\partial p}{\partial x_i} + (r + \frac{\partial r}{\partial x_i})(c_1 + c_0) + \frac{\partial r}{\partial x_i}(c_1 x_i + c_0 x_i') = 0 ,$$

or equivalently,

$$\frac{\partial p}{\partial x_i} + (r + \frac{\partial r}{\partial x_i} + \frac{\partial r}{\partial x_i}x_i)c_1 + (r + \frac{\partial r}{\partial x_i} + \frac{\partial r}{\partial x_i}x_i')c_0 = 0 .$$

According to the Verification Theorem, the latter identity is equivalent to the system

$$\frac{\partial p}{\partial x_i} + (r + \frac{\partial r}{\partial x_i})c_1 + rc_0 = 0 ,$$

$$\frac{\partial p}{\partial x_i} + rc_1 + (r + \frac{\partial r}{\partial x_i})c_0 = 0 ,$$

which we can write as a single equality:

(1.19) $(\dfrac{\partial p}{\partial x_i} + (r + \dfrac{\partial r}{\partial x_i})c_1 + rc_0 + 1)(\dfrac{\partial p}{\partial x_i} + rc_1 + (r + \dfrac{\partial r}{\partial x_i})c_0 + 1) = 1 .$

Thus the function (1.17) does not depend on x_i if and only if condition (1.19) is fulfilled. Therefore such a function exists if and only if equation (1.19) has a solution with respect to (c_0, c_1). So it remains to work out the consistency condition

$$(\frac{\partial p}{\partial x_i} + 1) \vee (\frac{\partial p}{\partial x_i} + r + 1)(\frac{\partial p}{\partial x_i} + r + \frac{\partial r}{\partial x_i} + 1) \vee (\frac{\partial p}{\partial x_i} + \frac{\partial r}{\partial x_i} + 1) = 1$$

as follows:

$$0 = \frac{\partial p}{\partial x_i}((\frac{\partial p}{\partial x_i} + r) \vee (\frac{\partial p}{\partial x_i} + r + \frac{\partial r}{\partial x_i}))(\frac{\partial p}{\partial x_i} + \frac{\partial r}{\partial x_i})$$

$$= \frac{\partial p}{\partial x_i}(1 + \frac{\partial r}{\partial x_i})(\frac{\partial r}{\partial x_i} + (\frac{\partial p}{\partial x_i} + r)(\frac{\partial p}{\partial x_i} + r + \frac{\partial r}{\partial x_i}))$$

$$= \frac{\partial p}{\partial x_i}(\frac{\partial r}{\partial x_i})'(\frac{\partial r}{\partial x_i} + r) = \frac{\partial p}{\partial x_i}(\frac{\partial r}{\partial x_i})'(1 + r) .$$

$\square$

Bochmann and Posthoff [1979] have also solved similar problems for properties (ii) and (iii) in Proposition 1.2 and for the property of being symmetric in the variables x_i and x_j.

Proposition 1.4. (Fügert [1975]*; cf. Posthoff [1978]). *Suppose $f \in \{0,1\}F(n+m)$. If the equation $f(X,Y) = 0$ has a unique solution with respect to $Y \in (\{0,1\}Fn)^m$, then this solution is*

(1.20) $$y_i = \frac{\partial^{m-1}f}{\partial y_1 \ldots \partial y_{i-1}\partial y_{i+1} \ldots \partial y_m}(X, 1) \qquad (i = 1, \ldots, m) .$$

COMMENT: The function in the right side of (1.20) depends on the $n+1$ variables (X, y_i).

PROOF: Let $y_i = g_i(X)$ $(i = 1, \ldots, m)$ be the unique solution. Then

$$f(X,Y) = 0 \Longleftrightarrow \bigvee_{i=1}^{m}(y_i + g_i(X)) = 0 ,$$

hence $f(X,Y) = \bigvee_{i=1}^{m}(y_i + g_i(X))$. Therefore

$$\frac{\partial f}{\partial y_1} = (g_1 \vee \bigvee_{i=2}^{m}(y_i + g_i)) + (g_1' \vee \bigvee_{i=2}^{m}(y_i + g_i))$$

and since $x \vee y = x + y + xy$, this implies

$$\frac{\partial f}{\partial y_1} = 1 + (g_1 + g_1') \bigvee_{i=2}^{m}(y_i + g_i) = \prod_{i=2}^{m}(y_i + g_i') ,$$

$$\frac{\partial^2 f}{\partial y_1 \partial y_2} = g_2' \prod_{i=3}^{m}(y_i + g_i') + g_2 \prod_{i=3}^{m}(y_i + g_i) = \prod_{i=3}^{m}(y_i + g_i') \,,$$

whence by easy induction we obtain the system

$$\frac{\partial^{m-1} f}{\partial y_1 \ldots \partial y_{i-1} \partial y_{i+1} \ldots \partial y_m} = y_i + g_i \qquad (i = 1, \ldots, m) \,,$$

which is equivalent to (1.20).   □

Another family of Boolean differential operators which has been studied in the literature is constructed as follows. For every $\Xi \in B^n$ and every Boolean function $f : B^{n+m} \longrightarrow B$, the Boolean function $d^n f/d\Xi : B^{n+m} \longrightarrow B$ is defined by

(1.21) $$\frac{d^n f}{d\Xi}(X, Y) = f(X, Y) + f(\Xi, Y) \qquad (\forall X \in B^n) \, (\forall Y \in B^m) \,.$$

We only mention here the following result:

Proposition 1.5. (Lapscher [1968]). *Let* $f : B^{n+m} \longrightarrow B$ *and* $h : B^n \longrightarrow B$ *be Boolean functions. Then the following conditions are equivalent:*

(i) there is a Boolean function $g : B^{1+m} \longrightarrow B$ *such that*

(1.22) $$f(X, Y) = g(h(X), Y) \qquad (\forall X \in B^n) \, (\forall Y \in B^m) \,;$$

(ii) there is a Boolean function $g : B^{1+m} \longrightarrow B$ *such that the identity*

(1.23) $$\frac{d^n f}{d\Xi} = a \frac{d^n h}{d\Xi}$$

holds for some $a \in \mathrm{BF}m$ *and some* $\Xi \in B^n$ *;*

(iii) there is a Boolean function $g^{1+m} : B^{1+m} \longrightarrow B$ *such that identity (1.23) holds for some* $a \in \mathrm{BF}m$ *and every* $\Xi \in B^n$ *.*

PROOF: (i)$\Longrightarrow$(iii): Identity (1.22) can be written in the form

$$f(X, Y) = a(Y)h(X) + b(Y) \,,$$

where $a, b : B^m \longrightarrow B$ are Boolean functions. Therefore

$$\frac{d^n f}{d\Xi}(X, Y) = a(Y)h(X) + b(Y) + a(Y)h(\Xi) + b(Y) = a(Y)(h(X) + h(\Xi)) \,.$$

(iii)$\Longrightarrow$(ii): Trivial.
(ii)$\Longrightarrow$(i): From

$$f(X) + f(\Xi) = a(Y)(h(X) + h(\Xi))$$

we infer

$$f(X) = a(Y)h(X) + (a(Y)h(\Xi) + f(\Xi)) .$$

$\square$

Fadini [1961] suggested the operator $\min_i$, defined by

(1.24)
$$\min_i f(X) = f(x_1, \ldots, x_{i-1}, 0, x_{i+1}, \ldots, x_n) \cdot$$
$$\cdot f(x_1, \ldots, x_{i-1}, 1, \ldots, x_n) ,$$

as a Boolean analogue of the concept of partial derivative. Note that (1.24) is actually of the form (1.1). The idea of considering (1.24) as a partial derivative was rediscovered by Posthoff [1978] and Bochmann and Posthoff [1979], who regarded the consistency condition $\prod_A f(A) = 0$ of a Boolean equation $f(X) = 0$ as being expressed by a derivative of higher order. Besides, these authors remarked that an equation $f(X) = 0$ has a unique solution with respect to x_i iff $\partial f/\partial x_i = 1$; this follows from Proposition 1.1 (i), since the equation is uniquely solvable with respect to x_i iff it is x_i−linear.

Yanushkevich [1994] devised matrix methods for the computation of Boolean derivatives. Bochmann [1977] described Boolean differentials in terms of graphs. Parfenov and Ryabinin [1997] introduced a couple of new derivatives and suggested a probabilistic approach to Boolean differential calculus. Serfati [1995] defined a total derivative Df by

(1.25)
$$(Df)(X) = f(X) + \prod_A f(A)$$

and proved that D is idempotent, $D(f) = 0$ iff f is a constant, $D(fg) = fDg \vee gDf$ and

(1.26)
$$D\left(\bigvee_A \Xi^A f_A\right) = \bigvee_A \Xi^A D(f_A)$$

for every $\Xi \in B^n$ and $(f_A)_{A \in \{0,1\}^n} \subset BFn$ (see also BFE, Theorem 4.1 and the present Lemma 5.1.2).

Several applications to automata theory have led to the concept of *sequential derivative* for functions $x : \mathbf{N} \longrightarrow \{0,1\}$, namely $\dot{x}$ is defined by $\dot{x}(t) = x(t) + x(t+1)$, where t is a discrete argument which measures time; see e.g. Bochmann [1977] and Bochmann and Posthoff [1979].

The paper by Ghilezan [1982], written in 1975, introduces a concept of derivative for functions

(1.27)
$$f : L^n \longrightarrow R ,$$

where L is a finite set and R is a commutative ring with unit. For every $a \in L$ and every $i \in \{1, \ldots, n\}$, one defines $\partial f_a/\partial x_i$ by

(1.28)
$$\frac{\partial f_a}{\partial x_i}(x_1, \ldots, x_n) = f(x_1, \ldots, x_{i-1}, a, x_{i+1}, \ldots, x_n) - f(x_1, \ldots, x_n) .$$

Properties (1.6), (1.8) and (1.9) are extended to this case, while the generalization of (1.5) states that for every function f and every $i \in \{1, \ldots, n\}$, the following conditions are equivalent: (i) f does not depend on x_i; (ii) $\exists a\, \partial f_a/\partial x_i = 0$;

(iii)$\forall a \, \partial f_a / \partial x_i = 0$. The derivatives of higher order are defined in a natural way and many computational rules are established, including two Taylor-like formulas; see also Ghilezan [1995]. An attempt to a further generalization is sketched in Ghilezan [1981a]. The paper by Ghilezan [1980b] introduces an appropriate concept of *total differential*, similar to the total differential of a Boolean function, which has been studied in the literature. The Ghilezan differentials are defined by

$$(1.29) \qquad\qquad df_a = \sum_{i=1}^{n} \frac{\partial f_a}{\partial x_i} dx_i \, ,$$

where $dx_i = a - x_i$ $(i = 1, \ldots, n)$, and has properties similar to the properties of the partial derivatives $\partial f_a / \partial x_i$. The total differentials of higher order are defined in a natural way.

Lee [1976] endows the Boolean algebra $\{0, 1\}^n$ with certain supplementary operations and extends many properties of Boolean functions to this new framework. In particular a Boolean-like differential calculus is developed in view of its applications to logical design.

2 An axiomatic approach

The axiomatic approach to Boolean differential calculus described in this section is due to Kühnrich [1986a].

Definition 2.1. Suppose $(B; \cdot, \vee, ', 0, 1)$ is a Boolean algebra and let $d : B \longrightarrow B$; the values $d(x)$ for $x \in B$ will usually be denoted simply by dx. Then d is called a (*Boolean*) *differential operator* over B provided the following identities hold:

$$(2.1) \qquad\qquad ddx = 0 \, ,$$

$$(2.2) \qquad\qquad dx' = dx \, ,$$

$$(2.3) \qquad\qquad d(xy) = xdy + ydx + dx \cdot dy \, .$$

$\square$

Example 2.1. The constant function 0 is a differential operator. It is, in fact, the single Boolean differential operator which is also a Boolean function. $\square$

Example 2.2. Recall that BFn stands for the Boolean algebra of all Boolean functions $f : B^n \longrightarrow B$ over a Boolean algebra B. For every $\Xi = (\xi_1, \ldots, \xi_n) \in B^n$, the function $d_\Xi : \mathrm{BFn} \longrightarrow \mathrm{BFn}$ defined by

$$(2.4) \qquad (d_\Xi f)(X) = f(x_1 + \xi_1, \ldots, x_n + \xi_n) + f(X) \qquad (\forall X \in B^n)$$

for every $f \in \mathrm{BFn}$, is a differential operator over BFn. The proof begins with the

remark that BFn is closed under composition of functions, while checking (2.1)-
(2.3) is routine. The operator (2.4) is known as *derivation along the direction*
Ξ. In particular the Boolean derivative $\partial/\partial x_i$ (cf. (1.2)) is obtained by taking
$\xi_j := \delta_j^i \; (j = 1, \ldots, n)$. $\qquad\qquad\qquad\qquad\qquad\qquad\qquad\qquad\qquad\qquad\qquad$ $\square$

A slightly more general variant of this example is obtained by taking $\{i_1, \ldots,$
$i_m\} \subseteq \{1, \ldots, n\}$, $\Xi = (\xi_1, \ldots, \xi_m) \in B^m$ and

$$(d_\Xi f)(X) = f(X) +$$
(2.4′)
$$+ \; f(x_1, \ldots, x_{i_1-1}, \xi_i, x_{i_1+1}, \ldots, x_{i_m-1}, \xi_m, x_{i_m+1}, \ldots, x_n) \;.$$

Proposition 2.1. *Every Boolean differential operator satisfies the following
properties:*

(2.5) $$d0 = d1 = 0 \;,$$

(2.6) $$d(x + y) = dx + dy \;,$$

(2.7) $$d(x \vee y) = x' dy + y' dx + dx \cdot dy \;,$$

(2.8) $$d(x \vee y) + d(xy) = dx + dy \;.$$

PROOF: We compute in turn

$$d(x \vee y) = d((x'y')') = d(x'y') = x' dy' + y' dx' + dx' \cdot dy' = x' dy + y' dx + dx \cdot dy \;,$$

$$d(xy) + d(x \vee y) = (x + x') dy + (y + y') dx + dx \cdot dy + dx \cdot dy = dy + dx \;,$$

$$d(x + y) = d((x \vee y)(x' \vee y'))$$
$$= (x \vee y)d(x' \vee y') + (x' \vee y')d(x \vee y) + d(x \vee y)d(x' \vee y')$$
$$= (x + y + xy)(xdy + ydx + dx \cdot dy) + (xy + 1)(x' dy + y' dx + dx \cdot dy) +$$
$$+ (x' dy + y' dx + dx \cdot dy)(xdy + ydx + dx \cdot dy)$$
$$= (xy + y + xy + y')dx + (x + xy + xy + x')dy +$$
$$+ (x + y + xy + xy + 1 + xy' + x + x'y + y + x' + y' + 1)dx \cdot dy$$
$$= dx + dy + (x(y + 1) + (x + 1)y + x + 1 + y + 1)dx \cdot dy = dx + dy \;,$$
$$d1 = d0' = d0 = d(x + x) = dx + dx = 0 \;.$$

$\qquad\qquad\qquad\qquad\qquad\qquad\qquad\qquad\qquad\qquad\qquad\qquad\qquad\qquad\qquad\qquad$ $\square$

Definition 2.2. By a *Boolean differential algebra of order* k is meant an algebra
$(B; \cdot, \vee, ', 0, 1, (d_i)_{i \in K})$, where $(B; \cdot, \vee, ', 0, 1)$ is a Boolean algebra, K is a set of
cardinality k, and $d_i \; (i \in K)$ are Boolean differential operators over B such that
$d_i d_j = d_j d_i$ for every $i, j \in K$. $\qquad\qquad\qquad\qquad\qquad\qquad\qquad\qquad\qquad\qquad$ $\square$

Remark 2.1. If J is a subset of K of cardinality $|J|$, then $(B; \cdot, \vee, ', 0, 1, (d_j)_{j \in J})$
is a Boolean differential algebra of order $|J|$. $\qquad\qquad\qquad\qquad\qquad\qquad\qquad$ $\square$

Example 2.3. If d is a Boolean differential operator over a Boolean algebra B, then $(B; \cdot, \vee, ', 0, 1, d)$ is a Boolean differential algebra of order 1. In particular every Boolean algebra is a Boolean differential algebra of order 1 with respect to the constant operator 0 (cf. Example 2.1).   □

Example 2.4. If B is a Boolean algebra, then BFn becomes a Boolean differential algebra of order $\mid B \mid^n$ with respect to the Boolean differential operators d_Ξ in Example 2.2 For if $\Xi, H \in B^n$, then

$$(d_H d_\Xi f)(X) = f(\ldots, x_i + \xi_i + \eta_i, \ldots) + f(\ldots, x_i + \eta_i, \ldots) + f(\ldots, x_i + \xi_i, \ldots) + f(X),$$

which is symmetric in Ξ and H, hence $d_H d_\Xi = d_\Xi d_H$.   □

Example 2.5. It follows from Examples 2.2, 2.4 and Remark 2.1 that BFn is a Boolean differential algebra of order n with respect to the Boolean derivatives $\partial/\partial x_1, \ldots, \partial/\partial x_n$.   □

Definition 2.3. A *constant* of a Boolean differential algebra $(B; \cdot, \vee, ', 0, 1, (d_i)$ $_{i \in K})$ is an element $c \in B$ such that $d_i c = 0$ for every $i \in K$.   □

Remark 2.2. It follows from (2.5), (2.2) and (2.3) that the set of constants of a Boolean differential algebra is a subalgebra. This set includes in particular the elements 0 and 1.   □

Now we extend Examples 2.2, 2.4 and 2.5 to the axiomatic level.

Definition 2.4. In a Boolean differential algebra B, for every $S \subseteq K$ define $\partial_S :$ $B \longrightarrow B$ as follows: $\partial_\emptyset = 0$ (the constant function) and $\partial_{\{i_1, \ldots, i_m\}} = d_{i_1} d_{i_2} \ldots d_{i_m}$ for $\{i_1, \ldots, i_m\} \subseteq K$.   □

The above definition makes sense due to the commutativity $d_i d_j = d_j d_i$.

Remark 2.3. It is easy to see that for every $S, T \subseteq K$,

$$\partial_S \partial_T = \partial_T \partial_S = \partial_{S \cup T} \text{ if } S \neq \emptyset \neq T \text{ and } S \cap T = \emptyset,$$
(2.9)
$$\text{else } 0 .$$

□

The sensitivity and delta operators introduced in the next definition are natural axiomatic counterparts of well-known concepts in the conventional Boolean differential calculus.

Definition 2.5. In a Boolean differential algebra, one associates with every $S \subseteq K$ the *sensitivity* operator σ_S and the *delta* operator δ_S defined by

(2.10)
$$\sigma_S x = \sum_{R \subseteq S} \partial_R x \text{ and } \delta_S = \bigvee_{R \subseteq S} \partial_R x .$$

□

Proposition 2.2. *Suppose* $(B; \cdot, \vee, ', 0, 1, (d_i)_{i \in K})$ *is a Boolean differential algebra. Then for every* $i \in K$, $S \subseteq K$, $k \in K \backslash S$ *and* $x \in B$,

$$(2.11) \qquad \sigma_\emptyset x = \delta_\emptyset x = \partial_\emptyset x = 0 \, ,$$

$$(2.12) \qquad \sigma_{\{i\}} x = \delta_{\{i\}} x = \partial_{\{i\}} x = d_i x \, ,$$

$$(2.13) \qquad \sigma_{S \cup \{k\}} x = \sigma_S x + \sigma_S d_k x + d_k x \, ,$$

$$(2.14) \qquad \delta_{S \cup \{k\}} x = \delta_S x \vee \delta_S d_k x \vee d_k x \, .$$

PROOF: Immediate from Definitions 2.4, 2.5, the usual convention for the empty sum (join) and the decomposition of a sum (join) into three parts according to the scheme

$$(2.15) \qquad R \subseteq S \cup \{k\} \Longleftrightarrow R \subseteq S \text{ or } R = S \cup \{k\} \text{ or } R = \{k\} \, .$$

$\square$

Proposition 2.3. *Suppose* $(B; \cdot, \vee, ', 0, 1, (d_i)_{i \in K})$ *is a Boolean differential algebra. Then for every* $k \in K$ *and* $T \subseteq S \subseteq K$, *every* $x, y \in B$ *and every constant* c,

$$(2.16) \qquad \partial_S c = 0 \, , \ \sigma_S c = 0 \, , \ \delta_S c = 0 \, ,$$

$$(2.17) \qquad \partial_S x' = \partial_S x \, , \ \sigma_S x' = \sigma_S x \, , \ \delta_S x' = \delta_S x \, ,$$

$$(2.18) \qquad \partial_S(cx) = c \partial_S x \, , \ \sigma_S(cx) = c \sigma_S x \, , \ \delta_S(cx) = c \delta_S x \, ,$$

$$(2.19) \qquad \partial_S(c \vee x) = c' \partial_S x \, , \ \sigma_S(c \vee x) = c' \sigma_S x \, , \ \delta_S(c \vee x) = c' \delta_S x \, ,$$

$$(2.20) \qquad \partial_S(x + y) = \partial_S x + \partial_S y \, , \ \sigma_S(x + y) = \sigma_S x + \sigma_S y \, ,$$

$$(2.21) \qquad d_k \partial_S x = \partial_S d_k x \, , \ d_k \sigma_S x = \sigma_S d_k x \, , \ d_k \delta_S x \leq \delta_S d_k x \, ,$$

$$(2.22) \qquad k \in S \Longrightarrow d_k \sigma_S x = \sigma_{S \backslash \{k\}} d_k x = \sigma_S x + \sigma_{S \backslash \{k\}} x + d_k x \, ,$$

$$(2.23) \qquad k \notin S \Longrightarrow d_k \sigma_S x = \sigma_{S \cup \{k\}} d_k x = \sigma_S x + \sigma_{S \cup \{k\}} x + d_k x \, ,$$

$$(2.24) \qquad \partial_S \partial_S x = 0 \, , \ \sigma_S \sigma_S x = 0 \, ,$$

$$(2.25) \qquad\qquad \sigma_S(xy) = x\sigma_S y + y\sigma_S x + \sigma_S x \cdot \sigma_S y \,,$$

$$(2.26) \qquad\qquad \sigma_T \sigma_S x = \sigma_S \sigma_T x \,.$$

PROOF: Properties (2.16)-(2.20) and the two equalities (2.21) are immediate. We prove the inequality (2.21) by induction on S. For $S := \emptyset$ it holds because $\delta_\emptyset = 0$. Now suppose the inequality is true for S, take $j \in K \backslash S$ and use in turn (2.4), (2.13), (2.7) and the inductive hypothesis. Then

$$(d_k \delta_{S \cup \{j\}} x)(\delta_{S \cup \{j\}} d_k x)' = d_k(\delta_S x \vee \delta_S d_j x \vee d_j x)(\delta_{S \cup \{j\}} d_k x)'$$

$$= ((\delta_S x \vee \delta_S d_j x)' d_k d_j x + (d_j x)' d_k(\delta_S x \vee \delta_S d_j x) +$$

$$+ d_k(\delta_S x \vee \delta_S d_j x) \cdot d_k d_j x)(\delta_{S \cup \{j\}} d_k x)'$$

$$= ((\delta_S x)'(\delta_S d_j x)' d_k d_j x + ((d_j x)' + d_k d_j x) \cdot$$

$$\cdot ((\delta_S x)' d_k \delta_S d_j x + (\delta_S d_j x)' d_k \delta_S x +$$

$$+ (d_k \delta_S x)(d_k \delta_S d_j x)))(\delta_S d_k x)'(\delta_S d_j d_k x)'(d_j d_k x)' = 0 \,.$$

Further, note that (2.1) implies

$$(2.27) \qquad\qquad k \in S \Longrightarrow d_k \partial_S x = \partial_S d_k x = 0 \,.$$

To prove (2.22) suppose $k \in S$. But

$$\sum_{R \subseteq S} \partial_R x = \sum_{R \subseteq S \backslash \{k\}} \partial_R x + \sum_{R \subseteq S, R \not\subseteq S \backslash \{k\}} \partial_R x \,,$$

and taking also into account (2.12), we obtain

$$\sigma_S x + \sigma_{S \backslash \{k\}} x + d_k x = \sum_{R \subseteq S, R \not\subseteq S \backslash \{k\}} \partial_R x + d_k x$$

$$= \sum_{k \in R \subseteq S} \partial_R x + d_k x = \sum_{T \subseteq S \backslash \{k\}} \partial_{T \cup \{k\}} x + d_k x$$

$$= \sum_{\emptyset \neq T \subseteq S \backslash \{k\}} \partial_{T \cup \{k\}} x + \partial_{\{k\}} x + d_k x$$

$$= \sum_{\emptyset \neq T \subseteq S \backslash \{k\}} \partial_T d_k x = \sum_{T \subseteq S \backslash \{k\}} \partial_T d_k x = \sigma_{S \backslash \{k\}} d_k x$$

and on the other hand, using in turn (2.9) and (2.27) via (2.12), then (2.6), we get

$$\sum_{T \subseteq S \backslash \{k\}} \partial_T d_k x = \sum_{T \subseteq S \backslash \{k\}} d_k \partial_T x + d_k \partial_S x$$

$$= d_k \Big(\sum_{T \subseteq S \setminus \{k\}} \partial_T x + \partial_S x \Big) = d_k \sum_{T \subseteq S} \partial_T x = d_k \sigma_S x \ .$$

To prove (2.23) suppose $k \notin S$. Then we obtain similarly

$$\sigma_S x + \sigma_{S \cup \{k\}} x + d_k x = \sum_{R \subseteq S \cup \{k\}, R \not\subseteq S} \partial_R x + d_k x$$

$$= \sum_{k \in R \subseteq S \cup \{k\}} \partial_R x + d_k x = \sum_{T \subseteq S} \partial_{T \cup \{k\}} x + d_k x$$

$$= \sum_{\emptyset \neq T \subseteq S} \partial_{T \cup \{k\}} x = \sum_{\emptyset \neq T \subseteq S} d_k \partial_T x$$

$$= d_k \Big(\sum_{\emptyset \neq T \subseteq S} \partial_T x \Big) = d_k \sigma_S x = \sum_{R \subseteq S} d_k \partial_R x \cdot \sigma_S d_k x$$

$$= \sum_{R \subseteq S} \partial_R d_k x = \sum_{R \subseteq S} \partial_R d_k x + \sum_{k \in R \subseteq S \cup \{k\}} \partial_R d_k x$$

$$= \sum_{R \subseteq S \cup \{k\}} \partial_R d_k x = \sigma_{S \cup \{k\}} d_k x \ .$$

Property (2.9) immediately implies $\partial_S \partial_S x = 0$ and using also (2.20) we obtain

$$\sigma_S \sigma_S x = \sum_{R \subseteq S} \partial_R \Big(\sum_{T \subseteq S} \partial_T x \Big) = \sum_{R, T \subseteq S} \partial_R \partial_T x$$

$$= \sum_{R \subset T \subseteq S} (\partial_R \partial_T x + \partial_T \partial_S x) = 0 \ .$$

Property (2.25) holds for $S := \emptyset$. Now suppose $S \subset K$ satisfies (2.25) and take $k \in K \setminus S$. Then properties (2.13), (2.3) and (2.20) imply

(2.28) $$x \cdot \sigma_{S \cup \{k\}} y = x \sigma_S y + x \sigma_S d_k y + x d_k y \ ,$$

(2.29) $$y \cdot \sigma_{S \cup \{k\}} x = y \sigma_S x + y \sigma_S d_k x + y d_k x \ ,$$

(2.30) $$\sigma_{S \cup \{k\}} x \cdot \sigma_{S \cup \{k\}} y = (\sigma_S x + \sigma_S d_k x + d_k x)(\sigma_S y + \sigma_S d_k y + d_k y)$$

$$= \sigma_S x \cdot \sigma_s y + \sigma_S x \cdot \sigma_S d_k y + \sigma_S x \cdot d_k y +$$

$$+ \sigma_S y \cdot \sigma_S d_k x + \sigma_S d_k x \cdot \sigma_S d_k y + \sigma_S d_k x \cdot d_k y +$$

$$+ d_k x \cdot \sigma_S y + d_k x \cdot \sigma_S d_k y + d_k x \cdot d_k y \ ,$$

$$\sigma_S d_k (xy) = \sigma_S (x d_k y + y d_k x + d_k x \cdot d_k y)$$

$$= \sigma_S (x d_k y) + \sigma_S (y d_k x) + \sigma_S (d_k x \cdot d_k y) \ ,$$

hence

(2.31)

$$\sigma_S d_k(xy)$$
$$= \sigma_S d_k y + d_k y \cdot \sigma_S x + \sigma_S x \cdot \sigma_S d_k y +$$
$$+ y\sigma_S d_k x + d_k x \cdot \sigma_S y + \sigma_S y \cdot \sigma_S d_k x +$$
$$+ d_k x \cdot \sigma_S d_k y + d_k y \cdot \sigma_S d_k x + \sigma_S d_k x \cdot \sigma_S d_k y \,,$$
$$\sigma_{S\cup\{k\}}(xy) = \sigma_S(xy) + \sigma_S d_k(xy) + d_k(xy) \,,$$

hence

(2.32)

$$\sigma_{S\cup\{k\}}(xy) = x\sigma_S y + y\sigma_S x + \sigma_S x \cdot \sigma_S y + \sigma_S d_k(xy) +$$
$$+ xd_k y + yd_k x + d_k x \cdot d_k y \,,$$

hence it follows, by comparing systems (2.32)&(2.31) and (2.28)&(2.20)&(2.30), that

$$\sigma_{S\cup\{k\}}(xy) = x\sigma_{S\cup\{k\}}y + y\sigma_{S\cup\{k\}}x + \sigma_{S\cup\{k\}}x \cdot \sigma_{S\cup\{k\}}y \,.$$

Property (2.26) holds for $S := \emptyset$. Now suppose $S \subset K$ satisfies (2.26), take $k \in K\backslash S$, $T \subseteq S\cup\{k\}$, and prove $\sigma_T \sigma_{S\cup\{k\}} = \sigma_{S\cup\{k\}}\sigma_T$. But properties (2.13) and (2.20) imply

(2.33)

$$\sigma_T \sigma_{S\cup\{k\}}x = \sigma_T \sigma_S x + \sigma_T \sigma_S d_k x + \sigma_T d_k x \,.$$

If $k \notin T$ then $T \subseteq S$, therefore (2.33), the inductive hypothesis, (2.21) and (2.13) imply

$$\sigma_T \sigma_{S\cup\{k\}}x = \sigma_S \sigma_T x + \sigma_S \sigma_T d_k x + d_k \sigma_T x$$
$$= \sigma_S \sigma_T x + \sigma_S d_k \sigma_T x + d_k \sigma_T x = \sigma_{S\cup\{k\}}\sigma_T x \,.$$

If $k \in T$ then T is of the form $T = V \cup \{k\}$ for some $V \subseteq S$. Now (2.13), (2.20), (2.6), (2.1), Remark 2.2 and (2.16) imply

$$\sigma_T \sigma_{S\cup\{k\}}x = \sigma_{V\cup\{k\}}(\sigma_S x + \sigma_S d_k x + d_k x)$$
$$= \sigma_V(\sigma_S x + \sigma_S d_k x + d_k x) + \sigma_V d_k(\sigma_S x + \sigma_S d_k x + d_k x) + d_k(\sigma_S x + \sigma_S d_k x + d_k x)$$
$$= \sigma_V \sigma_S x + \sigma_V \sigma_S d_k x + \sigma_V d_k x + \sigma_V d_k \sigma_S x + \sigma_V d_k \sigma_S d_k x + d_k \sigma_S x + d_k \sigma_S d_k x$$
$$= \sigma_V \sigma_S x + \sigma_V d_k x + d_k \sigma_S x$$

because $\sigma_V \sigma_S d_k x = \sigma_V d_k \sigma_S x$ by (2.21) and $\sigma_V d_k \sigma_S d_k x = 0 = d_k \sigma_S d_k x$ by (2.21) and (2.1). It follows by symmetry, the inductive hypothesis and (2.21) that

$$\sigma_{S\cup\{k\}}\sigma_T x = \sigma_{S\cup\{k\}}\sigma_{V\cup\{k\}}x = \sigma_S \sigma_V x + \sigma_S d_k x + d_k \sigma_V x$$
$$= \sigma_V \sigma_S x + d_k \sigma_S x + \sigma_V d_k x = \sigma_T \sigma_{S\cup\{k\}}x \,.$$

$\square$

Proposition 2.4. *The following identities hold in every Boolean differential algebra* $(B; \cdot, \vee, ', 0, 1, (d_i)_{i\in K})$ *and every* $S \subseteq K$:

(2.34)
$$\partial_S x = \sum_{R\subseteq S} \sigma_R x \quad \text{and} \quad \delta_S x = \bigvee_{R\subseteq S} \sigma_R x \,.$$

COMMENT: Compare to identities (2.10) in Definition 2.5.

PROOF: The first identity (2.34) holds for $S := \emptyset$ by Definitions 2.4 and 2.5. Now let S be a set which satisfies identity (2.34) and take again $k \in K \setminus S$. Then the same technique as in the previous proof yields

$$\partial_{S \cup \{k\}} x = d_k \partial_S x = d_k \sum_{R \subseteq S} \sigma_R x = \sum_{R \subseteq S} d_k \sigma_R x$$

$$= \sum_{R \subseteq S} \sigma_R d_k x = \sum_{R \subseteq S} (\sigma_{R \cup \{k\}} x + \sigma_R x + d_k x)$$

$$= \sum_{R \subseteq S} \sigma_{R \cup \{k\}} x + \sum_{R \subseteq S} \sigma_R x + \sum_{R \subseteq S} d_k x$$

$$= \sum_{k \in T \subseteq S \cup \{k\}} \sigma_T x + \sum_{k \notin R \subseteq S \cup \{k\}} \sigma_R x = \sum_{R \subseteq S \cup \{k\}} \sigma_R x \, ,$$

since $\sum_{R \subseteq S} d_k x = 0$ because it contains $2^{|S|}$ terms.

We use the same technique for the second identity (2.34). The inductive step uses (2.23) and the fact that

$$a \vee b \vee (a + b + c) = a \vee b \vee a(b + c)' \vee a'(b + c)$$

$$= a \vee b \vee (b + c) = a \vee b \vee bc' \vee b'c = a \vee b \vee c \, ,$$

hence

$$\delta_{S \cup \{k\}} x = \delta_S x \vee \delta_S d_k x \vee d_k x = \bigvee_{R \subseteq S} \sigma_R x \vee \bigvee_{R \subseteq S} \sigma_R d_k x \vee d_k x$$

$$= d_k x \vee \bigvee_{R \subseteq S} \sigma_R x \vee \bigvee_{R \subseteq S} (\sigma_R x + \sigma_{R \cup \{k\}} x + d_k x)$$

$$= \bigvee_{R \subseteq S} (d_k x \vee \sigma_R x \vee (d_k x + \sigma_R x + \sigma_{R \cup \{k\}} x))$$

$$= \bigvee_{R \subseteq S} (d_k x \vee \sigma_R x \vee \sigma_{R \cup \{k\}} x) = d_k x \vee \bigvee_{R \subseteq S} \sigma_R x \vee \bigvee_{R \subseteq S} \sigma_{R \cup \{k\}} x$$

$$= \sigma_{\{k\}} x \vee \bigvee_{k \notin R \subseteq S \cup \{k\}} \sigma_R x \vee \bigvee_{k \in T \subseteq S \cup \{k\}} \sigma_T x$$

$$= \bigvee_{R \subseteq S \cup \{k\}} \sigma_R x \, .$$

$\square$

Proposition 2.5. *Suppose d is a Boolean differential operator over a Boolean algebra $(B; \cdot, \vee, ', 0, 1)$ and define*

$$(2.35) \qquad \bar{d} x = x \vee dx \qquad (\forall x \in B) \, .$$

Then $(B; \cdot, \vee, ', 0, 1, \bar{d})$ is a closure algebra.

COMMENT: In other words, this means that $\overline{d}$ is a *Kuratowski closure operator*, that is, for every $x, y \in B$,

$$x \leq \overline{d}x , \tag{2.36}$$

$$\overline{d}\,\overline{d}x = \overline{d}x , \tag{2.37}$$

$$\overline{d}(x \vee y) = \overline{d}x \vee \overline{d}y , \tag{2.38}$$

$$\overline{d}0 = 0 . \tag{2.39}$$

PROOF: Property (2.36) is immediate and so is (2.39) in view of (2.5). Then

$$\overline{d}\,\overline{d}x = x \vee dx \vee d(x \vee dx)$$

$$= x \vee dx \vee x'dx \vee (dx)'dx \vee dx \cdot ddx = x \vee dx = \overline{d}x ,$$

$$\overline{d}(x \vee y) = x \vee y \vee d(x \vee y) = x \vee y \vee (x'dy + y'dx + dx \cdot dy)$$

$$= x \vee y \vee (x'dy + (y'dx((dx)' \vee (dy)') \vee (y \vee (dx)')dx \cdot dy))$$

$$= x \vee y \vee (x'dy + (y'dx(dy)' \vee ydx \cdot dy))$$

$$= x \vee y \vee x'dy(y \vee (dx)' \vee dy)(y' \vee (dx)' \vee (dy)') \vee (x \vee (dy)')(y'dx(dy)' \vee ydx \cdot dy)$$

$$= x \vee y \vee x'dy(y' \vee (dx)') \vee (dy)'y'dx = x \vee y \vee y'dy \vee dy(dx)' \vee (dy)'dx$$

$$= x \vee y \vee dy \vee dx = \overline{d}x \vee \overline{d}y .$$

$\square$

The paper by Kühnrich [1986a] studies this closure algebra in some detail, while Kühnrich [1986b] establishes deep connections with *cylindric algebras* (cf. Henkin, Monk and Tarski [1971]) and provides two representation theorems for Boolean differential algebras.

3 Boolean differential equations

In the previous two sections we have defined several kinds of Boolean derivatives and differentials and we have studied their properties. The next natural step is to study functional equations expressed in terms of these derivatives. As we are going to show in this section, the beginning of such a study does exist in the literature.

All the functions dealt with in this section are in BFn. Unless otherwise stated, the Boolean partial derivative $\partial f/\partial x_i$ has the meaning (1.2), while the partial derivatives of higher order are defined in a natural way. We refer to all of the equations under consideration as Boolean differential equations.

Proposition 3.1. α) *The Boolean differential equation*

(3.1)
$$\frac{\partial^m f}{\partial x_1 \ldots \partial x_m} = g \,,$$

where $m \leq n$, *has solutions if and only if* g *does not depend on the variables* $x_1, \ldots, x_m$.

β) *When this is the case, the solutions are exactly the functions of the form*

(3.2)
$$f = \sum_{S \subset M} h_S \prod_{i \in S} x_i + g x_1 \ldots x_m \,,$$

where $M = \{1, \ldots, m\}$ *and the functions* h_S ($S \subset M$) *do not depend on the variables* $x_1, \ldots, x_m$.

PROOF: Since the order of the variables is immaterial in the Boolean derivatives of higher order, it follows from (3.1) that for each $i \in \{1, \ldots, m,\}$, g is of the form $g = \partial g_i / \partial x_i$, hence it does not depend on x_i.

Now suppose g is independent of $x_1, \ldots, x_m$. Then it follows from (1.6) and (1.8) that (3.2) implies (3.1). Besides, the well-known expansion

$$f = \sum_{S \subseteq M} h_S \prod_{i \in S} x_i$$

implies

$$\frac{\partial^m f}{\partial x_1 \ldots \partial x_m} = h_M \,,$$

therefore (3.1) implies $h_M = g$, that is, (3.2) holds.   □

Corollary 3.1. (Bochmann [1977], Liang [1983]*, [1990]). *For each* $i \in \{1, \ldots, m\}$, *the Boolean differential equation*

(3.3)
$$\frac{\partial f}{\partial x_i} = g$$

is consistent if and only if g *does not depend on* x_i, *in which case the solutions are exactly the functions of the form*

(3.4)
$$f = g x_i + h \,,$$

where h *does not depend on* x_i.   □

Liang [1983]*, [1990] also counted the number of solutions of equation (3.3) in the case when the Boolean algebra reduces to $\{0, 1\}$.

Proposition 3.2. α) *The system of Boolean differential equations*

(3.5)
$$\frac{\partial f}{\partial x_i} = g_i \qquad (i = 1, \ldots, m) \,,$$

where $m \leq n$, *has solutions if and only if*

(3.6)
$$g_i \text{ does not depend on } x_i \qquad (i = 1, \ldots, m)$$

and

(3.7)
$$\frac{\partial g_i}{\partial x_j} = \frac{\partial g_j}{\partial x_i} \qquad (i, j = 1, \ldots, m \,;\, i \neq j) \,.$$

β) *When this is the case, the solutions are exactly the functions of the form*

(3.8)
$$f = \sum_{i=1}^{m} g_i x_i + \sum_{k=2}^{m} \sum_{i_1,\dots,i_k \in M} \frac{\partial^{k-1} g_{i_1}}{\partial x_{i_2} \dots \partial x_{i_k}} x_{i_1} \dots x_{i_k} + h \,,$$

where $M = \{1, \dots, m\}$, the indices $i_1, \dots, i_k$ in each term of (3.8) are pairwise distinct, and h is a function which does not depend on $x_1, \dots, x_m$.

COMMENT: System (3.5) was solved by Levchenkov [1999a] in the case $m :=
n := 2$.

PROOF: Relations (3.5) immediately imply (3.6) and also (3.7), in view of the commutativity $\partial^2 f/\partial x_i \partial x_j = \partial^2 f/\partial x_j \partial x_i$.

Conversely, suppose relations (3.6) and (3.7) hold. Then it is easily proved by induction on k that for every k–element subset $S = \{i_1, \dots, i_k\}$ of M, the partial derivative of order $k - 1$

$$\frac{\partial^{k-1} g_{i_1}}{\partial x_{i_2} \dots \partial x_{i_k}}$$

does not depend on the chosen permutation $(i_1, \dots, i_k)$ of S. Therefore the expression (3.8) is unambiguous. We are going to prove that f satisfies (3.5) if and only if it is of the form (3.8).

Corollary 3.1 shows that this is true for $m := 1$. The inductive step runs as follows. Split the condition that f is a solution of system (3.5) into two parts: f satisfies the first $m - 1$ equations (3.5), and $\partial f/\partial x_m = g_m$. In view of the inductive hypothesis, the former condition is equivalent to f being of the form

(3.9)
$$f = \sum_{i=1}^{m-1} g_i x_i + \sum_{k=2}^{m-1} \sum_{i_1,\dots,i_k \in M_1} \frac{\partial^{k-1} g_{i_1}}{\partial x_{i_2} \dots \partial x_{i_k}} x_{i_1} \dots x_{i_k} + h_1 \,,$$

where $M_1 = M \backslash \{m\}$ and h_1 does not depend on $x_1, \dots, x_{m-1}$. Now we impose the condition $\partial f/\partial x_m = g_m$ on the function (3.9):

$$\sum_{i=1}^{m-1} \frac{\partial g_i}{\partial x_m} x_i + \sum_{k=2}^{m-1} \sum_{i_1,\dots,i_k \in M_1} \frac{\partial^k g_{i_1}}{\partial x_{i_2} \dots \partial x_{i_k} \partial x_m} x_{i_1} \dots x_{i_k} + \frac{\partial h_1}{\partial x_m} = g_m \,,$$

whence Corollary 3.1 yields

(3.10)
$$h_1 = (g_m + \sum_{i=1}^{m-1}(\partial g_i/\partial x_m) x_i +$$
$$+ \sum_{k=2}^{m-1} \sum_{i_1,\dots,i_k \in M_1}(\partial^k g_{i_1}/\partial x_{i_2} \dots \partial x_{i_k} \partial x_m) x_{i_1} \dots x_{i_k}) x_m + h,$$

where h does not depend on x_m. Thus equation (3.5) is equivalent to system (3.9)-(3.10) and the latter system is equivalent to (3.8), because h is obtained from h_1 by taking $x_m := 0$, therefore h does not depend on $x_1, \dots, x_{m-1}, x_m$. □

Example 3.1. For $n := 3$ equation (3.8) becomes

$$f = g_1 x_1 + g_2 x_2 + g_3 x_3 + \frac{\partial g_1}{\partial x_2} x_1 x_2 + \frac{\partial g_2}{\partial x_3} x_2 x_3 + \frac{\partial g_3}{\partial x_1} x_3 x_1 +$$

$$+ \frac{\partial^2 g_1}{\partial x_2 \partial x_3} x_1 x_2 x_3$$

and variants of it can be obtained by using the relations

$$\frac{\partial g_1}{\partial x_2} = \frac{\partial g_2}{\partial x_1}, \quad \frac{\partial g_2}{\partial x_3} = \frac{\partial g_3}{\partial x_2}, \quad \frac{\partial g_3}{\partial x_1} = \frac{\partial g_1}{\partial x_3},$$

$$\frac{\partial^2 g_1}{\partial x_2 \partial x_3} = \frac{\partial^2 g_2}{\partial x_1 \partial x_3} = \frac{\partial^2 g_3}{\partial x_1 \partial x_2}.$$

□

Ghilezan [1976] solved equations of the form

$$\frac{\partial^m f_{a_1 \ldots a_m}}{\partial x_1 \ldots \partial x_m} = g$$

and systems of the form

$$\frac{\partial f_{a_i}}{\partial x_i} = g_i \qquad (i = 1, \ldots, m)$$

(cf. his derivatives (1.21)). The necessary and sufficient conditions for the consistency of the above system are quite similar to (3.6)-(3.7), while the form of the general solution is simpler than (3.8). As a matter of fact, the above proof of Proposition 3.2 is inspired by Ghilezan's proof. The paper by Ghilezan [1979a] solves equations of the form

$$g_1 \frac{\partial f_a}{\partial x} + g_2 \frac{\partial f_b}{\partial y} + g_3 \frac{\partial^2 f_{ab}}{\partial x \partial y} = g,$$

while the general linear equations with partial Ghilezan derivatives of higher order and constant coefficients is solved in Ghilezan [1981b]. The paper by Ghilezan and Udicki [1989] solves linear equations with partial Ghilezan derivatives up to the third order and arbitrary coefficients.

A result similar to Proposition 3.1 holds for the differential operator d_Ξ in Example 2.2.

Proposition 3.3. (Lapscher [1968]). *The differential equation*

(3.11) $$d_\Xi f = g$$

is consistent if and only if

(3.12) $$g(\Xi) = 0,$$

in which case the set of solutions has the representation

(3.13) $$f(X) = g(X) + h,$$

where the parameter h is a constant function.

COMMENT: The result is also valid, mutatis mutandis, in the case $\varXi \in B^m$, $m \leq n$ (cf. Example 2.2).

PROOF: It is plain that (3.11) implies (3.12). Conversely, suppose (3.12) holds. Then clearly (3.13) implies (3.11), while the latter equality can be written in the form $f(X) = g(X) + f(\varXi)$.   □

Liang [1983]*, [1990] studies differential equations with respect to the differential operator

$$(3.14) \qquad \frac{d^2 f}{dx_i x_j}(X) = f(X) + f(x_i = x_i + 1, x_j = x_j + 1) \,,$$

where the notation $f(x_i = a, x_j = b)$ means the element obtained from $f(X)$ when x_i is replaced by a and x_j is replaced by b.

Lemma 3.1 and Propositions 3.4 and 3.5 below are due to Liang [1983a]*, [1990].

Lemma 3.1. *The following identity holds:*

$$(3.15) \qquad \begin{aligned} \frac{d^2 f}{dx_i x_j} &= (f(x_i = 1, x_j = 0) + f(x_i = 0, x_j = 1))(x_i + x_j) + \\ &+ (f(x_i = 1, x_j = 1) + f(x_i = 0, x_j = 0))(x_i + x_j)' \,. \end{aligned}$$

PROOF: Set

$$(3.16) \qquad\qquad f_{ab} = f(x_i = a, x_j = b)$$

and recall that $x \vee y = x + y$ iff $xy = 0$. Then

$$\begin{aligned} \frac{d^2 f}{dx_i x_j} &= f_{11} x_i x_j + f_{10} x_i x_j' + f_{01} x_i' x_j + f_{00} x_i' x_j' + \\ &\quad + f_{11} x_i' x_j' + f_{10} x_i' x_j + f_{01} x_i x_j' + f_{00} x_i x_j \\ &= (f_{11} + f_{00})(x_i x_j + x_i' x_j') + (f_{10} + f_{01})(x_i x_j' + x_i' x_j) \,. \\ &= (f_{11} + f_{00})(x_i x_j \vee x_i' x_j') + (f_{10} + f_{01})(x_i x_j' \vee x_i' x_j) \,. \end{aligned}$$

□

Proposition 3.4. *The equation*

$$(3.17) \qquad\qquad \frac{d^2 f}{dx_i x_j} = 0$$

is consistent and its solutions are exactly the functions of the form

$$(3.18) \qquad\qquad f = (x_i + x_j)h_1 + (x_i + x_j)'h_0 \,,$$

where h_0 and h_1 are functions that do not depend on x_i and x_j.

PROOF: Using Lemma 3.1 and the notation (3.16), we see that (3.15) is equivalent to the identity

$$(f_{10} + f_{01})(x_i + x_j) + (f_{11} + f_{00})(x_i + x_j)' = 0 \,,$$

which holds if and only if

$$f_{10} + f_{01} = f_{11} + f_{00} = 0 \, .$$

The latter condition amounts to saying that f is of the form

$$f(X) = ax_i x_j + bx_i x_j' + bx_i' x_j + ax_i' x_j'$$

$$= a(x_i + x_j)' + b(x_i + x_j) \, ,$$

where the functions a and b do not depend on x_i and x_j. $\qquad \square$

Proposition 3.5. α) *The equation*

$$(3.19) \qquad\qquad \frac{d^2 f}{dx_i dx_j} = g$$

is consistent if and only if g is of the form

$$(3.20) \qquad\qquad g = (x_i + x_j)a + (x_i + x_j)'b \, ,$$

where a and b are functions that do not depend on x_i and x_j.

β) *When this is the case, the solutions are exactly the functions of the form*

$$(3.21) \qquad\qquad f(X) = (x_i + x_j)h_1 + (x_i + x_j)'h_0 + ax_i' x_j + bx_i' x_j' \, ,$$

where h_0 and h_1 are functions that do not depend on x_i and x_j.

PROOF: It is easily shown that $d^4 f/dx_i x_j x_i x_j = 0$. It follows that $d^2 g/dx_i x_j = 0$ is a necessary consistency condition, therefore g must be of the form (3.20), by Proposition 3.4.

Now suppose identity (3.20) holds. Then (3.21) implies

$$\frac{d^2 f}{dx_i x_j} = ax_i' x_j + bx_i' x_j' + ax_i x_j' + bx_i x_j = g \, .$$

Conversely, (3.19) implies

$$(f_{10} + f_{01})(x_i + x_j) + (f_{11} + f_{00})(x_i + x_j)' = (x_i + x_j)a + (x_i + x_j)'b$$

by Lemma 3.1. The last identity is equivalent to the conditions

$$f_{10} + f_{01} = a \text{ and } f_{11} + f_{00} = b \, ,$$

which amount to saying that f is of the form

$$f = f_{11}x_i x_j + f_{10}x_i x_j' + (f_{10} + a)x_i' x_j + (f_{11} + b)x_i' x_j'$$

$$= f_{11}(x_i + x_j)' + f_{10}(x_i + x_j) + ax_i' x_j + bx_i' x_j' \, .$$

$\qquad\qquad\qquad\qquad\qquad\qquad\qquad\qquad\qquad\qquad\qquad \square$

Independently of each other, Ghilezan and Liang have suggested that a Boolean analogue of integral calculus could be developed. Consider the simplest case pointed out in Corollary 3.1. If the general solution (3.4) of equation

$$(3.3) \qquad\qquad \frac{\partial f}{\partial x_i} = g$$

is written in the form

$$(3.22) \qquad\qquad \int g dx_i = g x_i + h \,,$$

then we may regard the family of functions (3.4) as the "primitive" of the function g, depending on an "additive constant" h. Besides, as emphasized by Liang in some detail, the "parameter" h can be determined by a set of "initial conditions", while Ghilezan pointed out the identities

$$(3.23) \qquad\qquad \int (g_1 + g_2) dx_i = \int g_1 dx_i + \int g_2 dx_i \,,$$

$$(3.24) \qquad\qquad \int c g dx_i = c \int g dx_i \,.$$

Serfati [1995] solved the differential equations $Df = \theta f$, $Df = g$ and $af \vee bDf = c$, where Df is the derivative (1.25) introduced by him.

See also A. G. Levin [1978].

11. Decomposition of Boolean functions

The synthesis problem in switching theory, that is, the problem of designing a circuit from given building blocks, has led since the 1950s to an extensive study of the decomposition of truth functions (usually referred to as Boolean functions). Roughly speaking, this problem consists in expressing an arbitrary truth function by means of other (convenient) truth functions. In BFE, Chapter 16, §1, we have reported a few methods and results in this line of research. We strongly suggest that this field deserves a separate monograph.

A natural generalization of this problem consists in replacing truth functions by arbitrary Boolean functions. We have already dealt with the decomposition of Boolean functions in Theorem 7.3.4, Corollaries 7.3.9, 7.3.10, and Propositions 10.1.2 and 10.1.5, and we continue in this chapter. We have selected results that do not require properties specific to truth functions (or equivalently, to simple Boolean functions). This choice is somewhat complementary to the potential monograph suggested above.

The aim of §1 is to point out that the decomposition of Boolean functions has a history "avant la lettre" starting as early as 1901. The second section illustrates how Boolean equations can be used in order to decompose Boolean functions.

1 A historical sketch

In this section we present five results on the decomposition of Boolean functions, obtained by five authors independently of each other: Whitehead in 1901, McKinsey in 1936, Povarov in 1954, Ashenhurst in 1957 and Fadini in 1972.

Whitehead's theorem, already reported in BFE, establishes a factorization of evanescible Boolean functions into "linear primes", analogous to the factorization of natural numbers into prime factors. Ashenhurst's paper is the starting point of the modern theory of decomposition of truth functions, referred to in the above introduction. The papers by Povarov, Ashenhurst and Fadini are somewhat in the same spirit.

Definition 1.1. A Boolean function $f : B^n \longrightarrow B$ is said to be: 1) *evanescible*, if the equation $f(X) = 0$ is consistent, and 2) a *linear prime* if it is of the form

p_Ξ for some $\Xi = (\xi_1, \ldots, \xi_n) \in B^n$, where

$$(1.1) \qquad p_\Xi(X) = \bigvee_{i=1}^{n} (x_i + \xi_i) \qquad (\forall X \in B^n) .$$

Theorem 1.1. (Whitehead [1901]; see also BFE, Lemma 13.1 and Theorem 13.3). *A Boolean function is evanescible if and only if it can be decomposed as a product of at most 2^n linear primes.* □

Proposition 1.1. (McKinsey [1936a]). *A Boolean function $f : B^{n+m} \longrightarrow B$ has a decomposition of the form*

$$(1.2) \qquad f(X,Y) = g(X) \bullet h(Y) ,$$

where $g : B^n \longrightarrow B$, $h : B^m \longrightarrow B$ are Boolean functions and $\bullet \in \{\cdot, \vee, +, \tilde{+}\}$,[1] if and only if for every $X, U \in B^n$ and every $Y, V \in B^m$,

$$(1.3) \qquad f(X,Y) \bullet f(U,V) = f(X,V) \bullet f(U,Y) .$$

PROOF: Necessity is immediate, because the associativity and commutativity of $\bullet$ imply that both sides of (1.3) equal $g(X) \bullet h(Y) \bullet g(U) \bullet h(V)$.

To prove sufficiency when $\bullet$ is $\cdot$, take $g(X) = \bigvee_C f(X, C)$ and $h(Y) = \bigvee_A f(A, Y)$, where C and A run over B^m and B^n, respectively. Then

$$g(X)h(Y) = \bigvee_{A,C} f(X,C)f(A,Y) = \bigvee_{A,C} f(X,Y)f(A,C)$$

$$= f(X,Y) \bigvee_{A,C} f(A,C) = f(X,Y)$$

because $\bigvee_{A,C} f(A,C) = \max f(X,Y)$.

To prove sufficiency when $\bullet$ is $+$, take $g(X) = f(X, O_2) + f(O_1, O_2)$ and $h(Y) = f(O_1, Y)$, where O_1 and O_2 are the zeros of B^n and B^m, respectively. Then

$$g(X) + h(Y) = f(X, O_2) + f(O_1, Y) + f(O_1, O_2)$$

$$= f(X,Y) + f(O_1, O_2) + f(O_1, O_2) = f(X,Y) .$$

The proof is completed by duality. □

Corollary 1.1. *A Boolean function $f^{n+m+p} \longrightarrow B$ has a decomposition of the form*

$$(1.4) \qquad f(X,Y,Z) = g(X,Z) \bullet h(Y,Z)$$

if and only if for every $X, U \in B^n$, every $Y, V \in B^m$ and every $D \in B^p$,

$$(1.5) \qquad f(X,Y,D) \bullet f(U,V,D) = f(X,V,D) \bullet f(U,Y,D) .$$

□

---

[1] $\tilde{+}$ is the dual of $+$.

Proposition 1.2. (Povarov [1954]). *The truth functions f, g, h, of $n+m$, $1+m$ and n variables, respectively, satisfy the identity*

(1.6) $$f(X,Y) = g(h(X),Y)$$

if and only if

(1.7) $$f(X,C) \in \{h(X), h'(X), 0, 1\} \qquad (\forall C \in \{0,1\}^m) .$$

PROOF: If (1.6) holds then $f(X,C) = g(h(X),C)$ is a truth function in the variable $h(X)$. Conversely, (1.7) implies

$$f(X,Y) = \bigvee_C f(X,C)Y^C = h'(X)\varphi_0(Y) \vee h(X)\varphi_1(Y) \vee \varphi_2(Y) ,$$

hence (1.6) holds for $g(x,Y) = x'\varphi_0(Y) \vee x\varphi_1(Y) \vee \varphi_2(Y)$. □

Ashenhurst [1957]* has studied the decompositions (1.6) of a truth function f in terms of the $2^n \times 2^m$ matrix $\| f(A,C) \|$.

Theorem 1.2. (Ashenhurst [1957]*). *Associate with a given truth function $f : \{0,1\}^{n+m} \longrightarrow \{0,1\}$ the $2^n \times 2^m$ matrix $\| f(A,C) \|$. Then the following conditions are equivalent:*

(i) there exists $g : \{0,1\}^{1+m} \longrightarrow \{0,1\}$ and $h : \{0,1\}^n \longrightarrow \{0,1\}$ such that identity (1.6) holds ;

(ii) the matrix $\| f(A,C) \|$ has at most two distinct rows ;

(iii) the matrix $\| f(A,C) \|$ has at most four distinct columns, the four possible types being $(\beta_A)_{A\in\{0,1\}^n}$, $(\beta'_A)_{A\in\{0,1\}^n}$, $(0)_{A\in\{0,1\}^n}$ and $(1)_{A\in\{0,1\}^n}$.

PROOF: (i)$\Longrightarrow$(ii): Since $f(A,C) = g(h(A),C)$ and $h(A) \in \{0,1\}$, the two possible rows are $(g(0,C))_{C\in\{0,1\}^m}$ and $(g(1,C))_{C\in\{0,1\}^m}$.

(ii)$\Longrightarrow$(i): Let $(\delta_C)_{C\in\{0,1\}^m}$ and $(\varepsilon_C)_{C\in\{0,1\}^m}$ be the two distinct rows. Define $h : \{0,1\}^n \longrightarrow \{0,1\}$ and $g : \{0,1\}^{1+m} \longrightarrow \{0,1\}$ by

(1.8) $$h(A) = 1 \text{ if } f(A,C) = \delta_C , \quad h(A) = 0 \text{ if } f(A,C) = \varepsilon_C ,$$

(1.9) $$g(x,C) = x\delta_C \vee x'\varepsilon_C .$$

Then

$$f(A,C) = \delta_C \Longrightarrow g(h(A),C) = g(1,C) = \delta_C ,$$
$$f(A,C) = \varepsilon_C \Longrightarrow g(h(A),C) = g(0,C) = \varepsilon_C ,$$

showing that $g(h(A),C) = f(A,C)$ in both cases.

(i)$\Longrightarrow$(iii): For each $C \in \{0,1\}^m$,

$$f(A,C) = g(h(A),C) \in \{h(A), h'(A), 0, 1\} .$$

(iii)$\Longrightarrow$(i): Define $h(A) = \beta_A$. Denote the four possible types of columns indicated in (iii) by I, II, III and IV, respectively. Define $g(x, C)$ to be x , x' , 0 and 1 according as column C is of type I, II, III and IV, respectively. Then $f(A, C) = g(h(A), C)$. □

Corollary 1.2. (Ashenhurst [1957]*). *If the equivalent conditions in Theorem 1.2 are fulfilled, then the functions h and g defined by* (1.8) *and* (1.9), *respectively, provide a decomposition* (1.6). □

Example 1.1. We are looking for a decomposition of the function

$$f(x_1, x_2, x_3, x_4) = x_1 x_2 x_3 x_4 \vee x_1 x_2' x_3' \vee x_1' x_3 x_4$$

with respect to the partition $X = \{x_1, x_2\}$, $Y = \{x_3, x_4\}$ of the set of variables. The rows of the matrix $\| f(A, C) \|$ are provided by

$$f(0, 0, Y) = f(0, 1, Y) = f(1, 1, Y) = x_3 x_4 , \quad f(1, 0, Y) = x_3' .$$

If we denote by (δ_C) and ε_C these two rows, then

$$\delta_{00} = \delta_{01} = \delta_{10} = 0 , \ \delta_{11} = 1 ,$$

$$\varepsilon_{00} = \varepsilon_{01} = 1 , \ \varepsilon_{10} = \varepsilon_{11} = 0 ,$$

hence formula (1.9) yields

$$g(x, Y) = \bigvee_C (x \delta_C \vee x' \varepsilon_C) Y^C$$

$$= x' x_3' x_4' \vee x' x_3' x_4 \vee x x_3 x_4 = x' x_3' \vee x x_3 x_4 .$$

It is convenient to use (1.8) in the form

$$h(A) = 1 \text{ if } f(A, C) = g(1, C) , \ h(A) = 0 \text{ if } f(A, C) = g(0, C) ,$$

which in our case yields

$$h(A) = 1 \iff f(A, C) = x_3 x_4 ,$$

hence

$$h(x_1, x_2) = x_1' x_2' \vee x_1' x_2 \vee x_1 x_2 = x_1' \vee x_2 ,$$

therefore the sought decomposition is

$$f(x_1, x_2, x_3, x_4) = g(x_1' \vee x_2, x_3, x_4) = x_1 x_2' x_3' \vee (x_1' \vee x_2) x_3 x_4 .$$

For a more general approach see Example 2.1. □

The next proposition is a generalization to arbitrary Boolean functions of a result established by Fadini [1972] for truth functions, in the same combinatorial spirit as Theorem 1.2.

Proposition 1.3. *Let* $f : B^{n+m} \longrightarrow B$ *and* $h_1, \ldots, h_r : B^n \longrightarrow B$ *be Boolean functions. Set* $H = (h_1, \ldots, h_r)$. *Then there is a Boolean function* $g : B^{r+m} \longrightarrow B$ *satisfying the identity*

$$(1.10) \qquad f(X,Y) = g(H(X),Y)$$

if and only if the correspondence

$$(1.11) \qquad H(A) \mapsto f(A,Y) \qquad (\forall A \in \{0,1\}^n)$$

is a function from $H(\{0,1\}^n)$ *to* BFm, *in which case the solutions of the functional equation* (1.10) *are characterized by*

$$(1.12) \qquad g(H(A),Y) = f(A,Y) \qquad (\forall A \in \{0,1\}^n) .$$

PROOF: If (1.10) holds, then for every $A, C \in \{0,1\}^n$,

$$H(A) = H(C) \Longrightarrow f(A,Y) = g(H(A),Y) = g(H(C),Y) = f(C,Y) .$$

Conversely, suppose the correspondence (1.11) is a function. Then the definition (1.12) of $g(H(A),Y)$ is unambiguous and the Verification Theorem yields a Boolean function g satisfying (1.11). $\qquad \square$

Example 1.2. (Fadini [1972]). Consider the truth function

$$f(X,Y) = x_1' x_3 x_4 \vee x_1 x_2 x_3' x_4 \vee x_1 x_2 x_3 x_5 \vee x_1' x_2 x_3 x_4 x_5' \vee x_1 x_2' x_3 x_4' x_5 ,$$

where $X = \{x_1, x_2, x_3\}$ and $Y = \{x_4, x_5\}$. Further, let

$$h_1(X) = x_1 x_2 \vee x_1' x_3 , \; h_2(X) = x_1' \vee x_3' , \; h_3(X) = x_2 x_3 .$$

Then

$$H(0,0,0) = H(0,1,0) = H(1,0,0) = (0,1,0) , \; H(0,0,1) = H(1,1,0) = (1,1,0) ,$$

$$H(0,1,1) = (1,1,1) , \; H(1,0,1) = (0,0,0) , \; h(1,1,1) = (1,0,1) ,$$

while

$$f(0,0,0,Y) = f(0,1,0,Y) = f(1,0,0,Y) = 0 ,$$

$$f(0,0,1,Y) = f(1,1,0,Y) = x_4 ,$$

showing that (1.11) is actually a map, and

$$f(0,1,1,Y) = x_4 , \; f(1,0,1,Y) = x_4' x_5' , \; f(1,1,1,Y) = x_5 .$$

Relations (1.12) become

$$g(0,1,0,Y) = 0 , \; g(1,1,0,Y) = x_4 , \; g(1,1,1,Y) = x_4 ,$$

$$g(0,0,0,Y) = x_4' x_5' , \; g(1,0,1,Y) = x_4 ,$$

while $g(0,0,1,Y)$, $g(0,1,1,Y)$ and $g(1,0,0,Y)$ remain arbitrary.

In other words, taking into account that $z_1 z_2 z_3' \vee z_1 z_2 z_3 \vee z_1 z_2' z_3 = z_1(z_2 \vee z_3)$, the solutions of (1.10) are

$$g(z_1, z_2, z_3, Y) = z_1(z_2 \vee z_3)x_4 \vee z_1' z_2' z_3' x_4' x_5' \vee$$

$$\vee z_1' z_2' z_3 p(x_4, x_5) \vee z_1' z_2 z_3 q(x_4, x_5) \vee z_1 z_2' z_3' r(x_4, x_5) \,,$$

where p, q, r are arbitrary.[2]

The reader is urged to check directly that $h_1' h_2' h_3 = h_1' h_2 h_3 = h_1 h_2' h_3' = 0$ (which explains the three arbitrary terms), and $h_1(h_2 \vee h_3) \vee h_1' h_2' h_3' x_4' x_5' = f$. Notice also that the equivalent conditions in Theorem 1.2 are not fulfilled in this case.   □

2 Decomposition via Boolean equations

In the previous section we have seen that, independently of each other, Povarov and Ashenhurst have studied decompositions of the form

(2.1) $f(X, Y) = g(h(X), Y) \,,$

which are known as *disjoint decompositions* of the function f. The more general case of *non-disjoint decompositions* that is, decompositions of the form

(2.2) $f(X, Y, Z) = g(h(X, Z), Y, Z)$

has been extensively studied in the literature.

This section is devoted to the decomposition of Boolean functions via the solution of several Boolean equations.

The first part of the section follows Rudeanu [1976b]. Although the original paper refers to non-disjoint decompositions (2.2), we present here the results for disjoint decompositions (2.1). This strategy is justified because, unlike what happens for truth functions, in the case of arbitrary Boolean functions the passage from disjoint decompositions to non-disjoint decompositions is a mere formal matter: just add the letter Z everywhere!

The second part of the section is mainly devoted to decompositions using the majority decision function and to the square root of a Boolean function.

Proposition 2.1. *Let $f : B^{n+m} \longrightarrow B$ and $h : B^n \longrightarrow B$ be Boolean functions. Then f has a decomposition of the form (2.1) if and only if*

(2.3)
$$(f(A, C) + f(A_1, C))(h(A) + h(A_1) + 1) = 0$$
$$(\forall A, A_1 \in \{0, 1\}^n \; ; \; \forall C \in \{0, 1\}^m) \,,$$

in which case the Boolean functions $g : B^{1+m} \longrightarrow B$ which satisfy (2.1) are given by the prescriptions

---

[2] This corrects a mistake in Fadini's paper.

$$(2.4.0) \qquad \bigvee_A h'(A)f(A,C) \le g(0,C) \le \prod_A (h(A) \vee f(A,C)) \ (\forall C \in \{0,1\}^m),$$

$$(2.4.1) \qquad \bigvee_A h(A)f(A,C) \le g(1,C) \le \prod_A (h'(A) \vee f(A,C)) \ (\forall C \in \{0,1\}^m).$$

PROOF: Taking into account that $x \vee y = x + y \iff xy = 0$, we can write the identity (2.1) in the following equivalent forms:

$$f(X,Y) + g(1,Y)h(X) + g(0,Y)h'(X) = 0,$$

$$\bigvee_A \bigvee_C (f(A,C) + g(1,C)h(A) + g(0,C)h'(A)) = 0,$$

$$\bigvee_A \bigvee_C (f'(A,C)(g(1,C)h(A) \vee g(0,C)h'(A))$$
$$\vee f(A,C)(g'(1,C)h(A) \vee g'(0,C)h'(A))) = 0,$$

and this equation can be decomposed as follows:

$$(2.5.0) \qquad \bigvee_A \bigvee_C (f'(A,C)h(A)g(1,C) \vee f(A,C)h(A)g'(1,C)) = 0,$$

$$(2.5.1) \qquad \bigvee_A \bigvee_C (f'(A,C)h'(A)g(0,C) \vee f(A,C)h'(A)g'(0,C)) = 0,$$

or equivalently,

$$(2.6.0) \qquad \begin{aligned} &g(1,C) \bigvee_A f'(A,C)h(A) \vee g'(1,C) \bigvee_A f(A,C)h(A) = 0 \\ &(\forall C \in \{0,1\}^m), \end{aligned}$$

$$(2.6.1) \qquad \begin{aligned} &g(0,C) \bigvee_A f'(A,C)h'(A) \vee g'(0,C) \bigvee_A f(A,C)h'(A) = 0 \\ &(\forall C \in \{0,1\}^m). \end{aligned}$$

It turns out that (2.6) is a system of 2×2^m equations, each of them in one unknown $g(\alpha, C)$, and which can be solved independently of each other. The consistency conditions are

$$(2.7.0) \qquad \left(\bigvee_A f'(A,C)h(A) \right)\left(\bigvee_A f(A,C)h(A) \right) = 0 \ (\forall C \in \{0,1\}^m),$$

$$(2.7.1) \qquad \left(\bigvee_A f'(A,C)h'(A) \right)\left(\bigvee_A f(A,C)h'(A) \right) = 0 \ (\forall C \in \{0,1\}^m),$$

and they can be put together in the equivalent system

$$\bigvee_{A,A_1} f(A,C)f'(A_1,C)(h(A)h(A_1) \vee h'(A)h'(A_1)) = 0 \ (\forall C \in \{0,1\}^m) \,.$$

Obvious considerations of symmetry enable us to write the last system in the form

$$\bigvee_{A,A_1} (f(A,C)f'(A_1,C) \vee f'(A,C)f(A_1,C))\cdot$$

$$\cdot (h(A)h(A_1) \vee h'(A)h'(A_1)) = 0 \qquad (\forall C \in \{0,1\}^m) \,,$$

which is precisely (2.3). If equations (2.6) are consistent, their solutions are (2.4).

$\square$

Proposition 2.2. *Let* $f : B^{n+m} \longrightarrow B$ *and* $g : B^{1+m} \longrightarrow B$ *be Boolean functions. Then* f *has a decomposition of the form (2.1) if and only if*

(2.8)
$$(\bigvee_C (f(A,C) + g(1,C)))(\bigvee_C (f(A,C) + g(0,C))) = 0$$
$$(\forall A \in \{0,1\}^n) \,,$$

in which case the Boolean functions $h : B^n \longrightarrow B$ *which satisfy (2.1) are given by the prescriptions*

(2.9)
$$\bigvee_C (f(A,C)+g(0,C)) \leq h(A) \leq \prod_C (f(A,C)+g(0,C)+1))$$
$$(\forall A \in \{0,1\}^n) \,.$$

PROOF: It was seen in the proof of Proposition 2.1 that the identity (2.1) is equivalent to the system (2.5). But this system can be written in the following equivalent forms:

$$\bigvee_A \bigvee_C (h(A)(f'(A,C)g(1,C) \vee f(A,C)g'(1,C))$$

$$\vee h'(A)(f'(A,C)g(0,C) \vee f(A,C)g'(0,C))) = 0 \,,$$

(2.10)
$$h(A) \bigvee_C (f(A,C) + g(1,C))$$
$$\vee h'(A) \bigvee_C (f(A,C) + g(0,C)) = 0 \qquad (\forall A \in \{0,1\}^n) \,.$$

But system (2.10) consists of 2^n equations, each of them in one unknown $h(A)$, and which can be solved independently of each other. The consistency condition for equation (2.10) is (2.8), while the solutions are given by (2.9). $\square$

As a matter of fact the problem of finding a decomposition of the form (2.1) can be specified in three ways: 1) given f and h, find g; 2) given f and g, find h; and 3) given f, find g and h. Propositions 2.1 and 2.2 solve problems 1) and 2), respectively. Proposition 2.1 can also be used for solving problem 3): first find h by solving equation (2.3), then construct g via (2.4). As a matter of principle, Propositions 2.1 and 2.2 provide all the solutions.

Example 2.1. Let us resume the function

(E1.1) $$f(x_1, x_2, x_3, x_4) = x_1 x_2 x_3 x_4 \vee x_1 x_2' x_3' \vee x_1' x_3 x_4$$

in Example 1.1 and look for all the decompositions (2.1) with $X := \{x_1, x_2\}$ and $Y := \{x_3, x_4\}$.

We have

$$f(0, 0, Y) = f(0, 1, Y) = f(1, 1, Y) = x_3 x_4 , \quad f(1, 0, Y) = x_3' ,$$

so that $f(A, Y) + f(A_1, Y) = 0$ unless $A \in \{(0,0), (0,1), (1,1)\}$ and $A_1 = (1, 0)$ (or conversely). Therefore system (2.3) reduces to

$$(\gamma_3 \gamma_4 + \gamma_3')(h(0, 0) + h(1, 0) + 1) = 0 ,$$

$$(\gamma_3 \gamma_4 + \gamma_3')(h(0, 1) + h(1, 0) + 1) = 0 ,$$

$$(\gamma_3 \gamma_4 + \gamma_3')(h(1, 1) + h(1, 0) + 1) = 0 ,$$

for all $\gamma_3, \gamma_4 \in \{0, 1\}$, which is equivalent to

$$h(0, 0) + h(1, 0) = h(0, 1) + h(1, 0) = h(1, 1) + h(1, 0) = 1 ,$$

hence the solutions are

(E1.2) $$h(1, 0) = a , \quad h(0, 0) = h(0, 1) = h(1, 1) = a' ,$$

where a is an arbitrary parameter. In other words,

(E1.3) $$h(x_1, x_2) = a x_1 x_2' \vee a'(x_1' x_2' \vee x_1' x_2 \vee x_1 x_2) = a x_1 x_2' \vee a'(x_1' \vee x_2) .$$

Therefore system (2.4) becomes

$$a \gamma_3 \gamma_4 \vee a' \gamma_3' \le g(0, \gamma_3, \gamma_4) \le (a' \vee \gamma_3 \gamma_4)(a \vee \gamma_3') ,$$

$$a' \gamma_3 \gamma_4 \vee a \gamma_3' \le g(1, \gamma_3, \gamma_4) \le (a \vee \gamma_3 \gamma_4)(a' \vee \gamma_3') ,$$

that is,

$$g(0, \gamma_3, \gamma_4) = a \gamma_3 \gamma_4 \vee a' \gamma_3' ,$$

$$g(1, \gamma_3, \gamma_4) = a' \gamma_3 \gamma_4 \vee a \gamma_3' .$$

In other words,

$$g(0, x_3, x_4) = a' x_3' x_4' \vee a' x_3' x_4 \vee a x_3 x_4 = a' x_3' \vee a x_3 x_4 ,$$

$$g(1, x_3, x_4) = a x_3' \vee a' x_3 x_4 ,$$

so that

(E1.4) $$g(x, x_3, x_4) = x'(a' x_3' \vee a x_3 x_4) \vee x(a x_3' \vee a' x_3 x_4) .$$

Formulas (E1.3) and (E1.4) provide all the decompositions (2.1) of the function (E1.1). Taking $a \in \{0, 1\}$, we obtain the solutions for which g and h are simple Boolean functions: they are just the solution in Example 1.1 and its dual.   □

Decompositions using the *majority decision* function

(2.11) $\text{maj}\,(x, y, z) = xy \vee xz \vee yz$

have been much studied in the literature, in view of their applications to logical design; cf. Ch.14, §3.

Recall the notation $f(x = a)$ for the function obtained from f by taking $x := a$.

Proposition 2.3. (Tohma [1964]). *Let f be a Boolean function and x one of its variables. Then f has a decomposition of the form*

(2.12) $f = \text{maj}\,(u, v, x)$

if and only if

(2.13) $f(x = 0) \leq f(x = 1)\,,$

in which case (2.12) has the general solution

(2.14.1) $u = f(x = 0) \vee p f(x = 1)\,,$

(2.14.2) $v = f(x = 0) \vee p' f(x = 1)\,,$

where p is an arbitrary function of the variables $\neq x$.

PROOF: The identity $f = uv \vee ux \vee vx$ is equivalent to the system

(2.15) $f(x = 0) = uv\ \&\ f(x = 1) = u \vee v\,,$

which implies (2.13). Conversely, suppose (2.13) holds. Then $f(x = 0) \vee f(x = 1) = f(x = 1)$, hence (2.14) implies (2.15) and conversely, any solution (u, v) of (2.15) is of the form (2.14) for $p := u$, because

$$f(x = 0) \vee u f(x = 1) = uv \vee u(u \vee v) = u\,,$$

$$f(x = 0) \vee u' f(x = 1) = uv \vee u'(u \vee v) = uv\,.$$

$\square$

The next proposition is a slight improvement of the main result in Rudeanu [1965].

Proposition 2.4. *The Boolean functional equation*

(2.16) $f = \text{maj}\,(u, v, w)$

has the reproductive solution

(2.17.1) $u = (p \vee q' \vee r')f \vee pq'r'f'\,,$

(2.17.2) $v = (p' \vee q \vee r')f \vee p'qr'f'\,,$

(2.17.3) $w = (p' \vee q' \vee r)f \vee p'q'rf'\,.$

PROOF: Taking into account the identity

(2.18) $\text{maj}'\,(x, y, z) = \text{maj}\,(x', y', z')$

(which is easy to check), the equation $f = uv \lor uw \lor vw$ can be written in the form

(2.19) $\qquad (u'v' \lor u'w' \lor v'w')f \lor (uv \lor uw \lor vw)f' = 0$

and it has the particular solution $u := v := w := f$. We are going to show that Löwenheim's reproductive solution amounts to formulas (2.17). In view of symmetry, it suffices to check (2.17.1). If we denote the left side of (2.19) by g, then Löwenheim's formula yields

$$u = fg(p,q,r) \lor pg'(p,q,r)$$

$$= (p'q' \lor p'r' \lor q'r')f \lor ((pq \lor pr \lor qr)f \lor (p'q' \lor p'r' \lor q'r')f')p$$

$$= (p'q' \lor p'r' \lor q'r'p \lor q'r'p' \lor pq \lor pr)f \lor pq'r'f'$$

$$= (p'q' \lor p'r' \lor pq'r' \lor p)f \lor pq'r'f' = (p \lor q' \lor r')f \lor pq'r'f' \ .$$

□

The papers by Tohma and Rudeanu also study the isotony of the solutions in the variable x.

Reischer and Simovici [1971] defined a *square root* of a Boolean function $f : B^n \longrightarrow B$ with respect to a variable x_i as a Boolean function $s : B^n \longrightarrow B$ which satisfies the identity

(2.20) $\qquad f(x_1,\ldots,x_n) = s(x_1,\ldots,x_{i-1},s(x_1,\ldots,x_n),x_{i+1},\ldots,x_n)$

and they found all the square roots of a simple Boolean function f. Note that (2.20) may be viewed as a decomposition of the form (2.2) with $Y := \emptyset$, $X := \{x_i\}$, $Z := \{x_1,\ldots,x_{i-1},x_{i+1},\ldots,x_n\}$ and $g = h$, therefore the general form of Propositions 2.1 and 2.2 (cf. the introduction to this section) can be applied to this case. However the next proposition seems to provide the simplest solution.

Proposition 2.5. (Rudeanu [1976b]). *The functional equation* (2.20) *is consistent if and only if*

(2.21) $\qquad f(x_i = 0) \leq f(x_i = 1) \ ,$

in which case the set of solutions is determined by formulas

(2.22) $\qquad s = f(x_i = 0) \lor f(x_i = 1)(x_ip \lor x_i'p') \ ,$

where p is an arbitrary function of the variables $\neq x_i$.

COMMENT: The consistency condition (2.21) is due to Reischer and Simovici [1971].

PROOF: Write the sought function s in the form $s = s_0x_i' \lor s_1x_i$, where s_0 and s_1 do not depend on x_i. Then the right side of (2.20) equals

$$s_1(s_0x_i' \lor s_1x_i) \lor s_0(s_0'x_i' \lor s_1'x_i) = s_1s_0x_i' \lor (s_1 \lor s_0)x_i$$

$$= s_1s_0x_i' \lor (s_1 \lor s_0)x_i \lor s_1s_0x_i = \mathrm{maj}\,(s_0,s_1,x_i) \ ,$$

whence the desired conclusion follows by Proposition 2.3.   □

Corollary 2.1. (Rudeanu [1976b]). *The square root can also be written in the form*

(2.23) $s = f(x_i = 0) \vee x_i t \vee x_i' t' f'(x_i = 0) f(x_i = 1)$,

where t is any function of the variables $\neq x_i$ which satisfies

(2.24) $t \leq f'(x_i = 0) f(x_i = 1)$.

PROOF: Work out formula (2.22) as follows:

$$s = f_0 \vee f_1(x_i p \vee x_i' p') = f_0 \vee f_0' f_1(x_i p \vee x_i' p')$$

$$= f_0 \vee x_i p f_0' f_1 \vee f_0' f_1 x_i'(p' \vee f_o \vee f_1') ,$$

and set $t = p f_0' f_1$.   □

For a further insight of equation (2.20), see isotony detectors in Ch.7, §2.

Another solution of the problem of square rooting a Boolean functions was obtained in terms of Boolean derivatives.

Lemma 2.1. *The following identity holds:*

(2.25) $(x_i + f)\dfrac{\partial f}{\partial x_i} + f = f(x_1, \ldots, x_{i-1}, f, x_{i+1}, \ldots, x_n)$.

PROOF: The left side of (2.25) equals

$$(x_i + f)(f(x_i = 1) + f(x_i = 0)) + x_i f(x_i = 1) + x_i' f(x_i = 0)$$

$$= x_i f(x_i = 0) + (f(x_i = 1) + f(x_i = 0))f + (x_i + 1)f(x_i = 0)$$

$$= f(x_i = 1)f + f(x_i = 0)f + f(x_i = 0) = f(x_i = 1)f + f(x_i = 0)f' .$$

                                                                □

Proposition 2.6. (Stiefel [1982]). *The equation (2.20) is consistent if and only if*

(2.26) $(x_i + f)\dfrac{\partial f}{\partial x_i} = 0$,

in which case the reproductive general solution of (2.20) is

(2.27) $s = f + (x_i + p)\dfrac{\partial f}{\partial x_i}$,

where p is any function such that

(2.28) $\dfrac{\partial p}{\partial x_i} \geq \dfrac{\partial f}{\partial x_i}$.

PROOF: In view of Lemma 2.1, the identity (2.20) is equivalent to

(2.29) $f = (x_i + s)\dfrac{\partial s}{\partial x_i} + s$,

which implies in turn

$$f + x_i \frac{\partial s}{\partial x_i} = s \frac{\partial s}{\partial x_i} + s = s \left(\frac{\partial s}{\partial x_i} \right)',$$

(2.30)
$$\frac{\partial f}{\partial x_i} = 1 \cdot \frac{\partial s}{\partial x_i} + \frac{\partial s}{\partial x_i} \cdot \frac{\partial s}{\partial x_i} + \frac{\partial s}{\partial x_i} = \frac{\partial s}{\partial x_i},$$

$$f + x_i \frac{\partial f}{\partial x_i} = f + x_i \frac{\partial s}{\partial x_i} = s \left(\frac{\partial s}{\partial x_i} \right)' = s \left(\frac{\partial f}{\partial x_i} \right)',$$

$$(x_i + f) \frac{\partial f}{\partial x_i} = x_i \frac{\partial f}{\partial x_i} + f \frac{\partial f}{\partial x_i} \cdot \frac{\partial f}{\partial x_i} = \frac{\partial f}{\partial x_i} \left(x_i + f \frac{\partial f}{\partial x_i} \right) = 0.$$

Conversely, suppose (2.26) holds. Then s given by (2.27) and (2.28) is a solution because we get in turn

$$\frac{\partial s}{\partial x_i} = \frac{\partial f}{\partial x_i} + 1 \cdot \frac{\partial f}{\partial x_i} + \frac{\partial p}{\partial x_i} \frac{\partial f}{\partial x_i} = \frac{\partial f}{\partial x_i},$$

$$(x_i + s) \frac{\partial s}{\partial x_i} + s$$

$$= (f + x_i \left(\frac{\partial f}{\partial x_i} \right)' + p \frac{\partial f}{\partial x_i}) \frac{\partial f}{\partial x_i} + f + x_i \frac{\partial f}{\partial x_i} + p \frac{\partial f}{\partial x_i}$$

$$= f \frac{\partial f}{\partial x_i} + p \frac{\partial f}{\partial x_i} + f + x_i \frac{\partial f}{\partial x_i} + p \frac{\partial f}{\partial x_i} = (x_i + f) \frac{\partial f}{\partial x_i} + f = f,$$

thus proving (2.29). Moreover, any solution s of (2.29) satisfies (2.27) with $p := s$. For, taking into account (2.30), we have

$$f + x_i \frac{\partial f}{\partial x_i} + s \frac{\partial f}{\partial x_i} = f + (x_i + s) \frac{\partial s}{\partial x_i} = s.$$

$\square$

Remark 2.1. It is easy to check directly the equivalence of the consistency conditions (2.21) and (2.26) found in Propositions 2.5 and 2.6, respectively. For the Verification Theorem transforms (2.26) into the system

$$f(x_i = 0) \frac{\partial f}{\partial x_i} = 0 \ \& \ f'(x_i = 1) \frac{\partial f}{\partial x_i} = 0$$

and each of the two conditions is equivalent to $f(x_i = 0) f'(x_i = 1) = 0$. $\square$

Example 2.2. Find the square root of the function

(E2.1) $$f(x_1, x_2, x_3, x_4) = x_1 x_2 \vee x_1 x_3 \vee (x_1 \vee x_3) x_4$$

with respect to x_4.

Using the notation in the proof of Corollary 2.1, we have

$$f_0 = x_1 x_2 \vee x_1 x_3 \ \& \ f_1 = x_1 \vee x_3,$$

$$f'_0 f_1 = (x'_1 \vee x'_2 x'_3)(x_1 \vee x_3) = x'_1 x_3 \vee x_1 x'_2 x'_3,$$

so that formulas (2.23) and (2.24) become

(E2.2) $$s = x_1x_2 \vee x_1x_3 \vee x_4t \vee (x_1'x_3 \vee x_1x_2'x_3')x_4't' ,$$

(E2.3) $$t \leq x_1'x_3 \vee x_1x_2'x_3' .$$

If we impose the supplementary condition that s be a simple Boolean function, then we write (E2.3) in the form

(E2.4) $$t = ax_1'x_2x_3 \vee bx_1'x_2'x_3 \vee cx_1x_2'x_3'   (a, b, c \in \{0, 1\})$$

and obtain 8 solutions.   □

Rozenfel'd [1974]* and Rozenfel'd and Silayev [1979] study decompositions of the form (1.8) and several generalizations by means of Boolean equations. [3]

Melter and Rudeanu [1980] study decompositions of the form $f = fgf$, where g is an injective Boolean function, which they call a *generalized inverse* of f. Several characterizations of generalized inverses are given.

---

[3] We have been unable to follow the computations of the latter paper.

12. Boolean-based mathematics

As was seen in Chapter 12 of BFE, several Boolean-based algebraic structures have been studied, that is, algebras having as support a Boolean algebra and whose operations are Boolean functions; homomorphisms connecting such algebras and expressed by Boolean functions have also been investigated. Chapter 13 of BFE deals with Boolean arithmetic and Boolean geometry. The former means the study of divisibility between Boolean functions. In Boolean geometry the rôle of the space is played by a Boolean algebra and one looks for analogues of the basic concepts of geometry; for instance, a "good" analogue of the distance function is the symmeric difference $d(x, y) = x + y$. As was seen in BFE, Chapter 14, Boolean analysis replaces the real line by a Boolean algebra, while the functions dealt with are Boolean functions.

It was seen in Ch.10 of the present book that Boolean differential calculus has much developed in the last years. Other recent results in Boolean-based mathematics are presented in this chapter. A continuation of Boolean geometry, with extensions to the geometry of 3−rings, is developed in §3. The other sections present the beginnings of new directions in the same spirit: §1 contains Boolean analogues of modus ponens and of the deduction theorem, §4 is devoted to a Boolean measure of central tendency, while §2 takes a step beyond the Boolean flag by constructing all semilattices defined by Post functions over a 3−Post algebra.

1 Mathematical logic

In this section we present Boolean analogues of modus ponens, the resolution rule and the deduction theorem.

The basic inference rule, known as *modus ponens*, is: "if x and $x \to y$ then y". In the Lindenbaum-Tarski algebra of the classical propositional calculus, modus ponens is expressed by the inequality $x \cdot (x \to y) \leq y$. Trillas and Cubillo [1996] have replaced theLindenbaum-Tarski algebra by an arbitrary Boolean algebra and they have taken two functions $f, g : B^2 \longrightarrow B$ in the rôles of conjunction ($\cdot$) and implication ($\to$), respectively. Their result is the determination of all simple Boolean functions f, g that satisfy the following generalization of modus ponens:

the identity

(1.1) $f(x, g(x, y)) \le y$

and $f(1, 1) = 1$. A similar problem is solved for the functional equation

(1.2) $f(x, g(x, y)) = xy$.

The results were further generalized to arbitrary Boolean functions f, g by Rudeanu [1998a], as we are going to show below.

Proposition 1.1. *The Boolean functions*

(1.3) $f(x, y) = axy \vee bxy' \vee cx'y \vee dx'y'$,

(1.4) $g(x, y) = pxy \vee qxy' \vee rx'y \vee sx'y'$,

satisfy identity (1.1) *if and only if the following relations hold:*

(1.5.1) $ab = cd = 0$,

(1.5.2) $b \le q \le a'$,

(1.5.3) $d \le s \le c'$.

PROOF: The inequality (1.1) is equivalent to

$$f(1, g(1, 0)) = 0 \ \& \ f(0, g(0, 0)) = 0$$

by the Verification Theorem, and since

(1.6) $\begin{aligned} f(x, g(x, y)) &= (ax \vee cx')(pxy \vee qxy' \vee rx'y \vee sx'y') \vee \\ &\vee (bx \vee dx')(p'xy \vee q'xy' \vee r'x'y \vee s'x'y') , \end{aligned}$

we see that (1.1) is equivalent to the system

(1.7.1) $aq \vee bq' = 0$,

(1.7.2) $cs \vee ds' = 0$.

The consistency conditions for the equations (1.7.1) and (1.7.2) are (1.5.1), while the solutions are (1.5.2) and (1.5.3), respectively.   □

Proposition 1.2. *The Boolean functions f and g satisfy* (1.1) *and* $f(1, 1) = 1$ *if and only if they are of the form*

(1.8) $f(x, y) = xy \vee cxy' \vee dx'y'$,

(1.9) $g(x, y) = pxy \vee rx'y \vee sx'y'$,

where

(1.10) $cd = 0 \ \& \ d \le s \le c'$.

PROOF: Taking $a := 1$ in Proposition 1.1, conditions (1.5) become $b = q = 0$&
(1.10), so that (1.3) and (1.4) reduce to (1.8) and (1.9), respectively.   □

Corollary 1.1. *There are 16 pairs of simple Boolean functions* (f, g) *satisfying*
(1.1) *and* $f(1, 1) = 1$, *namely*

(1.11)
$$f(x, y) = xy \vee cx'y = (x \vee c)y ,$$

(1.12)
$$g(x, y) = pxy \vee rx'y ,$$

and

(1.13)
$$f(x, y) = xy \vee dx'y' ,$$

(1.14)
$$g(x, y) = pxy \vee rx'y \vee x'y' ,$$

where the coefficients are arbitrary in $\{0, 1\}$.

PROOF: Taking $s := 0$ In Proposition 1.2, we obtain $d = 0$, so that conditions
(1.10) are fulfilled, while (1.8) and (1.9) reduce to (1.11) and (1.12), respectively.
Similarly, for $s := 1$ we obtain $c := 0$ and the solutions (1.13), (1.14).   □

Remark 1.1. The 16 solutions determined in Corollary 1.1 are the 14 solutions
found by Trillas and Cubillo, plus the solutions $(xy, 0)$ and $(y, 0)$, which do not
seem to have a modus ponens interpretation.   □

Proposition 1.3. *The Boolean functions* f *and* g *satisfy the identity*

(1.2)
$$f(x, g(x, y)) = xy$$

if and only if they are of the form

(1.15)
$$f(x, y) = axy \vee a'xy' \vee cx'y \vee dx'y' ,$$

(1.16)
$$g(x, y) = axy \vee a'xy' \vee rx'y \vee sx'y' ,$$

where

(1.17.1)
$$cd = 0 ,$$

(1.17.2)
$$d \le r \le c' ,$$

(1.17.3)
$$d \le s \le c' .$$

PROOF: In view of (1.6), identity (1.2) is equivalent to the system

(1.18.1)
$$ap \vee bp' = 1 ,$$

(1.18.2)
$$aq \vee bq' = 0 ,$$

(1.18.3)
$$cr \vee dr' = 0 ,$$

(1.18.4) $$cs \vee ds' = 0 \ .$$

The consistency conditions for the above equations in p, q, r and s are $a \vee b = 1$, $ab = 0$ and (1.17.1) (twice), respectively. The first two conditions are equivalent to $b = a'$. Therefore (1.18.1) and (1.18.2) become $ap \vee a'p' = 1$ and $aq \vee a'q' = 0$ and hence they are equivalent to $p = a$ and $q = a'$, respectively. We have thus proved that (1.3) and (1.4) reduce to (1.15) and (1.16), respectively, while (1.18.3) and (1.18.4) can be written in the form (1.17.2) and (1.17.3), respectively.   □

Proposition 1.4. *The Boolean functions f and g satisfy (1.2) and $f(1,1) = 1$ if and only if they are of the form*

(1.19) $$f(x,y) = xy \vee cx'y \vee dx'y' \ ,$$

(1.20) $$g(x,y) = xy \vee rx'y \vee sx'y' \ ,$$

where the coefficients satisfy (1.17).

PROOF: Take $a := 1$ in Proposition 1.3.   □

Corollary 1.2. (Trillas and Cubillo [1996]) . *There are 6 pairs of simple Boolean functions satisfying (1.2) and $f(1,1) = 1$, namely*

(1.21) $$f(x,y) = y \ , \ g(x,y) = xy \ ,$$

(1.22) $$f(x,y) = xy \vee x'y' \ , \ g(x,y) = x' \vee y \ ,$$

and

(1.23) $$f(x,y) = xy \ , \ g(x,y) = xy \vee rx'y \vee sx'y' \ ,$$

where r, s are arbitrary in $\{0,1\}$.

PROOF: Taking $c := 1$ in Proposition 1.4, conditions (1.17) become $d = r = s = 0$ and we obtain the solution (1.21). For $c := 0$ conditions (1.17) reduce to $d \le rs$, therefore either $d := 1$, which implies $r = s = 1$ and this yields the solution (1.22), or $d := 0$, which produces the 4 solutions (1.23).   □

Similarly, the *resolution rule* "if $x \to y$ and $x \vee z$, then $y \vee z$" has the Boolean translation $(x \to y) \cdot (x \vee z) \le y \vee z$. Martin and Nipkow [1988] suggested the generalization $(x \to y)w \le y \vee z$, which is equivalent to $(x' \vee y)wy'z' = 0$, that is, $x'y'z'w = 0$. Therefore the solutions are $w \le x \vee y \vee z$.

Now let us recall the *deduction theorem*

(1.24) $$X \cup \{y\} \vdash z \Longleftrightarrow X \vdash y \vee z \ ,$$

where X is a set of sentences and y, z are sentences. Several generalizations of this basic theorem have been obtained, which are of the form

(1.25) $$X \cup \{y\} \, R \, z \Longleftrightarrow X \, R \, h(y,z) \ ,$$

where R is a relation from sets of sentences to sentences and h is a binary syntactic operator; cf. Porte [1982] and Corcoran [1985].

In the remainder of this section we follow Rudeanu [1990] in obtaining a theorem of the above type, stated within the framework of an arbitrary Boolean algebra B and generalizing the Boolean-algebra counterpart of the deduction theorem for the classical propositional calculus.

The result will be obtained under two mild conditions. The first one is the finiteness of the set X. As a matter of fact this is equivalent to dealing with singletons, because the set $X = \{x_1, \ldots, x_n\} \subseteq B$ can be replaced by the Boolean conjunction $x = x_1 \ldots x_n$. So (1.25) will reduce to

$$(1.26) \qquad \{xy\}\, R\, z \Longleftrightarrow \{x\}\, R\, h(y, z)$$

and since in (1.26) the sets are singletons, we can further replace (1.26) by

$$(1.27) \qquad xy\, \rho\, z \Longleftrightarrow x\, \rho\, h(y, z)\,,$$

where $\rho \subseteq B^2$, $h : B^2 \longrightarrow B$ and $x, y, z \in B$. The second natural restriction is to require that h be a Boolean function and ρ a *Boolean relation*, that is,

$$(1.28) \qquad x\, \rho\, y \Longleftrightarrow f(x, y) = 0$$

for some Boolean function f. Let us put all this in a formal way.

Definition 1.1. By a *Boolean deduction theorem* we mean an equivalence of the form (1.27), where ρ is a Boolean relation and h is a Boolean function.   □

Thus our aim is to determine all Boolean deduction theorems; in other words, all Boolean pairs (ρ, h) satisfying (1.27). In particular we will see that there are several such pairs of the form $(\rho, \rightarrow)$, but only one such that ρ is a partial order: the pair $(\leq, \rightarrow)$ corresponding to the classical deduction theorem. As a matter of fact we will solve a functional Boolean equation more general than (1.27) and as a by-product we will recapture an old result of Bernstein [1924] on associative Boolean operations.

Lemma 1.1. *Suppose* $f, h \in BF2$ *and* ρ *is defined by* (1.28), $\rho \neq \emptyset$. *Then* (1.27) *holds if and only if* f *and* h *satisfy the identity*

$$(1.29) \qquad f(xy, z) = f(x, h(y, z))\,.$$

PROOF: By the Verification Theorem.   □

As a matter of fact we are going to solve the more general Boolean functional equation

$$(1.30) \qquad f(g(x, y), z) = f(x, h(y, z))$$

in the unknown functions f, g, h. We set

$$(1.31) \qquad f(x, y) = axy \vee bxy' \vee cx'y \vee dx'y'\,,$$

$$(1.32) \qquad g(x, y) = jxy \vee kxy' \vee mx'y \vee nx'y'\,,$$

$$(1.33) \qquad h(x, y) = qxy \vee rxy' \vee sx'y \vee tx'y'\,,$$

Lemma 1.2. *Let f,g,h be given by* (1.31)–(1.33). *The Boolean functional equation* (1.30) *is equivalent to the following system of equations:*

(1.34) $$(a+c)(j'\vee m)q\vee((a+b)j\vee(b+c)j'\vee(a+d)m\vee(c+d)m')q' = 0 \,,$$

(1.35) $$((a+b)j\vee(a+d)j'\vee(b+c)m\vee(c+d)m')r\vee(b+d)(j'\vee m)r' = 0 \,,$$

(1.36) $$(a+c)(k'\vee n)s\vee((a+b)k\vee(b+c)k'\vee(a+d)n\vee(c+d)n')s' = 0 \,,$$

(1.37) $$((a+b)k\vee(a+d)k'\vee(b+c)n\vee(c+d)n')t\vee(b+d)(k'\vee n)t' = 0 \,.$$

PROOF: Identity (1.30) is equivalent to the system of equalities obtained from it by giving to the vector (x, y, z) the 2^3 values $(0, 0, 0), \ldots, (1, 1, 1)$. We get the system

(1.38.0) $$f(n, 0) = f(0, t) \,,$$

(1.38.1) $$f(n, 1) = f(0, s) \,,$$

(1.38.2) $$f(m, 0) = f(0, r) \,,$$

(1.38.3) $$f(m, 1) = f(0, q) \,,$$

(1.38.3) $$f(k, 0) = f(1, t) \,,$$

(1.38.5) $$f(k, 1) = f(1, s) \,,$$

(1.38.6) $$f(j, 0) = f(1, r) \,,$$

(1.38.7) $$f(j, 1) = f(1, q) \,,$$

and we notice that if in equations (1.38.0), (1.38.1), (1.38.4) and (1.38.5) we replace the unknowns n, k, t, s by m, j, r, q, respectively, we obtain equations (1.38.2), (1.38.3), (1.38.6) and (1.38.7), respectively, therefore it suffices to process the former system of equations.

Equation (1.38.0) can be written in the form $bn + dn' = ct + dt'$, or else

$$(bn + dn' + c)t + (bn + dn' + d)t' = 0 \,;$$

but $bn + dn' + c = (b + c)n + (d + c)n'$ and $dn' + d = dn$, hence we obtain

$$((b + c)n \vee (d + c)n')t \vee (b + d)nt' = 0$$

and similarly, equations (1.38.1), (1.38.4) and (1.38.5) can be written in the form

$$(a+c)ns \vee ((a+d)n \vee (c+d)n')s' = 0 \,,$$

$$((a+b)k \vee (a+d)k')t \vee (b+d)k't' = 0 \,,$$

$$(a+c)k's \vee ((a+b)k \vee (b+c)k')s' = 0 \,.$$

But the last system of four equations is equivalent to (1.36) & (1.38). It follows by the symmetry mentioned above that the system (1.38.2), (1.38.3), (1.38.6) and (1.38.7) is equivalent to (1.34) & (1.35). □

Proposition 1.5. *Let* f,g,h *be given by* (1.31)–(1.33). *The Boolean functional equation* (1.30) *is consistent and its solutions are constructed as follows:* a,b,c,d *are taken arbitrarily; for each* (a,b,c,d), *both* (j,m) *and* (k,n) *run over the set of solutions of the equation*

(1.39)
$$(ab(c' \vee d') \vee a'b(c \vee d))v' \vee (cd(a' \vee b') \vee c'd'(a \vee b))w$$
$$\vee (ab'(c'\vee d)\vee a'b(c\vee d'))vw \vee (cd'(a'\vee b)\vee c'd(a\vee b'))v'w' = 0 \,,$$

in the unknowns v,w, *while for each* (a,b,c,d,j,m,k,n), *the elements* q, r, s, t *are determined by equations* (1.34)–(1.37).

PROOF: In view of Lemma 1.2, the solutions (f,g,h) of equation (1.30) are obtained by solving equations (1.34)-(1.37) with respect to the unknowns q, r, s, t. Since each unknown appears in exactly one equation, the consistency condition for the system is simply the logical conjunction of the consistency conditions of the four equations, that is,

$$(a+c)((a+b)jm \vee (b+c)j' \vee (a+d)m \vee (c+d)j'm') = 0 \,,$$

$$(b+d)((a+b)jm \vee (a+d)j' \vee (b+c)m \vee (c+d)j'm') = 0 \,,$$

$$(a+c)((a+b)kn \vee (b+c)k' \vee (a+d)n \vee (c+d)k'n') = 0 \,,$$

$$(b+d)((a+b)kn \vee (a+d)k' \vee (b+c)n \vee (c+d)k'n') = 0 \,.$$

The first two equations are equivalent to (1.39) with $v := j$ and $w := m$, while the last two equations express (1.39) with $v := k$ and $w := n$. Equation (1.39) is consistent since it has the particular solution $v := 1$, $w := 0$. □

Now we confine to equation (1.29). Proposition 1.5 yields the following result.

Lemma 1.3. *Let* f *and* h *be given by* (1.31) *and* (1.33), *respectively. Then equation* (1.29) *is consistent if and only if*

(1.40)
$$a'b'c \vee b'c'd \vee abc' \vee bcd' = 0 \,,$$

in which case the solutions are characterized by the following equations:

(1.41)
$$((a+b) \vee (c+d))q' = 0 \,,$$

(1.42)
$$((a+b) \vee (c+d))r = 0 \,,$$

(1.43)
$$(a+c)s \vee ((b+c) \vee (c+d))s' = 0 \,,$$

(1.44) $((a+d) \vee (c+d))t \vee (b+d)t' = 0$.

PROOF: Equation (1.29) is obtained from (1.30) by taking $g(x,y) := xy$, that is, $j := 1$ and $k := m := n := 0$. This specialization reduces equations (1.34)-(1.37) in Lemma 1.2 to (1.41)-(1.44). Equation (1.39) is anyway satisfied for $v := 1$, $w := 0$, and $v := w := 0$ must be another solution. This yields

$$0 = ab(c' \vee d') \vee a'b'(c \vee d) \vee cd'(a' \vee b) \vee c'd(a \vee b')$$

$$= abc' \vee abcd' \vee a'b'c \vee a'b'c'd \vee cd'a'b' \vee cd'b \vee cd'ab \vee c'db' \,,$$

which coincides with (1.40).   □

Theorem 1.1. *Let f and h be given by (1.31) and (1.33), respectively, and let ρ be defined by (1.28). Then the following conditions are equivalent:*

   (i) *the deduction theorem (i.e., $\rho \neq \varnothing$ and (1.27) holds);*

   (ii) *equations*

(1.45) $ab \vee a'b'c \vee b'c'd \vee bcd' = 0$

and (1.41)–(1.44) are fulfilled.

PROOF: In view of Lemmas 1.1 and 1.3, condition (i) is equivalent to (1.40)-(1.44) plus $abcd = 0$. The last equality transforms (1.40) into

$$0 = a'b'c \vee b'c'd \vee abc' \vee bcd'(a \vee a') \vee abcd$$

$$= a'b'c \vee b'c'd \vee ab \vee a'bcd' = a'b'c \vee b'c'd \vee ab \vee bcd' \,.$$

                                                                         □

Corollary 1.3. *The set of all pairs (ρ, h) satisfying the deduction theorem (1.27) is obtained as follows: (a, b, c, d) runs over the set of solutions to (1.45), while for each (a, b, c, d), the elements q, r, s, t run over the solutions of (1.41)–(1.44).*

PROOF: Equation (1.45) is consistent: take e.g. $b := c := d := 0$.   □

   We are now looking for significant specializations, namely $h(x,y) := x \to y$, then $\rho :=$ a partial order.

Proposition 1.6. *A consistent Boolean relation ρ (i.e., $\rho \neq \varnothing$) satisfies the specialized deduction theorem*

(1.46) $xy \rho z \iff x \rho (y \to z)$

if and only if it is of the form

(1.47) $x\rho y \iff b \leq (x \to y) \leq a'$,

for some fixed $a, b \in B$ with $ab = 0$.

PROOF: Apply Theorem 1.1 with $h(y) := x \to y = x' \lor y$, that is, $q := s := t := 1$, $r := 0$; cf. (1.33). Then conditions (1.41) and (1.42) are fulfilled, while (1.43) and (1.44) reduce to $a + c = a + d = c + d = 0$. It is plain that the last system is equivalent to $c = d$ & $a = c$, which reduces condition (1.45) to $ab = 0$, while (1.31) becomes

$$f(x, y) = axy \lor bxy' \lor ax'y \lor ax'y' = a(x' \lor y) \lor bxy',$$

so that (1.28) becomes (1.47).   □

Proposition 1.7. *The following conditions are equivalent for the relation ρ in Theorem 1.1:*

(i) ρ is reflexive ;

(ii) ρ is a quasi-order ;

(iii) $a = c = d = 0$, i.e., $x\rho y \Longleftrightarrow bx \leq y$.

PROOF: (i)⟺(iii): Reflexivity means $f(x, x) = 0$, that is, $ax \lor dx' = 0$, or equivalently, $a = d = 0$. But this implies $c = 0$, because (1.45) reduces to $b'c \lor bc = 0$.

(iii)⟹(ii): ρ is reflexive by the previous equivalence and we prove ρ is transitive. If $bx \leq y$ and $by \leq z$, then

$$bxz' = bxyz' \lor bxy'z' \leq byz' \lor bxy' = 0.$$

(ii)⟹(i): Trivial.   □

Theorem 1.2. *The unique Boolean deduction theorem (1.27) for which ρ is a partial order is the classical deduction theorem*

(1.48) $$xy \leq z \Longleftrightarrow x \leq y \to z.$$

PROOF: Since (1.48) is a specialization of (1.27), it remains to prove that if (1.27) holds and ρ is a partial order, then ρ is $\leq$ and $h(x, y) = x \to y$.

In view of Proposition 1.7, the assumption on ρ implies $a = c = d = 0$ plus antisymmetry. The latter property can be written in the form

$$bxy' = 0 \text{ \& } byx' = 0 \Longrightarrow xy' \lor x'y = 0,$$

which, in view of the Verification Theorem, is equivalent to

$$xy' \lor x'y \leq bxy' \lor bx'y;$$

taking $y := x'$ we obtain $1 \leq bx \lor bx'$, that is, $b = 1$.

We have thus proved that ρ coincides with $\leq$. On the other hand, relations (1.41)-(1.44), which hold by Theorem 1.1, reduce to $q' = 0$, $r = 0$, $s' = 0$, $t' = 0$, therefore $h(x, y) = xy \lor x'y \lor x'y' = x' \lor y$.   □

Finally we turn to another specialization of equation (1.30), namely $h := g := f$. The functional equation

(1.49) $$f(f(x, y), z) = f(x, f(y, z))$$

has an algebraic meaning: it expresses the associativity of the operation f.

Proposition 1.8. (Bernstein [1924]; cf. BFE, Theorem 12.3). *The Boolean operation f given by (1.31) is associative if and only if*

(1.50)
$$a'd \vee (b+c)(a' \vee d) = 0 \,.$$

PROOF: We apply Lemma 1.2 with $a = j = q$, $b = k = r$, $c = m = s$, $d = n = t$. Conditions (1.34)-(1.37) become

$$(a+c)(a' \vee c)a \vee ((a+b)a \vee (b+c)a' \vee (a+d)c \vee (c+d)c')a' = 0 \,,$$

$$((a+b)a \vee (a+d)a' \vee (b+c)c \vee (c+d)c')b \vee (b+d)(a' \vee c)b' = 0 \,,$$

$$(a+c)(b' \vee d)c \vee ((a+b)b \vee (b+c)b' \vee (a+d)d \vee (c+d)d')c' = 0 \,,$$

$$((a+b)b \vee (a+d)b' \vee (b+c)d \vee (c+d)d')d \vee (b+d)(b' \vee d)d' = 0 \,,$$

that is,

$$(b+c)a' \vee a'cd \vee a'c'd = 0 \,,$$

$$a'bd \vee bc'd \vee b'd(a' \vee c) = 0 \,,$$

$$a'c(b' \vee d) \vee a'bc' \vee a'c'd = 0 \,,$$

$$a'bd \vee a'b'd \vee (b+c)d = 0 \,,$$

which becomes (1.50) after obvious absorptions.   □

We also mention Levchenkov [1999b], who discusses several logical paradoxes in terms of solutions of Boolean equations.

2 Post-based algebra

In this section we initiate Post-based algebra by determining all semilattice structures on a 3−Post algebra L for which the semilattice operation is a Post function.

So we take $r := 3$. In this case the chain of constants is $e_0 = 0 < e_1 < e_2 = 1$ and we will use the simpler notation $e_1 = e$. The disjunctive representation of an element x (cf. Definition 5.1.1) becomes

(2.1)
$$x = x^1 e \vee x^2 \,.$$

If we adopt the code $C_r = \{0, 1, 2\}$, then the canonical expansion of a Post function $f : L^2 \longrightarrow L$ given in Theorem 5.2.1 can be written

(2.2)
$$f(x,y) = c_{00}x^0 y^0 \vee c_{01}x^0 y^1 \vee c_{02}x^0 y^2 \vee c_{10}x^1 y^0 \vee$$
$$\vee c_{11}x^1 y^1 \vee c_{12}x^1 y^2 \vee c_{20}x^0 y^2 \vee c_{21}x^2 y^1 \vee c_{22}x^2 y^2 \,,$$

where

(2.3)
$$c_{ij} = f(e_i, e_j) \qquad (i, j \in \{0, 1, 2\}) \,.$$

Lemma 2.1. *Let L be a $3-Post$ algebra. A Post function $f : L^2 \longrightarrow L$ is commutative and idempotent if and only if it is of the form*

$$
\begin{aligned}
f(x,y) = ex^1y^1 \vee x^2y^2 \vee \\
\vee a(x^0y^1 \vee x^1y^0) \vee b(x^0y^2 \vee x^2y^0) \vee c(x^1y^2 \vee x^2y^1) \, .
\end{aligned}
\tag{2.4}
$$

PROOF: Apply Corollary 5.2.2 to the function (2.2) via (2.3): idempotency is equivalent to $c_{00} = 1$, $c_{11} = e$ and $c_{22} = 1$, while commutativity amounts to $c_{ij} = c_{ji}$. □

Proposition 2.1. *Let L be a $3-Post$ algebra and $f : L^2 \longrightarrow L$ a Post function. The following conditions are equivalent:*

(i) $(L; f)$ is a semilattice;

(ii) the function f is of the form (2.4), where

$$
a^0b^0 \vee a^1c^1 \vee b^2c^2 = 1 \, ;
\tag{2.5}
$$

(iii) the function f is of the form (2.4), where

$$
b = a^2 \vee ea^1t^1 \vee t^2 \, ,
\tag{2.6.1}
$$

$$
c = a \vee (a^0 \vee e)t^2 \vee e(a^1 \vee a^0((t^0 \vee t^1)))s^1 \vee (a^0 \vee t^2)s^2 \, ,
\tag{2.6.2}
$$

where a, s, t are arbitrary.

PROOF: (i)$\Longleftrightarrow$(ii): In view of Lemma 2.1, it remains to prove that the Post function (2.4) satisfies the associativity condition $f(x, f(y, z)) = f(f(x, y), z)$ if and only if relation (2.5) holds.

Write (2.4) in the form

$$
\begin{aligned}
f(x,y) = (0 \cdot x^0 \vee ax^1 \vee bx^2)y^0 \vee \\
\vee (ex^1 \vee ax^0 \vee cx^2)y^1 \vee (1 \cdot x^2 \vee bx^0 \vee cx^1)y^2
\end{aligned}
\tag{2.7}
$$

and recall that $0^0 = e^1 = 1^2 = 1$, else $(e_i)^h = 0$ (cf. Remark 5.1.1). Then (2.7) and Lemma 5.1.2 imply

$$
f^0(y,z) = (y^0 \vee a^0y^1 \vee b^0y^2)z^0 \vee (a^0y^0 \vee c^0y^2)z^1 \vee (b^0y^0 \vee c^0y^1)z^2 \, ,
\tag{2.8.0}
$$

$$
f^1(y,z) = (a^1y^1 \vee b^1y^2)z^0 \vee (y^1 \vee a^1y^0 \vee c^1y^2)z^1 \vee (b^1y^0 \vee c^1y^1)z^2 \, ,
\tag{2.8.1}
$$

$$
f^2(y,z) = (a^2y^1 \vee b^2y^2)z^0 \vee (a^2y^0 \vee c^2y^2)z^1 \vee (y^2 \vee b^2y^0 \vee c^2y^1)z^2 \, .
\tag{2.8.2}
$$

Setting

$$
f(x, f(y, z)) = g(x, y, z) \, ,
$$

it follows from (2.7) and (2.8) that

$$
(2.9.0) \quad
\begin{aligned}
g(x,0,z) &= (ax^1 \vee bx^2)(z^0 \vee a^0 z^1 \vee b^0 z^2) \vee \\
&\vee (ex^1 \vee ax^0 \vee cx^2)(a^1 z^1 \vee b^1 z^2) \vee (x^2 \vee bx^0 \vee cx^1)(a^2 z^1 \vee b^2 z^2) \;,
\end{aligned}
$$

$$
(2.9.1) \quad
\begin{aligned}
g(x,1,z) &= (ax^1 \vee bx^2)(a^0 z^0 \vee c^0 z^2) \vee (ex^1 \vee ax^0 \vee cx^2) \cdot \\
&\cdot (a^1 z^0 \vee z^1 \vee c^1 z^2) \vee (x^2 \vee bx^0 \vee cx^1)(a^2 z^0 \vee c^2 z^2) \;,
\end{aligned}
$$

$$
(2.9.2) \quad
\begin{aligned}
g(x,2,z) &= (ax^1 \vee bx^2)(b^0 z^0 \vee c^0 z^1) \vee (ex^1 \vee ax^0 \vee cx^2) \cdot \\
&\cdot (b^1 z^0 \vee c^1 z^1) \vee (x^2 \vee bx^0 \vee cx^1)(b^2 z^0 \vee c^2 z^1 \vee z^2) \;.
\end{aligned}
$$

On the other hand, commutativity implies

$$
f(f(x,y),z) = f(z,f(x,y)) = f(z,f(y,x)) = g(z,y,x) \;,
$$

hence associativity amounts to $g(x,y,z) = g(z,y,x)$, or equivalently, $g(\alpha,\beta,\gamma) = g(\gamma,\beta,\alpha)$ for $\alpha,\beta,\gamma \in \{0,1,2\}$, by Corollary 5.2.2. Therefore, taking in turn $(\alpha,\gamma) := (0,1),(0,2),(1,2)$, relations (2.9) yield the following system of conditions equivalent to associativity:

$$
aa^1 \vee ba^2 = a \;,
$$

$$
a = aa^0 \vee ea^1 \vee ca^2 \;,
$$

$$
ac^1 \vee bc^2 = ab^0 \vee eb^1 \vee cb^2 \;,
$$

$$
ab^1 \vee bb^2 = b \;,
$$

$$
ac^1 \vee bc^2 = ba^0 \vee ca^1 \vee a^2 \;,
$$

$$
b = bb^0 \vee cb^1 \vee b^2 \;,
$$

$$
ab^0 \vee eb^1 \vee cb^2 = ba^0 \vee ca^1 \vee a^2 \;,
$$

$$
ac^0 \vee ec^1 \vee cc^2 = c \;,
$$

$$
c = bc^0 \vee cc^1 \vee c^2 \;.
$$

Taking into account the identities $xx^0 = 0$, $xx^1 = x^1 e$ and $xx^2 = x^2$, which follow from (2.1), we see that the above nine conditions, which express associativity, coincide with conditions (E2.1)–(E.2.9), dealt with in Example 5.3.2, where it was proved that this system of conditions is equivalent to (E2.16), which we have re-numbered here as (2.5)

(ii)$\Longleftrightarrow$(iii): This was proved in Example 5.3.7.   □
See also Example 5.3.8.

3 Geometry

As was shown in BFE, Chapter 13, §§2,3 , there are several proposals for convenient definitions of Boolean analogues of certain basic concepts of geometry. The first part of this section follows Melter and Rudeanu's approach to the concepts of distance, area, isometry and circle; cf. Melter and Rudeanu [1976]. The second part of the section presents a Boolean analogue of the Steiner problem, due to Ting and Zhao [1991].

Beside Boolean geometry, there is also a geometry studied by Foster [1951], Zemmer [1956] and Batbedat [1971], in which the rôle of the space is played by a p–ring, that is, a ring satisfying the identities $x^p = x$ and $px = 0$, where p is a prime number. Foster and Zemmer have shown that if $(B; \cdot, +, 0, 1)$ is a Boolean ring, then the set

(3.1) $$R = \{(x_1, x_2) \in B^2 \mid x_1 x_2 = 0\}$$

is a 3–ring with respect to the operations

(3.2)
$$(x_1, x_2) + (y_1, y_2)$$
$$= (x_1 + y_1 + x_1 y_2 + x_2 y_1 + x_2 y_2 , \ x_2 + y_2 + x_1 y_2 + x_2 y_1 + x_1 y_1) ,$$

(3.3) $$(x_1, x_2)(y_1, y_2) = (x_1 y_1 + x_2 y_2, x_1 y_2 + x_2 y_1) ;$$

it is immediately seen that the set of idempotents of R is $\{(x, 0) \mid x \in B\}$, which can be identified with B. Since we want to construct two parallel geometries, taking as space B^2 and R, respectively, we need a few technical preliminaries.

A *polynomial* of the ring R is understood in the usual ring-theoretical sense; it will be termed *simple* if it has an expression which involves no constant of R, except possibly 1. So, polynomials and simple polynomials are the specializations to R of the universal-algebra concepts of algebraic function and polynomial, respectively[1], just like Boolean functions and simple Boolean functions, respectively. As a matter of fact, there is more than a similitude: it is easily seen that every polynomial (simple polynomial) $p : R^n \longrightarrow R$ is of the form $p = (f \times g)|_{R^n}$, where $f, g : B^{2n} \longrightarrow B$ are Boolean (simple Boolean) functions and $f \times g$ is defined by

(3.4) $$(f \times g)(X, Y, \ldots) = (f(X, Y, \ldots), g(X, Y, \ldots)) \ (\forall X, Y, \ldots \in B^2) .$$

By a *Boolean-valued polynomial* we mean a polynomial p with values in the Boolean algebra B; this makes sense in view of the above identification of B with the subset of idempotents $(x, 0)$ of R.

Notation. Unless otherwise stated, in the first part of this section the elements of B^2 will be denoted by $X = (x_1, x_2), Y = (y_1, y_2), \ldots$, while the notation $A = (\alpha_1, \alpha_2), C = (\gamma_1, \gamma_2), D = (\delta_1, \delta_2), \ldots$, is reserved for the elements of $\{0, 1\}^2$. The letter S stands for each of the sets B^2 and R.   □

We need an extension to R of the Verification Theorem; cf. BFE, Theorems 2.13, 2.14 and Corollary.

Lemma 3.1. *The following conditions are equivalent for two Boolean functions* $f, g : B^{2n} \longrightarrow B$:

(i) $f|_{R^n} = g|_{R^n}$;

(ii) $f(A, C, \ldots) = g(A, C, \ldots) \ (\forall A, C, \ldots \in R \cap \{0, 1\}^2)$.

PROOF: Suppose (ii). Then for every $A, C, \ldots \in \{0, 1\}^2$ we have

---

[1] See the note on terminology in Ch.2, §3.

$$f(A, C, \ldots) \vee \alpha_1 \alpha_2 \vee \gamma_1 \gamma_2 \vee \ldots = g(A, C, \ldots) \vee \alpha_1 \alpha_2 \vee \gamma_1 \gamma_2 \vee \ldots ,$$

because if e.g. $A \notin R$, then $\alpha_1 \alpha_2 = 1$, hence both sides of the above relation equal 1. Therefore

$$f(X, Y, \ldots) \vee x_1 x_2 \vee y_1 y_2 \vee \ldots = g(X, Y, \ldots) \vee x_1 x_2 \vee y_1 y_2 \vee \ldots$$

for all $X, Y, \ldots \in B^2$, which implies (i). The converse is trivial.   □

Lemma 3.2. *The following conditions are equivalent for two Boolean functions* $f, g : B^{2n} \longrightarrow B$ *such that the equation* $f(X, Y, \ldots) = 0$ *has a solution in* R^n:

(i) $f(X, Y, \ldots) = 0 \Longrightarrow g(X, Y, \ldots) = 0 \qquad (\forall X, Y, \ldots \in R)$;

(ii) $g(X, Y, \ldots) \leq f(X, Y, \ldots) \qquad (\forall X, Y, \ldots \in R)$;

(iii) $g(A, C, \ldots) \leq f(A, C, \ldots) \qquad (\forall A, C, \ldots \in R \cap \{0,1\}^2)$.

PROOF: From the corresponding Boolean result, using the same technique as in the proof of Lemma 3.1.   □

Corollary 3.1. *Let* $f, g : B^{2n} \longrightarrow B$ *be Boolean functions and assume that one of the equations* $f(X, Y, \ldots) = 0$, $g(X, Y, \ldots) = 0$ *has a solution in* R^n. *Then these equations have the same solutions in* R^n *if and only if* $f|_{R^n} = g|_{R^n}$.   □

Remark 3.1. The above results can be extended from R to any subset of B^2 which is characterized by a simple Boolean equation $\varphi(x_1, x_2) = 0$.   □

Definition 3.1. A function $d : S \longrightarrow B$ is called a *distance function* of the space S ($:= B^2$ or R), provided it satisfies the following conditions: for every $X, Y, Z \in S$,

(D1) $$d(X, Y) = 0 \Longleftrightarrow X = Y ,$$

(D2) $$d(X, Y) = d(Y, X) ,$$

(D3) $$d(X, Y) \leq d(X, Z) \vee d(Z, Y) .$$

□

Proposition 3.1. *The following conditions are equivalent for a Boolean function* $d : B^4 \longrightarrow B$ *(for a simple Boolean-valued polynomial* $d : R^2 \longrightarrow B$):

(i) d *is a distance function of* S ;

(ii) d *fulfils* (D1) ;

(iii) d *is the function*

(3.5) $$d(X, Y) = (x_1 + y_1) \vee (x_2 + y_2) .$$

PROOF: Consider first the case $S := B^2$.

(ii)$\Longleftrightarrow$(iii): This is a particular case of a theorem of Whitehead (see e.g. BFE, Theorem 6.6), provided we interpret (D1) as expressing the fact that equation $d(X, Y) = 0$ has the unique solution $X = Y$ with respect to X.

(iii)$\Longrightarrow$(i): Routine, using the consensus inequality.

(i)$\Longrightarrow$(ii): Trivial.

Now take $S := R$. Clearly the implications (iii)$\Longrightarrow$(i) $\Longrightarrow$(ii) are still valid, so that it remains to prove (ii) $\Longrightarrow$(iii). Suppose d fulfils (D1) and let $f : B^4 \longrightarrow B$ be a simple Boolean function such that $f|_{R^2} = d$. If we succeed to prove that $f(A,C) = (\alpha_1 + \gamma_1) \vee (\alpha_2 + \gamma_2)$ for every $A, C \in R \cap \{0,1\}^2$, it will follow from Lemma 3.1 that $f(X,Y) = (x_1 + y_1) \vee (x_2 + y_2)$ for every $X, Y \in R$, that is (iii). If the above points A, C are distinct, then $(\alpha_1 + \gamma_1) \vee (\alpha_2 + \gamma_2) = 1$, while $f(A,C) = d(A,C) \neq 0$, hence $f(A,C) = 1$ because f is a simple Boolean function (cf. BFE, Theorem 1.7). If $A = C$ then $(\alpha_1 + \gamma_1) \vee (\alpha_2 + \gamma_2) = 0 = d(A,C) = f(A,C)$. $\qquad\square$

Corollary 3.2. *The distance between any two distinct points $A, C \in \{0,1\}$ is 1.* $\qquad\square$

Notation. Unless otherwise stated, in the remainder of this section the letter d will stand for the distance function (3.5) $\qquad\square$

The following results will also be needed in the sequel.

Lemma 3.3. *The following conditions are equivalent in the space S $(:= B^2$ or $R)$ for a Boolean function $\circ : B^2 \to B$:*

$$(3.6) \qquad\qquad d(X,Y) \leq d(X,Z) \circ d(Z,Y) \qquad (\forall X, Y, Z \in S) ;$$

$$(3.7) \qquad\qquad\qquad x \vee y \leq x \circ y \qquad (\forall x, y \in B) .$$

PROOF: (3.7)$\Longrightarrow$(3.6): By (D3).

(3.6)$\Longrightarrow$(3.7): Write down the inequality (3.6) for the following elements of S: $X := (0,0)$, $Y := (1,0)$ and $Z \in \{(0,0),(1,0),(0,1)\}$. Using Corollary 3.2, we obtain $1 \leq 0 \circ 1$, $1 \leq 1 \circ 0$ and $1 \leq 1 \circ 1$, respectively. Setting $x \circ y = axy \vee bxy' \vee cx'y \vee dx'y'$, the above inequalities become in fact $c = 1$, $b = 1$ and $a = 1$, respectively; therefore $x \circ y = x \vee y \vee dx'y' = x \vee y \vee d$. $\qquad\square$

Definition 3.2. For every three points $X, Y, Z \in S$ $(:= B^2$ or $R)$, set $d_1 = d(Y,Z)$, $d_2 = d(X,Z)$, and $d_3 = d(X,Y)$. A function $\alpha : S^3 \longrightarrow S$ is called an *area function* of the space S provided there exists a Boolean function $a : B^3 \longrightarrow B$ satisfying the following conditions:

$$(A0) \qquad\qquad \alpha(X,Y,Z) = a(d_1,d_2,d_3) \qquad (\forall X,Y,Z \in S) ,$$

$$(A1) \qquad a(0,d_2,d_3) = a(d_1,0,d_3) = a(d_1,d_2,0) \qquad (\forall X,Y,Z \in S) ,$$

$$(A2) \qquad\qquad a \text{ is symmetric in } d_1, d_2, d_3 ,$$

$$(A3) \qquad\qquad\qquad a(1,1,1) = 1 .$$

$\qquad\square$

Remark 3.2. For eny $k \in B$ there exist $X, Y \in S$ such that $d(X, Y) = k$, say $X = (k, 0)$ and $Y = (0, 0)$. Therefore $a(d_1, d_2, d_3)$ is actually defined on B^3. $\quad\square$

Remark 3.3. It follows from (A0) that

(3.8)
$$d(X_i, X_j) = d(Y_i, Y_j) \; (i, j \in \{1, 2, 3\}) \Longrightarrow$$
$$\Longrightarrow \alpha(X_1, X_2, X_3) = \alpha(Y_1, Y_2, Y_3) \, .$$

$\quad\square$

Lemma 3.4. *For every* $X, Y, Z \in S$,

(3.9)
$$d_1 d_2 d_3 = d_1 + d_2 + d_3 \, .$$

PROOF: In view of the Verification Theorem or of Lemma 3.1, it suffices to prove (3.9) for $A, C, D \in S \cap \{0, 1\}^2$. If A, B, C are pairwise distinct, then Corollary 3.2 reduces (3.9) to $1 \cdot 1 \cdot 1 = 1 + 1 + 1$. If $A = B = C$ then (3.9) reduces similarly to $0 \cdot 0 \cdot 0 = 0 + 0 + 0$, while in the remaining cases, say $A = B \neq C$, the desired property is reduced to $1 \cdot 1 \cdot 0 = 1 + 1 + 0$. $\quad\square$

Proposition 3.2. *The following conditions are equivalent for a Boolean function* $\alpha : B^6 \longrightarrow B$ *(for a Boolean-valued polynomial* $\alpha : R^3 \longrightarrow B$*):*

(i) α *is an area function ;*

(ii) α *fulfils* (A0), (A1) *and* (A3) *;*

(iii) α *fulfils* (A0) *with*

(3.10)
$$a(d_1, d_2, d_3) = d_1 d_2 d_3 \, ;$$

(iv) α *fulfils* (A0) *with*

(3.11)
$$a(d_1, d_2, d_3) = d_1 + d_2 + d_3 \, .$$

PROOF: (i)$\Longrightarrow$(ii): Trivial.

(ii)$\Longrightarrow$(iii): Setting

$$a(d_1, d_2, d_3) = p d_1 d_2 d_3 + q_1 d_2 d_3 + q_2 d_1 d_3 + q_3 d_1 d_2 + r_1 d_1 + r_2 d_2 + r_3 d_3 + s \, ,$$

the hypotheses imply the identities

$$q_1 d_2 d_3 + r_2 d_2 + r_3 d_3 + s = 0 \, ,$$

$$q_2 d_1 d_3 + r_1 d_1 + r_3 d_3 + s = 0 \, ,$$

$$q_3 d_1 d_2 + r_1 d_1 + r_2 d_2 + s = 0 \, ,$$

$$p + q_1 + q_2 + q_3 + r_1 + r_2 + r_3 + s = 1 \, .$$

The first equation, written for $d_2 := d_3 := 0$, yields $s = 0$, hence for $d_2 := 1$, $d_3 := 0$ it implies $r_2 = 0$. We obtain similarly $r_1 = 0$ and $r_3 = 0$. It follows that taking $d_1 := d_2 := d_3 := 1$, the first three identities reduce to $q_1 = 0$, $q_2 = 0$ and $q_3 = 0$, respectively. Therefore the last equality becomes $p = 1$.

(iii)$\Longrightarrow$(iv): By Lemma 3.4.

(iv)$\Longrightarrow$(i): Immediate. $\square$

Corollary 3.3. *The area function can be written in the form*

$$(3.12) \qquad \alpha(X, Y, Z) = (y_1+z_1)(y_2+z_2)+(x_1+z_1)(x_2+z_2)+(x_1+y_1)(x_2+y_2).$$

PROOF: From (3.11), (2.5) and $a \vee b = a + b + ab$. $\square$

Corollary 3.4. *In the case of the space R, the area function can be written in the form*

$$(3.13) \qquad \alpha(X, Y, Z) = y_1 z_2 + y_2 z_1 + x_1 z_2 + x_2 z_1 + x_1 y_2 + x_2 y_1 .$$

$\square$

Note that formula (3.13) can be given a form which coincides, up to a factor, with a well-known formula in analytic geometry:

$$\alpha(X, Y, Z) = \begin{vmatrix} 1 & 1 & 1 \\ x_1 & y_1 & z_1 \\ x_2 & y_2 & z_2 \end{vmatrix} .$$

The next result exhibits the form of motions in Boolean geometry.

Proposition 3.3. *The following conditions are equivalent for a Boolean transformation $\mu = \varphi \times \psi : B^2 \longrightarrow B^2$:*

(i) μ preserves distances ;

(ii) μ is a bijection which preserves distances ;

(iii) μ is of the form

$$(3.14.1) \qquad \varphi(X) = \varphi_0 + \varphi_1 x_1 + \varphi_2 x_2 ,$$

$$(3.14.2) \qquad \psi(X) = \psi_0 + \psi_1 x_1 + \psi_2 x_2 ,$$

where φ_h, ψ_h ($h = 0, 1, 2$) are constants satisfying

$$(3.15) \qquad \varphi_1 \psi_2 + \varphi_2 \psi_1 = 1 .$$

PROOF: (i)$\Longrightarrow$(iii): The general form of the transformation $\mu = \varphi \times \psi$ is

$$(3.16.1) \qquad \varphi(X) = \varphi_0 + \varphi_1 x_1 + \varphi_2 x_2 + \varphi_2 x_1 x_2 ,$$

$$(3.16.2) \qquad \psi(X) = \psi_0 + \psi_1 x_1 + \psi_2 x_2 + \psi_3 x_1 x_2 ,$$

thus implying

$$\mu((0,0)) = (\varphi_0, \psi_0), \mu((1,0)) = (\varphi_0 + \varphi_1, \psi_0 + \psi_1), \mu((0,1)) = (\varphi_0 + \varphi_2, \psi_0 + \psi_2).$$

Using (iii) and Corollary 3.2, we obtain

$$\varphi_1 \vee \psi_1 = (\varphi_0 + (\varphi_0 + \varphi_1)) \vee (\psi_0 + (\psi_0 + \psi_1)) = d((\varphi_0, \psi_0), (\varphi_0 + \varphi_1, \psi_0 + \psi_1))$$

$$= d(\mu((0,0)), \mu((1,0))) = d((0,0), (1,0)) = 1$$

and similarly $\varphi_2 \vee \psi_2 = 1$ and $(\varphi_1 + \varphi_2) \vee (\psi_1 + \psi_2) = 1$. Writing down these relations in ring form and summing we obtain (3.15).

We continue by writing down the distances between the above three points and

$$\mu((1,1)) = (\varphi_0 + \varphi_1 + \varphi_2 + \varphi_3, \psi_0 + \psi_1 + \psi_2 + \psi_3),$$

hence we obtain as above

$$(\varphi_1 + \varphi_2 + \varphi_3) \vee (\psi_1 + \psi_2 + \psi_3) = 1,$$

$$(\varphi_2 + \varphi_3) \vee (\psi_2 + \psi_3) = 1,$$

$$(\varphi_1 + \varphi_3) \vee (\psi_1 + \psi_3) = 1.$$

Hence, using again the ring form and summation, we obtain

$$\varphi_3 + \psi_3 + \varphi_1\psi_2 + \varphi_2\psi_1 + \varphi_3\psi_3 = 1,$$

which, in view of (3.15), reduces to $\varphi_3 + \psi_3 + \varphi_3\psi_3 = 0$, that is, $\varphi_3 \vee \psi_3 = 0$ and finally $\varphi_3 = \psi_3 = 0$.

(iii)$\Longrightarrow$(ii): Since (3.15) implies $1 \leq \varphi_1 \vee \psi_1$ and similarly $\varphi_2 \vee \psi_2 = 1$, we infer

$$d(\mu(X), \mu(Y)) = (\varphi_1 x_1 + \varphi_2 x_2 + \varphi_1 y_1 + \varphi_2 y_2) \vee (\psi_1 x_1 + \psi_2 x_2 + \psi_1 y_1 + \psi_2 y_2)$$

$$= \varphi_1(x_1 + y_1) + \varphi_2(x_2 + y_2) + \psi_1(x_1 + y_1) + \psi_2(x_2 + y_2) +$$

$$+ (\varphi_1(x_1 + y_1) + \varphi_2(x_2 + y_2))(\psi_1(x_1 + y_1) + \psi_2(x_2 + y_2))$$

$$= (\varphi_1 + \psi_1 + \varphi_1\psi_1)(x_1 + y_1) + (\varphi_2 + \psi_2 + \varphi_2\psi_2)(x_2 + y_2) +$$

$$+ (\varphi_1\psi_2 + \varphi_2\psi_1)(x_1 + y_1)(x_2 + y_2)$$

$$= (x_1 + y_1) + (x_2 + y_2) + (x_1 + y_1)(x_2 + y_2) = (x_1 + y_1) \vee (x_2 + y_2) = d(X, Y).$$

Moreover, given any $Y \in B^2$, the equation $\mu(X) = Y$ can be written in the form

(3.17.1) $\varphi_1 x_1 + \varphi_2 x_2 = \varphi_0 + y_1,$

(3.17.2) $\psi_1 x_1 + \psi_2 x_2 = \psi_0 + y_2;$

according to a theorem of Parker and Bernstein [1955] (see also BFE, Theorem 6.10), condition (3.15), which can be written in the form

$$\begin{vmatrix} \varphi_1 & \varphi_2 \\ \psi_1 & \psi_2 \end{vmatrix} = 1,$$

ensures that system (3.17) has a unique solution.

(ii)$\Longrightarrow$(i): Trivial. $\square$

Proposition 3.4. *Let $f, g : B^2 \longrightarrow B$ be Boolean functions. The following conditions are equivalent for the transformation $\mu = (\varphi \times \psi)|_R$:*

(i) $\mu : R \longrightarrow R$ and μ preserves distances ;

(ii) $\mu : R \longrightarrow R$ and μ is a bijection which preserves distances ;

(iii) μ is of the form

(3.18.1)
$$\varphi(X) = \varphi_1\varphi_2 + \varphi_1 x_1 + \varphi_2 x_2 ,$$

(3.18.2)
$$\psi(X) = \psi_1\psi_2 + \psi_1 x_1 + \psi_2 x_2 ,$$

where φ_1, φ_2, ψ_1, ψ_2 are constants satisfying (3.15).

PROOF: (i)$\Longrightarrow$(iii): The transformation μ is of the form (3.16), where we may assume without loss of generality that $\varphi_3 = \psi_3 = 0$, because the restriction of $\varphi \times \psi$ to R does not change. Moreover, relations $\varphi_1 \vee \psi_1 = \varphi_2 \vee \psi_2 = 1$ and (3.15) hold by exactly the same proof as for (i)$\Longrightarrow$(iii) in Proposition 3.3. It will be convenient to write (3.15) in the form

(3.15')
$$\varphi_1\psi_2 \cdot \varphi_2\psi_1 \vee (\varphi_1' \vee \psi_2')(\varphi_2' \vee \psi_1') = 0 .$$

Now $\mu(X) \in R$ ($\forall X \in R$), that is, $\varphi(X)\psi(X) = 0$ ($\forall X \in R$) and this is equivalent to $\varphi(A)\psi(A) = 0$ for $A \in \{(0,0), (1,0), (0,1)\}$ by Lemma 3.1. We thus obtain the equivalent formulation

(3.19.1)
$$\varphi_0\psi_0 = 0 ,$$

(3.19.2)
$$\varphi_0\psi_1 + \varphi_1\psi_0 + \varphi_1\psi_1 = (\varphi_0 + \varphi_1)(\psi_0 + \psi_1) = 0 ,$$

(3.19.3)
$$\varphi_0\psi_2 + \varphi_2\psi_0 + \varphi_2\psi_2 = (\varphi_0 + \varphi_2)(\psi_0 + \psi_2) = 0 ,$$

Multiplying (3.19.2) by φ_2 and (3.19.3) by φ_1 and summing, we obtain

(3.20.1)
$$\varphi_0(\psi_1\varphi_2 + \psi_2\varphi_1) + \varphi_1\varphi_2(\psi_1 + \psi_2) = 0 ;$$

but (3.15) implies

(3.21.1)
$$\varphi_1\varphi_2(\psi_1 + \psi_2) = \varphi_1\varphi_2\psi_1\psi_2' \vee \varphi_1\varphi_2\psi_1'\psi_2$$
$$= \varphi_1\varphi_2\psi_1 \vee \varphi_1\varphi_2\psi_2 = \varphi_1\varphi_2(\psi_1 \vee \psi_2) = \varphi_1\varphi_2$$

and using (3.15), we see that (3.20.1) reduces to $\varphi_0 + \varphi_1\varphi_2 = 0$. Similarly we get $\psi_0 = \psi_1\psi_2$.

(iii)$\Longrightarrow$(ii): It follows from (3.18) and (3.15') that $\varphi_0\psi_0 = \varphi_1\varphi_2\psi_1\psi_2 = 0$,

$$\varphi_0\psi_1 + \varphi_1\psi_0 + \varphi_1\psi_1 = \varphi_1\varphi_2 + \varphi_1\psi_1\psi_2 + \varphi_1\psi_1$$

$$= \varphi_1\psi_1(\varphi_2 + \psi_2 + 1) = \varphi_1\psi_1(\varphi_2\psi_2 \vee \varphi_2'\psi_2') = 0$$

and similarly we can verify (3.19.3). Thus μ satisfies (3.19), hence $\mu(X) \in R$ for every $X \in R$. We also know from Proposition 3.3 that μ preserves distances.

Finally we have to show that if $Y \in R$ then the unique solution X of system (3.17), where $\varphi_0 = \varphi_1\varphi_2$ and $\psi_0 = \psi_1\psi_2$, belongs to R. But according to the

Parker-Bernstein theorem, this solution is provided by Cramer's rule, hence, taking into account (3.21), we obtain

$$x_1 x_2 = (\varphi_1\varphi_2\psi_2 + \psi_2 y_1 + \varphi_2\psi_1\psi_2 + \varphi_2 y_2)(\varphi_1\psi_1\psi_2 + \varphi_1 y_2 + \varphi_1\varphi_2\psi_1 + \psi_1 y_1)$$

$$= \varphi_1\varphi_2\psi_2 y_2 + \varphi_1\psi_1\psi_2 y_1 + \psi_1\psi_2 y_1 + \varphi_2\psi_1\psi_2 y_1 + \varphi_1\varphi_2 y_2 + \varphi_1\varphi_2\psi_1 y_2$$

$$= \psi_1\psi_2(\varphi_1 + \varphi_2 + 1)y_1 + \varphi_1\varphi_2(\psi_1 + \psi_2 + 1)y_2$$

$$= (\psi_1\psi_2 + \psi_1\psi_2)y_1 + (\varphi_1\varphi_2 + \varphi_1\varphi_2)y_2 = 0 \ .$$

(ii)$\Longrightarrow$(i): Trivial.   □

We now introduce the following natural analogue of the concept of circle.

Definition 3.3. Given a point $C \in R$ and an element $r \in B$, the *circle* of *center* C and *radius* r is the set

$$(3.22) \qquad\qquad C(r) = \{X \in R \mid d(C, X) = r\} \ .$$

□

Proposition 3.5. *A subset of the space R is a circle if and only if it is the set of solutions to an equation of the form*

$$(3.23) \qquad\qquad ax_1 + bx_2 = c \ ,$$

where a and b are constants of B satisfying

$$(3.24) \qquad\qquad a \vee b = 1 \ .$$

The center of the circle is the point $(b', a') \in R$ and the radius is $r = a + b + c$.

PROOF: The equation $d(C, X) = r$ can be successively written in the following equivalent forms:

$$(c_1 + x_1) + (c_2 + x_2) + (c_1 + x_1)(c_2 + x_2) = r \ ,$$

$$c_1 + c_2 + x_1 + c_2 x_1 + x_2 + c_1 x_2 = r \ ,$$

$$c_2' x_1 + c_1' x_2 = c_1 + c_2 + r \ .$$

So the equation is of the form (3.23) for $a := c_2'$, $b := c_1'$ and $c := c_1 + c_2 + r = a + b + r$. We have $a \vee b = (c_1 c_2)' = 1$.   □

Corollary 3.5. *For every $C \in R$ and every $r \in B$, the circle $C(r)$ is not empty and its equation (3.23) is uniquely determined.*

PROOF: For instance, the point $(ac, a'c) \in R$ satisfies (3.23): since $a'b' = 0$, we have $a \cdot ac + b \cdot a'c = ac + a'c = c$.   □

Unfortunately, for every three points X, Y, Z located on a circle (3.23), the area (3.13) of the triangle XYZ is 0. In view of Lemma 3.2, the proof is done by showing that

$$\alpha(X, Y, Z) \le (ax_1 + bx_2 + c) \vee (ay_1 + by_2 + c) \vee (az_1 + bz_2 + c)$$

for all $X, Y, Z \in R \cap \{0, 1\}^2$.

In the remainder of this section we follow Ting and Zhao [1991]. Their geometry, different from the one above, falls essentially within the Ellis-Blumenthal line, begun in the early fifties; cf. BFE, Chapter 13.

Definition 3.4. Given a Boolean algebra B, the associated *Boolean geometry* regards B as a *space* endowed with the ring sum as distance $d(x, y) = x + y$ between the points $x, y \in B$. The *linear subspace* generated by two points $a, b \in B$ is the interval

(3.27)
$$< a, b > = [ab, a \vee b] .$$

The *diameter* of the subspace $< a, b >$ is $a + b$.   □

Proposition 3.6. *The diameter $a + b$ is the greatest distance between any two points $x, y \in < a, b >$.*

PROOF: If $x, y \in < a, b >$ then $ab \leq x, y \leq a \vee b$, hence $x', y' \leq a' \vee b'$, therefore

$$xy', x'y \leq (a \vee b)(a' \vee b') = ab' \vee a'b = a + b ,$$

hence $d(x, y) = xy' \vee x'y \leq a + b$. The upper bound $a + b$ is actually reached, for instance

$$d(ab, a \vee b) = ab \cdot a'b' \vee (a' \vee b')(a \vee b) = a + b .$$

□

Definition 3.5. The *general Steiner problem* for $2n$ elements $c_1, \ldots, c_n, x_1, \ldots, x_n \in B$ is to find $\min \varphi(x)$, where $\varphi(x) = \bigvee_{i=1}^{n} c_i d(x_i, x)$. The elements $c_1, \ldots, c_n$ are termed *weights*.

Proposition 3.7. *Set*

(3.28)
$$a = \bigvee_{i=1}^{n} c_i x_i , \quad b = \bigvee_{i=1}^{n} c_i x_i' .$$

Then the range of the function φ is the subspace $< a, b >$, while the solution set for the general Steiner problem is the subspace $< a', b >$.

PROOF: The first statement follows from the fact that $a = \varphi(0)$ and $b = \varphi(1)$. Now an element $x \in B$ minimizes the function φ if and only $\varphi(x) = ab$. But this equation can be written in the form $ax \vee bx' \leq ab$, or equivalently, $ab'x \vee a'bx' = 0$, therefore the solution set is described by $a'b \leq x \leq a' \vee b$.   □

Proposition 3.8. *The following conditions are equivalent for a given system of weights $c_1, \ldots, c_n$:*

(i) the general Steiner problem has a unique solution for every n elements $x_1, \ldots, x_n$;

(ii) the system $(c_1, \ldots, c_n)$ is orthonormal .

PROOF: The subspace $< a', b >$ is a singleton if and only if $a' = b$ (cf. (3.28)) that is,

$$(3.29) \qquad \prod_{i=1}^{n}(c_i' \vee x_i') = \bigvee_{i=1}^{n} c_i x_i' \, .$$

Now condition (i) states that (3.29) is an identity. This happens if and only if

$$\prod_{i=1}^{n}(c_i' \vee \alpha_i) = \bigvee_{i=1}^{n} c_i \alpha_i \qquad (\forall \alpha_1, \ldots, \alpha_n \in \{0,1\})$$

and this is equivalent to

$$(3.30) \qquad \prod_{i \in CM} c_i' = \bigvee_{i \in M} c_i \qquad (\forall M \subseteq \{0,1\}^n) \, ,$$

where $CM = \{0,1\}^n \backslash M$. It is easily seen that (3.30) expresses the orthonormality of $(c_1, \ldots, c_n)$. $\qquad \square$

The paper by Ting and Zhao [1991] contains a detailed study of the general Steiner problem.

See also Melter and Rudeanu [1993]; cf. Ch.14, §7.

4 Statistics

Melter and Rudeanu [1981] have introduced Boolean analogues of the concepts of measure of central tendency and dispersion, for which they have obtained a Chebyshev-like inequality, as we are going to show in this section.

The values which are to be averaged by a measure of central tendency, like the mean or the median, are usually real numbers. The following Boolean analogue seems natural.

Definition 4.1. By a *Boolean measure of central tendency* for n variables over a Boolean algebra B we mean a symmetric isotone simple Boolean function $\beta : B^n \longrightarrow B$ such that

$$(4.1) \qquad x_1 \leq x_2 \leq \ldots \leq x_n \Longrightarrow \beta(x_1, \ldots, x_n) = x_{[n/2]+1} \, ,$$

where [ ] is the greatest integer function. $\qquad \square$

Let us determine the functions with these properties.

Lemma 4.1. *The only isotone symmetric simple Boolean functions $\beta : B^n \longrightarrow B$ are the constant functions $0, 1$ and the fundamental symmetric Boolean functions*

$$(4.2) \qquad \beta_k(x_1, \ldots, x_n) = \bigvee_{i_1, \ldots, i_k} x_{i_1} \ldots x_{i_k} \qquad (k = 1, \ldots, n) \, ,$$

where $\bigvee_{i_1,\ldots,i_k}$ denotes disjunction over all $k-$element subsets $\{i_1,\ldots,i_k\}$ of $\{1,\ldots,n\}$.

PROOF: Clearly the constant functions 0,1 and the fundamental symmetric Boolean functions are isotone symmetric simple Boolean functions. To prove the converse, recall (see e.g. BFE, Theorem 11.3) that a Boolean function β is isotone if and only if it can be written in the form

$$\beta(x_1,\ldots,x_n) = a_0 \vee \bigvee_{p=1}^{n} \bigvee_{i_1,\ldots,i_p} a_{i_1\ldots i_p} x_{i_1} \ldots x_{i_p} .$$

Then $a_0 = \beta(0,\ldots,0) \in \{0,1\}$ because β is a simple Boolean function. For $a_0 = 1$ we obtain the constant function $\beta = 1$. Now suppose $a_0 = 0$. It follows that for each set $\{i_1,\ldots,i_p\}$ we have $a_{i_1\ldots i_p} = f(\xi_1,\ldots,\xi_n)$, where $\xi_i = 1$ if $i \in \{i_1,\ldots,i_p\}$, else $\xi_i = 0$. Hence for each p, the symmetry of β implies that the coefficients $a_{i_1\ldots i_p}$ are equal, that is, β is of the form

$$\beta(x_1,\ldots,x_n) = \bigvee_{p=1}^{n} a_p \bigvee_{i_1,\ldots,i_p} x_{i_1} \ldots x_{i_p}$$

and moreover, $a_p = \beta(\xi_1,\ldots,\xi_n) \in \{0,1\}$ for every p. If all $a_p = 0$ then β is the constant function $\beta = 0$. Otherwise $\beta = \beta_k$, where $a_k = 1$ and $a_j = 0$ for $j < k$. □

Proposition 4.1. For every $n \in \mathbf{N}$, the only Boolean measure of central tendency for n variables is $\beta_{[(n-1)/2]+1}$.

PROOF: If $x_1 \leq x_2 \leq \ldots \leq x_n$, formula (4.2) yields

$$\beta_k(x_1,\ldots,x_n) = x_{n-k+1} x_{n-k+2} \ldots x_n = x_{n-k+1} ,$$

therefore condition (4.1) reduces to $n - k = [n/2]$. If $n = 2h$ this means $k = h$, while for $n = 2h + 1$ we obtain $k = h + 1$. It is easy to check that in each case $k = [(n - 1)/2] + 1$. □

Example 4.1. The first five functions of this kind are

$$\beta(x_1) = x_1 ,$$

$$\beta(x_1, x_2) = x_1 \vee x_2 ,$$

$$\beta(x_1, x_2, x_3) = x_1 x_2 \vee x_1 x_3 \vee x_2 x_3 ,$$

$$\beta(x_1, x_2, x_3, x_4) = x_1 x_2 \vee x_1 x_3 \vee x_1 x_4 \vee x_2 x_3 \vee x_2 x_4 \vee x_3 x_4 ,$$

$$\beta(x_1, x_2, x_3, x_4, x_5) = x_1 x_2 x_3 \vee x_1 x_2 x_4 \vee x_1 x_2 x_5 \vee x_1 x_3 x_4 \vee x_1 x_3 x_5 \vee$$

$$\vee x_1 x_4 x_5 \vee x_2 x_3 x_4 \vee x_2 x_3 x_5 \vee x_2 x_4 x_5 \vee x_3 x_4 x_5 .$$

□

The following analogue of the dispersion function also seems natural.

Definition 4.2. By a *Boolean dispersion function* of n variables over a Boolean algebra B we mean a Boolean function $\delta : B^n \longrightarrow B$ such that

$$(4.3) \qquad \delta(x_1, \ldots, x_n) = 0 \Longleftrightarrow x_1 = x_2 = \ldots = x_n .$$

□

Proposition 4.2. *The only Boolean dispersion function of n variables is*

$$(4.4) \qquad \delta(x_1, \ldots, x_n) = (x_1 + x_2) \vee (x_1 + x_3) \vee \ldots \vee (x_1 + x_n) ,$$

which can also be written in the form

$$(4.5) \qquad \delta(x_1, \ldots, x_n) = (x_1 \vee \ldots \vee x_n)(x_1' \vee \ldots \vee x_n') .$$

PROOF: $(4.3) \Longleftrightarrow (4.4)$: From

$$x_1 = x_2 = \ldots = x_n \Longleftrightarrow (x_1 + x_2) \vee (x_1 + x_3) \vee \ldots \vee (x_1 + x_n) = 0 ,$$

via the Verification Theorem (see e.g. Corollary to Theorem 2.14 in BFE).
  $(4.4) \Longleftrightarrow (4.5)$: The equality of the right sides is immediate. □

Remark 4.1. a) If $x_1 \leq x_2 \leq \ldots \leq x_n$ then

$$\delta(x_1, \ldots, x_n) = x_1' x_n = x_1 + x_n ,$$

where $x + y$ may be viewed as the distance between x_1 and x_n; cf. BFE, Chapter 13, §§2,3, and the present §3.
  b) δ is symmetric.
  c) $\delta(x_1, \ldots, x_n) = 1 \Longleftrightarrow \prod_{i=1}^{n} x_i = 0$ & $\bigvee_{i=1}^{n} x_i = 1$.
  d) $\delta(x_1 + y, \ldots, x_n + y) = \delta(x_1, \ldots, x_n)$. □

We are now in a position to prove an analogue of the Chebyshev inequality.

Proposition 4.3. *The following inequalities hold for all $n \in \mathbf{N}$ and all $k \in \{1, \ldots, n\}$:*

$$(4.6) \qquad x_j + \beta_k(x_1, \ldots, x_n) \leq \delta(x_1, \ldots, x_n) \qquad (j = 1, \ldots, n) .$$

PROOF: Taking into account that by multiplying out

$$\prod_{i_1, \ldots, i_k} (x_{i_1}' \vee \ldots \vee x_{i_k}')$$

we obtain a disjunction of terms in which every literal is complemented, we have

$$x_j + \beta_k(x_1, \ldots, x_n)$$

$$= x_j \prod_{i_1,\ldots,i_k} (x'_{i_1} \vee \ldots \vee x'_{i_k}) \vee x'_j \bigvee_{i_1,\ldots,i_k} x_{i_1} \ldots x_{i_k}$$

$$\leq x_j \bigvee_{i=1}^{n} x'_i \vee x'_j \bigvee_{i=1}^{n} x_i \leq (x_1 \vee \ldots \vee x_n)(x'_1 \vee \ldots \vee x'_n) = \delta(x_1,\ldots,x_n) .$$

$\square$

Mayor and Martin [1999] define a *locally internal aggregation function* to be a continuous isotone real function $f : \mathbf{R}^n \longrightarrow \mathbf{R}$ such that $f(z_1,\ldots,z_n) \in \{z_1,\ldots,z_n\}$ for all $z_1,\ldots,z_n \in \mathbf{R}$; this property implies isotony. Taking the restriction of such a function to $\{0,1\}^n$ establishes a mapping Φ from the set of all locally internal aggregation functions to the set of all non-constant isotone truth functions. A mapping Ψ in the opposite sense is obtained by associating with each non-constant isotone truth function β a real function defined as follows: interpret the variables of β as real variables and replace conjunction and disjunction by min and max, respectively. Clearly $\Phi \circ \Psi$ is the identity mapping, implying that Φ is surjective and Ψ is injective. Mayor and Martin claim that Φ is also an injection. The case $n := 2$ is studied in some detail.

Note that by restricting Φ to symmetric locally internal aggregation functions and taking into account the isomorphism between truth functions and simple Boolean functions, one obtains a parallel theory involving the Boolean measures of central tendency.

13. Miscellanea

This chapter collects results that can hardly be classified within the other chapters of this book.

There is an extension of truth equations which is much studied nowadays: MVL equations, that is, equations over a finite linear set, which may be regarded as the set of truth values of a multiple-valued logic. In §1 we present a related earlier result of Itoh on equations over a lattice of the form L^I, where L is a finite chain. An even older result of Schröder, establishing the reproductive solution of an equation $f(x) = 0$ over a relation algebra, is also included.

The basic features of the theory of Boolean equations, recaptured for Post equations, have been extended beyond lattice theory, namely to functionally complete algebras, i.e., to algebras in which every function is algebraic. This is reported in §2.

There are two subjects in §3. One of them is the class of generalized Boolean functions $f : B^n \longrightarrow B$, introduced and studied by Ţăndăreanu. The other subject is the study of isotone, not necessarily Boolean functions $f : B^n \longrightarrow B$, following Takagi, Nakamura and Nakashima.

The class of generalized Boolean functions, as well as the class of Boolean functions and several subclasses of it, can be characterized by certain functional Boolean equations, as shown in §4.

The idea dealt with in §5 is to associate with every property $P(f)$ defined for Boolean functions $f : B^n \longrightarrow B$ and every $X \in B^n$, a "local property" $P(f, X)$ in such a way that f has property $P(f)$ iff $P(f, X)$ holds for every $X \in B^n$. This is studied for the properties of isotony and injectivity, with applications to extremal solutions of Boolean equations.

1 Equations in MVL and relation algebras

We have briefly mentioned in BFE, Appendix, some results on algebraic equations in certain non-Boolean lattices. A part of this work has been extensively included in Chapters 1, 3, 4 and 5 of this book. The present section is another addition to this subject, in the same sketchy style as in the BFE Appendix.

While in Ch. 9 we have briefly referred to truth equations (which in the literature are usually known as Boolean equations), the study of switching functions (i.e., truth functions, usually known as Boolean functions) has developed tremendously under the impetus of applications to circuits, which has led to specific problems. As we have already already suggested in BFE, an updated monograph on truth (switching) fgunctions and equations would be most welcome.

As mentioned in the Appendix to BFE, the study of many-valued functions $f : \{0, 1, \ldots, r - 1\}^n \longrightarrow \{0, 1, \ldots, r - 1\}$ and many-valued equations was begun by Itoh, Gotō and Carvallo, independently of each other. We have seen in Ch.5 of this book the extent of the theory of Post functions and equations. But the study of MVL (multiple-valued logic) functions and equations is now incorporated in a theory which applies MVL algebra to the study of modern electronic technology. This generalization of switching theory is rapidly growing nowadays. See Davio, Deschamps and Thayse [1978], Bochmann and Posthoff [1981]*, Yanushkevich [1998] and numerous research papers published in journals such as MVL - An International Journal, and/or presented at the International Symposia on MVL.

Having in mind the above context, we present below the results obtained by Itoh [1955]*, [1956]* on the equation in one unknown over a lattice of the form L^I, where L is a finite chain and $I \neq \emptyset$ an arbitrary set (so this framework is more general than the MVL algebra L). The last proposition, due to Schröder, refers to the equation in one unknown over an algebra of of binary relations. Finally we mention the theorems obtained by Rybakov on equations over certain free pseudoboolean algebras.

We begin with Itoh's results. The framework is the familiar pointwise structure of bounded distributive lattice of L^I induced by the finite chain $(L; \cdot, \vee, 0, 1)$. Define

(1.1) $$\delta : L \times L^I \longrightarrow L^I , \ \delta(a, x)(i) = \delta^a_{x(i)} ,$$

for every $a \in L$, $x \in L^I$, $i \in I$, where δ^a_b is the Kronecker delta.

Lemma 1.1. *For every* $x, y \in L^I$,

(1.2) $$x = y \Longleftrightarrow \bigvee_{a \in L} \delta(a, x)\delta(a, y) = 1 .$$

PROOF: For every $i \in I$,

$$(\bigvee_{a \in L} \delta(a, x)\delta(a, y))(i) = 1 \Longleftrightarrow \bigvee_{a \in L} \delta^a_{x(i)}\delta^a_{y(i)} = 1$$

$$\Longleftrightarrow \exists a \in L \ x(i) = a = y(i) \Longleftrightarrow x(i) = y(i) .$$

□

Corollary 1.1. *Every system of algebraic equations in the algebra* $(L^I; \cdot, \vee, 0, 1, \delta)$ *is equivalent to a single algebraic equation of the form* $f = 1$. □

Clearly an algebraic equation $f(x) = 1$ can be written in the form

(1.3) $$\bigvee_{a \in L} c_a \delta(a, x) = 1 ,$$

where $c_a = f(a)$ $(\forall a \in L)$.

Lemma 1.2. *An element $x \in L^I$ satisfies (1.3) if and only if*

$$(1.4) \qquad c_{x(i)}(i) = 1 \qquad (\forall i \in L) .$$

PROOF: Since $\delta^a_{x(i)} \in \{0,1\}$ for all a and all x, we have

$$(1.5) \qquad (\bigvee_{a \in L} c_a \delta(a,x))(i) = \bigvee_{a \in L} c_a(i) \delta^a_{x(i)} = c_{x(i)}(i) .$$

$\square$

Proposition 1.1. *The following conditions are equivalent:*

(i) equation (1.3) is consistent ;

(ii) $\forall i \in I \, \exists a \in A \; c_a(i) = 1$;

(iii) $\bigvee_{a \in L} \delta(1, c_a) = 1$.

PROOF: (i)$\Longrightarrow$(ii): By Lemma 1.2.

(iii)$\Longleftrightarrow \forall i \in I \; \bigvee_{a \in L} \delta^1_{c_a(i)} = 1 \Longleftrightarrow$(ii) .

(ii)$\Longrightarrow$(i): For every $i \in I$ choose $a \in L$ such that $c_a(i) = 1$ and set $x(i) = a$. Then x is a solution of (1.3) by Lemma 1.2. $\square$

Proposition 1.2. *Suppose ξ_a, $a \in L$, are solutions of (1.3) such that*

$$(1.6) \qquad \bigvee_{b \in L} \delta(a, \xi_b) = \delta(1, c_a) \qquad (\forall a \in L) .$$

Then formula

$$(1.7) \qquad x = \bigvee_{a \in L} \xi_a \delta(a, t) ,$$

where t is an arbitrary parameter in L^I, defines the general solution of equation (1.3).

PROOF: Let $g(t)$ denote the right side of (1.7). Then

$$(1.8) \qquad g(t)(i) = \bigvee_{a \in L} \xi_a(i) \delta^a_{t(i)} = \xi_{t(i)}(i) \qquad (\forall i \in I) ,$$

and on the other hand condition (1.6) becomes

$$\bigvee_{b \in L} \delta^a_{\xi_b(i)} = \delta^1_{c_a(i)} \qquad (\forall i \in I)$$

or equivalently,

$$(1.9) \qquad \forall a \in L \; \forall i \in L \; c_a(i) = 1 \Longleftrightarrow \exists b \in L \; \xi_b(i) = a .$$

Taking $a := g(t)(i)$, we have $a = \xi_{t(i)}(i)$ by (1.8), hence for this a the right side of (1.9) is true, therefore (1.9) implies that $c_{g(t)(i)}(i) = 1$, showing that $g(t)$ is a solution of equation (1.3) by Lemma 1.2.

Conversely, let x be an arbitrary solution of equation (1.3). Take $i \in I$. Then $c_{x(i)}(i) = 1$, again by Lemma 1.2. It follows by (1.9) that $\xi_b(i) = x(i)$ for some

$b \in L$. Hence for every $i \in I$ we can choose an element $t(i) \in L$ such that $\xi_{t(i)}(i) = x(i)$. We have thus defined an element $t \in L^I$ such that

$$g(t)(i) = (\bigvee_{a \in L} \xi_a \delta(a,t))(i) = \bigvee_{a \in L} \xi_a(i)\delta^a_{t(i)} = \xi_{t(i)}(i) = x(i)$$

for all $i \in I$, therefore $g(t) = x$.   □

The *algebra of binary relations* over a set S is $(\mathcal{P}(S^2); \cap, \cup, ^-, \varnothing, S^2, \circ, \Delta_S, ^{-1})$, where $(\mathcal{P}(S^2); \cap, \cup, ^-, \varnothing, S^2)$ is the Boolean algebra of all subsets of S^2, $\circ$ is the composition of relations, Δ_S is the equality relation and ρ^{-1} is defined by $x\rho^{-1}y \iff y\rho x$. More generally, a *relation algebra* is an algebra $(L; \cdot, \vee, ', 0, 1, \circ, e, ^{-1})$ of type $(2,2,1,0,0,2,0,1)$, where $(L; \cdot, \vee, ', 0, 1)$ is a Boolean algebra and several axioms are assumed, which generalize certain obvious identities of the algebra of binary relations. Thus the algebra of binary relations is a relation algebra, but the converse does not hold.

It is immediately seen that, if $\rho \subseteq S^2$, then $S^2 \circ \rho \circ S^2$ is the universal relation S^2 if $\rho \neq \varnothing$, otherwise it is the empty relation $\varnothing$. More generally, it is easy to prove the identity $1 \circ x \circ 1 = (\delta^0_x)'$ in any relation algebra. This property is the basis for the next result.

Proposition 1.3. (Schröder [1890-1905], vol.3, §11). *Let a be a solution of an equation $f(x) = 0$ in a relation algebra. Then formula*

(1.10) $$x = a \cdot (1 \circ f(t) \circ 1) + t \cdot (1 \circ f(t) \circ 1)' ,$$

where t is an arbitrary parameter, defines the reproductive solution of equation $f(x) = 0$.

PROOF: Let $g(t)$ denote the right side of (1.10). For every t,

$$f(t) = 0 \implies g(t) = a \cdot 0 + t \cdot 1 = t \implies f(g(t)) = f(t) = 0 ,$$

$$f(t) \neq 0 \implies g(t) = a \cdot 1 + t \cdot 0 = a \implies f(g(t)) = f(a) = 0 .$$

□

We reproduce here the summaries of the papers by V. I. Levin [1975a]*, [1975b]*.

"We consider equations of infinite-valued logic with one unknown that contain three possible logical operations: disjunction, conjunction and negation. We solve some typical equations and inequalities in this class. We show how on the basis of these solutions we can solve an equation of general type."

"We consider equations of infinite-valued logic that contain an algebraic operation of time shift in addition to the logical operations of disjunction, conjunction and negation. We solve typical equations and inequalities of this class having one unknown. We outline the solution of a general linear or nonlinear equation with one unknown."

Rybakov [1986a], [1986b], [1992] has provided algorithms for checking the consistency and solving equations over the free pseudoboolean algebras associated with the modal system S4 and with Heyting's intuitionistic propositional calculus.

2 Equations in functionally complete algebras

Unification theory in algebras is one of the means of improving the efficiency of resolution-based deduction techniques, which have developed under the impetus of artificial intelligence. Unification means, in fact, solving equations $f(X) = g(X)$ in certain algebras: a solution X "unifies" f with g. Unification theory manipulates particular solutions, parametric solutions (cf. BFE, Definition 2.4) and general solutions, under the names *ground solutions, unifiers* and *most general unifiers*, respectively; see e.g. Büttner [1987] and the literature quoted there.

Unification theory developed independently of the theory of Boolean equations until Martin and Nipkow [1986], [1988] rediscovered Löwenheim's reproductive solution associated with a particular solution and an algorithm much in the spirit of Theorem 9.3 in BFE, both within the framework of a Boolean ring. Then Nipkow and Büttner succeeded in extending four basic features of Boolean and Post equations to other algebraic structures, within and beyond lattice theory. These four features are the following: the reduction of any system to a single equation $f(X) = 0$, the consistency condition, the method of successive elimination of variables, and Löwenheim's reproductive solution, as we are going to see below.

Definition 2.1. A finite algebra A is said to be *functionally complete (primal)* if for every $n \in \mathbf{N}$, any function $f : A^n \longrightarrow A$ is algebraic (a polynomial). □

Thus every primal algebra is functionally complete. As we know, the Boolean algebra $\{0, 1\}$ (the r−element Post algebra C_r) is the unique primal Boolean algebra (r−Post algebra); see e.g. the Corollary of Theorem 1.11 in BFE and the present Proposition 5.2.3, respectively.

Proposition 2.1. (Post [1921]). *A finite algebra A, card(A) ≥ 2, is functionally complete if and only if there are two elements $0, 1 \in A$, $0 \neq 1$, and two algebraic functions $+, \cdot : A^2 \longrightarrow A$ such that for every $x \in A$,*

(2.1) $$x + 0 = 0 + x = x \ \& \ x \cdot 0 = 0 \cdot x = 0 \ \& \ x \cdot 1 = x \, ,$$

and for every $a \in A$, the function $\kappa_a : A \longrightarrow A$ defined by

(2.2) $$\kappa_a(x) = 1 \text{ if } x = a, \text{ else } 0 \, ,$$

is algebraic.[1]

PROOF: If A is functionally complete, choose $0, 1 \in A$, $0 \neq 1$, and define $+, \cdot$ by extending arbitrarily conditions (2.1). Then $+, \cdot$ and κ_a, $a \in A$, are algebraic functions.

Conversely, suppose the algebra A is endowed with two algebraic functions satisfying (2.1) and the functions κ_a defined by (2.2) are algebraic. Set $\sum_{i=1}^{1} x_i = x_1$, $\sum_{i=1}^{m} x_i = (\sum_{i=1}^{m-1} x_i) + x_m$ and define similarly $\prod_{i=1}^{m} x_i$. For every $C = (c_1, \ldots, c_n) \in A^n$ define $\kappa_C : A^n \longrightarrow A$ by

---

[1] This algebraic structure was rediscovered by Prešić; cf. Ch.1 §2.

$$(2.3) \qquad \kappa_C(x_1, \ldots, x_n) = \prod_{i=1}^{n} \kappa_{c_i}(x_i) \, .$$

Then the functions κ_C are algebraic and

$$(2.4) \qquad \kappa_C(X) = 1 \text{ if } X = C \, , \text{ else } 0 \, .$$

It follows that any function $f : A^n \longrightarrow A$ is algebraic, because it can be represented in the form

$$(2.5) \qquad f(X) = \sum_{C \in A^n} f(C) \cdot \kappa_C(X) \, .$$

$\square$

Remark 2.1. It is easy to see that primal algebras have a quite similar characterization: just replace "algebraic function" by "polynomial" and add the requirement that every $a \in A$ can be expressed as a polynomial. $\square$

In Propositions 2.2–2.5 below, due to Nipkow [1988], we use the same notation as in Proposition 2.1.

Proposition 2.2. *In a functionally complete algebra, every system of algebraic equations and/or negated equations*

$$(2.6.i) \qquad g_i(X) = h_i(X) \qquad (i \in I) \, ,$$

$$(2.6.j) \qquad g_j(X) \neq h_j(X) \qquad (j \in J) \, ,$$

where I and J are finite sets, is equivalent to a single equation of the form

$$(2.7) \qquad f(X) = 0 \, .$$

PROOF: The functions

$$(2.8) \qquad eq(x, y) = \sum_{a \in A} \kappa_a(x) \kappa_a(y) \, ,$$

$$(2.9) \qquad neq(x, y) = \sum_{a,b \in A; a \neq b} \kappa_a(x) \kappa_b(y) \, ,$$

$$(2.10) \qquad vel(x, y) = \sum_{(a,b) \neq (0,0)} \kappa_{(a,b)}(x, y) \, ,$$

satisfy $eq(x, x) = 1$, $eq(x, y) = 0$ for $x \neq y$, $neq(x, x) = 0$, $neq(x, y) = 1$ for $x \neq y$, $vel(0, 0) = 0$, $vel(x, y) = 1$ for $(x, y) \neq (0, 0)$. Define

$$(2.11) \qquad \text{Vel}_{k=1}^{1} x_k = x_1 \text{ \& } \text{Vel}_{k=1}^{m} x_k = vel(\text{Vel}_{k=1}^{m-1} x_k, x_m) \, .$$

Then system (2.6) can be rewritten in the form

$$(2.6'.i) \qquad neq(g_i(X), h_i(X)) = 0 \qquad (i \in I) \, ,$$

$$(2.6'.j) \qquad eq(g_j(X), h_j(X)) = 0 \qquad (j \in J) \, ,$$

and a system of equations of the form

(2.12) $$f_k(X) = o \qquad (k = 1, \ldots, m)$$

is equivalent to the single equation

(2.13) $$\mathrm{Vel}_{k=1}^m f_k(X) = 0 .$$

$\square$

Proposition 2.3. *In a functionally complete algebra, set*

(2.14) $$\mathrm{con}(x, y) = (\sum_{a \neq 0} \kappa_a(x))(\sum_{a \neq 0} \kappa_a(y)) ,$$

(2.15) $$\mathrm{Con}_{k=1}^1 x_k = x_1 \ \& \ \mathrm{Con}_{k=1}^m x_k = \mathrm{con}(\mathrm{Con}_{k=1}^{m-1} x_k, x_m) .$$

Then equation (2.7) *is consistent if and only if*

(2.16) $$\mathrm{Con}_{C \in A^n} f(C) = 0$$

for one, and hence all of the possible total orderings of the set A^n.

PROOF: If $x = 0$ or $y = 0$ then $\mathrm{con}(x, y) = 0$, else $\mathrm{con}(x, y) = 1$. Hence $\mathrm{Con}_{k=1}^m y_k = 0$ iff $y_k = 0$ for some k. $\square$

The next corollary and proposition are the basis for the method of successive elimination of variables.

Corollary 2.1. *Take* $x \in X$ *and set* $Y = X \backslash \{x\}$. *Then equation* (2.7) *is consistent if and only if the equation*

(2.17) $$\mathrm{Con}_{a \in A} f(a, Y) = 0$$

is consistent. Moreover, (2.7) *implies* (2.17). $\square$

Proposition 2.4. *Suppose* $Y = G(X)$ *is a reproductive solution of equation* (2.17). *Then formulas* $x = t(p, Q)$, $Y = G(Q)$, *where we have set*

(2.18) $$t(x, Y) = d(f(x, Y), x, \mathrm{Max}_{a \in A} a \cdot \kappa_0(f(a, G(Y)))) ,$$

(2.19) $$d(x, y, z) = y \cdot \kappa_0(x) + z \cdot \sum_{a \neq 0} \kappa_a(x) ,$$

(2.20) $$\max(x, y) = d(x, y, x) ,$$

(2.21) $$\mathrm{Max}_{k=1}^1 x_k = x_1 \ \& \ \mathrm{Max}_{k=1}^m x_k = \max(\mathrm{Max}_{k=1}^{m-1} x_k, x_m) ,$$

and (p, Q) *is the parameter set, define the reproductive solution of equation* (2.7).

PROOF: Note first that $d(x, y, z) = y$ if $x = 0$, else z. Hence $\max(x, y) = y$ if $x = 0$, else x; in particular $\max(0, y) = y$ and $\max(x, 0) = x$. Therefore $\mathrm{Max}_{k=1}^m x_k$ is 0 if all $x_k = 0$, otherwise it is the first element $x_i \neq 0$.

Now if $f(p, Q) = 0$, then $t(p, Q) = p$ by (2.18) and Q satisfies (2.17) by Corollary 2.1(i), hence $Q = G(Q)$. This also implies $f(t(p, Q), G(Q)) = f(p, Q) =$

0 under the assumption $f(p, Q) \neq 0$. But in this case

(2.22) $t(p, Q) = \text{Max}_{a \in A} a \cdot \kappa_0(f(a, G(Q)))$;

setting $A_1 = \{a \in A \mid f(a, G(Q)) = 0\}$, we have to prove that $t(p, Q) \in A_1$. But

(2.23) $\{a \cdot \kappa_0(f(a, G(Q))) \mid a \in A\} = A_1 \cup \{0\}$.

Now apply (2.22) and (2.23). If $A_1 \backslash \{0\} \neq \emptyset$ then $t(p, Q)$ is an element of this set. Otherwise, since $A_1 \neq \emptyset$ by hypothesis, it follows that $A_1 = \{0\}$, hence $t(p, Q) = 0 \in A_1$.   □

Remark 2.2. The function (2.19) may be called a *discriminator*, because the value $d(x, y, z)$ depends on the answer to the question: is x distinct from 0? Such discriminators play an important rôle in universal algebra; see Burris and Sankappanavar [1981].   □

Proposition 2.5. *Suppose $C = (c_1, \ldots, c_n)$ is a particular solution of equation (2.7) over a functionally complete algebra. Then formulas*

(2.24) $x_i = d(f(P), p_i, c_i)$ $(i = 1, \ldots, n)$,

where $P = (p_1, \ldots, p_n)$ is the parameter set, define the reproductive solution of equation (2.7).

PROOF: We use the property of the discriminator (2.19). If $f(P) = 0$ then formulas (2.24) yield $x_i = p_i$ $(i = 1, \ldots, n)$. This also implies

$$f(d(f(P), p_1, c_1), \ldots, d(f(P), p_n, c_n)) = f(p_1, \ldots, p_n) = 0 .$$

If $f(P) \neq 0$ then

$$f(d(f(P), p_1, c_1), \ldots, d(f(P), p_n, c_n)) = f(c_1, \ldots, c_n) = 0 .$$

                                    □

Nipkow [1988] gives applications to equations in the Post algebra C_r and beyond lattice theory, to matrix rings. For Boolean unification techniques in predicate calculus see Martin and Nipkow [1989]. Propositions 2.2–2.5, valid in particular for primal algebras, have been extended to direct powers of primal algebras and to varieties generated by primal algebras, with examples in certain specific 3–rings; cf. Nipkow [1990]. All of these papers pay special attention to the complexities of the algorithms they suggest.

Büttner [1987] suggests a promising approach to solving arbitrary (i.e., not necessarily algebraic) equations over a finite algebra. Namely, the signature of the algebra is enriched so as to obtain a functionally complete algebra, and the original equation becomes an algebraic equation which is solved by unification theory techniques. An example refers to the four-element Boolean algebra. Unfortunately, we have been unable to understand completely this paper.

3 Generalized Boolean functions and non-Boolean functions

The fundamental paper by Jónsson and Tarski [1951], [1952] studies functions whose arguments and values are in a Boolean algebra and which are not necessarily Boolean. Their properties turn out to be useful in the theory of cylindric algebras, which is an algebraic counterpart of predicate calculus; cf. Henkin, Monk and Tarki [1971].

The present monograph is outside this line of research. We deal with non-Boolean functions defined over a Boolean algebra, only in this and the next section. Namely, we present the theory of generalized Boolean functions, due to Ţăndăreanu, and a slight generalization of the results obtained by Takagi, Nakamura and Nakashima [1997] on isotone functions $f : B^n \longrightarrow B$.

Throughout this section B is again an arbitrary Boolean algebra, A is a finite set, $\{0,1\} \subseteq A \subset B$, and n is a positive integer.

The basic definitions, examples and Propositions 3.1 and 3.2 were given by Ţăndăreanu [1981].

Definition 3.1. We denote by $G(A)$ the set of all functions $g : A \times B \longrightarrow B$ such that $g(0,0) = g(1,1) = 1$ and the set $\{g(a,x) \mid a \in A\}$ is orthonormal for every $x \in B$. □

Here are a few examples of functions in $G(A)$.

Example 3.1. The function g defined by

(3.1) $g(a,x) = x^a$ if $a \in \{0,1\}$, else 0 .

□

Example 3.2. Fix an element $a_0 \in A$ and define

(3.2) $g(a,x) = \delta_x^a$ (Kronecker) if $x \in A$, else 0 .

If $a, b \in A$ and $a \neq b$, then

$$g(a,x)g(b,x) = \delta_x^a \delta_x^b = 0 \text{ if } x \in A, \text{ else } = \delta_{a_0}^a \delta_{b_0}^b = 0 ,$$

while

$$\bigvee_{a \in A} g(a,x) = \bigvee_{a \in A} \delta_x^a = 1 \text{ if } x \in A, \text{ else } = \bigvee_{a \in A} \delta_{a_0}^a = 1 .$$

□

Example 3.3. Suppose $B = \{0, a, a', 1\}$ and $A = \{0, a, 1\}$. Define g by the prescriptions

(3.3.1) $g(0,x) = \delta_x^0$,

(3.3.2) $g(a,x) = 0$ if $x \in \{0,1\}$, else x ,

(3.3.3) $g(1, x) = x$ if $x \in \{0, 1\}$, else x' .

Then $g(0, 0) = g(1, 1) = 1$. To prove orthogonality, note that $g(a, 0) = 0 = g(1, 0)$, while for $x \neq 0$ we have $g(0, x) = 0$ and $g(a, x)g(1, x) = 0$ if $x = 1$, else $= xx' = 0$. Finally, if $x \in \{0, 1\}$ then

$$g(0, x) \vee g(1, x) = \delta_x^0 \vee x = 1 ,$$

while if $x \notin \{0, 1\}$ then

$$g(a, x) \vee g(1, x) = x \vee x' = 1 .$$

□

Remark 3.1. For any fixed $a \in A$, the function in Example 3.1 is Boolean, whereas the function in Example 3.2 is not Boolean. For setting

$$h(x) = g(a, 1)x \vee g(a, 0)x' = \delta_1^a x \vee \delta_0^a x' ,$$

two cases may occur. If $a \in \{0, 1\}$ then $h(x) \in \{x, x'\}$, therefore $h(x) \notin \{0, 1\}$ for $x \notin \{0, 1\}$, while $g(a, x) \in \{0, 1\}$ for every x. If $a \notin \{0, 1\}$ then $h(x) = 0$ for every x, while $g(a, a) = 1$.

The functions (3.3.1), (3.3.2) and (3.3.3) are not Boolean. For

$$g(0, 1)x \vee g(0, 0)x' = x' , \text{ while } g(0, x) = 0 \text{ if } x \neq 0 ;$$

$$g(a, 1)x \vee g(a, 0)x' = 0 , \text{ while } g(a, x) = x \text{ if } x \notin \{0, 1\} ;$$

$$g(1, 1)x \vee g(1, 0)x' = x , \text{ while } g(1, x) = x' \text{ if } x \notin \{0, 1\} .$$

□

Remark 3.2. If $g \in G(A)$ and $a, b \in \{0, 1\}$, then $g(a, b) = \delta_b^a$. For $a \neq b$ implies $g(a, b) = g(a, b)g(b, b) = 0$. □

Proposition 3.1. *Suppose $g \in G(A)$ and for every $(a_1, \ldots, a_n) \in A$ let $b_{a_1 \ldots a_n}$ and $c_{a_1 \ldots a_n}$ be elements of B. Setting*

$$e(x_1, \ldots, x_n) = \bigvee_{a_1, \ldots, a_n \in A} e_{a_1 \ldots a_n} g(a_1, x_1) \ldots g(a_n, x_n) ,$$

for $e \in \{b, c\}$, the following identities hold:

(3.4) $$\bigvee_{a_1, \ldots, a_n \in A} g(a_1, x_1) \ldots g(a_n, x_n) = 1 ,$$

(3.5)
$$b(x_1, \ldots, x_n) \vee c(x_1, \ldots, x_n)$$
$$= \bigvee_{a_1, \ldots, a_n \in A} (b_{a_1 \ldots a_n} \vee c_{a_1 \ldots a_n})g(a_1, x_1) \ldots g(a_n, x_n) ,$$

$$(3.6) \quad b(x_1,\ldots,x_n)c(x_1,\ldots,x_n)$$
$$= \bigvee\nolimits_{a_1,\ldots,a_n \in A} b_{a_1\ldots a_n} c_{a_1\ldots a_n} g(a_1,x_1)\ldots g(a_n,x_n) ,$$

$$(3.7) \quad (b(x_1,\ldots,x_n))' = \bigvee_{a_1,\ldots,a_n \in A} (b_{a_1\ldots a_n})' g(a_1,x_1)\ldots g(a_n,x_n) .$$

PROOF: Same as in the conventional case $A := \{0,1\}$, $g(a,x) = x^a$; see e.g. BFE, Theorem 1.5, and also the present Proposition 5.2.1. $\qquad\Box$

Definition 3.2. Let $g \in G(A)$. We denote by GBFn(g) the set of all functions $f : B^n \longrightarrow B$ that satisfy the identity

$$(3.8) \quad f(x_1,\ldots,x_n) = \bigvee_{a_1,\ldots,a_n \in A} f(a_1,\ldots,a_n)g(a_1,x_1)\ldots g(a_n,x_n)$$

and we set GBFn(A) $= \bigcup_{g\in G(A)}$ GBFn(g). The elements of GBFn(g) and GBFn(A) are called $g-$*generalized Boolean functions* and $A-$*generalized Boolean functions*, respectively, or simply *generalized Boolean functions*. $\qquad\Box$

Example 3.4. The following function is a $g-$generalized Boolean function, where g is the function in Example 3.3:

$$(3.9) \quad f(0) = a , \; f(a) = a' , \; f(a') = f(1) = 1 .$$

For, setting $h(x) = ag(0,x) \vee a'g(a,x) \vee g(1,x)$, we get $h(0) = a$, $h(1) = 1$ and $h(x) = x \vee x' = 1$ for $x \notin \{0,1\}$. Therefore $h(x) = f(x)$ for all x.

The function f is not Boolean, because $f(1)a \vee f(0)a' = a \neq a' = f(a)$. $\qquad\Box$

Remark 3.3. (Țăndăreanu [1983]). If $g \in G(A)$ then GBFn(g) is the Boolean subalgebra of B^{B^n} generated by the constant functions and the functions of the form $f(x_1,\ldots,x_n) = g(a,x_i)$, for $a \in A$ and $i \in \{1,\ldots,n\}$. The proof uses Proposition 3.1 and is quite similar to the well-known proof for the algebra BFn of Boolean functions; see e.g. BFE, Definition 1.13 and Theorem 1.6'. $\qquad\Box$

Remark 3.4. (Țăndăreanu [1982]). It follows from (3.4) that any function obtained from a $g-$generalized Boolean function by introduction of fictitious variables is also a $g-$generalized Boolean function. $\qquad\Box$

Lemma 3.1. *Suppose* $b \in A$ *and let* g *be the function in Example* 3.2. *Define* $f : B^n \longrightarrow B$ *by* $f(x_1,\ldots,x_n) = g(b,x_1)$. *Then* $f \in$ GBFn(A).

PROOF: Define $f_1 : B \longrightarrow B$ by $f_1(x) = g(b,x)$. In view of Remark 3.4 it suffices to prove that $f_1 \in$ GBF1(A). But

$$\bigvee_{a\in A} f_1(a)g(a,x) = \bigvee_{a\in A} g(b,a)g(a,x) = \bigvee_{a\in A} \delta_a^b g(a,x) = g(b,x) = f_1(x).$$

$\qquad\Box$

Proposition 3.2. BFn $\subset$ GBFn$(A) \subset B^{B^n}$.

PROOF: BFn $\subseteq$ GBFn because every Boolean function is a g–generalized Boolean function, where g is the function in Example 3.1. The inclusion GBFn(A) $\subseteq B^{B^n}$ is trivial. The function $f : B^n \longrightarrow B$ defined by $f(x_1,\ldots,x_n) = 0$ if $x_1,\ldots,x_n \in A$, else 1, is not a generalized Boolean function: otherwise (3.8) would imply $f(x_1,\ldots,x_n) = 0$ identically. Therefore GBFn$(A) \subset B^{B^n}$. The first inclusion is also strict, because e.g. the function f in Lemma 3.1 is not Boolean: otherwise $f_1(x) = f(x,\ldots,x) = g(b,x)$ would be a Boolean function, in contradiction with Remark 3.1. $\qquad\square$

The next Propositions 3.3–3.5 can be found in Ţăndăreanu [1982].

Proposition 3.3. *For every $f \in$ GBFn(A) and every $x_1,\ldots,x_n \in B$,*

$$(3.10) \qquad \prod_{a_1,\ldots,a_n \in A} f(a_1,\ldots,a_n) \le f(x_1,\ldots,x_n) \le \bigvee_{a_1,\ldots,a_n \in A} f(a_1,\ldots,a_n) .$$

PROOF: The second inequality is a consequence of (3.8); by applying it to f' we obtain the first inequality. $\qquad\square$

Proposition 3.4. *In $\{0,1\} \subseteq C \subset A$ then:*

(i) every function $g \in G(C)$ has a unique extension to a function $\bar{g} \in G(A)$, *and*

(ii) $G(C) \subset \{g|_C \mid g \in C(A)\}$.

PROOF: (i) The function $\bar{g} : A \times B \longrightarrow B$ defined by

$$(3.11) \qquad\qquad \bar{g}(a,x) = g(a,x) \text{ if } a \in C , \text{ else } 0 ,$$

is clearly an extension in $G(A)$ of g. Now suppose $h \in G(A)$ is another extension of g. Then for every $c \in C\backslash A$,

$$h(c,x) = \prod_{b \in A\backslash\{c\}} h'(b,x) \le \prod_{b \in C} h'(b,x) = \prod_{b \in C} g'(b,x) = 0 ,$$

therefore $h = \bar{g}$.

(ii) $G(C) \subseteq \{g|_C \mid g \in C(A)\}$ follows from (i). To prove the inclusion is proper, consider again the function g in Example 3.2, where $a_0 \notin C$. Then $g|_C \notin G(C)$ because $\bigvee_{c \in C} g(c,a_0) = \bigvee_{c \in C} \delta^c_{a_0} = 0$. $\qquad\square$

Proposition 3.5. *If $\{0,1\} \subseteq C \subset A$ then GBFn$(C) \subset$ GBFn(A) .*

PROOF: If $f \in$ GBFn(C) then Proposition 3.4 implies

$$f(x_1,\ldots,x_n) = \bigvee_{c_1,\ldots,c_n \in C} f(c_1,\ldots,c_n)g(c_1,x_1)\ldots g(c_n,x_n)$$

$$= \bigvee_{c_1,\ldots,c_n \in C} f(c_1,\ldots,c_n)\bar{g}(c_1,x_1)\ldots\bar{g}(c_n,x_n)$$

$$= \bigvee_{a_1,\ldots,a_n \in A} f(a_1,\ldots,a_n)\overline{g}(a_1,x_1)\ldots\overline{g}(a_n,x_n) \,,$$

proving that $f \in \mathrm{GBFn}(A)$. Thus $\mathrm{GBFn}(C) \subseteq \mathrm{GBFn}(A)$. To prove the inclusion is strict, take the function f in Lemma 3.1 with $b \notin C$. In view of Remark 3.4, in order to show that $f \notin \mathrm{GBFn}(C)$ it suffices to take $f_1 : B \longrightarrow B$, $f_1(x) = g(b,x)$, and prove that $f_1 \notin \mathrm{GBF1}(C)$. But

$$\bigvee_{c \in C} f_1(c)g(c,b) = \bigvee_{c \in C} \delta_c^b g(c,x) = 0 \neq 1 = g(b,b) = f_1(b) \,.$$

$\square$

Another theorem in Ţăndăreanu [1982] studies the behaviour of $\mathrm{GBFn}\ (A_1)$ and $\mathrm{GBFn}(A_2)$ with respect to union and intersection.

Yongcai Liu [1988] devises a matrix method for generating the functions $g \in G(A)$ and counting them.

Now we turn to the case $n := 1$. The next proposition is a partial converse to Proposition 3.3.

Proposition 3.6. *The following conditions are equivalent for a function* $f :$ $B \longrightarrow B$:

(i) $f \in \mathrm{GBF1}(A)$;

(ii) for every $x \in B$,

(3.12)
$$\prod_{a \in A} f(a) \leq f(x) \leq \bigvee_{a \in A} f(a) \,;$$

(iii) f *satisfies* (3.12) *for every* $x \in B\backslash A$.

COMMENT: This is a slight refinement of Theorem 1 in Ţăndăreanu [1984a].
PROOF: (i)$\Longrightarrow$(ii): By Proposition 3.3.

(ii)$\Longrightarrow$(iii): Trivial.

(iii)$\Longrightarrow$(i): For every $x \in B$, Theorem 4.8 in BFE ensures the existence of an orthonormal solution $\{h(a,x) \mid a \in A\}$ to the equation $\bigvee_{a \in A} f(a)h(a,x) = f(x)$. Define

$$g(a,x) = \delta_x^a \text{ if } x \in A \,, \text{ else } h(a,x) \,.$$

Then it is easy to check that $g \in G(A)$ and $\bigvee_{a \in A} f(a)g(a,x) = f(x)$. $\square$

Corollary 3.1. *If* $f : B \longrightarrow B$ *satisfies* $f(\alpha) = \alpha$ ($\alpha = 0,1$), *then* $f \in$ $\mathrm{GBF1}(\{0,1\})$. $\square$

Example 3.5. (Ţăndăreanu [1984a]). Let $B = \{0,a,a',1\}$. Then the function $f : B \longrightarrow B$ defined by

$$f(x) = 1 \text{ if } x \in \{0,1\} \,, \text{ else } x \,,$$

is by no means (that is, for no subset A) a generalized Boolean function. For $f(0)f(1) = 1 \nleq f(a)$, $f(0)f(1)f(a) = 1 \nleq f(a')$ and $f(0)f(1)f(a') = a' \nleq f(a)$. $\square$

Țăndăreanu [1984a] gives also a necessary and sufficient condition on f for the existence of a unique function $g \in G(A)$ such that $f \in \mathrm{GBF1}(g)$.

Another specialization studied in Țăndăreanu [1984b], [1985a] is the case $A := \{0, 1\}$ (which is still more general than the case of Boolean functions, due to the function g).

Remark 3.5. (Țăndăreanu [1985b]). The set of isotone functions $f : B \longrightarrow B$ and the set of antitone functions $f : B \longrightarrow B$ are strictly included in $\mathrm{GBF1}(\{0, 1\})$. For the isotony of f implies $f(0)f(1) = f(0) \le f(x) \le f(1) = f(0) \vee f(1)$, hence $f \in \mathrm{GBF1}(\{0, 1\})$ by Proposition 3.6. The inclusion is proper because a function in $\mathrm{GBF1}(\{0, 1\})$ need not fulfil $f(0) \le f(1)$. There is a similar proof for antitone functions.   □

Remark 3.6. (Melter and Rudeanu [1982]). Associate with every $f : B^n \longrightarrow B$ the Boolean function βf defined by

$$(3.13) \qquad (\beta f)(X) = \bigvee_{A \in \{0,1\}^n} f(A) X^A \qquad (\forall X \in B^n) \,,$$

which has the property $\beta f|_{\{0,1\}^n} = f|_{\{0,1\}^n}$. Then $\beta : B^{B^n} \longrightarrow \mathrm{BFn}$ is a surjective Boolean homomorphism and in fact it is a retract of the inclusion mapping $\iota : \mathrm{BFn} \hookrightarrow B^{B^n}$. Therefore BFn is isomorphic to $B^{B^n}/\ker\beta$.   □

Remark 3.7. For every $g \in G(\{0, 1\})$, the restriction $\beta : \mathrm{GBFn}(g) \longrightarrow \mathrm{BFn}$ of the morphism constructed in Remark 3.6 is an isomorphism, because the unique f such that βf is given by (3.13), is

$$f(x_1, \ldots, x_n) = \bigvee_{\alpha_1, \ldots, \alpha_n \in \{0,1\}} f(\alpha_1, \ldots, \alpha_n) g(\alpha_1, x_1) \ldots g(x_n, \alpha_n) \,.$$

□

The theory of Boolean partial derivatives (cf. Ch.10, §1) can be extended to generalized Boolean functions: Țăndăreanu [1985a] defines $\partial f / \partial x_i$ by 10.(1.4) and proves properties 10.(1.5)-(1.9) within the framework of $\mathrm{GBFn}(\{0, 1\})$.

A more recent field of research, referred to as *set-valued functions*, is the study of functions $f : \mathcal{P}(\mathbf{r})^n \longrightarrow \mathcal{P}(\mathbf{r})$ subject to no supplementary condition, where $\mathbf{r} = \{0, 1, \ldots, r - 1\}$ is the set of truth values of a many-valued logic. Completeness properties in the Post-Rosenberg line and the approximation of arbitrary such functions by Boolean functions are among the problems studied in the field, which has important applications; see e.g. the survey Ngom, Reischer, Simovici and Stojmenović [1979] and Ch.14, §3. In particular Takagi, Nakamura and Nakashima [1997] studied isotone functions. We present below a generalization of their results to an arbitrary Boolean algebra.

For every $a, x \in B$ set

$$(3.14) \qquad\qquad [a, x) = 1 \text{ if } a \le x \,, \text{ else } 0 \,,$$

and for every $a \in B$ define $[a) : B \longrightarrow B$ by $[a)(x) = [a, x)$. Further, for every $a_1, \ldots, a_n \in B$ define $[a_1) \ldots [a_n) : B^n \longrightarrow B$ by $[a_1) \ldots [a_n)(x_1, \ldots, x_n) = [a_1, x_1) \ldots [a_n, x_n)$.

Remark 3.8. The above functions are $\{0,1\}$–valued.   □

Taking into account Remark 3.8 one proves easily

$$(3.15) \qquad [a,x) \leq [a,y) \; \forall a \in B \Longleftrightarrow x \leq y \,,$$

$$(3.16) \qquad [a_1,x_1)\ldots[a_n,x_n) = 1 \text{ if } a_k \leq x_k \; (k=1,\ldots,n) \,, \text{ else } 0 \,,$$

$$(3.17) \qquad \begin{aligned} &[a_1,x_1)\ldots[a_n,x_n) \leq [b_1,x_1)\ldots[b_n,x_n) \; \forall(x_1,\ldots,x_n)\\ &\Longleftrightarrow (b_1,\ldots,b_n) \leq (a_1,\ldots,a_n) \end{aligned}$$

(to prove $\Longrightarrow$ in (3.15) take $a := x$ and to prove $\Longrightarrow$ in (3.17) take $x_k := a_k$ $(k = 1,\ldots,n))$.[2]

Definition 3.3. For every $a \in B$ and every $f : B^n \longrightarrow B$ define $[a,f) : B^n \longrightarrow B$ by $[a,f)(x_1,\ldots,x_n) = [a,f(x_1,\ldots,x_n))$ and $g_{af} : B^n \longrightarrow B$ by

$$(3.18) \qquad g_{af}(x_1,\ldots,x_n) = \bigvee\{[a_1,x_1)\ldots[a_n,x_n) \mid a \leq f(a_1,\ldots,a_n)\}$$

(recall the convention $\bigvee\varnothing = 0$).   □

Remark 3.9. It follows from Remark 3.8 that, if the set in the right side of (3.18) is not empty, then (3.18) becomes

$$(3.19) \qquad g_{af}(x_1,\ldots,x_n) = \max\{[a_1,x_1)\ldots[a_n,x_n) \mid a \leq f(a_1,\ldots,a_n)\} \,,$$

showing that Definition 3.3 makes sense although we have not assumed the Boolean algebra B to be complete.   □

Definition 3.4. Following a usual abuse of notation we denote the functions of the form $[a_1)\ldots[a_n)$ by $[a_1,x_1)\ldots[a_n,x_n)$ and call them *order terms*, and we denote $[a,f)$ by $[a,f(x_1,\ldots,x_n))$.   □

The discrepancy between Definition 3.4 and the conventional concept of a Boolean term, which also includes products of less than n literals, is only apparent: any product of less than n factors $[a_i,x_i)$ can be written as a product of n factors by using the identity $[0,x) = 1$.

Remark 3.10. The functions g_{af} are isotone. For, in view of Remark 3.9, $(x_1,\ldots,x_n) \leq (y_1,\ldots,y_n)$ and $g_{af}(x_1,\ldots,x_n) = 1$ imply $[a_1,x_1)\ldots[a_n,x_n) = 1$ for some $a_1,\ldots,a_n \in B$ satisfying $a \leq f(a_1,\ldots,a_n)$, hence $a_k \leq x_k \leq y_k$ $(k = 1,\ldots,n)$, therefore $[a_1,y_1)\ldots[a_n,y_n) = 1$, showing that $g_{af}(y_1,\ldots,y_n) = 1$.   □

Proposition 3.7. *A function* $f : B^n \longrightarrow B$ *is isotone if and only if* $[a,f) = g_{af}$ *for every* $a \in B$.

---

[2] This corrects a mistake in the original proof.

PROOF: Suppose f is isotone and take $a \in B$. Further, fix an arbitrary $(a_1, \ldots, a_n) \in B^n$ and prove that $[a, f(a_1, \ldots, a_n)) = g_{af}(a_1, \ldots, a_n)$. If $a \leq f(a_1, \ldots, a_n)$ then

$$[a, f(a_1, \ldots, a_n)) = 1 = [a_1, a_1) \ldots [a_n, a_n) = g_{af}(a_1, \ldots, a_n)$$

by Remark 3.9. If $a \nleq f(a_1, \ldots, a_n)$ then $[a, f(a_1, \ldots, a_n)) = 0$ and we are going to prove that $g_{af}(a_1, \ldots, a_n) = 0$. Otherwise $g_{af}(a_1, \ldots, a_n) = 1$, hence Remark 3.9 implies that $[b_1, a_1) \ldots [b_n, a_n) = 1$ for some $b_1, \ldots, b_n \in B$ such that $a \leq f(b_1, \ldots, b_n)$. But $b_k \leq a_k$ ($k = 1, \ldots, n$) by (3.16), hence $f(b_1, \ldots, b_n) \leq f(a_1, \ldots, a_n)$, therefore $a \leq f(a_1, \ldots, a_n)$, a contradiction.

Conversely, suppose $[a, f) = g_{af}$ for every $a \in B$ and take $a_k, b_k \in B$ ($k = 1, \ldots, n$) such that $(a_1, \ldots, a_n) \leq (b_1, \ldots, b_n)$. In view of (3.15), in order to prove $f(a_1, \ldots, a_n) \leq f(b_1, \ldots, b_n)$, it suffices to show that $[a, f(a_1, \ldots, a_n)) \leq [a, f(b_1, \ldots, b_n))$ for every $a \in B$. Suppose $[a, f(a_1, \ldots, a_n)) = 1$. Then $g_{af}(a_1, \ldots, a_n) = 1$, hence $[c_1, a_1) \ldots [c_n, a_n) = 1$ for some $c_1, \ldots, c_n \in B$ such that $a \leq f(c_1, \ldots, c_n)$. But $c_k \leq a_k$ ($k = 1, \ldots, n$), hence $c_k \leq b_k$ ($k = 1, \ldots, n$), therefore $[c_1, b_1) \ldots [c_n, b_n) = 1$, proving that

$$[a, f(b_1, \ldots, b_n)) = g_{af}(b_1, \ldots, b_n) = 1 .$$

□

In the case of finite Boolean algebras, a theory of implicants and prime implicants for order terms can be constructed which recaptures the essentials of the conventional theory for Boolean functions; the proof techniques are the same.

4 Functional characterizations of classes of functions over B

As reported in BFE, Exercise 1.12, the first system of functional equations characterizing Boolean functions was given by McColl [1877-80]. The idea was developed by Melter and Rudeanu [1982], [1983], [1984b] and extended by Ţăndăreanu [1984b] to generalized Boolean functions, as we are going to show here. The section will end with a few words about several classes of truth functions that can be characterized by functional equations; cf. Ekin, Foldes, Hammer and Hellerstein [2000].

Let again B denote an arbitrary Boolean algebra.

Lemma 4.1. (Ţăndăreanu [1984b]). *Suppose $f, h : B \longrightarrow B$ and $h(\alpha) = \alpha$ ($\alpha = 0, 1$). Then the following identities are equivalent:*

(4.1) $f(x) = f(1)h(x) \vee f(0)h'(x) ;$

(4.2) $f(x) + f(0) = (f(1) + f(0))h(x) ;$

(4.3) $f(x) + f(y) = ((f(1) + f(0))(h(x) + h(y))$;

(4.4) $f(x) + f(y) \leq h(x) + h(y)$;

(4.5) $f(x)f^{\alpha}(x) \leq f(\alpha) \leq f(x) \vee (h^{\alpha}(x))'$ $(\alpha = 0, 1)$.

PROOF: (4.1)$\Longrightarrow$(4.2): From

$$f(1)h(x) \vee f(0)h'(x) = f(1)h(x) + f(0)(h(x) + 1) .$$

(4.2)$\Longrightarrow$(4.3)$\Longrightarrow$(4.4): Obvious.
(4.4)$\Longrightarrow$(4.5): Taking $y := \alpha$ we obtain

$$f(x) + f(\alpha) \leq h(x) + \alpha = (h^{\alpha}(x))' ,$$

hence

$$f(x)f'(\alpha)h^{\alpha}(x) \vee f'(x)f(\alpha)h^{\alpha}(x) = 0 .$$

(4.5)$\Longrightarrow$(4.1): Taking the meet of each side with $h^{\alpha}(x)$, we obtain $f(x)h^{\alpha}(x) = f(\alpha)h^{\alpha}(x)$ $(\alpha = 0, 1)$. $\qquad\Box$

Proposition 4.1. (Melter and Rudeanu [1982]). *The following conditions are equivalent for a function $f : B \longrightarrow B$:*

(i) f is a Boolean function ;

(ii) $f(x) + f(y) = (f(1) + f(0))(x + y)$ $(\forall x, y \in B)$;

(iii) $f(x) + f(y) \leq x + y$ $(\forall x, y \in B)$;

(iv) $f(x)x^{\alpha} \leq f(\alpha) \leq f(x) \vee (x^{\alpha})'$ $(\forall x, y \in B)$ $(\alpha = 0, 1)$;

(v) $x^{\alpha} f(x) = x^{\alpha} f(\alpha)$ $(\forall x \in B)$ $(\alpha = 0, 1)$.

COMMENT: The generalization of condition (v) to n variables, that is, $X^A f(X) = X^A f(A)$ $(\forall A)$, is due to McColl [1877-80], who first proved its equivalence with f being a Boolean function; cf. BFE, Exercise 1.12.
PROOF: (i)$\Longleftrightarrow$(ii)$\Longleftrightarrow$(iii) $\Longleftrightarrow$(iv): Take $h(x) := x$ in Lemma 4.1.
(iv)$\Longrightarrow$(v): Take the meet of each side with x^{α}.
(v)$\Longrightarrow$(i): Take the join of the two identities. $\qquad\Box$

Proposition 4.2. (Țăndăreanu [1984b]). *A function $f : B \longrightarrow B$ is in GBF1 ($\{0, 1\}$) (cf. Definition 4.2) if and only if there exists $h : B \longrightarrow B$ satisfying $h(\alpha) = \alpha$ $(\alpha = 0, 1)$ and the equivalent identities in Lemma 4.1.*

PROOF: According to a result establieshed in Țăndăreanu [1984b], $f \in$ GBF1 ($\{0, 1\}$) if and only if $f = \beta f \circ h$, where $(\beta f)(x) = f(1)x \vee f(0)x'$ $(\forall x \in B)$ and $h(\alpha) = \alpha$ $(\alpha = 0, 1)$ (see also Remark 3.6). But the former condition amounts to (4.1). $\qquad\Box$

Remark that condition (iii) in Proposition 4.1 is of the form

(4.6) $\qquad\qquad \varphi(x, y, f(x), f(y)) = 0 \qquad (\forall x, y \in B)\,,$

where $\varphi : B^4 \longrightarrow B$ is the Boolean function $\varphi(x, y, z, t) = (x + y)'(z + t)$. This justifies the study undertaken in the subsequent proposition and remarks.

Proposition 4.3. (Melter and Rudeanu [1983]). *The functional equation (4.6) is satisfied by any Boolean function $f : B \longrightarrow B$ if and only if the Boolean function φ is of the form*

(4.7) $\qquad\qquad \varphi(x, y, z, t) = (pxy \vee qx'y')z't \vee (rxy \vee sx'y')zt'\,.$

PROOF: Equation (4.6) is fulfilled by any Boolean function $f : B \longrightarrow B$ if and only if

$$\varphi(x, y, ax \vee bx', ay \vee by') = 0 \qquad (\forall x, y, a, b \in B)$$

and this condition is equivalent to the system of identities

(4.8.0) $\qquad\qquad\qquad \varphi(x, y, 0, 0) = 0\,,$

(4.8.1) $\qquad\qquad\qquad \varphi(x, y, x', y') = 0\,,$

(4.8.2) $\qquad\qquad\qquad \varphi(x, y, x, y) = 0\,,$

(4.8.3) $\qquad\qquad\qquad \varphi(x, y, 1, 1) = 0\,,$

by the Verification Theorem. Set

$$\varphi(x, y, z, t) = \varphi_0(x, y)z't' \vee \varphi_1(x, y)z't \vee \varphi_2(x, y)zt' \vee \varphi_3(x, y)zt\,,$$

where $\varphi_k : B^2 \longrightarrow B$ $(k = 0, 1, 2, 3)$ are Boolean functions. But $\varphi_0 = \varphi_3 = 0$ identically by (4.8.0) and (4.8.3), respectively, so that (4.8.1) and (4.8.2) reduce to

$$\varphi_1(x, y)xy' \vee \varphi_2(x, y)x'y = 0\,,$$
$$\varphi_1(x, y)x'y \vee \varphi_2(x, y)xy' = 0\,,$$

and this system is equivalent to the single identity

$$\varphi_1(x, y)(x + y) \vee \varphi_2(x, y)(x + y) = 0\,,$$

or equivalently,

$$\varphi_k(x, y) \leq (x + y)' = xy \vee x'y' \qquad (k = 1, 2)\,.$$

In other words, the φ_ks are of the form $\varphi_1(x, y) = pxy \vee qx'y'$ and $\varphi_2(x, y) = rxy \vee sx'y'$. $\qquad\square$

Remark 4.1. We can restate Proposition 4.3 as follows: the functional equations of the form (4.6), where φ is a Boolean function, satisfied by any Boolean function $f : B \longrightarrow B$, are exactly those of the form

(4.9) $$(pxy \lor qx'y')f'(x)f(y) \lor (rxy \lor sx'y')f(x)f'(y) = 0 \ .$$

In particular every truth function $f : \{0,1\} \longrightarrow \{0,1\}$ satisfies exactly 10 distinct equations of the form (4.9). They are obtained by assigning to the vector (p,q,r,s) values in $\{0,1\}^4$ such that for every $\alpha, \beta, \gamma, \delta \in \{0,1\}$ exactly one of the values $(\alpha, \beta, \gamma, \delta)$ and $(\gamma, \delta, \alpha, \beta)$ is assigned (because interchanging x and y in (4.9) is immaterial).   □

In general an equation of the form (4.9) has non-Boolean solutions f as well. Taking $p := q := r := s := 1$ yields the functional equation

(4.10) $$(x + y)'(f(x) + f(y)) = 0 \ ,$$

which coincides with (iii) in Proposition 4.1 and therefore all of its solutions are Boolean. It follows from the Verification Theorem applied in the Boolean algebra B^B that (4.10) is the unique Boolean functional equation (4.6) which characterizes Boolean functions.

Melter and Rudeanu [1983] proved that the solutions of equation (4.6) form a Boolean ring of functions. They also introduced the class of *upper semi-Boolean* functions and *lower semi-Boolean* functions, defined by the functional equations $xf(x) = xf(1)$ and $x'f(x) = x'f(0)$, respectively (cf. condition (v) in Proposition 4.1). Here is a sample result: the upper semi-Boolean functions form a ring isomorphic to the Cartesian product of B and B^B/I_B, where $I_B = \{f : B \longrightarrow B \mid x'f(x) = 0\}$. Two characterizations of upper semi-Boolean functions are also provided.

Another technique introduced by Melter and Rudeanu [1982] in order to study classes of functions from B^B consists in characterizing the functions we are interested in by certain orthonormal quadruples. This idea may also be regarded as a functional characterization (although not by a functional equation).

Define the *orthonormal characteristic* of a function $f : B \longrightarrow B$ as the vector (f_0, f_1, f_2, f_3), where

(4.11) $$f_0 = f'(0)f'(1), f_1 = f'(0)f(1), f_2 = f(0)f'(1), f_3 = f(0)f(1) \ ,$$

and refer to the vector (f_1, f_2, f_3) as the *orthogonal characteristic* of the function f.

The latter term is justified by the fact that (4.11) implies

(4.12) $$f(0) = f_2 \lor f_3 \ \& \ f(1) = f_1 \lor f_3$$

and on the other hand $f_0 = f_1'f_2'f_3'$, showing that (4.11) is the unique orthonormal extension of the vector (f_1, f_2, f_3), which justifies the former term.

It is easy to check the identity

(4.13) $$f_1 x \lor f_2 x' \lor f_3 = f(1)x \lor f(0)x'(= (\beta f)(x)) \ ,$$

whence we obtain the following result.

Proposition 4.4. (Melter and Rudeanu [1982]). *A function $f \in B^B$ is Boolean if and only if it satisfies the identity*

(4.14) $$f(x) = f_1 x \vee f_2 x' \vee f_3 \,,$$

where (f_1, f_2, f_3) is an orthogonal vector. When this is the case, (f_1, f_2, f_3) is the orthogonal characteristic of f.

PROOF: If the function f is Boolean, then (4.13) implies (4.14). Conversely, if identity (4.14) is satisfied for an orthogonal vector (f_1, f_2, f_3), then the function f is Boolean function and $f(0) = f_3$, $f(1) = f_1 \vee f_3$, hence $f(0)f(1) = f_3$ and $f'(0)f(1) = f_3'f_1 = f_1$; similarly $f_2 = f(0)f'(1)$.   □

The proofs of the next Propositions 4.5−4.7 is left to the reader.

Proposition 4.5. *The orthonormal characteristic (4.11) of a function $f \in$ BF1 satisfies the following properties:*

(i) *f is isotone $\Longleftrightarrow f_2 = 0$,*

(ii) *f is antitone $\Longleftrightarrow f_1 = 0$,*

(iii) *f is constant $\Longleftrightarrow f_1 = f_2 = 0$,*

(iv) *f is injective $\Longleftrightarrow f_0 = f_3 = 0$,*

(v) *f is simple $\Longleftrightarrow f_i \in \{0, 1\} \ (i = 0, 1, 2, 3)$.*   □

Proposition 4.6. *Let (f_1, f_2, f_3, f_4) and (g_1, g_2, g_3, g_4) be the orthonormal characteristics of the Boolean functions f and g, respectively. Then the following conditions are equivalent:*

(i) *$f \leq g$;*

(ii) *$f_1 \leq g_1 \vee g_2$ & $f_2 \leq g_2 \vee g_3$ & $f_3 \leq g_3$;*

(iii) *$g_0 \leq f_0$ & $g_1 \leq f_0 \vee f_1$ & $g_2 \leq f_0 \vee f_2$.*   □

Proposition 4.7. (Ţăndăreanu [1984]). *Suppose $f \in B^B$ and $g \in G(\{0, 1\})$. Then $f \in$ GBF1$(\{0, 1\})$ if and only if it satisfies the identity*

(4.15) $$f(x) = f_1 g(1, x) \vee f_2 g(0, x) \vee f_3 \,,$$

where (f_1, f_2, f_3) is an orthogonal vector. When this is the case, (f_1, f_2, f_3) is the orthogonal charasteristic of f.   □

For the background of the next result see Ch.7, §1.

Proposition 4.8. *Suppose B is a complete Boolean algebra. Then BF1 is a double Moore family of the Boolean algebra B^B and the associated closure operator $\overline{\beta}$ and interior operator $\underline{\beta}$ are constructed as follows:*

(4.16) $$(\overline{\beta}f)(x) = (\bigvee_{a \in B} af(a))x \vee (\bigvee_{a \in B} a'f(a))x' \,,$$

(4.17) $$(\underline{\beta}f)(x) = (\prod_{a \in B} (a' \vee f(a)))x \vee (\prod_{a \in B} (a \vee f(a)))x' \,.$$

PROOF: Clearly $\overline{\beta}f$ is a Boolean function and $f \leq \overline{\beta}f$. If $g \in \mathrm{BF1}$ and $f \leq g$ then $\bigvee_{a \in B} af(a) \leq f(1) \leq g(1)$ and similarly $\bigvee_{a \in B} a'f(a) \leq g(0)$, therefore $(\overline{\beta}f)(x) \leq g(x)$. Therefore $\overline{\beta}$ is the required closure operator and the proof is completed by duality. □

Remark 4.2. $\overline{\beta}$ is a complete-join homomorphism and $\underline{\beta}$ is a complete-meet homomorphism, by Proposition 7.1.1. □

As a matter of fact, the case of a complete Boolean algebra is studied in some detail.

Ekin, Foldes, Hammer and Hellerstein [2000] consider several classes of truth functions, defined in terms of their disjunctive normal forms, and characterize these classes by functional truth equations. For instance, a *Horn function* is defined by the existence of a disjunctive normal form $f = t_1 \vee \ldots \vee t_m$ with the property that each term t_i has at most one negated variable [3]; it is proved that a truth function f is Horn if and only if it satisfies the identity $f(XY) \leq f(X) \vee f(Y)$. The authors also investigate the most general conditions under which a class $\mathbf{K}$ of truth functions can be characterized by a set of identities. A necessary condition is that $\mathbf{K}$ be closed under permutations and identifications of variables. The condition is also sufficient if $\mathbf{K}$ is closed under addition of fictitious variables.

5 Local properties of Boolean functions and extremal solutions of Boolean equations

The idea developed in this section is that with every property $P(f)$ defined for any function $f : B^n \longrightarrow B$ one can associate a corresponding "local" property $P(f, X)$ defined for any f and any $X \in B^n$ in such a way that the "global" property $P(f)$ is satisfied if and only if $P(f, X)$ holds for every $X \in B^n$. The idea was introduced and studied in Rudeanu [1975a], [1975b] and [1976a][4] in the case of Boolean functions and for the properties of injectivity, isotony and with applications to extremal solutions of Boolean equations, respectively. We present below these papers.

The notations A, C stand for vectors in $\{0, 1\}^n$, while D will denote a vector in $\{0, 1\}^m$.

Lemma 5.1. *The following conditions are equivalent for* $f, h, k \in \mathrm{BF}n$ *and* $X \in B^n$:

(5.1)
$$\bigvee_A (f(A)h(A) \vee f'(A)k(A))X^A = 0 \, ;$$

---

[3] Do not confuse with Horn formulas in the sense of Ch.5, §3.

[4] Unfortunately, there are many annoying misprints in this paper.

$$(5.2) \qquad f(X) = \bigvee_A (f(A)h'(A) \vee f'(A)k(A))X^A \ ;$$

$$(5.3) \qquad \bigvee_A k(A)X^A \le f(X) \le \bigvee_A h'(A)X^A \ ;$$

$$(5.4) \qquad f(X) = \bigvee_A (f(A) \vee k(A))X^A = \bigvee_A f(A)h'(A)X^A \ .$$

PROOF: $(5.3) \Longleftrightarrow (5.4)$: From $f(X) = \bigvee_A f(A)X^A$.

$(5.3) \Longleftrightarrow (5.1)$: Write the first inequality (5.3) in the form

$$0 = (\bigvee_A k(A)X^A)(\bigvee_A f'(A)X^A) = \bigvee_A k(A)f'(A)X^A$$

and similarly for the second inequality (5.3).

$(5.2) \Longrightarrow (5.1)$: Take the meet of both sides of (5.2) with $f(A)h(A)X^A$, then with $f'(A)k(A)X^A$.

$(5.1) \Longrightarrow (5.2)$: For

$$\bigvee_A (f(A)h'(A) \vee f'(A)k(A))X^A = \bigvee_A f(A)h'(A)X^A$$

$$= \bigvee_A (f(A)h'(A) \vee f(A)h(A))X^A = \bigvee_A f(A)X^A = f(X) \ .$$

$\square$

Lemma 5.2. *Suppose* $\varphi, \psi \in BF(m+n)$ *and* $\forall X \in B^n \ \exists Y \in B^m \ \varphi(X,Y) = 0$. *Then for every* $X \in B^n$ *the following conditions are equivalent:*

$$(5.5) \qquad \forall Y \in B^m \ \varphi(X,Y) = 0 \Longrightarrow \psi(X,Y) = 0 \ ;$$

$$(5.6) \qquad \bigvee_A (\bigvee_D \varphi'(A,D)\psi(A,D))X^A = 0 \ .$$

PROOF: Since X is fixed, φ and ψ become functions of Y. Therefore the Verification Theorem shows that (5.5) is equivalent to $\varphi'(X,D)\psi(X,D) = 0$ $(\forall D \in \{0,1\}^m$ and this can be written in the following equivalent forms:

$$0 = \bigvee_D \varphi'(X,D)\psi(X,D) = \bigvee_D (\bigvee_A \varphi'(A,D)X^A)(\bigvee_A \psi(A,D)X^A$$

$$= \bigvee_D \bigvee_A \varphi'(A,D)\psi(A,D)X^A = \bigvee_A (\bigvee_D \varphi'(A,D)\psi(A,D))X^A \ .$$

$\square$

Corollary 5.1. *If φ is a simple Boolean function, then (5.5) is equivalent to*

(5.7)
$$\bigvee_A (\bigvee_D \{\psi(A, D) \mid \varphi(A, D) = 0\}) X^A = 0 .$$

□

Corollary 5.2. *If ψ is a simple Boolean function, then (5.5) is equivalent to*

(5.8)
$$\bigvee_A (\bigvee_D \{\varphi'(A, D) \mid \psi(A, D) = 1\}) X^A = 0 .$$

□

Definition 5.1. A function $f : B^n \longrightarrow B$ is said to be *locally injective at the point* $X \in B^n$, while X is called a *point of local injectivity for* f, provided

(5.9)
$$\forall Y \in B^n \ f(X) = f(Y) \Longrightarrow X = Y ;$$

we may omit "locally" and/or "local".

□

Clearly f is injective if and only if it is locally injective at every point $X \in B^n$.

Proposition 5.1. *Suppose $f : B^n \longrightarrow B$ is a Boolean function and $X \in B^n$. Then each of the following conditions is equivalent to X being a point of local injectivity for f:*

(5.10)
$$\bigvee_A (f(A) \bigvee_{C \neq A} f(C) \vee f'(A) \bigvee_{C \neq A} f'(C)) X^A = 0 ;$$

(5.11)
$$f(X) = \bigvee_A (f(A) \prod_{C \neq A} f'(C) \vee f'(A) \bigvee_{C \neq A} f'(C)) X^A ;$$

(5.12)
$$\bigvee_A (\bigvee_{C \neq A} f'(C)) X^A \leq f(X) \leq \bigvee_A (\prod_{C \neq A} f'(C)) X^A ;$$

(5.13)
$$f(X) = \bigvee_A (f(A) \vee \bigvee_{C \neq A} f'(C)) X^A = \bigvee_A f(A) (\prod_{C \neq A} f'(C)) X^A .$$

PROOF: The condition (5.9) of local injectivity is of the form (5.5) for $m := n$, $\varphi(X, Y) := f(X) + f(Y)$ and $\psi(X, Y) := \bigvee_{i=1}^n (x_i + y_i)$. Then

$$\bigvee_C \{\varphi'(A, C) \mid \psi(A, C) = 1\} = \bigvee_C \{f(A) f(C) \vee f'(A) f'(C) \mid C \neq A\}$$

$$= f(A) \bigvee_{C \neq A} f(C) \vee f'(A) \bigvee_{C \neq A} f'(C) ,$$

therefore Corollary 5.2 shows that $(5.9) \Longleftrightarrow (5.10)$. Finally conditions $(5.10)-(5.13)$ are equivalent by Lemma 5.1 with $h(A) := \bigvee_{C \neq A} f(C)$ and $k(A) := \bigvee_{C \neq A} f'(C)$.

□

Corollary 5.3. *A Boolean function $f : B^n \longrightarrow B$ possesses points of local injectivity if and only if*

(5.14)
$$\bigvee_A (f(A) \prod_{C \neq A} f'(C) \vee f'(A) \prod_{C \neq A} f(C)) = 1 \,.$$

PROOF: Immediate from the consistency condition

$$\prod_A (f(A) \bigvee_{C \neq A} f(C) \vee f'(A) \bigvee_{C \neq A} f'(C)) = 0 \,.$$

$\square$

Certain results already known in the literature can be recaptured from the previous theory. Thus e.g. we have the following

Corollary 5.4. *There are no injective Boolean functions of more than one variable, while the injective Boolean functions of one variable are those of the form $f(x) = x + b$.*

PROOF: In view of Proposition 5.1, the condition that every $X \in B^n$ is a point of local injectivity becomes

$$f(A) \bigvee_{C \neq A} f(C) \vee f'(A) \bigvee_{C \neq A} f'(C) = 0 \qquad (\forall A \in \{0,1\}^n) \,,$$

or equivalently, $f(A)f(C) \vee f'(A)f'(C) = 0$ whenever $A \neq C$. The latter equality means $f(C) = f'(A)$ and if $n > 1$ there exist three distinct points A, C, E, which cannot fulfil this condition. For $n := 1$ the unique condition is $f(1) = f'(0)$, which yields $f(x) = x + f(0)$.

$\square$

More generally, it was seen in BFE that Boolean functions of one variable have many special properties. Here is another example:

Corollary 5.5. *A function $f \in$ BF1 possesses points of local injectivity if and only if it is injective.*

PROOF: Condition (5.14) in Corollary 5.3 becomes $f(1) = f'(0)$. $\square$

We now pass to the study of local isotony

Definition 5.2. A function $f : B^n \longrightarrow B$ is said to be *locally upper (lower) isotone* at the point $X \in B^n$, while X is called a *point of local upper (lower) isotony for f*, provided the condition

(5.15)
$$\forall Y \in B^n : X \leq Y \Longrightarrow f(X) \leq f(Y)$$

holds (provided the condition

(5.16)
$$\forall Y \in B^n : Y \leq X \Longrightarrow f(Y) \leq f(X)$$

holds); we may omit "locally" and/or "local". $\square$

Remark 5.1. For every isotone function $f : B^n \longrightarrow B$, $I = (1, \ldots, 1)$ is a point of upper isotony, while $O = (0, \ldots, 0)$ is a point of lower isotony. $\square$

Proposition 5.2. *Suppose* $f : B^n \longrightarrow B$ *is a Boolean function and* $X \in B^n$. *Then each of the following conditions is equivalent to* X *being a point of upper isotony for* f:

(5.17)
$$\bigvee_A (f(A) \vee f'(C)) X^A = 0 ;$$

(5.18)
$$f(X) = \bigvee_A (\prod_{C \geq A} f(C)) X^A ;$$

(5.19)
$$f(X) \leq \bigvee_A (\prod_{C > A} f(C)) X^A .$$

PROOF: To express (5.15) in the form (5.17), we apply Corollary 5.1 for $\varphi(X, Y) := \bigvee_{i=1}^n x_i y_i'$ and $\psi(X, Y) := f(X) f'(Y)$. Then

$$\bigvee_C \{\psi(A, C) \mid \varphi(A, C) = 0\} = \bigvee_C \{f(A) f'(C) \mid A \leq C\}$$

$$= \bigvee_{C > A} f(A) f'(C) = f(A) \bigvee_{C > A} f'(C) ,$$

showing that equation (5.7), which corresponds to (5.15) in the rôle of (5.5), actually becomes (5.17).

Furthermore, (5.17) can be written in the form

$$f(A) X^A \leq \prod_{C > A} f(C) \qquad (\forall A \in \{0, 1\}^n)$$

and this system of inequalities is equivalent to (5.19).

Clearly (5.18) implies (5.19). Conversely, suppose (5.19) holds. Then for each $A \in \{0, 1\}^n$, taking the meet of both sides with X^A we obtain $f(A) X^A \leq (\prod_{C > A} f(C)) X^A$, or equivalently,

$$f(A) X^A = f(A) X^A (\prod_{C > A} f(C)) X^A = (\prod_{C \geq A} f(C)) X^A ,$$

which proves (5.18).   □

We leave to the reader the dual study of lower isotony.

Definition 5.3. A function $f : B^n \longrightarrow B$ is said to be *locally isotone at the point* $X \in B^n$, while X is called a *point of local isotony for* f, provided X is both a point of upper isotony and a point of lower isotony; we may omit "locally" and/or "local".   □

Remark 5.2. It follows immediately from Definitions 5.2 and 5.3 that the following conditions are equivalent for a function $f : B^n \longrightarrow B$:

  (i) f is upper isotone at every point $X \in B^n$;

  (ii) f is lower isotone at every point $X \in B^n$;

(iii) f is locally isotone at every point $X \in B^n$;

(iv) f is isotone .   □

Proposition 5.3. *Suppose $f : B^n \longrightarrow B$ is a Boolean function and $X \in B^n$. Then each of the following conditions is equivalent to X being a point of local isotony for f:*

(5.20)
$$\bigvee_A (f(A) \bigvee_{C>A} f'(C) \vee f'(A) \bigvee_{C<A} f(C))X^A = 0 ;$$

(5.21)
$$f(X) = \bigvee_A (\prod_{C \geq A} f(C))X^A = \bigvee_A (\bigvee_{C \leq A} f(C))X^A ;$$

(5.22)
$$\bigvee_A (\bigvee_{C<A} f(C)X^A \leq f(X) \leq \bigvee_A (\prod_{C>A} f(C))X^A .$$

PROOF: Immediate from Proposition 5.2 and its analogue for points of lower isotony, left to the reader.   □

Corollary 5.6. *A Boolean function $f : B^n \longrightarrow B$ possesses points of local isotony if and only if $f(O) \leq f(I)$.*

PROOF: The consistency condition for equation (5.20) is

$$\prod_A (f(A) \bigvee_{C>A} f'(C) \vee f'(A) \bigvee_{C<A} f(C)) = 0 ,$$

which is equivalent to

$$1 = \bigvee_A (f(A) \prod_{C>A} f(C) \vee f'(A) \prod_{C<A} f'(C))$$

$$= \bigvee_A \prod_{C \geq A} f(C) \vee \bigvee_A \prod_{C \leq A} f'(C) = f(I) \vee f'(O) .$$

□

The concepts of lower (upper) local injectivity, local strict isotony, local antitony and local constancy are also introduced and studied.

Certain applications to graph theory (see e.g. Hammer and Rudeanu [1968], Chapter X, §§ 2, 3, 5) have led to the study of *extremal solutions* of truth equations. By this term we mean *maximal solutions* of an equation $f(X) = 0$ or *minimal solutions* of an equation $f(X) = 1$, in both cases f being an isotone truth function. In the sequel we deal with maximal solutions.

Suppose momentarily that f is an isotone truth function. Then a maximal solution of the equation $f(X) = 0$ is defined by the condition

(M1) $f(X) = 0 \ \& \ \forall Y (X < Y \Longrightarrow f(Y) = 1) ,$

which can also be written in the form

(M2) $$f(X) = 0 \ \& \ \forall i \in \{1, \ldots, n\} \ (x_i = 0 \Longrightarrow f(X_i, 1) = 1) \,,$$

where we have set $X_i = (x_1, \ldots, x_{i-1}, x_{i+1}, \ldots, x_n)$ and $f(X) = f(X_i, x_i)$, for $i = 1, \ldots, n$. While (M2) is very convenient in practice, a more symmetric form of (M1) is

(M3) $$f(X) = 0 \ \& \ \forall Y \ (X < Y \Longrightarrow f(X) < f(Y)) \,.$$

Whereas conditions (M1)–(M3) are easily shown to be equivalent for an isotone truth function f, this is no longer true in the general case, studied in the paper Rudeanu [1976a], which we present in the sequel. This means that from now on, unless otherwise stated, $f : B^n \longrightarrow B$ is an arbitrary Boolean function; thus only the implications (M1)$\Longrightarrow$(M2) and (M1)$\Longrightarrow$(M3) are guaranteed.

Proposition 5.4. *Suppose $f : B^n \longrightarrow B$ is a Boolean function and $X \in B^n$. Then each of the following conditions is equivalent to (M2):*

(5.23) $$f(X) = 0 \ \& \ x_i' f'(X_i, 1) = 0 \qquad (i = 1, \ldots, n) \,;$$

(5.24) $$x_i = f'(X_i, 1) \ \& \ x_i' f(X_i, 0) = 0 \qquad (i = 1, \ldots, n) \,;$$

(5.25) $$\bigvee_A (f(A) \vee \bigvee \{f'(A_i, 1) \mid \alpha_i = 0\}) X^A = 0 \,.$$

PROOF: (M2)$\Longleftrightarrow$(5.23): By the Verification Theorem.
(5.23)$\Longleftrightarrow$(5.24): Write (5.23) in the form

$$x_i f(X_i, 1) = x_i' f(X_i, 0) = x_i' f'(X_i, 1) = 0 \qquad (i = 1, \ldots, n)$$

and note that the equality between the first and the third term is equivalent to $x_i = f'(X_i, 1)$.
(5.23)$\Longleftrightarrow$(5.25): For $f(X) = \bigvee_A f(A) X^A$ and

$$\bigvee_{i=1}^n x_i' f'(X_i, 1) = \bigvee_A (\bigvee_{i=1}^n \alpha_i' f'(A_i, 1)) X^A \,.$$

$\square$

Lemma 5.3. *The following relation holds:*

(5.26) $$\prod_A (f(A) \vee \bigvee_{C > A} f'(C)) = \prod_A f(A) \,.$$

PROOF: Multiply the parentheses in decreasing order of the number of 1's in A. $\square$

Corollary 5.7. *Every consistent Boolean equation $f(X) = 0$ has solutions satisfying (M2).*

PROOF: The consistency condition for equation (5.25) is satisfied because, taking into account Lemma 5.3, we obtain

$$\prod_A (f(A) \vee \bigvee \{f'(A_i, 1) \mid \alpha_i = 0\}) = \prod_A f(A) = 0 .$$

□

Corollary 5.8. *Suppose* $f : B^n \longrightarrow B$ *is an isotone Boolean function, say*

$$f(X) = x_i g_i(X_i) \vee h_i(X_i) \qquad (i = 1, \ldots, n) .$$

Then each of the following conditions are equivalent to (M2):

(5.27) $$f(X) = 0 \, \& \, x_i' g_i'(X_i) = 0 \qquad (i = 1, \ldots, n) ;$$

(5.28) $$x_i = g_i'(X_i) \, \& \, h_i(X_i) = 0 \qquad (i = 1, \ldots, n) .$$

COMMENT: Rudeanu [1966b] proved that in the case of a truth function f, condition (M2) is equivalent to $x_i = g_i'(X_i)$ $(i = 1, \ldots, n)$ and it implies $h_i(X_i) = 0$ $(i = 1, \ldots, n)$; cf. Rudeanu and Hammer [1968], Theorem VII.6. See also Davio [1970].

PROOF: Note that $f(X_i, 0) = h_i(X_i)$ and $f(X_i, 1) = g_i(X_i) \vee h_i(X_i)$. Therefore (5.23) can be written in the form $x_i g_i = h_i = x_i' g_i' h_i' = 0$, which is clearly equivalent to $x_i g_i = h_i = x_i' g_i' = 0$, for $i = 1, \ldots, n$; but the latter condition is (5.27) and it is also equivalent to $x_i = g_i' \, \& \, h_i = 0$ $(i = 1, \ldots, n)$, which is (5.28).   □

Proposition 5.5. *Suppose* $f : B^n \longrightarrow B$ *is a Boolean function and* $X \in B^n$. *Then the following conditions are equivalent to* (M3):

(5.29) $$f(I)X^I = 0 \, \& \, \forall Y \, (X < Y \Longrightarrow f(X) < f(Y)) ;$$

(5.30) $$f(X) = 0 \, \& \, \bigvee_{A \neq I} (\bigvee_{C > A} f'(C)) X^A = 0 ;$$

(5.31) $$\bigvee_A (f(A) \vee \bigvee_{C > A} f'(C)) X^A = 0 .$$

PROOF: Condition (M3) is the logical conjunction of $f(X) = 0$, which can be written in the form

(5.32) $$f(I)X^I = \bigvee_{A \neq I} f(A) X^A = 0 ,$$

and the property which in Rudeanu [1975b] was called " X is a point of upper strict isotony" and it was shown that it is equivalent to

(5.33) $$\bigvee_{A \neq I} f(A) X^A = \bigvee_{A \neq I} (\bigvee_{C > A} f'(C)) X^A = 0 .$$

Now it is clear that each of the conditions (5.29)–(5.31) is equivalent to (5.32)&
(5.33). $\qquad\qquad\qquad\qquad\qquad\qquad\qquad\qquad\qquad\qquad\qquad\qquad$ □

Corollary 5.9. *Every consistent Boolean equation $f(X) = 0$ has solutions satisfying* (M3).

PROOF: The consistency condition for equation (5.31) is satisfied by Lemma
5.3. $\qquad\qquad\qquad\qquad\qquad\qquad\qquad\qquad\qquad\qquad\qquad\qquad\qquad$ □

At this point let us recall that in Ch.6, §4 we have discussed the method
of successive elimination of variables for solving Boolean equations. It is easily
seen by induction that Definition 6.4.1 of the *eliminants* of a Boolean function
f amounts to

(5.34)
$$f_k(x_k,\ldots,x_n) = \prod_{\alpha_1,\ldots,\alpha_{k-1}} f(\alpha_1,\ldots,\alpha_{k-1},x_k,\ldots,x_n)$$
$$(k = 1,\ldots,n+1)\,,$$

yielding in particular $f_1 = f$ and $f_{n+1} = \prod_A f(A)$. Therefore the well-known
tree-like construction of the set of all the solutions (cf. Corollary 6.4.1; see also
BFE, Chapter 2, §4) can be written in the form

(5.35)
$$\prod_{\alpha_1,\ldots,\alpha_{j-1}} f(\alpha_1,\ldots,\alpha_{j-1},0,x_{j+1},\ldots,x_n)$$
$$\leq x_j \leq \bigvee_{\alpha_1,\ldots,\alpha_{j-1}} f'(\alpha_1,\ldots,\alpha_{j-1},1,x_{j+1},\ldots,x_n)$$
$$(j = 1,\ldots,n)\,.$$

This suggests one more concept of maximal solution:

(M4)
$$x_j = \bigvee_{\alpha_1,\ldots,\alpha_{j-1}} f'(\alpha_1,\ldots,\alpha_{j-1},1,x_{j+1},\ldots,x_n)\ (j = 1,\ldots,n)\,.$$

Proposition 5.6. *If the Boolean equation $f(X) = 0$ is consistent, then* (M4)
$\Longrightarrow$(M2).

PROOF: Since the vector X constructed in (M4) satisfies $f(X) = 0$, it follows
that $x_i' f(X_i,0) = 0$ and $x_i \leq f'(X_i,1)$, hence Proposition 5.4 implies that it
remains to prove $f'(X_i,1) \leq x_i$ for all i. But

$$x_i' f'(X_i,1) = (\prod_{\alpha_1,\ldots,\alpha_{i-1}} f(\alpha_1,\ldots,\alpha_{i-1},1,x_{i+1},\ldots,x_n))\cdot$$

$$\cdot(\bigvee_{\alpha_1,\ldots,\alpha_{i-1}} f'(\alpha_1,\ldots,\alpha_{i-1},1,x_{i+1},\ldots,x_n)x_1^{\alpha_1}\ldots x_{i-1}^{\alpha_{i-1}}) = 0\,.$$

$\qquad\qquad\qquad\qquad\qquad\qquad\qquad\qquad\qquad\qquad\qquad\qquad\qquad$ □

As usual, the case $n := 1$ is very peculiar.

Proposition 5.7. *If the equation $ax \vee bx' = 0$ is consistent, then for each $i \in \{2,3,4\}$, a' is the unique element satisfying* (Mi).

PROOF: Property (M4) becomes $x = a'$. In view of Proposition 5.5, property (M3) is equivalent to (5.30), which becomes $f(x) = 0$ & $f'(1)x^0 = 0$. Proposition 5.4 shows that property (M2) is equivalent to (5.23), which reduces to $f(x) = 0$ & $x'f'(1) = 0$. Therefore each of the conditions (M2) and (M3) is equivalent to the system $ax \lor bx' = 0$ & $x'a' = 0$, which can be written in the form $ax \lor (a' \lor b)x' = 0$. But this equation is equivalent to $ax \lor a'x' = 0$ (because $ab = 0$, hence $b \leq a'$), therefore it has the unique solution $x = a'$.   □

14. Applications

The monograph by Hammer and Rudeanu [1968] is devoted to pseudo-Boolean programming, that is, to Boolean techniques for solving optimization problems in 0−1 variables. It includes numerous applications, in particular to graph theory, among which we mention here the utilization of Boolean matrices and of free Boolean algebras in path problems, Boolean proofs of certain graph-theoretical properties and the use of Boolean equations for the determination of independent sets, dominated sets, kernels and chromatic decompositions.

Applications of Boolean equations to switching circuits are described in BFE, Chapter 16, §3, where other applications are also briefly mentioned.

The present chapter contains applications of Boolean equations to graph theory, the algebraic theory of automata, the synthesis of circuits and fault detection (including applications of Boolean derivatives), and databases. There is also an application of Post functions and equations to a marketing problem. Other applications are collected in the last section.

1 Graph theory

The applications in this section refer to kernels, chromatic decompositions, the König-Egerváry property and the planarity of graphs.

First we recall several definitions and facts concerning kernels and related concepts in graphs. Let (V, U) be a finite *directed graph*, where V is the *vertex set* and $U \subseteq V \times V$ is the set of *arcs* or *directed edges*. A subset $S \subseteq V$ is said to be: *independent* or *internally stable* if $S^2 \cap U = \emptyset$, i.e., there is no arc between two vertices of S, *dominated* or *externally stable* if for every $i \in V \backslash S$ there is $j \in S$ such that $(i, j) \in U$; and a *kernel*, if it is both independent and dominated. The following properties are well known. Every independent (dominated) set is included in a maximal independent (includes a minimal dominated) set, where the terms "maximal" and "minimal" refer to set inclusion. Every kernel is both a maximal independent set and a minimal dominated set. A graph may have several kernels, a single kernel or none.

The *adjacency matrix* $A = \| a_{ij} \|$ of the graph is the square matrix of order n defined by $a_{ij} = 1$ if $(i, j) \in U$, else $a_{ij} = 0$, where we have set, without loss of generality, $V = \{1, \ldots, n\}$. On the other hand, every subset $S \subseteq V$ is determined by its *characteristic vector* $X = (x_1, \ldots, x_n)$, defined by $x_i = 1$ if $i \in S$, else $x_i = 0$. As reported in Hammer and Rudeanu [1968], the following equations characterize independent sets, maximal independent sets, dominated sets, minimal dominated sets and kernels, respectively:

$$(1.1) \qquad \bigvee_{i=1}^{n} \bigvee_{j=1}^{n} a_{ij} x_i x_j = 0 \,,$$

$$(1.2) \qquad x_i = a'_{ii} \prod_{j=1, j \neq i}^{n} (a'_{ij} a'_{ji} \vee x'_j) \qquad (i = 1, \ldots, n) \,,$$

$$(1.3) \qquad \prod_{i=1}^{n} \bigvee_{j=1}^{n} (a_{ij} \vee \delta_{ij}) x_j = 1 \,,$$

$$(1.4) \qquad x_i = \bigvee_{j=1}^{n} \prod_{h=1, h \neq j}^{n} (a'_{hj} \delta'_{hj} \vee x'_j) \qquad (i = 1, \ldots, n) \,,$$

$$(1.5) \qquad x_i = \prod_{j=1}^{n} (a'_{ij} \vee x'_j) \qquad (i = 1, \ldots, n) \,,$$

where δ_{ij} stands for the Kronecker delta. Note that kernels are also characterized by each of the following systems of equations: (1.1) and (1.3), (1.1) and (1.4), (1.2) and (1.3), (1.2) and (1.4). As specified in Hammer and Rudeanu (op. cit.), several authors have contributed to the above results, including Rudeanu [1966a], [1966b]. Let us add now that several procedures have been suggested which in many cases reduce the problem of determining the kernel(s) of a graph to the same problem in a graph which is simpler than the original one; cf. Rudeanu [1964] and Tinhofer [1972].

It can be seen by comparing systems (1.2) and (1.5) or by a direct argument that the maximal independent sets of a graph coincide with the kernels of the *symmetric graph* which is obtained from the original one by adding an arc (j, i) to each arc $(i, j) \in U$ (unless we already had $(j, i) \in U$). Căzănescu and Rudeanu [1978] devise a computer program for solving system (1.5) and report favourable computational experience. The same paper suggests a generalization of the concepts discussed above to hypergraphs.

The determination of all maximal independent sets of a graph yields solutions to some other problems as well. Obviously one of them is the determination of the *independence number*, which means the maximum cardinality of an independent set. Direct methods for computing the independence number have been much studied in the literature. For instance, Hertz [1997] uses pseudo-Boolean and Boolean methods in order to reduce this problem to the same problem for a graph of smaller dimension.

Recall that a *chromatic decomposition* of a graph is a partition of the set of vertices into independent subsets; the *chromatic number* of the graph is the minimum cardinality of such a partition, while the partition is called a *minimal chromatic decomposition*. More generally, a covering of least cardinality with independent sets will be called a *minimal covering*.

Remark 1.1. Let $\{I_1, I_2, \ldots, I_m\}$ be a covering of the vertices of a graph with maximal independent sets. Define $C_1 = I_1$ and $C_{k+1} = I_k \backslash (C_1 \cup \ldots \cup C_k)$. Then it is plain that the non-empty C_ks form a chromatic decomposition.   □

Lemma 1.1. *If $\{I_1, \ldots, I_m\}$ is a minimal covering with maximal independent sets, then the chromatic decomposition obtained in Remark 1.1 is minimal.*

PROOF: Let $p \leq m$ be the cardinality of the chromatic decomposition in Remark 1.1. Suppose there exists a chromatic decomposition of cardinality $q < p$. Then by embedding each set of the latter chromatic decomposition into a maximal independent set, one obtains a covering of V with r maximal independent sets, $r \leq q$. But $r < m$, a contradiction.   □

Corollary 1.1. *If, moreover, $\{I_1, \ldots, I_m\}$ is a partition, then it is a minimal chromatic decomposition.*

PROOF: The procedure in Remark 1.1 leaves $\{I_1, \ldots, I_m\}$ invariant.   □

On the other hand there are many algorithms for solving the following problem: given a set X and a family $\mathcal{F} \subseteq \mathcal{P}(X)$ that covers X, find a minimal covering of X with subsets from $\mathcal{F}$. In particular there are Boolean and pseudo-Boolean procedures for solving this problem; cf. Hammer and Rudeanu [1968].

The above results suggest the following method for finding the chromatic number of a graph: determine all maximal independent sets, then a minimal covering with maximal independent sets, from which the construction in Remark 1.1 yields a minimal chromatic decomposition. It is true that this devious procedure may be tedious in certain cases. However there are also cases when it works quite well, as will be shown below.

Proposition 1.1. (Rudeanu [1969]). *Let $G = (V, U)$ be a graph which contains a vertex v_0 that belongs to exactly one maximal independent set V_0. Let G' be the subgraph $G' = (V \backslash V_0, U|_{(V \backslash V_0)^2})$. Then:*

(i) There exists a minimal chromatic decomposition of G containing the set V_0.

(ii) Formula

(1.6)
$$V = V_0 \cup V_1 \cup \ldots \cup V_h$$

is a chromatic decomposition of G if and only if

(1.7)
$$V \backslash V_0 = V_1 \cup \ldots \cup V_h$$

is a chromatic decomposition of G'.

(iii) The decomposition (1.6) is minimal for G if and only if the decomposition (1.7) is minimal for G'.

COMMENT: Clearly every vertex of a finite graph is contained in at least one maximal indepndent set.

PROOF: (i) Note that V_0 belongs to every covering of V with maximal independent sets and apply Remark 1.1 to a minimal such covering, taking $C_1 = I_1 = V_0$. Clearly the resulting chromatic decomposition is minimal.

(ii) Obvious.

(iii) Suppose (1.6) is a minimal chromatic decomposition of G. It follows by (ii) that (1.7) is a chromatic decomposition of G'; suppose it is not minimal. Then, if

$$V \setminus V_0 = W_1 \cup \ldots \cup W_k$$

is a chromatic decomposition of G' with $k < h$, it follows, again by (ii), that

$$V = V_0 \cup W_1 \cup \ldots \cup W_k$$

is a chromatic decomposition of G, of cardinality $k + 1 < h + 1$. Contradiction.

Conversely, suppose (1.7) is a minimal chromatic decomposition of G'. Then (1.6) is a minimal covering of V with maximal independent sets, otherwise the minimality of (1.7) would be contradicted. Therefore (1.6) is a minimal chromatic decomposition by Corollary 1.1.   □

Corollary 1.2. *Let γ be the chromatic number of G. Then the chromatic number of G' is $\gamma - 1$.*   □

Example 1.1. (Rudeanu [1969]). Consider the graph $V = \{1, \ldots, 7\}$, $U = U_1 \cup U_2$,

$$U_1 = \{(1,2), (1,3), (2,3), (2,4), (3,5), (4,3)\} ,$$

$$U_2 = \{(5,1), (5,4), (6,4), (6,7), (7,1), (7,4), (7,5)\} .$$

Then equations (1.2) become

(1.8.1) $x_1 = x_2' x_3' x_5' x_7' ,$

(1.8.2) $x_2 = x_1' x_3' x_4' ,$

(1.8.3) $x_3 = x_1' x_2' x_4' x_5' ,$

(1.8.4) $x_4 = x_2' x_3' x_5' x_6' x_7' ,$

(1.8.5) $x_5 = x_1' x_3' x_4' x_7' ,$

(1.8.6) $x_6 = x_4' x_7' ,$

(1.8.7) $x_7 = x_1' x_4' x_5' .$

Systems of equations of this form are conveniently solved by a bifurcation process and it is recommended to begin by setting to 1 the variable corresponding to the longest equation.

Thus we begin with the case $x_4 = 1$. Then (1.8.4) implies $x_2 = x_3 = x_5 = x_6 = x_7 = 0$. These values transform equations (1.8.2)–(1.8.7) into identities, while (1.8.1) reduces to $x_1 = 1$. We have thus found a solution of system (1.8) and the case $x_4 = 1$ is over. This means that $V_0 = \{1, 4\}$ is the unique maximal independent set containing vertex 4. Therefore it is preferable to give up finding the other solutions of system (1.8) and apply Proposition 1.1 instead.

The reduced graph $G' = (V', U')$ is given by $V' = \{2, 3, 5, 6, 7\}$ and $U' = \{(2, 3), (3, 5), (5, 1), (6, 7), (7, 5)\}$. The system (1.2) corresponding to this graph can be written directly from (1.8):

(1.9.2)
$$x_2 = x_3' \,,$$

(1.9.3)
$$x_3 = x_2' x_5' \,,$$

(1.9.5)
$$x_5 = x_3' x_7' \,,$$

(1.9.6)
$$x_6 = x_7' \,,$$

(1.9.7)
$$x_7 = x_5' x_7' \,.$$

We begin the splitting process on system (1.9) with the case $x_5 = 1$. Then (1.9.5) implies $x_3 = x_7 = 0$ and these values transform equations (1.9.3), (1.9.5) and (1.9.7) into identities, while (1.9.2) and (1.9.6) reduce to $x_2 = 1$ and $x_6 = 1$, respectively. We have found a solution of system (1.9) and the case $x_5 = 1$ is over. This means that $V_0' = \{2, 5, 6\}$ is the unique maximal independent set of G' that contains vertex 5. Therefore we break the solving process and apply Proposition 1.1.

The reduced graph $G'' = (V'', U'')$ is given by $V'' = \{3, 7\}$ and $U'' = \emptyset$. Thus G'' is the empty graph and it has the unique maximal independent set $V_0'' = V''$ (which is also the unique kernel). Therefore Proposition 1.1 implies that

$$V = V_0 \cup V_0' \cup V_0'' = \{1, 4\} \cup \{2, 5, 6\} \cup \{3, 7\}$$

is a minimal chromatic decomposition, hence the chromatic number is $\gamma = 3$. $\square$

Anthonisse [1968] remarked that a given vertex belongs to a unique maximal independent set if and only if the set of all vertices that are not adjacent to the given vertex is independent. He devised a computer program for performing this test.

Now we quote from Simeone [1979] the following preliminaries to the next proposition. A *matching* of a graph $G = (V, U)$ is a subset $M \subseteq U$ such that no two arcs of M have a common vertex; a vertex of G is said to be *matched* by M of *free* according as it is the extremity of an arc in M or not. A *transversal* is a subset $T \subseteq V$ such that every arc has at least an extremity in T. A pair (S, M) where $S \subseteq V$ and M is a matching, is called a *rake* provided every edge of M

has an extremity in S.[1] The greatest cardinality $\mu(G)$ of a matching of G and the least cardinality $\beta(G)$ of a transversal of G are related by $\mu(G) \leq \beta(G)$. A graph for which $\mu(G) = \beta(G)$ is said to have the *König-Egerváry property*. The following conditions are known to be equivalent:

(i) G has the König-Egerváry property ;

(ii) there exists a (necessarily *minimal*, i.e., of least cardinality) transversal T and a (necessarily *maximal*, i.e., of greatest cardinality) matching M such that the pair (T, M) is a rake ;

(iii) for every minimal transversal T and every maximal matching M, the pair (T, M) is a rake .

Simeone (op.cit.) associates with every graph G and every maximal matching M, a quadratic truth function $f(G, M)$ defined as follows. Suppose $M = \{e_1, \ldots, e_q\}$, $p = \text{card}(V) - 2q$ and $n = p+q$; then the *set F of free vertices* has cardinality p. For each $e_i \in M$, associate the literal x_i with one of the extremities of e_i and the literal x'_i with the other extremity. Further, associate a literal x_i $(i = q+1, \ldots, n)$ with each free vertex. Finally, denoting by $\xi(v)$ the literal associated with vertex v, set

$$(1.10) \qquad \begin{aligned} f(G, M)(x_1, \ldots, x_n) &= \bigvee\{\xi(u)\xi(v) \mid (u, v) \in U \backslash M\} \vee \\ &\vee \bigvee\{\xi'(w) \mid w \in F\} . \end{aligned}$$

The function (1.10) depends in fact on the numbering of M and F and on the choices between x_i and x'_i as well, but this dependency is immaterial for the next proposition.

Proposition 1.2. (Simeone [1979]). *Let G be a graph and M a maximal matching of it. Then G has the König-Egerváry property if and only if the truth equation $f(G, M)(X) = 0$ is consistent.*

PROOF: Suppose $A \in \{0, 1\}^n$ is a solution of the equation. Let T be the set of those vertices v for which $\xi(v) := 0$ in the solution. We prove that T is a transversal and (T, M) is a rake, which will imply that G has the required property.

Take $(u, v) \in U$. If $(u, v) \in U \backslash M$ then in A we have $\xi(u)\xi(v) := 0$, hence $u \in T$ or $v \in T$. If $(u, v) \in M$ then we have e.g. $\xi(u) = x_i$ and $\xi(v) = x'_i$ for some i, therefore exactly one of the vertices u, v is in T.

Conversely, suppose G has the König-Egerváry property. Let T be a minimal transversal; then (T, M) is a rake. For every $v \in V$ take $\xi(v) := 0$ if $v \in T$, else $\xi(v) := 1$. We prove that the assignment A defined in this way satisfies $f(G, M)(A) = 0$.

Since every arc of M has exactly one extremity in T, the assignment is consistent with the labelling (x_i, x'_i) of the arcs of M. Since T is a transversal, it follows that for every $(u, v) \in U$ (!) we have $u \in T$ or $v \in T$, which means that in A we have $\xi(u)\xi(v) := 0$. The transversality also implies that every vertex in T is matched, hence $F \subseteq V \backslash T$, which means that in A we have $\xi(v) := 1$ for every $v \in F$. Therefore $f(G, M)(A) = 0$. □

---

[1] This corrects a mistake in op.cit., page 140.

Example 1.2. The graph in Example 1.1 has the obvious maximal matching $M = \{(4,3),(5,1),(6,7)\}$, for which $F = \{2\}$. Consider the labelling $\xi(4) = x_1$, $\xi(3) = x_1'$, $\xi(5) = x_2$, $\xi(1) = x_2'$, $\xi(6) = x_3$, $\xi(7) = x_3'$, $\xi(2) = x_4$. Then the equation $f(G,M)(x_1,x_2,x_3,x_4) = 0$ becomes

$$x_2'x_4 \vee x_2'x_1' \vee x_4x_1' \vee x_4x_1 \vee x_1'x_2 \vee x_2x_1 \vee x_3x_1 \vee x_3'x_2' \vee x_3'x_1 \vee x_3'x_2 \vee x_4' = 0\,,$$

which is inconsistent because it implies the inconsistent equation $x_4x_1' \vee x_4x_1 \vee x_4' = 0$. Therefore this graph does not have the König-Egerváry property.

As a matter of fact, the König-Egerváry property can be tested directly, using Boolean techniques presented in Hammer and Rudeanu [1968], which enable the determination of all maximal matchings and all minimal transversals. In this case one finds out that there is a single minimal transversal, namely $\{4,1,3,7\}$, hence $\beta(G) = 4$, while $\mu(G) = 3$.   □

As shown in Simeone [1979], the opposite way is also valid: with every quadratic truth equation is associated a graph such that the equation is consistent if and only if the associated graph has the König-Egerváry property. Even more, this technique can be extended to arbitrary truth equations and hypergraphs.

Takenaka [1978] rediscovered the characterization of kernels by the dual form of system (1.5),

$$x_i' = \bigvee_{j=1}^{n} a_{ij}x_j \qquad (i = 1,\ldots,n)\,,$$

which she interpreted as representing the Boolean matrix equation $AX = X'$. She then proved, by a thorough analysis of all possible situations, that if for every odd number k, each diagonal element a_{ii}^k of the k-th power A^k is equal to 0, then equation $AX = X'$ is consistent. In graph-theoretic terms, this means that every graph without odd circuits has a kernel, a result which is known as Richardson's theorem.

The aim of the paper by Yanpei Liu [1987a], [1987b] is to characterize the planarity of a graph by the consistency of an associated Boolean equation. The paper is highly technical and no concrete example is given.

2 Automata theory

In most of this section we follow the by paper Cerny and Marin [1977], which uses Boolean functions and equations in the study of automata. We adopt the semi-informal style of the paper, but we handle Boolean equations in our algebraic way, while Cerny and Marin use map techniques.

Remark 2.1. Consider a system of Boolean equations

$$\varphi_1(X,Z) = 1 \ \& \ \varphi_2(X,Z,Y) = 1\,,$$

where the vector variables X, Z and Y have dimensions n, p and q, respectively. Then the result of eliminating Z from the above system is

(2.1)
$$\bigvee_{C \in \{0,1\}^p} \varphi_1(X, C)\varphi_2(X, C, Y) = 1 .$$

$\square$

Definition 2.1. A *Mealy automaton* is a system $\mathbf{M} = (\mathbf{X}, \mathbf{Y}, \mathbf{Z}; \delta, \mu)$, where $\mathbf{X}, \mathbf{Y}$ and $\mathbf{Z}$ are finite sets of elements called *inputs* or *input states*, *outputs* or *output states*, and *(internal) states*, respectively, while $\delta : \mathbf{X} \times \mathbf{Z} \longrightarrow \mathbf{Z}$ and $\mu : \mathbf{X} \times \mathbf{Z} \longrightarrow \mathbf{Y}$ are called the *transition function* and the *output function*, respectively. $\square$

This definition is intended as a mathematical model for a technical device which works at discrete-time instants denoted $0, 1, \ldots, n, \ldots$. If $X \in \mathbf{X}$ and $Z \in \mathbf{Z}$ denote the input and the internal state, respectively, at time n, then $\mu(X, Z)$ is the output at time n, while $Z^+ = \delta(X, Z)$ denotes the internal state at time $n + 1$.

Definition 2.2. *Behavioural characteristics.*

a) *Stability.* $\mathbf{M}$ is *stable for* $(X, Z) \in \mathbf{X} \times \mathbf{Z}$ if, whenever the stimulus X is maintained (see d), after a finite number (k) of transitions, $\mathbf{M}$ reaches a *stable state* Z_s, that is, $Z_s = \mu(X, Z_s)$; otherwise $\mathbf{M}$ is said to be *oscillatory for* (X, Z). If $\mathbf{M}$ is stable (oscillatory) for (X, Z) for some $X \in \mathbf{X}$ and all $Z \in \mathbf{Z}$, then $\mathbf{M}$ is said to be *stable (oscillatory) at* X.

b) *Steady state.* For a given input X, the stable states and the states Z such that $\mathbf{M}$ is oscillatory for (X, Z), are called *steady states* ; all the other states are then *transient*.

c) *Simple oscillations.* If $\mathbf{M}$ is oscillatory for some (X, Z), then it has *simple oscillations at* X if there is no other state $Z_1 \neq Z$ such that $\mathbf{M}$ would be stable for (X, Z_1).

d) It will be understood that $\mathbf{M}$ operates in the following way: α) the input X cannot change unless $\mathbf{M}$ is in a steady state, and β) the state Z and hence the response Y are of interest to the external environment only when $\mathbf{M}$ has reached steady-state conditions.

e) $\mathbf{M}$ is *zero-transition* provided its transitions from an unstable state lead directly to a steady state. If, moreover, $\mathbf{M}$ is stable, then it is called *normal*.

f) *Combinational behaviour.* If $\mathbf{M}$ is in a steady state at some $X \in \mathbf{X}$, then it is said that $\mathbf{M}$ has a *combinational behaviour at* X with respect to the output provided a unique response Y is associated with X , independently of the state Z; otherwise $\mathbf{M}$ has a *sequential behaviour at* X. $\square$

Since the sets $\mathbf{X}, \mathbf{Y}, \mathbf{Z}$ are finite, they can be described by a binary coding, say

(2.2) $$\mathbf{X} \subseteq \{0,1\}^n , \ \mathbf{Z} \subseteq \{0,1\}^p , \ \mathbf{Y} \subseteq \{0,1\}^q ,$$

(2.3) $\delta = (\delta_1, \ldots, \delta_p)$, $\delta_j : \mathbf{X} \times \mathbf{Z} \longrightarrow \{0, 1\}$ $(j = 1, \ldots, p)$,

(2.4) $\mu = (\mu_1, \ldots, \mu_q)$, $\mu_k : \mathbf{X} \times \mathbf{Z} \longrightarrow \{0, 1\}$ $(k = 1, \ldots, q)$,

which opens the way to the use of Boolean techniques.

Definition 2.3. *Characteristic functions* (CF) *related to* **M**.
   a) *State transition* CF:

(2.5)
$$f_\delta(X, Z, Z^+) = \prod_{j=1}^{p}((z_i^+ \delta_i(X, Z) \vee (z_i^+)' \delta_i'(X, Z))$$
   if $X \in \mathbf{X}$ and $Z, Z^+ \in \mathbf{Z}$, else 0 .

   b) *Stable-state* CF:

(2.6)
$$f_\delta^s(X, Z^+) = \prod_{j=1}^{p}(z_i^+ \delta_i(X, Z^+) \vee (z_i^+)' \delta_i'(X, Z^+))$$
   if $X \in \mathbf{X}$ and $Z^+ \in \mathbf{Z}$, else 0 .

   c) *Next-state* CF:

(2.7)
$$f_\delta^z(X, Z^+) = \bigvee_{C \in \{0,1\}^p} f_\delta(X, C, Z^+) .$$

   d) *Steady-state* CF is the function f_δ^c such that the equation $f_\delta^c(X, Z^+) = 1$, $X \in \mathbf{X}$, has as solutions the steady states $Z^+ \in \mathbf{Z}$ that can be reached by some input sequence terminated at X.

   e) *Circuit* CF, or *overall circuit* CF, or *CCF* for short, is the function $f_\mathbf{M}$ such that the equation $f_\mathbf{M}(X, Y) = 1$, $X \in \mathbf{X}$, has as solutions the steady ouput responses of **M**.

   f) The CCF of the *output generator* $Y = \mu(X, Z)$ is

(2.8)
$$f_\mu(X, Z, Y) = \prod_{k=1}^{q}(y_k \mu_k(X, Z) \vee y_k' \mu_k'(X, Z)) .$$

$\square$

Remark 2.2. *Properties of* CF:
   a) The equation $f_\delta(X, Z, Z^+) = 1$ in the unknown Z^+ has the unique solution $Z^+ = \delta(X, Z)$.

   b) The solutions of the equation $f_\delta^s(X, Z^+) = 1$ in the unknown Z^+ coincide with the stable states that can be reached by some input sequence terminated at X.

   c) The solutions of the equation $f_\delta^z(X, Z^+) = 1$ in the unknown Z^+ consist of all the states **M** may pass through from any $Z_0 \in \mathbf{Z}$; this includes both transient and steady states.

   d) The inequalities

(2.9) $$f_\delta^s(X, Z^+) \le f_\delta^c(X, Z^+) \le f_\delta^z(X, Z^+)$$

hold identically. The 0-transition automata are characterized by the identity

$f_\delta^c(X, Z^+) = f_\delta^z(X, Z^+)$, while $f_\delta^s(X, Z^+) = f_\delta^c(X, Z^+)$ characterizes stable automata.

e) A steady output response of **M** means $Y = \mu(X, Z^+)$, where (X, Z^+) is a steady state. In terms of Definition 2.3 this means $f_\mu(X, Z, Y^+) = 1$ and $f_\delta^c(X, Z^+) = 1$. It then follows by Remark 2.1 that

$$(2.10) \qquad f_\mathbf{M}(X, Y) = \bigvee_{C \in \{0,1\}^p} f_\delta^c(X, C) f_\mu(X, C, Y) .$$

$\square$

Remark 2.3. It follows immediately from Definitions 2.2.f and 2.3.e that **M** has a combinational behaviour if and only if the equation $f_\mathbf{M}(X, Y) = 1$ has a unique solution at $X \in \mathbf{X}$. $\square$

Remark 2.4. It follows from Remark 2.3 that, if the equation $f_\delta^c(X, Z^+) = 1$ has a unique solution for $x \in \mathbf{X}$, then **M** has a combinational behaviour at X. For let $C \in \mathbf{Z}$ be that solution; then in view of (2.10) the equation $f_\mathbf{M}(X, Y) = 1$ reduces to $f_\mu(X, C, Y) = 1$, which has the unique solution $y_k = \mu_k(X, C)$ ($k = 1, \ldots, q$). $\square$

Example 2.1. Let $\mathbf{X} := \mathbf{Y} := \{0, 1\}^2$ and define δ by

$$(2.11.1) \qquad \delta_1(X, Z) = x_1 x_2'(z_1 \vee z_2') \vee x_1' x_2 z_1 \vee x_1 x_2 z_2 ,$$

$$(2.11.2) \qquad \delta_2(X, Z) = x_1 x_2' z_1' z_2 \vee x_1' x_2(z_1' \vee z_2) \vee x_1 x_2 z_2 .$$

For the convenience of writing set $Z^+ = (w_1, w_2)$. Then $f_\delta(X, Z, Z^+) = \varphi_1 \varphi_2$, where

$$\varphi_1 = w_1(x_1 x_2'(z_1 \vee z_2') \vee x_1' x_2 z_1 \vee x_1 x_2 z_2) \vee$$
$$\vee w_1'(x_1' x_2' \vee x_1 x_2' z_1' z_2 \vee x_1' x_2 z_1' \vee x_1 x_2 z_2') ,$$
$$\varphi_2 = w_2(x_1 x_2' z_1' z_2 \vee x_1' x_2(z_1' \vee z_2) \vee x_1 x_2 z_2) \vee$$
$$\vee w_2'(x_1' x_2' \vee x_1 x_2'(z_1 \vee z_2') \vee x_1' x_2 z_1 z_2' \vee x_1 x_2 z_2') ,$$

hence

$$f_\delta(X, Z, Z^+) = w_1' w_2'(x_1' x_2' \vee x_1 x_2 z_2') \vee w_1 w_2'(x_1 x_2'(z_1 \vee z_2') \vee$$
$$(2.12) \qquad \vee x_1' x_2 z_1 z_2') \vee w_1' w_2(x_1 x_2' z_1' z_2 \vee x_1' x_2 z_1') \vee w_1 w_2(x_1' x_2 z_1 z_2 \vee$$
$$x_1 x_2 z_2) ,$$

$$f_\delta^s(X, Z^+) = w_1' w_2'(x_1' x_2' \vee x_1 x_2) \vee w_1 w_2'(x_1 x_2' \vee x_1' x_2) \vee$$
$$(2.13) \qquad \vee w_1' w_2(x_1 x_2' \vee x_1' x_2) \vee w_1 w_2 x_2$$

and using Remark 7.3.2 and (2.13) we obtain

$$f_\delta^z(X, Z^+) = \text{ind}_Z f_\delta(X, Z, Z^+) = f_\delta^s(X, Z^+) ,$$

whence Remark 2.2.d implies

$$(2.14) \qquad f_\delta^s(X, Z^+) = f_\delta^c(X, Z^+) = f_\delta^z(X, Z^+) .$$

$\square$

The structure $(\mathbf{X}, \mathbf{Z}; \delta)$ is usually called a *semiautomaton*. However it can be regarded as an automaton by adding $\mathbf{Y=Z}$ and $\mu = \delta$. Then formulas (2.14) show that the automaton in Example 2.1 is normal; cf. Definition 2.2.e.

Example 2.2. Let us complete Example 2.1 by taking $\mathbf{Y} := \{0,1\}$ and the output function

$$(2.15) \qquad \mu(X, Z) = x_2 \vee x_1(z_1 \vee z_2) .$$

Then

$$(2.16) \qquad f_\mu(X, Z, Y) = y(x_2 \vee x_1(z_1 \vee z_2)) \vee y'x_2'(x_1' \vee z_1'z_2') ,$$

hence (2.10), (2.13) and (2.14) imply

$$f_\mathbf{M}(X, Y) = f_\delta^s(X, 0, 0)f_\mu(X, 0, 0, Y) \vee f_\delta^s(X, 0, 1)f_\mu(X, 0, 1, Y) \vee$$

$$\vee f_\delta^s(X, 1, 0)f_\mu(X, 1, 0, Y) \vee f_\delta^s(X, 1, 1)f_\mu(X, 1, 1, Y)$$

$$= (x_1'x_2' \vee x_1x_2)(yx_2 \vee y'x_2') \vee (x_1x_2' \vee x_1'x_2)(y(x_1 \vee x_2) \vee y'x_1'x_2) \vee$$

$$\vee (x_1x_2' \vee x_1'x_2)(y(x_1 \vee x_2) \vee y'x_1'x_2') \vee x_2(y(x_1 \vee x_2) \vee y'x_1'x_2')$$

$$= yx_1x_2 \vee y'x_1'x_2' \vee (x_1 \vee x_2)(y(x_1 \vee x_2) \vee y'x_1'x_2') ,$$

$$(2.17) \qquad f_\mathbf{M}(X, Y) = (x_1 \vee x_2)y \vee x_1'x_2'y' .$$

Since the equation $f_\mathbf{M}(X, Y) = 1$ has the unique solution $y = x_1 \vee x_2$, $\mathbf{M}$ has a combinational behaviour. □

We now pass to the converse problem of determining an automaton which operates in a prescribed way.

Definition 2.4. The *output characteristic function* (OCF for short) is the function f such that the solutions of the equation $f(X, Y) = 1$, $X \in \mathbf{X}$, are those output states Y that define the range of responses within which an automaton is allowed to operate, in order to satisfy some conditions.

In particular if $\mathbf{M}$ operates within these limits for all $X \in \mathbf{X}$, it is said that $\mathbf{M}$ *realizes* f. □

A paraphrase of the above definition states that $\mathbf{M}$ realizes f provided

$$(2.18) \qquad f_\mathbf{M}(X, Y) \le f(X, Y) .$$

It seems to us that a more natural definition would have required equality instead of $\le$.

For instance, suppose it is required that the output should be a partially defined vector truth function $Y \in [\Phi, \Phi \vee \Gamma]$. This means

$$\varphi_j(X) \le y_j \le \varphi_j(X) \vee \gamma_j(X) \qquad (j = 1, \ldots, q) ,$$

therefore the associated OCF is

$$f(X,Y) = \prod_{j=1}^{q} (\varphi_j'(X) \vee y_j)(y_j' \vee \varphi_j(X) \vee \gamma_j(X))$$

$$= \prod_{j=1}^{q} (\varphi_j'(X)y_j' \vee \varphi_j'(X)\gamma_j(X) \vee y_j\varphi_j(X) \vee y_j\gamma_j(X))$$

$$= \prod_{j=1}^{q} (y_j\varphi_j(X) \vee y_j'\varphi_j'(X) \vee \varphi_j'(X)\gamma_j(X)) .$$

A *cascade realization* of an automaton **M** means that **M** consists of two "sub-automata" M_1 and M_2 such that the inputs of M_1 are those of **M**, the inputs of M_2 are are those of **M** together with the outputs of M_1, while M_2 produces the outputs of **M**. Stated formally, this amounts to $M_1 = (X, Y_1, Z; \delta_1, \mu_1)$, $M_2 = (X \times Y_1, Y, Z; \delta_2, \mu_2)$, while

$$f_{\mathbf{M}}(X,Y) = 1 \text{ iff } f_{\mathbf{M}_1}(X,Y_1) = 1 \text{ and } f_{\mathbf{M}_2}((X,Y_1),Y) = 1 .$$

Suppose $X = \{0,1\}^n$, $Y = \{0,1\}^q$ and $Y_1 = \{0,1\}^r$. Then Remark 2.1 implies

(2.19)
$$f_{\mathbf{M}}(X,Y) = \bigvee_{D \in \{0,1\}^r} f_{\mathbf{M}_1}(X,D) f_{\mathbf{M}_2}(X,D,Y) .$$

The problem of obtaining a cascade realization of an automaton **M** may be understood either as Problem 2.1 or as Problem 2.2 below.

Problem 2.1. Given **M** and M_2, find M_1.

Proposition 2.1. *Let f be the OCF of* **M**. *Then the OCF of* M_1 *is*

(2.20)
$$f_1(X,Y_1) = \prod_{C \in \{0,1\}^q} (f_{\mathbf{M}_2}'(X,Y_1,C) \vee f(X,C)) .$$

PROOF: In view of (2.18) and (2.19), the hypothesis can be successively written in the following equivalent forms: for every X, every Y and every Y_1,

$$\prod_{D \in \{0,1\}^r} (f_{\mathbf{M}_1}'(X,D) \vee f_{\mathbf{M}_2}'(X,D,Y)) \vee f(X,Y) = 1 ,$$

$$\prod_{D \in \{0,1\}^r} (f_{\mathbf{M}_1}'(X,D) \vee f_{\mathbf{M}_2}'(X,D,Y) \vee f(X,Y)) = 1 ,$$

$$f_{\mathbf{M}_1}'(X,Y_1) \vee f_{\mathbf{M}_2}'(X,Y_1,Y) \vee f(X,Y) = 1 ,$$

$$\prod_{C \in \{0,1\}^q} (f_{\mathbf{M}_1}'(X,Y_1) \vee f_{\mathbf{M}_2}'(X,Y_1,C) \vee f(X,C)) = 1 ,$$

$$f_{\mathbf{M}_1}'(X,Y_1) \vee f_1(X,Y_1) = 1 ,$$

and applying again (2.18), the last identity is equivalent to (2.20).   □

Corollary 2.1. *If* M_2 *is combinational, then* $f_1(X, Y_1) = f(X, C)$, *where* C *is the unique solution of the equation* $f(X, Y_1, Y) = 1$.   □

Problem 2.2. Given **M** and M_1, find M_2.

Proposition 2.2. *Let* f *be the OCF of* **M**. *Then the OCF of* M_2 *is*

$$(2.21) \qquad f_2(X, Y_1, Y) = f'_{M_1}(X, Y_1) \vee f(X, Y).$$

PROOF: Condition (2.20) can be written in the form

$$f_{M_2}(X, Y_1, Y) \leq f'_{M_1}(X, Y_1) \vee f(X, Y),$$

which, in view of (2.18), is equivalent to (2.21).   □

The *parallel realization* of an automaton is treated in a similar way.

Example 2.3. The automaton **M** in Examples 2.1–2.2 has $X = Z = \{0, 1\}^2$ and $Y = \{0, 1\}$. The output characteristic function of $Y(X) = x_1 \vee x_2$ is $f(X, Y) = f_M(X, Y)$ given by (2.17), therefore **M** realizes f by (2.18).

Now let us try to obtain a cascade decomposition of **M** with $Y_1 = Y$ and M_2 an automaton such that $f_{M_2}(X, Y_1, Y) = f_\mu(X, Y_1, Y)$; cf. (2.16). Setting $Y_1 = (v_1, v_2)$, we have

$$f'_{M_2}(X, Y_1, Y) = yx'_2(x'_1 \vee v'_1 v'_2) \vee y'(x_2 \vee x_1(v_1 \vee v_2)),$$

hence Proposition 2.1 yields

$$f_1(X, Y_1) = (f'_{M_2}(X, Y_1, 0) \vee f(X, 0))(f'_{M_2}(X, Y_1, 1) \vee f(X, 1))$$

$$= (x_2 \vee x_1(v_1 \vee v_2) \vee x'_1 x'_2)(x'_2(x'_1 \vee v'_1 v'_2) \vee x_1 \vee x_2)$$

$$= (x_2 \vee x_1(v_1 \vee v_2) \vee x'_1)(x'_1 \vee v'_1 v'_2 \vee x_1 \vee x_2),$$

$$(2.22) \qquad f_1(X, Y_1) = x'_1 \vee x_2 \vee v_1 \vee v_2.$$

We can take as M_1 the normal (semi)automaton in Example 2.1. For in view of Definition 2.3.d,e and the identification $Y_1 = Z$, the CCF of the latter automaton is $f_{M_1}(X, Y_1) = f^c_\delta(X, Y_1)$, while (2.14), (2.13) and (2.22) imply

$$f^c_\delta(X, Y_1) = f^s_\delta(X, Y_1) \leq f_1(X, Y_1).$$

Now let us solve Problem 2.2. Suppose M_1 has the CCF

$$(2.23) \qquad f_{M_1}(X, Y_1) = v'_1 v'_2(x_1 x_2 \vee x'_1 x'_2) \vee v_1 v'_2 x_1 \vee v'_1 v_2(x_1 \vee x_2) \vee v_1 v_2 x_1 x_2.$$

Then Proposition 2.2 yields

$$f_{M_2}(X, Y_1, Y) = v'_1 v'_2(x_1 x'_2 \vee x'_1 x_2) \vee v_1 v'_2 x'_1 \vee$$

$$\vee v'_1 v_2 x'_1 x'_2 \vee v_1 v_2(x'_1 \vee x'_2) \vee (x_1 \vee x_2)y \vee x'_1 x'_2 y'.$$

It is plain that the former function f_μ given by (2.16) satisfies the realization theorem: $f_\mu(X, Y_1, Y) \leq f_{M_2}(X, Y_1, Y)$.   □

In the last part of their paper, Cerny and Marin study degenerate sequential circuits and apply the Löwenheim reproductive solution of an equation $f(X, Y) = 1$ to determine all direct-transition circuits which realize f.[2]

For instance, they consider the OCF f_1 given by (2.22). The equation $f_1(X, Y_1) = 1$ in the unknown Y_1 has the particular solution $v_1 := 0$, $v_2 := 1$, for which Löwenheim's formula of the reproductive solution

$$Y_1 = P f_1(X, P) \vee \Xi f_1'(P)$$

(see e.g. BFE, dual of Theorem 2.11) becomes

(2.24.1) $$v_1 = p_1(x_1' \vee x_2 \vee p_1 \vee p_2) = p_1 \,,$$

(2.24.2) $$v_2 = p_2(x_1' \vee x_2 \vee p_1 \vee p_2) \vee x_1 x_2' p_1' p_2' = p_2 \vee x_1 x_2' p_1' \,,$$

where $P : \{0,1\}^4 \longrightarrow \{0,1\}^2$ is an arbitrary parameter.

Taking e.g. $p_1 := x_1 x_2' \vee x_1' x_2$, $p_2 := 0$, yields $v_1 = x_1 x_2' \vee x_1' x_2$, $v_2 = 0$, that is, the semiautomaton $\delta_1(X) = x_1 x_2' \vee x_1' x_2$, $\delta_2(X) = 0$. This coincidence is not accidental: the above value of P satisfies $f_1(X, P) = 1$ and the solution (2.24) is reproductive. For the same reason, if one takes $P := \delta$ given by (2.11) one recaptures the semiautomaton in Example 2.1. Of course, the same result is obtained by a direct computation.

See also the short survey by Cerny [1977].

It seems appropriate to mention here the paper Prusinkiewicz [1976]. The author considers an asynchronous linear logical circuit which consists of n elements. The system is described by a square matrix $\| a_{ik} \|$ of order n, where $a_{ik} = 1$ if the output of element i is an input of element k, else $a_{ik} = 0$. Let $B = A + I$, where I is the unit matrix of order n and $+$ means the sum (mod 2) performed componentwise. Let $Y = \| y_k \|_{n \times 1}$ and $X = \| x_i \|_{n \times 1}$, where x_i is the sum (mod 2) of those inputs of the whole system that are also inputs of element i. Then Y represents a stable set iff it satisfies the equation $BY = X$ over the field $\{0,1\}$.

3 Synthesis of circuits

Boolean equations have been used in the literature for the synthesis of combinational and sequential circuits, as was briefly shown in BFE, Chapter 16, §§1 and 2, respectively. The present Ch. 11, this section and Appendix 2 may be regarded as an addition to that sketchy survey.

The conventional problem of constructing a combinational circuit from given gates amounts to the problem of decomposing a Boolean function into functions

---

[2] Unfortunately, we have been unable to understand completely the physical aspects of this last part of the paper.

from a given library. This central problem has been extensively studied in the literature. Appendix 2 is a tentative bibliography on this subject, while in Ch. 11 we have pointed out the rôle of Boolean equations in the decomposition of Boolean functions.

This section comprises two parts. The first one sketches a Boolean-equation approach to the synthesis of combinational circuits using CMOS technologies; cf. Davio and Rudeanu [1993]. The second part resumes and completes the solution devised by Davio and presented in BFE, for the general synthesis problem of sequential circuits.

The paper by Davio and Rudeanu [1993][3] begins with a presentation of the class of autodual negative truth functions and its connections with CMOS technologies. Then the paper concentrates on the *complemented majority function* (CMF for short)

$$(3.1) \qquad \mathrm{maj}'(x, y, z) = y'z' \vee x'z' \vee x'y' ,$$

which is *negative*, i.e., decreasing in all of its variables; besides, the reader can easily check that (3.1) is actually the complement of the well-known *majority function*

$$(3.2) \qquad \mathrm{maj}(x, y, z) = yz \vee xz \vee xy ,$$

therefore (3.1) is also *autodual* or *selfdual*, i.e.,

$$(\mathrm{maj}'(x', y', z'))' = \mathrm{maj}'(x, y, z) .$$

Since $\mathrm{maj}'(x, y, y) = x'y'$ and the Sheffer stroke $\mathrm{NOR}(x, y) = x'y'$ is universal, it follows that CMF is also *universal*, that is, every truth function can be realized in terms of the function (3.1) only.

In order to obtain a decomposition of an arbitrary truth function f using only CMF, variables and the constants 0,1, one can adopt the following recursive strategy. Suppose f is a function of n variables; choose a variable x. Use Proposition 3.1 below in order to decompose f by a CMF in terms of x and of two auxiliary functions g and h, which are both decreasing in x. Then apply twice the next Proposition 3.2 in order to decompose each of the functions g and h by a CMF in terms of x and of two new auxiliary functions that do not depend on x. Thus the variable x is eliminated and the fact that Propositions 3.1 and 3.2 provide the general solutions of the corresponding problems increases the chances of a rapid convergence of the iterative process.

Proposition 3.1. *The Boolean functional equation*

$$(3.3) \qquad f = \mathrm{maj}'(g, h, x)$$

has solutions (g, h) with both g and h decreasing in x. The set of these solutions has the representation

$$(3.4.1) \qquad g(x = 0) = f'(x = 0) \vee p_0 q_0' \vee f'(x = 1)(p_0 \vee p_1 \vee q_0' \vee q_1') ,$$

---

[3] Unfortunately, there are many annoying misprints in this paper.

(3.4.2) $g(x = 1) = f'(x = 1)(p_1 \vee q'_1 \vee f(x = 0)(p_0 \vee q'_0))$,

(3.4.3) $h(x = 0) = f'(x = 0) \vee p'_0 q_0 (f(x = 1) \vee p'_1 q_1)$,

(3.4.4) $h(x = 1) = f'(x = 1)(p'_1 \vee q_1)(f'(x = 0) \vee p'_0 p'_1 q_0 q_1)$,

where p_0, p_1, q_0, q_1 are arbitrary Boolean functions of the variables other than x.

PROOF: For any Boolean function φ, let us set $\varphi(x = \alpha) = \varphi_\alpha$ ($\alpha \in \{0, 1\}$). Then equation (3.3) is equivalent to the system

(3.5) $$f_0 = g'_0 \vee h'_0 ,$$

(3.6) $$f_1 = g'_1 h'_1 ,$$

while the conditions imposed on g and h are expressed by

(3.7) $$g_1 g'_0 \vee h_1 h'_0 = 0 .$$

Equation (3.5) can be written in the form $f_0 g_0 h_0 \vee f'_0(g'_0 \vee h'_0) = 0$ and has the particular solution $g_0 := h_0 := f'_0$. We work out Löwenheim's reproductive solution:

$$g_0 = f'_0(f_0 r_0 s_0 \vee f'_0(r'_0 \vee s'_0)) \vee r_0(f_0(r'_0 \vee s'_0) \vee f'_0 r_0 s_0)$$

$$= f'_0(r'_0 \vee s'_0) \vee f_0 r_0 s'_0 \vee f'_0 r_0 s_0 ,$$

(3.8.1) $$g_0 = f'_0 \vee r_0 s'_0 ,$$

(3.8.2) $$h_0 = f'_0 \vee r'_0 s_0 ,$$

and we obtain similarly the reproductive solution of (3.6):

(3.9.1) $$g_1 = f'_1(r_1 \vee s'_1) ,$$

(3.9.2) $$h_1 = f'_1(r'_1 \vee s_1) .$$

Now we introduce (3.8) and (3.9) into (3.7) and obtain

(3.10) $$f'_1 f_0((r_1 \vee s'_1)(r'_0 \vee s_0) \vee (r'_1 \vee s_1)(r_0 \vee s'_0)) = 0 ,$$

or equivalently,

(3.10′) $$f'_1 f_0(r_0 r_1 s'_0 s'_1 \vee r'_0 r'_1 s_0 s_1)' = 0 ,$$

which we treat as an equation in the unknowns r_0, r_1, s_0 and s_1. The Löwenheim formula applied with the particular solution $r_0 := r_1 := 1$, $s_0 := s_1 := 0$ yields

$$r_0 = p_0 \vee f'_1 f_0((p_1 \vee q'_1)p'_0 \vee (p_1 \vee q'_1)q_0 \vee (p'_1 \vee q_1)q'_0) ,$$

(3.11.1)
$$r_0 = p_0 \vee f_1' f_0 (p_1 \vee q_1' \vee q_0') \,,$$

(3.11.2)
$$r_1 = p_1 \vee f_1' f_0 (p_0 \vee q_1' \vee q_0') \,,$$

$$s_0 = q_0 (f_1 \vee f_0' \vee p_0 p_1 q_0' q_1' \vee p_0' p_1' q_0 q_1) \,,$$

(3.11.3)
$$s_0 = q_0 (f_1 \vee f_0' \vee p_0' p_1' q_1) \,,$$

(3.11.4)
$$s_1 = q_1 (f_1 \vee f_0' \vee p_0' p_1' q_0) \,.$$

Finally we obtain (3.4) by introducing (3.11) into (3.8) and (3.9). This computation is left to the reader.   □

Proposition 3.2. *If the Boolean function f is decreasing in x, then equation (3.3) has solutions (g, h) with both g and h independent of x. The set of these solutions has the representation*

(3.12.1)
$$g = f'(x = 0) \vee t f'(x = 1) \,,$$

(3.12.2)
$$h = f'(x = 0) \vee t' f'(x = 1) \,,$$

where t is an arbitrary Boolean function of the variables other than x.

PROOF: We have seen that equation (3.3) is equivalent to system (3.5),(3.6). The condition that g and h do not depend on x transforms the above system into

$$g' \vee h' = f_0 \ \& \ g' h' = f_1 \,.$$

But it is well known that the equivalent system

$$gh = f_0' \ \& \ g \vee h = f_1' \,,$$

where $f_0' \leq f_1'$, has the general solution

$$g = f_0' \vee t f_1' \ \& \ h = f_0' \vee t' h_1' \,.$$

□

The two steps for eliminating x can be joined into a single one by a third proposition, which we omit here.

Example 3.1. Let us decompose the majority function

(3.13)
$$f := \mathrm{maj}(x, y, z) = yz \vee xz \vee xy \,.$$

We have $f_0 = yz$ and $f_1 = y \vee z$, hence formulas (3.4) yield

$$g_0 = y' \vee z' \vee p_0 q_0' \vee y' z' (p_0 \vee p_1 \vee q_0' \vee q_1') = y' \vee z' \vee p_0 q_0' \,,$$

$$g_1 = y' z' (p_1 \vee q_1' \vee yz(p_0 \vee q_0')) = y' z' (p_1 \vee q_1') \,,$$

$$h_0 = y' \lor z' \lor p_0'q_0(y \lor z \lor p_1'q_1) = y' \lor z' \lor p_0'q_0 \, ,$$
$$h_1 = y'z'(p_1' \lor q_1)(y' \lor z' \lor p_0'p_1'q_0q_1) = y'z'(p_1' \lor q_1) \, .$$

Taking $p_0 := p_1 := q_0 := q_1 := 0$, we obtain $g_0 = h_0 = y' \lor z'$ and $g_1 = h_1 = y'z'$, that is,

$$g = h = x'(y' \lor z') \lor xy'z' = x'(y' \lor z') \lor y'z' = \mathrm{maj}(x,y,z) \, .$$

The process is over with the result

(3.14) $\mathrm{maj}(x,y,z) = \mathrm{maj}'(x, \mathrm{maj}'(x,y,z), \mathrm{maj}'(x,y,z)) \, .$

$\square$

Remark 3.1. It is useful to have a library of CMF decompositions for several usual functions. This can easily be constructed by using e.g. the identity $x'y' = \mathrm{maj}'(x,y,1)$. $\square$

Remark 3.2. If $f(x=0) = f'(x=1)$, i.e., $f_0 = f_1'$, then formulas (3.4) are simplified as follows:

(3.15) $g_0 = f_1 \lor \rho, \, g_1 = f_1'\rho, \, h_0 = f_1 \lor \rho', \, h_1 = f_1'\rho' \, ,$

where $\rho = p_0 \lor p_1 \lor q_0' \lor q_1'$ is an arbitrary parameter. $\square$

Example 3.2. Let us decompose the function

(3.16) $s = x + y + z \, .$

We have $s_1 = yz \lor y'z'$ and $s_0 = s_1'$, hence we can apply Remark 3.2. Taking $\rho := s_1$ we get $g_0 = s_1, g_1 = 0, h_0 = 1, h_1 = s_1'$, that is, $g = x's_1, h = x' \lor s_1'$ and $s = \mathrm{maj}'(x, x's_1, x' \lor s_1')$.

Further, we apply Proposition 3.2 with $f := x's_1 = x'(yz \lor y'z')$. Then $f_0 = yz \lor y'z'$ and $f_1 = 0$. Re-denote $g := \gamma$ and $h := \delta$, so that $\gamma = yz' \lor y'z \lor t$ and $\delta = yz' \lor y'z \lor t'$. Taking $t := y$ we obtain $\gamma = y \lor z$ and $\delta = y' \lor z'$, therefore $x's_1 = \mathrm{maj}'(x, y \lor z, y' \lor z')$.

Apply again Proposition 3.2 with $f := x' \lor s_1' = x' \lor yz' \lor y'z$. Now $f_0 = 1$ and $f_1 = yz' \lor y'z$. Re-denote $g := \lambda$ and $h := \mu$, so that $\lambda = t(yz' \lor y'z)$ and $\mu = t'(yz' \lor y'z)$. Taking $t := y$ we obtain $\lambda = yz$ and $\mu = y'z'$, therefore $x' \lor s_1' = \mathrm{maj}'(x, yz, y'z')$.

Now we put together the above three results:

(3.17) $s = \mathrm{maj}'(x, \mathrm{maj}'(x, y \lor z, y' \lor z'), \mathrm{maj}'(x, yz, y'z')) \, .$

We have eliminated x, to the effect that it remains to decompose the functions $y \lor z, y' \lor z', yz$ and $y'z'$. This task is left to the reader; cf. Remark 3.1. $\square$

We have thus synthesized a binary adder; for the functions in Examples 3.2 and 3.1 represent the sum and the carry, respectively.

In the second part of this section we resume and complete an application of Boolean equations to the synthesis of sequential circuits, already presented in

BFE, Chapter 16, §2. We go directly to the Boolean-equation point; the reader is referred to BFE for the general context.

A *flip-flop* is a device which memorizes an internal state $p \in \{0, 1\}$, as dictated by an input signal $I \in \{0, 1\}^m$, where m depends on the type of the flip-flop, $m \in \{1, 2, 3\}$. More exactly, time is quantified $(t = 1, 2, \ldots)$ and the flip-flop operates according to an equation $p^+ = c(p, I)$, known as the *characteristic equation* or the *next-state equation*, where p^+ denotes the value of p at the next moment.

On the other hand, the synthesis problem imposes on each flip-flop a condition of the form $p^+ = f(p)$, where f is a truth function which depends on the specific synthesis problem; in particular it may also depend on other variables than p.

Now we eliminate p^+ from the above two equations and obtain a new condition

$$(3.18) \qquad\qquad c(p, I) = f(p) ,$$

known as the *input equation*. We write it in the form

$$(3.18') \qquad\qquad c(p, I) = ap \vee bp' (= ab + bp') ,$$

where the coefficients a and b depend on the specific synthesis problem to be solved; in particular a and b may depend on certain variables other than p.

Every solution $I = \varphi(a, b, p)$ of equation (3.18) amounts to expressing I as a combinational circuit connecting p and the variables in a and b; this ensures that the flip-flop will operate in the desired way $p^+ = f(p)$. The idea in Davio [1968] was to solve equation (3.18) for each type of flip-flop and for arbitrary a and b. Therefore in order to solve a specific synthesis problem one just has to introduce the functions a and b corresponding to that problem into the solution $I = \varphi(a, b, p)$. Davio solved the input equations for the D, T, JK, RS and RST flip-flops; cf. BFE, Table 16.6. Later on and independently of Davio, the idea was re-found by Bochmann and Posthoff [1979], who dealt with a larger class of flip-flops, including also the JR, SK and DV types.

We resume below the results in BFE, Table 16.6, while for the last three types of flip-flops we obtain solutions simpler than those of Bochmann and Posthoff.

Here are the characteristic equations of the flip-flops and the input equations:

$$D : \; p^+ = i = ab \vee bp' ,$$

$$T : \; p^+ = t + p = ap \vee bp' ,$$

$$JK : \; p^+ = jp' \vee k'p = jp' + k'p = ap + bp' ,$$

$$RS : \; p^+ = s \vee r'p = ap \vee pb' \; \& \; rs = 0 ,$$

$$RST : \; p^+ = s \vee tp' \vee r't'p = ap \vee bp' \; \& \; rs = rt = st = 0 ,$$

$$JR : \; p^+ = r'(j \vee p) = ap \vee bp' ,$$

$$SK : \; p^+ = s \vee k'p = ap \vee bp' ,$$

$$DV : \; p^+ = v'p \vee dv = v'p + dv = ap + bp' .$$

Proposition 3.3. *The following formulas define the reproductive solutions of the characteristic equations (3.18'):*

$$D: \; i = ap \vee bp' \,,$$

$$T: \; t = a'p \vee bp' \,,$$

$$JK: \; j = bp' \vee \lambda p, \, k = a'p \vee \mu p' \,,$$

$$RS: \; r = a'p \vee \lambda b'p' \,, \, s = bp' \vee \mu ap \,,$$

$$RST: \; r = \lambda a'p \vee \mu b'p' \,, \, s = \nu ap \vee \pi bp' \,, \, t = \lambda' a'p \vee \pi' bp' \,,$$

$$JR: \; j = bp' \vee \lambda \,, \, r = a'p \vee b'p'(\lambda \vee \mu) \,,$$

$$SK: \; k = a'p \vee \lambda \,, \, s = bp' \vee ap(\lambda \vee \mu) \,,$$

$$DV: \; d = (a \vee p')\lambda \vee bp'\lambda' \,, \, v = (a' \vee \lambda\mu)p \vee (b \vee \lambda'\mu)p' \,,$$

where the parameters are arbitrary functions of the variables of a and b.

PROOF: (sketch). The D-equation is already solved. The RST and SK equations are left to the reader. The other input equations can be written in the following forms:

$$(3.19) \qquad\qquad t = ap + bp' + p = a'p + bp' = a'p \vee bp' \,,$$

$$(3.20) \qquad\qquad (a + k')p + (b + j)p' = 0 \iff (a + k')p = (b + j)p' = 0 \,,$$

$$(3.21) \qquad\qquad (s \vee r'p)(a'p \vee b'p') \vee s'(r \vee p')(ap \vee bp') \vee rs = 0 \,,$$

$$(3.22) \qquad\qquad r'(jp' \vee p)(a'p \vee b'p') \vee (r \vee j'p')(ap \vee b'p') = 0 \,,$$

$$(3.23) \qquad\qquad v'p + dv + (ap + bp')(v + v') = 0 \,.$$

Equation (3.19) provides the unique solution of the T-equation. The reproductive solution of the JK-equation is obtained by solving separately each equation (3.20).

By eliminating s from equation (3.21) we obtain the equation $(ap \vee bp')r \vee a'pr' = 0$, whose reproductive solution is

$$r = (a'p \vee b'p')\lambda \vee a'p\lambda' = a'p \vee b'p'\lambda \,,$$

which transforms equation (3.21) into

$$s(a'p \vee b'p') \vee s'bp' = 0 \,,$$

whose reproductive solution is $s = bp' \vee ap\mu$.

Similarly, the elimination of r from (3.22) yields $bp'j' = 0$, whose reproductive solution $j = bp' \vee \lambda$ transforms (3.22) into

$$r'(a'p \vee b'\lambda p') \vee r(ap \vee bp') = 0 \, ,$$

whose reproductive solution is

$$r = (a'p \vee b'p')\mu \vee (a'p \vee b'\lambda p')\mu' = a'p \vee b'p'(\mu \vee \lambda) \, .$$

We process (3.23) in a similar way; the elimination of v results in the equation $a'pd + bp'd' = 0$. □

Corollary 3.1. *The input equations have the following particular solutions:*

$$\text{JK}: \ j = b, \ k = a' \, ,$$

$$\text{RS}: \ r = a'p, \ s = bp' \, ,$$

$$\text{JR}: \ j = b, \ r = a'p \, ,$$

$$\text{SK}: \ k = a', \ s = bp' \, ,$$

$$\text{DV}: \ d = bp', \ v = a' \vee p' \, .$$

PROOF: For JK take $\lambda := b$, $\mu := a'$. For RS take $\lambda := \mu := 0$. For JR take $\lambda := bp$, $\mu := 0$. For SK take $\lambda := a'p'$, $\mu := 0$. For DV take $\lambda := 0$, $\mu := 1$. □

Remark 3.3. 1) In certain particular cases it is possible to obtain solutions that are simpler than those in Corollary 3.1. For instance, if $a = b$ then taking $\lambda := \mu := 1$ in the solution of the RS-equation we get $r = a'$, $s = b$.

2) It is easy to make one of the three RST inputs equal to 0, but this amounts to using in fact an RS flip flop.

3) See also Example 3.3. □

The above solutions of the most general input equations are usually ignored in the literature, the equations being solved in the particular case dealt with.

Example 3.3. In a serial binary adder the carry c obeys the law $c^+ = xy \vee cx \vee cy$, where x and y are binary digits. Independently of each other, Even and Meyer [1969] and Papaioannou and Barrett [1975] solved the input equations for the synthesis of such an adder using an RS flip-flop. The former authors obtained the general solution for these particular equations, while the latter authors obtained the solution $r = x'y'c$, $s = xyc'$, by a rather complicated method. Yet $c^+ = (x \vee y)c \vee xyc'$, so that the above solutions are immediately obtained from Proposition 3.3 and Corollary 3.1, respectively, with $a := x \vee y$ and $b := xy$. Note however that in this case we can take $\lambda := x'y'$, $\mu := xy$ in the solution provided by Proposition 3.3 and obtain the simpler particular solution $r := x'y'$, $s := xy$ (cf. Remark 3.3). □

Example 3.4. Levchenkov [1999a], [1999c] realizes the recurrence $p^+ = xy' \vee xp$ by using an RS flip-flop for which he finds the solutions

$$r = x'p \vee \alpha(x' \vee y)p', \ s = xy'p' \vee \beta xp \, ,$$

[4]which in fact is precisely the solution provided by Proposition 3.3 with $a := x$ and $b := xy'$. □

---

[4] This corrects the misprint $x'y'p'$ in the paper.

It should also be mentioned that Wang [1959] and Even and Meyer [1969] studied sequential Boolean equations in view of their applications to the synthesis of switching circuits.

It seems appropriate to finish this section by mentioning the applications of set-valued functions, referred to in Ch.13, §3. We quote from Ngom, Reischer, Simovici and Stojmenović [1997], where this field is called *set-valued algebra*: "Set logic algebra (SLA) is a special class of multiple-valued logic algebra. It was proposed first as a new foundation of biological molecular computing in [1].[5] Introductory presentations of this new computer paradigm can be found in [2,3]. More formal algebraic aspects of SLA have been studied in [4-13]. As an ultra higher-valued logic system, a set logic system offers a new solution to the interconnection problems that occur in highly parallel VLSI systems. The fundamental concept inherent to a set logic system is multiplex commuting or logic values multiplexing: this means the simultaneous transmission of logic values. This basic concept enables the realization of superchips free from interconnection problems. Parallel processing with multiplexable information carriers makes it possible to construct large-scale highly parallel systems with reduced interconnections. Since the multiplexing of logic values increases the information density, several binary functions can be executed in parallel in a single module. Therefore a great reduction of interconnections can be achieved using optimal multiplexing scheme (see [14,15]) . Possible approaches to the implementation of the set logic system are based on frequencies multiplexing, waves multiplexing and molecules multiplexing, and are called carrier computing systems. For a general perspective on the applications of multivalued logic see [16]."

4 Fault detection in combinational circuits

In this section we resume, perhaps more explicitly, the problem of fault detection in combinational circuits, which we have already presented in BFE, Chapter 16, §3. Then we refer briefly to research that has been done in the meantime.

Consider a circuit with input $X = (x_1, \ldots, x_n) \in \{0,1\}^n$ and the output $y \in \{0,1\}$ described by an equation

(4.1) $$y = f(X) \,.$$

Then the operation of each gate of the circuit is also described by an equation of the form

(4.2) $$z = h(X) \,,$$

where z is the output of that gate and the function h depends on the gate.

The above equations describe the ideal operation of the circuit. Suppose, however, that a certain gate, say the one described by equation (4.2), may present a malfunctioning, while the other gates are reliable; this is known as the *single-fault* assumption. We disregard the trivial case when f or h would be a constant.

---

[5] The numbers within brackets refer to the 56-item bibliography of the paper.

By a *malfunctioning* we mean the fact that for a certain input Ξ, both z and y are wrong, that is, $z \neq h(\Xi)$ and $y \neq f(\Xi)$. Hence the single-fault assumption implies that the real operation of the circuits is described by an equation of the form

(4.3) $$y = g(z, X) ,$$

where the equations (4.1)–(4.3) are related by the identity

(4.4) $$f(X) = g(h(X), X) .$$

Suppose further that the possible errors of gate z are of the form *stuck-at*-0 and *stuck-at*-1. This means there exists $\alpha \in \{0, 1\}$ such that $z = \alpha$ no matter what the input X is. Then for every $X \in \{0, 1\}$ we have $y = g(\alpha, X)$. Now the complete description of the malfunctioning amounts to the existence of an input Ξ such that $\alpha \neq h(\Xi)$ and $g(\alpha, \Xi) \neq g(h(\Xi), \Xi)$. The latter condition is equivalent to $\partial g/\partial z(\Xi) = 1$ (cf. Ch.10, §1).

We are looking for a test which should enable one to identify the malfunctioning of the gate, although its output z is not directly observable. A sufficient condition for the existence of such a test is that the systems of truth equations

(4.5) $$h(A) = 0 \ \& \ \frac{\partial g}{\partial z}(A) = 1$$

and

(4.6) $$h(C) = 1 \ \& \ \frac{\partial g}{\partial z}(C) = 1$$

be consistent. In that case the experimenter should apply to the circuits the inputs A and C; the block z has a malfunctioning if and only if one of the corresponding outputs of the circuit is in error.

To prove this, suppose first that $h(A) = 0$ and $\partial g/\partial z(A) = 1$ and the output is not $f(A)$. Then

(4.7) $$f(A) = g(0, A) \neq g(1, A)$$

by (4.4), hence the output is $g(1, A)$ by (4.3), which proves that z is stuck at 1. A quite similar argument shows that if the output for C is wrong, then z is stuck at 0. Conversely, suppose e.g. that z is stuck at 1. Then (4.4) and (4.5) imply (4.7), hence the corresponding output given by (4.3) is wrong.

Example 4.1. Consider a circuit having three binary inputs x_1, x_2, x_3 and three NAND gates: $z_1 = \text{NAND}(x_1, x_3)$, $z_2 = \text{NAND}(x_2, x_3)$ and the output gate $y = \text{NAND}(z_1, z_2)$; recall that $\text{NAND}(x, y) = x' \vee y'$. Let us apply the test described above to $z := z_1$. Then $X = (x_1, x_2, x_3)$, $h(X) = x_1' \vee x_3'$,

$$g(z, X) = z' \vee z_2' = z' \vee x_2 x_3 ,$$

$$\frac{\partial g}{\partial z}(X) = 1 + x_2 x_3 = x_2' \vee x_3' ,$$

while systems (4.5) and (4.6) become

(4.8) $$x_1' \vee x_3' = 0 \ \& \ x_2' \vee x_3' = 1$$

and

(4.9) $\qquad\qquad\qquad x_1' \vee x_3' = 1 \ \& \ x_2' \vee x_3' = 1$,

respectively. System (4.8) has the unique solution $X := (1,0,1)$, while the solutions to (4.9) are $X := (0,0,1)$ and the four vectors for which $x_3 := 0$. So we apply the inputs (1,0,1) and, say, (0,0,1). $\square$

Another method, briefly referred to in BFE, does not require the single-fault assumption, but still needs the hypothesis that all possible faults are of the form stuck-at-0 and stuck-at-1. Let $z_1, \ldots, z_m$ be the gate outputs that may be in error. Associate with every $i \in \{1, \ldots, \}$ the following two variables:

(4.10) $\qquad\qquad\qquad a_i = 1$ if z_i is stuck at 1 , else 0 ,

(4.11) $\qquad\qquad\qquad b_i = 1$ if z_i is stuck at 0 , else 0 .

Then $a_i b_i = 0$ for all i and the real output of gate i is described by

(4.12) $\qquad\qquad\qquad z_{ic} = a_i \vee b_i' h_i(X)$,

where $h_i(X)$ describes the correct operation of output z_i; cf. (4.2). Hence the output y depends in fact on the variables $x_1, \ldots, x_n, z_{1c}, \ldots, z_{mc}$. Taking into account formulas (4.12) for $i := 1, \ldots, m$, we obtain an equation of the form

(4.13) $\qquad\qquad\qquad y = \varphi(x_1, \ldots, x_n, a_1, b_1, \ldots, a_m, b_m)$.

Now the experimenter applies a sequence of inputs and observes the corresponding outputs given by (4.13). This transforms (4.13), together with conditions

(4.14) $\qquad\qquad\qquad a_i b_i = 0 \qquad (i = 1, \ldots, m)$,

into a system of equations with respect to the unknowns $a_1, b_1, \ldots, a_m, b_m$. The diagnosis amounts to solving this system.

Example 4.2. Let us resume the circuit in Example 4.1. In order to find out whether gates z_1 and z_2 are correct, we introduce the variables (4.10), (4.11) and (4.12) for $i = 1, 2$, and calculate the function φ according to (4.13) and taking into account (4.14), written in the form $a_i \leq b_i'$ or $b_i \leq a_i'$:

$$y = z_{1c}' \vee z_{2c}' = a_1'(b_1 \vee x_1 x_3) \vee a_2'(b_2 \vee x_2 x_3) ,$$

(4.15) $\qquad\qquad\qquad y = b_1 \vee b_2 \vee a_1' x_1 x_3 \vee a_2' x_2 x_3$.

By applying in turn the inputs (1,1,1), (1,0,1), (0,1,0), then the other inputs, the outputs are

(4.16) $\qquad\qquad y_1 = a_1' \vee a_2' , \ y_2 = a_1' \vee b_2 , \ y_3 = b_1 \vee a_2' , \ y_4 = b_1 \vee b_2$.

It follows from (4.16) and (4.14) that $y_1 \geq y_2, y_3 \geq y_4$, therefore the following three cases are possible: I) $y_1 = 1 , y_4 = 0$; II) $y_1 = y_2 = y_3 = y_4 = 0$; III) $y_1 = y_2 = y_3 = y_4 = 1$.

In case I $b_1 = b_2 = 0$, hence $y_2 = a_1'$ and $y_3 = a_2'$, which implies $a_1' \vee a_2' = y_1 = 1$, therefore the following subcases are possible: I.1) $a_1 = a_2 = 0$, hence y_1 and y_2 operate correctly; I.2) $a_1 = 1$, $a_2 = 0$, hence y_1 is stuck at 1, while y_2 operates correctly; I.3) $a_1 = 0$, $a_2 = 1$, hence y_1 operates correctly, while y_2 is stuck at 1.

In case II $a_1' = a_2' = b_1 = b_2 = 0$, hence both y_1 and y_2 are stuck at 1.

In case III it follows that at least one of the outputs y_1 and y_2 is wrong (because the correct operation means $a_1 = a_2 = b_1 = b_2 = 0$, that is, $y_1 = y_2 = y_3 = 1$ and $y_4 = 0$), but no other conclusion can be drawn.   □

Boolean differential calculus in fault detection has also been used by Ku and Masson [1975] (including maultiple faults), Thomasson and Page [1976] (combined with probabilities), Lee [1976] (who extends Boolean differential calculus to vector Boolean algebra), Nguyen [1976]* (including multiple faults), Bozoyan [1978] (using also probabilities and the numerical concept of *activity*), Ubar [1979] (including multiple faults) and Petrosyan [1982]* (using also *activities*). The second method described above was extended by Breuer, Chang and Su [1976] and Cerny [1978]. Papaioannou and Barrett [1975] determined optimal tests for fault detection by applying pseudo-Boolean programming; cf. Hammer and Rudeanu [1968]. See also the references quoted in the above papers.

5 Databases

Helm, Marriott and Odersky [1991], [1995] developed query optimization techniques to bridge the gap between the high-level query language required in spatial database systems and the simpler query language supported by the underlying spatial data-structure. The optimization comprises two main steps. First, it approximates multivariate Boolean constraints by a sequence of univariate constraints. In the second step each univariate Boolean constraint is approximated by a range query over a domain involving simpler shapes. In this section we present the first step and a sketch of the second one.

The authors work in a field $(\mathcal{B}; \cap, \cup,^-, \varnothing, U)$ of subsets of a certain universe U; we will denote intersection $\cap$ by concatenation. Objects in the database are to be retrieved from conjunctive systems of Boolean constraints of the form $f(X_1,\ldots,X_n) = \varnothing$ and $g(X_1,\ldots,X_n) \neq \varnothing$. To be specific, one is faced with the problem of solving an elementary GSBE (cf. Definition 5.5.3)

(5.1)
$$f(X_1,\ldots,X_n) = \varnothing \ \& \ g_1(X_1,\ldots,X_n) \neq \varnothing \ \& \ldots$$
$$\& \ g_m(X_1,\ldots,X_n) \neq \varnothing$$

and this can be done by successive elimination of variables. If the Boolean algebra is big enough, in particular if it is infinite, then it follows from Proposition 5.5.5 that system (5.1) is consistent if and only if all the atomic GSBE's

(5.2)
$$f(X_1,\ldots,X_n) = \varnothing \ \& \ g_k(X_1,\ldots,X_n) \neq \varnothing \qquad (k = 1,\ldots,m)$$

are consistent. Taking also into account Proposition 5.5.4, it follows that for each $i \in \{1, \ldots, n\}$, the result of eliminating the variable X_i from system (5.1) is

$$
\begin{aligned}
(5.3) \quad & f(X_i = \emptyset) f(X_i = U) = \emptyset \ \& \ f'(X_i = \emptyset) g_k(X_i = \emptyset) \cup \\
& \cup f'(X_i = U) g_k(X_i = U) \neq \emptyset \qquad (k = 1, \ldots, m) \,.
\end{aligned}
$$

In other words, (5.3) is the necessary and sufficient condition on $X_1, \ldots, X_{i-1}$, $X_{i+1}, \ldots, X_n$, for the existence of X_i such that $(X_1, \ldots, X_n)$ be a solution of (5.1).

Consider the following example. "Let us assume smugglers are 'importing' prohibited goods into a given country C and wish to know where to site their distribution operation. The goods must be imported at some border town T and transported into some destination area A in C. Assume further that, while it is relatively easy to enter the country C, there are massive police patrols along the country's internal state boundaries. The transport of the prohibited goods is safe as long as no internal state boundary is crossed. Hence, the smugglers also want to find a road R from T to A, which does not cross a state boundary between T and A, that is, which proceeds entirely within some state B. Assuming the smugglers have access to a spatial database, they could formalize their problem" as follows: $A \subset C$, $B \subseteq C$, $T \not\subseteq C$, $R \subseteq T \cup B \cup A$, $RT \neq \emptyset$ and $RA \neq \emptyset$. Assume we are given A and C, while T, R and B must be found. Then we get the following formalization of the problem:

$$
(5.4) \qquad B\overline{C} \cup R\overline{T}\,\overline{B}\,\overline{A} = \emptyset \ \& \ T\overline{C} \neq \emptyset \ \& \ RT \neq \emptyset \ \& \ RA \neq \emptyset \,.
$$

Then B is determined by the inequalities

$$
(5.5.1) \qquad R\overline{T}\,\overline{A} \subseteq B \subseteq C
$$

under the condition $RT\overline{A}\,\overline{C} = \emptyset$. Since $\overline{A}\,\overline{C} = \overline{C}$, system (5.4) reduces to

$$
(5.4') \qquad R\overline{T}\,\overline{C} = \emptyset \ \& \ T\overline{C} \neq \emptyset \ \& \ RT \neq \emptyset \ \& \ RA \neq \emptyset \,.
$$

Then R is determined by the conditions

$$
(5.5.2) \qquad R \subseteq T \cup C \ \& \ RT \neq \emptyset \ \& \ RA \neq \emptyset \,.
$$

Now we must apply formula (5.3). We have $f(R = U) = \overline{T}\,\overline{C}$ and $f(R = \emptyset) = \emptyset$. The function g corresponding to $RT \neq \emptyset$ is given by $g(R = U) = T$ and $g(R = \emptyset) = \emptyset$, hence (5.3) becomes $(T \cup C)T \neq \emptyset$, that is $T \neq \emptyset$, which in fact follows from (5.5.2). The condition $RA \neq \emptyset$ yields $g(R = U) = A$ and $g(R = \emptyset) = \emptyset$, so that (5.3) reduces to $(T \cup C)A \neq \emptyset$, which follows from $AC = A \neq \emptyset$. Therefore the elimination of R does not introduce further conditions, so that system (5.4') reduces to the inequality

$$
(5.5.3) \qquad T\overline{C} \neq \emptyset \,,
$$

which is already solved.

We have thus solved system (5.4) in the recursive form (5.5).

Another example is somehow a variant of the previous one, but in quite a serious context and in a more technical formulation. It is taken from the authors' work on parsing visual languages and diagrams and concerns the recognition of state-charts: "A state-chart consists of states, which are represented by rectangles, and transitions, which are represented by arrows consisting of a head and a shaft. In the parser, diagrams are 'recognized' if the objects within it satisfy particular relationships. For example, a transition to a local state can be recognized if the objects c, a, h, t satisfy the following relationships: rectangle a within rectangle c, the head h of the arrow touches a but its shaft t does not. In our parser, objects are stored as tuples in a spatial database and recognition corresponds to querying the database. Assume that the tuples are stored in the relations *Rectangle*, *Head* and *Shaft* and each relation has a single attribute *extent*, which contains the geometric extent of the object. A local state can be recongnized with the query: *find* $a, c \in$ *Rectangle and* $h \in$ *Head and* $t \in$ *Shaft such that* $A = a \cdot$ *extent,* $C = c \cdot$ *extent,* $H = h \cdot$ *extent and* $T = t \cdot$ *extent satisfy* $A \subset C, H \subseteq C, HA \neq \emptyset, HT \neq \emptyset, T \nsubseteq C$ *and* $TA = \emptyset$." Thus we have the constraints

$$(5.6) \qquad A\overline{C} = \emptyset \ \& \ \overline{A}C \neq \emptyset \ \& \ H\overline{C} = \emptyset$$
$$\& \ HA \neq \emptyset \ \& \ HT \neq \emptyset \ \& \ T\overline{C} \neq \emptyset \ \& \ AT = \emptyset \,.$$

We begin with the variable C:

$$(5.7.1) \qquad A \cup H \subseteq C \ \& \ \overline{A}C \neq \emptyset \ \& \ T\overline{C} \neq \emptyset \,,$$

hence the elements in (5.3) are $f(C = U) = \emptyset$, $f(C = \emptyset) = A \cup H$ and $g(C = U) = \overline{A}$, $g(C = \emptyset = \emptyset$ for $\overline{A}C \neq \emptyset$, while $g(C = U) = \emptyset$, $g(C = \emptyset) = T$ for $T\overline{C} \neq \emptyset$. Therefore (5.3) yields $\overline{A} \neq \emptyset$ and $\overline{A}HT \neq \emptyset$. But $\overline{A} \neq \emptyset$ follows from (5.7.1), hence the result of the elimination of C from (5.6) is

$$(5.6') \qquad \overline{H}\,\overline{A}T \neq \emptyset \ \& \ HA \neq \emptyset \ \& \ HT \neq \emptyset \ \& \ AT = \emptyset \,.$$

Now we obtain

$$(5.7.2) \qquad A \subseteq \overline{T} \ \& \ HA \neq \emptyset \ \& \ \overline{H}\,\overline{A}T \neq \emptyset \,,$$

hence $f(A = U) = T$, $f(A = \emptyset) = \emptyset$ and $g(A = U) = H$, $g(A = \emptyset) = \emptyset$ for $HA \neq \emptyset$, while $g(A = U) = \emptyset$, $g(A = \emptyset) = \overline{H}T$ for $\overline{H}\,\overline{A}T \neq \emptyset$. Therefore $\overline{T}H \neq \emptyset$ and $\overline{H}T \neq \emptyset$ by (5.3) and the result of eliminating A from (5.6') is

$$(5.7.3) \qquad H\overline{T} \neq \emptyset \ \& \ \overline{H}T \neq \emptyset \ \& \ HT \neq \emptyset \,,$$

which coincides with its solution.

The second step consists in approximating the solution obtained in the first step by a range query over a domain involving simpler shapes. In most cases the authors work with a Boolean algebra of the form $\mathcal{P}(\mathbf{R}^k)$ and the simpler shape which approximates a region $X \in \mathcal{P}(\mathbf{R}^k)$ is its *bounding-box*

$$(5.8) \qquad [X] = \{Y \in \mathbf{R}^k \mid \inf X \leq Y \leq \sup X\}$$

or bb for short, where $\inf X$ and $\sup X$ refer to the standard Cartesian ordering of $\mathbf{R}^k$. For instance, if $k := 2$, then $[X]$ is the least rectangle which has the sides parallel to the axes and includes X. In the general case the sets (5.8) may be regarded as the *intervals* of the poset $\mathbf{R}^k$ and together with $\emptyset$ and $\mathbf{R}^k$ they form a complete lattice with respect to set inclusion; let us call it the *bounding-box lattice* or bb-lattice. The meet operation in this lattice is intersection and this property extends to arbitrary meets, but the join operation ⊔ is not the intersection; in fact $[X] \cup [Y] \subseteq [X] \sqcup [Y]$. Notice that the bb-lattice is not distributive.

The polynomials of the bb-lattice are called *bounding-box functions*, or bbf for short. One says that a bbf F *approximates* a Boolean function f *from below*, written $F \uparrow f$, if the identity

$$F([X_1], \ldots, [X_n]) \subseteq [f(X_1, \ldots, X_n)]$$

holds, while F *approximates* f *from above*, written $F \downarrow f$, provided the opposite inclusion holds identically. The *best lower bounding-box approximation* L_f of a Boolean function f is the greatest bbf which approximates f from below, while the *best upper bounding-box approximation* U_f of f is defined dually. The authors claim[6] that, if f is a simple Boolean function, then $L_f = \bigsqcup_{X \leq f} [X]$, while U_f is obtained from the Blake canonical form of f by removing all negative literals, then simplifying and finally converting into bb form. For instance, if $f = Y \cup X \cup \overline{Z}$, then $L_f = [Y]$ and $U_f = [Y] \sqcup [X][V]$.

In the first example treated above, the solution (5.5) is approximated by

(5.9) $[B] \subseteq [C] \,\&\, [R] \subseteq [T] \sqcup [C] \,\&\, [R][T] \neq \emptyset \,\&\, [R][A] \neq \emptyset$,

while in the second example the solution (5.7) is approximated by

(5.10) $[A] \sqcup [H] \subseteq [C] \,\&\, [H][A] \neq \emptyset \,\&\, [H][T] \neq \emptyset$.

6 Marketing

In this section we present (with a slight improvement) an example given by Bordat [1975], [1978] to illustrate the possibility of applying Post equations to certain marketing problems.

A business company delivers books in three variants: 0) Deluxe edition, 1) standard edition, and 2) pocket edition, to five categories of customers: 0) regular, 1) from the Administration, 2) of Deluxe editions, 3) of standard and pocket editions, and 4) foreigner. Besides, the company considers the following thresholds for the amount of an order: 0) less than 10 F, 1) less than 500 F, and 2) less than 1000 F.

On the other hand, there are several options as concerns: reduction of price (namely: 1) 0%, 2) 10%, and 3) 20%), postal charges (namely: 1) no, and 2) yes), and way of payment (namely: 1) cash on delivery, 2) within 30 days, and 3) within 90 days).

As a matter of principle the three criteria (edition, customer and amount of order) yield $3 \times 5 \times 3 = 45$ possible types of order. Yet the company has

---

[6] We have been unable to understand completely their proofs.

considered only 14 types of order for which it has established the corresponding options. Now the 14 rules can be described by Post functions, while the converse problem, i.e., find all possible types of order that correspond to a given option, can be solved by means of Post equations.

Let x, y, z be three variables denoting edition, customer and amount of order, respectively. Then x, y, z take values in the sets $\{0, 1, 2\}, \{0, 1, 2, 3, 4\}$ and $\{0, 1, 2\}$, respectively, but the three sets have different significations. Further, let f_1, f_2, f_3 be three variables denoting reduction of price, postal charges and way of payment, respectively, including the value $f_i = 0$, with the meaning "no option". Then f_1, f_2, f_3 take values in the sets $\{0, 1, 2, 3\}, \{0, 1, 2\}$ and $\{0, 1, 2, 3\}$, respectively, but again these sets have different significations. Each of the variables f_1, f_2, f_3 depends on x, y, z, so that we have in fact three functions

$$f_1 : \{0, 1, 2\} \times \{0, 1, 2, 3, 4\} \times \{0, 1, 2\} \longrightarrow \{0, 1, 2, 3\} ,$$

$$f_2 : \{0, 1, 2\} \times \{0, 1, 2, 3, 4\} \times \{0, 1, 2\} \longrightarrow \{0, 1, 2\} ,$$

$$f_3 : \{0, 1, 2\} \times \{0, 1, 2, 3, 4\} \times \{0, 1, 2\} \longrightarrow \{0, 1, 2, 3\} .$$

Now we regard the above sets as the Post algebras C_3, C_5 and C_4 and we embed the codomain of f_2 into C_4, so that we obtain the functions

(6.0) $f_i : C_3 \times C_5 \times C_3 :\longrightarrow C_4$ $(i = 1, 2, 3)$.

The variables x, y, z are determined by their disjunctive components (x^0, x^1, x^2), $(y^0, y^1, y^2, y^3, y^4)$ and (z^0, z^1, z^2), respectively (cf. Definition 5.1.1). Moreover, since the set of all functions of the form $f : X \longrightarrow C_4$ is endowed with a structure of 4$-$Post algebra with respect to the operations defined pointwise, our functions have expansions of the form 5.(1.7). Thus the 14 rules of this example, initially given in the form of a table, have the following algebraic expressions:

(6.1)
$$f_1 = 1(y^0(z^2)' \vee x^2y^3(z^2)') \vee 2(y^0z^2 \vee y^1(z^2)' \vee x^0y^2z^0 \vee$$
$$x^0y^4(z^2)' \vee (x^0)'y^4 \vee x^1y^3(z^2)' \vee x^2y^3z^2) \vee 3(y^1z^2 \vee x^0y^2z^1 \vee$$
$$\vee x^0y^2z^2 \vee x^0y^4z^2 \vee x^1y^3z^2) ,$$

(6.2)
$$f_2 = 1(y^0(z^2)' \vee y^0z^2 \vee y^1(z^2)' \vee y^1z^2 \vee x^0y^2z^2 \vee x^1y^3(z^2)' \vee$$
$$x^1y^3z^2 \vee x^2y^3(z^2)' \vee x^2y^3z^2) \vee 2(x^0y^2z^0 \vee x^0y^2z^1 \vee$$
$$\vee x^0y^4(z^2)' \vee x^0t^4z^2 \vee (x^0)'y^4)$$

(remember that $f_2 \in \{0, 1, 2\}$),

(6.3)
$$f_3 = 1y^0(z^2)' \vee 2(y^0z^2 \vee (x^0)'y^4 \vee x^0y^4(z^2)' \vee x^1y^3(z^2)' \vee$$
$$x^1y^3z^2 \vee x^2y^3(z^2)' \vee x^2y^3z^2) \vee 3(y^1(z^2)' \vee y^1z^2 \vee x^0y^0z^0 \vee$$
$$\vee x^0y^2z^0 \vee x^0y^2z^1 \vee x^0y^2z^2 \vee x^0y^4z^2) .$$

Formulas (6.1)$-$(6.3) can be simplified by introducing absorbed terms and by using the identities $x^k \vee (x^k)' = z^k \vee (z^k)' = 2$ and $y^k \vee (y^k)' = 4$. Thus

$$f_1 = f_1 \vee 1y^0z^2 \vee 1x^2y^3z^2 \vee 2y^1z^2 \vee 2x^0y^2z^1 \vee 2x^0y^2z^2 \vee 2x^0y^4z^2 \vee 2x^1y^3z^2$$

$$= 1(y^0 \vee x^2y^3) \vee 2(y^0z^2 \vee y^1 \vee x^0y^2 \vee x^0y^4 \vee (x^0)'y^4 \vee x^1y^3 \vee x^2y^3z^2) \vee 3(\ldots) ,$$

$$(6.4) \quad \begin{aligned} f_1 &= 1(y^0 \vee x^2 y^3) \vee 2(y^0 z^2 \vee y^1 \vee x^0 y^2 \vee y^4 \vee x^1 y^3 \vee x^2 y^3 z^2) \vee \\ &\quad \vee 3(y^1 z^2 \vee x^0 y^2 z^1 \vee x^0 y^2 z^2 \vee x^0 y^4 z^2 \vee x^1 y^3 z^2) \,, \end{aligned}$$

$$f_2 = 1(y^0 \vee y^1 \vee x^0 y^2 z^2 \vee x^1 y^3 \vee x^2 y^3) \vee 2(x^0 y^2 z^0 \vee x^0 y^2 z^1 \vee x^0 y^4 \vee (x^0)' y^4) \,,$$

$$(6.5) \quad f_2 = 1(y^0 \vee y^1 \vee x^0 y^2 z^2 \vee x^1 y^3 \vee x^3 y^3) \vee 2(x^0 y^2 (z^2)' \vee y^4) \,,$$

$$\begin{aligned} f_3 &= 1y^0 (z^2)' \vee 2(y^0 z^2 \vee (x^0)' y^4 \vee x^0 y^4 (z^2)' \vee x^1 y^3 \vee x^2 y^3) \vee \\ &\quad \vee 3(y^1 \vee x^0 y^2 \vee x^0 y^4 z^2) \vee 1y^0 z^2 \vee 2x^0 y^4 z^2 \\ &= 1y^0 \vee 2(y^0 z^2 \vee (x^0)' y^4 \vee x^0 y^4 \vee x^1 y^3 \vee x^2 y^3) \vee 3(\ldots) \,, \end{aligned}$$

$$(6.6) \quad f_3 = 1y^0 \vee 2(y^0 z^2 \vee y^4 \vee x^1 y^3 \vee x^2 y^3) \vee 3(y^1 \vee x^0 y^2 \vee x^0 y^4 z^2) \,.$$

To see how this works, consider an example. Suppose a regular customer purchases books for a total price between 500 F and 1000 F. This means $y = 0$ and $z = 2$, while x is not determined. Then $y^0 = z^2 = 1$, while $y^k = 0$ for $k \neq 2$ and $z^0 = z^1 = 0$. Then (6.4)–(6.6) imply $f_1 = 1 \vee 2 = 2$, $f_2 = 1$, $f_3 = 1 \vee 2 = 2$, that is: a reduction of 10%, no postal charges and payment within 30 days. This is one of the 14 rules.

As another example, consider a customer purchasing Deluxe editions, who orders a pocket edition for which the price is less than 10 F. This means $x = 2, y = 2, z = 0$, that is, $x^j = y^j = z^k = 0$ for $j \neq 2$ and $k \neq 0$. Now it follows from formulas (6.4)–(6.6) that $f_1 = f_2 = f_3 = 0$, which means no option. In other words, the above type of order has not been included among the 14 rules.

Now let us illustrate the converse problem. Suppose one asks whether there exist orders consistent with the options: 20% reduction, no postal charges, payment within 90 days. This amounts to finding values of x, y, z for which

$$(6.7) \quad f_1 = 3 \,, \; f_2 = 1 \,, \; f_3 = 3 \,.$$

We claim that system (6.7) is equivalent to

$$(6.8.1) \quad y^1 z^2 \vee x^0 y^2 z^1 \vee x^0 y^2 z^2 \vee x^0 y^4 z^2 \vee x^1 y^3 z^2 = 1 \,,$$

$$(6.8.2) \quad y^0 \vee y^1 \vee x^0 y^2 z^2 \vee x^1 y^3 \vee x^2 y^3 = 1 \,,$$

$$(6.8.3) \quad y^1 \vee x^0 y^2 \vee x^0 y^4 z^2 = 1 \,,$$

because it is clear that $f_1 = 3$ and $f_3 = 3$ are equivalent to (6.8.1) and (6.8.3), respectively, while the equivalence between $f_2 = 1$ and (6.8.2) is also based on the fact that the coefficients of 1 and 2 in (6.5) are orthogonal.

We multiply (6.8.2) by (6.8.3) and obtain $y^1 \vee x^0 y^2 z^2 = 1$. Further multiplication by (6.8.1) yields

(6.9) $$y^1 z^2 \vee x^0 y^2 z^2 = 1 .$$

Thus system (6.8) is equivalent to equation (6.9), which decomposes into $z^2 = 1$ and $y^1 \vee x^0 y^2 = 1$. The latter equation has the solutions $y^1 := 1$ and $x^0 := y^2 := 1$. This means the following solutions in terms of (x, y, z):

(6.10) $$(\text{arbitrary}, 1, 2) , (0, 2, 2) .$$

Therefore the following types of order yield the options given above:
   a) customer from the Administration, price less than 1000 F, and
   b) Deluxe edition, customer of Deluxe edition, price less than 1000 F.

7 Other applications

The first application in this section refers to Voronoi diagrams in Boolean algebras and is taken from Melter and Rudeanu [1993]. The present approach eliminates a redundancy in the original paper.

Given a set $S = \{P_1, \ldots, P_n\}$ of distinct points in a metric space, with each point $P_i \in S$ is associated the *Voronoi region* V_i, which consists of all the points P of the space that are closer to P_i than to any other point of S. The *Voronoi diagram* determined by S is defined as the family of sets $\{V_1, \ldots, V_n\}$. For example, if P_1, P_2, P_3 are the vertices of a non-degenerate triangle in the plane, then the boundaries of the Voronoi regions are rays determined by the perpendicular bisectors of the sides of the triangle. In general the Voronoi diagram is found by an algorithm. Voronoi diagrams are frequently used as a tool in *computer vision*.

Several authors have considered Voronoi diagrams for other metrics, e.g., the city block distance

$$d_4([X_1, Y_1], [X_2, Y_2]) = |X_1 - X_2| + |Y_1 - Y_2| .$$

[7] In fact one can consider metrics which take values in a set whose algebraic structure is different from that of real numbers; see e.g. BFE, Chapters 13 and 14, the present Ch.12, §3, as well as two papers on Boolean distance for graphs.[8] This is the context in which it seems natural to consider Voronoi diagrams for Boolean algebras; the metric will be the usual one, that is, the ring sum $x + y = xy' \vee x'y$.

Definition 7.1. Let $S = \{a_1, \ldots, a_n\}$ be a set of n distinct elements of a Boolean algebra B. For each $i \in \{1, \ldots, n\}$, the set

(7.1) $$V_i = \{x \in B \mid x + a_i \le x + a_j , j \in \{1, \ldots, n\} \setminus \{i\}\}$$

---

[7] See R.Klein, Lecture Notes Comput. Sci. No.40, Springer-Verlag 1989.
[8] See Harary, Melter, Tomescu and Peled, Discrete Math. 39(1982), 123-127, and Melter and Tomescu, Rev. Roumaine Math. Pures Appl. 29(1984), 407- 415.

is called the *Voronoi region* of a_i. The *Voronoi diagram* determined by S is the set $\{V_1, \ldots, V_n\}$. □

Proposition 7.1. *For every $i \in \{1, \ldots, n\}$, the Voronoi diagram is determined by*

$$(7.2) \qquad a_i \bigvee_{j=1}^{n} a_j' \leq x \leq a_i \vee \prod_{j=1}^{n} a_j' \,.$$

PROOF: For every $i, j \in \{1, \ldots, n\}$, we have

$$x + a_i \leq x + a_j \iff (xa_i' \vee x'a_i)(xa_j \vee x'a_j') = 0$$

$$\iff xa_i'a_j \vee x'a_ia_j' = 0 \iff a_ia_j' \leq x \leq a_i \vee a_j' \,.$$

□

Corollary 7.1. *For every $i \in \{1, \ldots, n\}$, the Voronoi region V_i is the interval $[v_i, v_i \vee \prod_{j=1}^{n} v_j']$, where we have set*

$$(7.3) \qquad v_i = a_i \bigvee_{j=1}^{n} a_j' \qquad (i = 1, \ldots, n) \,.$$

PROOF: We have

$$v_i \vee \prod_{j=1}^{n} v_j' = \prod_{j=1}^{n}(v_i \vee v_j')$$

$$= \prod_{j=1}^{n}(a_i(\prod_{k=1}^{n} a_k)' \vee a_j' \vee \prod_{k=1}^{n} a_k) = \prod_{j=1}^{n}(a_i \vee a_j') = a_i \vee \prod_{j=1}^{n} a_j' \,.$$

□

The above corollary justifies the following

Definition 7.2. The system $\{v_1, \ldots, v_n\}$ defined by (7.3) will be called the *Voronoi marker* of the set $S = \{a_1, \ldots, a_n\}$. □

Proposition 7.2. α) *A set $V = \{v_1, \ldots, v_n\}$ is the Voronoi marker of a set S if and only if*

$$(7.4) \qquad \prod_{j=1}^{n} v_i = 0 \,.$$

β) *When this is the case, the sets S for which V is a Voronoi marker coincide with those of the form*

$$(7.5) \qquad a_i = v_i \vee v \prod_{j=1}^{n} v_j' \qquad (i = 1, \ldots, n)$$

for some $v \in B$.

PROOF: Clearly (7.3) implies (7.4). Now suppose (7.4) holds. Then (7.5) implies in turn, via (7.4),

$$\prod_{i=1}^{n} a_i = \prod_{i=1}^{n} v_i \vee v \prod_{j=1}^{n} v_j' = v \prod_{j=1}^{n} v_j',$$

$$a_i(\prod_{j=1}^{n} a_j)' = (v_i \vee v \prod_{j=1}^{n} v_j')(v' \vee \bigvee_{j=1}^{n} v_j) = v_i,$$

which is (7.3). Conversely, if (7.3) holds, then taking $v := \prod_{j=1}^{n} a_j$, we get

$$v_i \vee v \prod_{k=1}^{n} v_k' = a_i(\prod_{j=1}^{n} a_j)' \vee (\prod_{j=1}^{n} a_j) \prod_{k=1}^{n} (a_k' \vee \prod_{j=1}^{n} a_j)$$

$$= a_i(\prod_{j=1}^{n} a_j)' \vee \prod_{j=1}^{n} a_j = a_i \vee \prod_{j=1}^{n} a_j = a_i.$$

□

Remark 7.1. It follows from Corollary 7.1 that a Voronoi region V_i is a singleton if and only if $\prod_{j=1}^{n} v_j' = 0$. Therefore this happens if and only if all the Voronoi regions are singletons. □

Research in the same geometric spirit is carried out in Melter [1988]. We quote from the introduction: "Paths in the plane that move through lattice points can be described by finite sequences whose elements are taken from $\{1, 2, \ldots, n\}$, where n is fixed. In particular, if in going from point to point the path is restricted to move either upward or to the right, then it can be associated with an element of a finite Boolean algebra. Paths which have the same initial point will also have the same terminal point if and only if their Boolean-algebra representations have the same level, i.e., the same number of ones." Consequently, if B stands for the Boolean algebra in question, the author seeks to determine those functions $f : B^n \longrightarrow B$ (not a priori Boolean) which have the property that when all arguments have the same level, then the image also has this level. The result is that a function has this property if and only if it is a projection. Necessity is established by a combinatorial argument, while sufficiency is trivial.

Another problem is solved by Dincă and Ţăndăreanu [1981]. Their starting point belongs to numerical analysis: it is known that a sufficient condition for the convergence of the Gauss-Seidel method for solving a system of linear equations $Ax = b$ is that $|a_{ii}| > \sum_{j\neq i} |a_{ij}|$ (ordinary sum) for all i. This motivates the following problem: given a real matrix A, is it possible to find a permutation of its rows and/or columns, such that the elements of the leading diagonal of the transformed matrix have the greatest absolute value in the corresponding rows? A variant of this problem occurs in many books on computer programming.

Dincă and Ţăndăreanu (op.cit) solve a more general problem. Let $A = \| a_{ij} \|$ be a square matrix of order n, with no hypothesis on the nature of its entries;

suppose, however, that some of its elements are marked. Find all possible permutations, if any, of the rows and/or of the columns of A, such that all the elements of the leading diagonal of the transformed matrix be marked. As a matter of fact, they prove that if the problem has a solution, then the desired transformation can be obtained by just a row permutation, and a column permutation is also sufficient. Suppose e.g. one looks for a row permutation. For each $i \in \{1, \ldots, n\}$, let R_i be the set of indices j such that a_{ij} is a marked element, and consider the truth equation

$$(7.7) \qquad \bigvee_{i=1}^{n} \left(\prod_{j \in R_i} x'_{ij} \vee \bigvee_{j,k \in R_i} x_{ij} x_{ik} \delta'_{jk} \right) = 0 ,$$

where δ stands for the Kronecker delta.

Proposition 7.3. *There exists a row permutation φ which solves the above problem if and only if the truth equation (7.7) is consistent. When this is the case, the solutions φ are associated with the solutions of equation (7.7) by the following rule: $\varphi(i)$ is the unique $j \in R_i$ such that $x_{ij} = 1$.*

COMMENT: The similar result for column permutations is left to the reader.
PROOF: Remark first that two elements a_{ij} and a_{ik} are left on the same row by every row permutation and every column permutation of the matrix A. Therefore, if the problem has a solution π, then exactly one element from each set R_i is moved to the leading diagonal. Now define $x_{ij} = 1$ if a_{ij} is moved to the leading diagonal by the permutation π, else $x_{ij} = 0$. It follows that for each $i \in \{1, \ldots, n\}$ there is a unique $j \in R_i$ such that $x_{ij} = 1$, hence the elements x_{ij} satisfy equation (7.7).

Conversely, suppose the elements $x_{ij} \in \{0, 1\}$ satisy equation (7.7). This means that for each i there is a unique $j \in R_i$ such that $x_{ij} = 1$. In other words, the prescription $\varphi(i)$ in statement $\beta)$ defines an injection $\varphi : \{1, \ldots, n\} \longrightarrow \{1, \ldots, n\}$, hence φ is in fact a bijection, therefore one can identify it with a row permutation. But we have just seen that for each $i \in \{1, \ldots, n\}$ there is exactly one marked element a_{ij} such that $x_{ij} = 1$; it follows that $\varphi(i) = j$, which means that φ moves a_{ij} on row j, while a_{ij} remains in column j.   □

Remark 7.2. Equation (7.7) implies the "subequation" $\bigvee_{i=1}^{n} \prod_{j \in R_i} x'_{ij} = 0$. It seems convenient to begin by solving this subequation, written in the form

$$(7.8) \qquad \prod_{i=1}^{n} \bigvee_{j \in R_i} x_{ij} = 1 .$$

Equations of this form have extensively been used in the literature (see e.g. Hammer and Rudeanu [1968]) and in our case the solving process is easier if we take into account the orthogonality conditions expressed by the other terms of equation (7.7). Each solution of equation (7.8) reduces equation (7.7) to a quadratic truth equation (see Ch.9, §3) or to the inconsistency 1=0.   □

Example 7.1. Let A be a square matrix of order 4, whose marked elements are located on cells (1,2), (1,3), (2,1), (2,4), (3,2), (3,3), (4,3) and (4,4). Then $R_1 = \{2,3\}$, $R_2 = \{1,4\}$, $R_3 = \{2,3\}$, and $R_4 = \{3,4\}$.

In this case equation (7.7) reads

(7.9)
$$(x'_{12}x'_{13} \vee x_{12}x_{13}) \vee (x'_{21}x'_{24} \vee x_{21}x_{24} \vee x_{21}x_{23}) \vee (x'_{32}x'_{33} \vee$$
$$x_{32}x_{33} \vee x_{31}x_{33} \vee x_{31}x_{34} \vee x_{33}x_{34}) \vee (x'_{43}x'_{44} \vee x_{43}x_{44} \vee$$
$$x_{42}x_{44}) = 0\,,$$

while equation (7.8) becomes in turn

$$(x_{12} \vee x_{13})(x_{21} \vee x_{24})(x_{32} \vee x_{33})(x_{43} \vee x_{44}) = 1\,,$$

$$(x_{12}x_{33} \vee x_{13}x_{32})(x_{21}x_{43} \vee x_{21}x_{44} \vee x_{24}x_{43}) = 1\,,$$

(7.10)
$$x_{12}x_{33}x_{21}x_{44} \vee x_{13}x_{32}x_{21}x_{44} = 1\,.$$

The truth equation (7.10) has two families of solutions. One of them is defined by $x_{12} := x_{33} := x_{21} := x_{44} := 1$, which reduces equation (7.9) to

$$x_{13} \vee x_{24} \vee x_{23} \vee x_{32} \vee x_{31} \vee x_{31}x_{34} \vee x_{34} \vee x_{43} \vee x_{42} = 0\,,$$

whose unique solution is obtained by taking all the variables occurring in it equal to 0.

We obtain in this way the row permutation $(2,1,3,4)$ and from the other family of solutions of equation (7.10) one obtains similarly the permutation $(3,2,1,4)$.

□

The problem of maximizing a linear pseudo-Boolean function $\sum_{i=1}^{n} c_i x_i$ (ordinary sum) under constraints of the form $x_i + x_j \geq 1$, $x_h + x_k \leq 1$ (ordinary sum), $x_r \leq x_s$, has been investigated in the literature. Crama, Hammer, Jaumard and Simeone [1987] remarked that since the above constraints can be expressed as a quadratic truth equation, the parametric solution of this equation transforms the above pseudo-Boolean problem into the problem of maximizing another (non-linear) pseudo-Boolean function whose variables (the former parameters!) are subject to no constraints, and the number of variables does not increase by this transformation. This may be advantageous in many cases.

Here are the author's summaries of two papers by Dimitrov: [1980a]* and [1980b]*, respectively.

"Differential operators, called directed, are defined and studied. This allows one to determine the conditions for calculating the value of a given function in a specific direction. It is shown that the directed differential operators represent a peculiar minimal form for defining the corresponding undirected operators. Their use in computer-directed algorithms leads to economy of computer memory and CPU time."

"In connection with computing directed Boolean differential operators, algorithms are offered for directional differentiation of Boolean functions, defined by the 'truth' vector or cubed cover. The algorithms are illustrated by examples."

Nurlybaev [1990]* studies "recognition algorithms of 'conjunction sorting' type that reduce to a solution of systems of Nelson-type Boolean equations" (cf. MR, 93d: 06014).

Élyashberg, Moskovina and Gribov [1971] use the solution of certain truth equations in order to establish the structure of complex molecules from their vibration spectra.

Here is the summary of the book by Shapiro [1984]*:

"We show the possibility of constructing algorithms for the solution of a broad class of logical problems using the propositional algebra. We consider questions of the diagnosis, analysis and synthesis of relay-switching circuits, scheduling problems, problems of calculators, automata, etc."

The monograph by Obukhov and Pavlov [1992], having the title "Logical equations and applied problems", is a poor book, at least as concerns its mathematical part.

Appendix 1. Errata to BFE

Dr. Yukio Moriwaki, the translator into Japanese of BFE, has kindly called my attention to a mistake in Theorem 4.9. It should be corrected as follows. Add to the hypotheses the orthonormality of $(a_i^j)_j$ for each i. The computation for property (i) runs as follows:

$$\bigcup_{i=1}^{n} a_i^j x^i = \bigcup_{i=1}^{n} a_i^j \bigcup_{k=1}^{m} b^k x_k^i = \bigcup_{i=1}^{n} \bigcup_{k=1}^{m} a_i^j x_k^i \bigcup_{h=1}^{n} a_h^k x_k^h$$

$$\bigcup_{i=1}^{n} \bigcup_{k=1}^{m} \bigcup_{h=1}^{n} a_i^j a_i^k x_k^i x_k^h \bigcup_{i=1}^{n} \bigcup_{k=1}^{m} a_i^j a_i^k x_k^i = \bigcup_{i=1}^{n} a_i^j x_j^i = b^j .$$

Other corrections:

Page 35, Corollary 1: Boolean function $\mapsto$ Boolean algebra
Page 35, Lemma 1.2: $f^* : B^{*n} \longrightarrow B$
Page 56, Definition 2.6: (1.13) $\mapsto$ (2.13)
Page 90, line 14 from bottom: minimal $\mapsto$ least
Page 90, line 10 from bottom: $(x_2, \ldots, x_n) \mapsto f_1(x_2, \ldots, x_n)$
Page 102, Table 3.1: the first two solutions are (1,1) and (0,0)
Page 102, line 10 from bottom: can be given only the values 0 and 1
Page 112, line 13 from bottom: theorem 3.2 $\mapsto$ theorem 3.3
Page 115, Corollary: be a particular
Page 125, formula (4.10): $y_\ell^j \mapsto y^j$
Page 156, line 15: this $\mapsto$ a more general
Page 191, the first Remark 7.2″ is in fact Remark 7.2′
Page 233, line 4 from bottom: proposition 9.6
Page 239, Theorem 9.7: $\mathbf{B} \mapsto \mathbf{B}_2$ (twice)
Page 239, formula (9.47): $h_1 \mapsto h$
Page 279, line 15: lunges $\mapsto$ hinges
Page 315, line 6: This line of research and its generalizations is
Page 346, equation (15.1): $g(X) = 1$
Page 409, line 14 from bottom: input $f \mapsto$ output f
Page 425, Livovschi: [1970] $\mapsto$ [1971]
Update the bibliography: reference Cerny and Marin [1973] was published in IEEE Trans. Computers C-23 (1974), no.5, 455-465.
Add Cardoso 288 to the Author Index

Appendix 2. Decomposition of Boolean functions and applications: a bibliography

AKERS, S.B.

1. On a theory of Boolean functions. SIAM J., 7 (1959), 487-498.

ARNOLD, R.F.

1. Group methods in combinational switching theory. Ph.D. Thesis, Univ. Michigan, 1963.

ASHENHURST, R.L.

1. Non-disjoint decomposition. Harvard Comput. Lab. Report BL-4, 1953.

2. The decomposition of switching functions. Proc. Intern. Symp. Theory Switching, 1957, Cambridge., Mass., 74-116. Cambridge Harvard Univ. Press, Ann. Comput. Lab. Harvard Univ., 29 (1957).

BIBILO, P. N.

1. Functional decomposition with fixed functions (Russian). Vescī Akad. Navuk BSSR Ser. Fiz.-Mat. Navuk 1980, no.4, 33-38, 140.

2. Probability of the existence of a multiple decomposition of a completely determined Boolean function (Russian). Veshchi Akad. Nauk BSSR Ser. Fiz. i Mat. Nauk 142 (5) (1981), 120-121.

BIBILO, P. N.; ENIN, S. V.

1. Decomposition of a Boolean function with a minimum number of significant arguments of a subfunction. Engrg. Cybernet. 18(3) (1980), 75-81.

2. Joint decompositions of a system of Boolean functions. Engrg. Cybernet. 18(2) (1980), 96-102.

BOCHMANN, D.; DRESIG, F.; STEINBACH, B.

1. A new decomposition method for multilevel circuit design. Proc. EDAC, February 1991; 374-377.

BOCHMANN, G. V.; ARMSTRONG, W.W.

1. Properties of Boolean functions with a tree decomposition. BIT 14 (1974), 1-13.

BRAYTON, R. K.; RUDELL, R.; SANGIOVANNI-VINCENTELLI, A.; WANG, A. R.

1. MIS: a multiple-level logic optimization system. IEEE Trans. CAD, CAD-6 (1987), 1062-1081.

BUTAKOV, E. A.

1. Methods of synthesis of switching circuits from threshold elements (Russian). Ènergiya, Moskva 1970.

CARVALLO, M.
  1. Logique à trois valeurs, logique à seuil. Gauthier-Villars, Paris 1968.

CERNY, E.; MARIN, M. A.
  1. A computer algorithm for the synthesis of memoryless logic circuits. IEEE Trans. Comput. C-23 (1974), 455-465.
  2. An approach to unified methodology of combinational switching circuits. IEEE Trans. Comput. C-26 (1977), 745-756.

CHANG, S.; MAREK-SADOWSKA, M.
  1. Technology mapping via transformation of function graphs. Proc. ICCD, October 1992; 159-162.

CURTIS, H. A.
  1. A new approach to the design of switching circuits. Van Nostrand, Princeton, NJ, 1962.

DAVIO, M.; DESCHAMPS, J.-P.; THAYSE, A.
  1. Discrete and switching functions. McGraw Hill & Georgi Publ. Co., New York/St. Saphorin, 1978.

DAVIO, M.; QUISQUATER, J. J.
  1. Affine cascades. Philips Res. Rep. 29 (1974), 193-213.

DAVIO, M.; RUDEANU, S.
  1. Boolean design with autodual negative gates. Rev. Roumaine Sci. Tech. Sér. Electrotech. & Energ. 39 (1993), 241-251, 475-489.

DESCHAMPS, J.
  1. Obtention des ensembles principaux de décomposition simple disjointe d'une fonction booléenne comme solution d'une équation booléenne. Application aux fonctions incomplètes. Rev. Roumaine Math. Pures Appl. 22 (1977), 613-631.

DESCHAMPS, J.; LAPSCHER, F.
  1. Présentation et optimisation d'un programme de recherche des décompositions disjointes d'une fonction booléenne. Rev. Roumaine Math. Pures Appl. 24 (1979), 893-931.

DESCHAMPS, J.-P.
  1. On a theory of discrete functions. Part III. Decomposition of discrete functions. Philips Res. Rep. 29 (1974), 193-213.
  2. Application de la notion de fermeture à l'étude des décompositions des fonctions booléennes. Thèse, Univ. Sci. du Languedoc, Montpellier, 1974.
  3. Binary simple decompositions of discrete functions. Digital Process. 1 (1975), 123-140.

DUCA, I.; DUCA, M.; OPREA, G.
  1. Sur la décomposition disjointe simple des fonctions booléennes. Politehn. Univ. Bucharest Sci. Bull. Ser. A Appl. Math. Phys. 61 (1999), no.3-4, 53-65.

ELLIS, D.
  1. Remarks on Boolean functions. II. J. Math. Soc. Japan 8 (1956), 363- 368.

ELSPAS, B.; STONE, H.

1. Decomposition of group functions and the synthesis of multi-cascades. IEEE Conf. Rec. 8-th Ann. Symp. Switching and Automata Theory, 1967; 184-196.

ENIN, S. V.; BIBILO, P. N.

1. The redundancy of arguments and the decomposition of Boolean functions (Russian). Automat. i Vychislit. Tekhn. (Riga) 1978, no.4, 16-21; 91.

2. Joint decomposition of a system of vector Boolean functions. Automat. Control Comput. Sci. 13 (1979), 14-20.

FADINI, A.

1. Algoritmo per la construzione di una funzione booleana composta mediante un'assegnata famiglia di funzioni booleane. Ricerca (Napoli), (2) 23, (1972), gennaio-aprile, 23-30.

GIVONE, D. D.

1. A tabular method to determine simple decompositions of switching functions. Thesis Math. Sci, Cornell Univ., 1963.

GREENE, C.; TAKEUTI, G.

1. On the decomposition of Boolean polynomials. J. Fac. Sci. Univ. Tokyo Sect IA Math. 24(1974), 23-28.

GUPTA, S.C.

1. A method for finding simple disjunctive decompositions with one free variable. Comput. Electr. Engrg. 7 (1980), 141-146.

2. Decomposition of four-variable Boolean functions. Comput. Electr. Engrg. 8 (1981), 41-48.

HARTMANIS, J.; STEARNS, R.

1. Algebraic structure of sequential machines. Prentice Hall, Englewood Hills 1966.

HIGH, S. L.

1. Complex disjunctive decomposition of incompletely specified Boolean functions. IEEE Trans. Comput. C-22 (1973), 103-110.

HU, S.-T.

1. On the decomposition of switching functions. Lockheed Missiles and Space Co., Techn. Rep. No. LMSD-6-90-61-15, Sunnyvale, Calif., June 1961.

KARP, R. M.

1. Functional decomposition in switching circuit design. SIAM J. 11 (1963), 291-335.

KARP, R. M.; McFARLIN, F.E.; ROTH, J. P.; WILLIS, J. R.

1. A computer program for the synthesis of combinational switching circuits. Proc. 2-nd AIEE Symp. Switching Circuit Theory and Logical Design, October 1961; 152-152.

KOHAVI, Z.

1. Switching and finite automata theory. McGraw Hill, New York 1970.

KOLP, C.

1. The synthesis of multivalued cellular cascades and the decomposition of

group functions. IEEE Trans. Comput. C-21 (1972), 489-492.

KUNTZMANN, J.
1. Algèbre de Boole. Dunod, Paris 1965.

KUZNETSOV, A. V.
1. On repetition-free switching circuits and repetition-free superpositions of the functions of the algebra of logic (Russian). Trudy Mat. Inst. V. A. Steklova 51 (1958), 186-225.

LAPSCHER, F.
1. Propriétés des fonctions booléennes admettant certaines décompositions disjointes. Actes Congrès AFIRO, Lille, juin 1966; 386-390.
2. Quelques propriétés des décompositions simples. Congrès AFIRO, Nancy, mai 1967.
3. Decomposition of Boolean functions. IFIP Congress Edinburgh, August 5-10, 1968.
4. Application de la notion de fermeture à l'étude des fonctions booléennes. Thèse, Univ. Grenoble, 1968.
5. Décompositions simples de fonctions booléennes. Automatisme, 1968 (?).
6. Sur la recherche des décompositions disjointes d'une fonction booléenne. Rev. Française Autom. Inform. Rech. Opér. 6 (1972), 92-112.

LEE, G.; SAUL, J.
1. Synthesis of LUT-type FPGAs using AND/OR/EXOR representations. Proc. SASIMI, November 1996; 74-77.

LIVOVSCHI, L.
1. Utilizarea ecuaţiilor booleene în probleme de reprezentare. Stud. Cerc. Mat. 22 (1970), 39-49.

MATSUNAGA, Y.
1. An attempt to factor logic functions using exclusive-or decompositions. Proc SASIMI, November 1996; 78-83.

MIYATA, F.
1. An extension of the method of Cohn and Lindman. IEEE Trans. Electronic Comput. EC-13 (1964), 625-629.

McKINSEY, J. C. C.
1. Reducible Boolean functions. Bull. Amer. Math. Soc. 42 (1956), 263-267.

McNAUGHTON, R.
1. Unate truth functions. IRE Trans. EC-10 (1961), 1-6.

MUZIO, J.; MILLER, D.
1. Decomposition of ternary switching functions. Intern. Symp. Multiple-Valued Logic, 1973; 156-165.

PAVLOVSKIĬ, A. I.
1. On the problem of iterative decomposition of Boolean functions (Russian). Dokl. Akad. Nauk BSSR 21 (1977), 879-881; 955.

PICHAT, E.
1. Décompositions simples disjointes de fonctions booléennes données par

leurs monômes premiers. Rev. Française Traitement Inf. 8 (1965), 63-66.

2. Décompositions des fonctions booléennes. Thèse 3-ème cycle, Fac. Sci. Grenoble, janvier 1966.

3. Décompositions et écritures minimales des fonctions booléennes. Bull. Math. Soc. Sci. Math. R.S. Roumanie 12(60) (1968).

4. Décompositions simples de fonctions booléennes. Rev. Inform. Rech. Opér. 2 (1968), no. 7, 51-70.

PIECHA, J.

1. Test prostej alternatywnej dekompozycji funkcji logicznich. Arch. Automat. i Telemech. 19 (1974), 207-215.

POVAROV, G.N.

1. On functional separability of Boolean functions (Russian). Dokl. Akad. Nauk SSSR 94 (1954), 801-803.

2. On the functional decomposition of Boolean functions (Russian). Dokl. Akad. Nauk SSSR 123 (1958), 774.

3. A mathematical theory for the synthesis of networks with one input and k outputs. Proc. Intern. Symp. Theory Switching. 30 (1959), 74-94.

POZDNYAKOV, Y. M.; MASHCHENKO, S. O.

1. Optimization of decomposition (Russian). Issled. Operatsii i ASU 128 (1981), no.18, 27-35.

PRATHER, R.E.

1. Three-variable multiple output tree circuits. IEEE Trans. EC-15 (1966), no.1, February, 3-13.

RIGHI, R.

1. Le funzioni di commutazione in genere e quelle simmetrici in particolare. Ingegneria feroviara 10 (1955), 719-737.

ROTH, J. P.

1. Minimization over Boolean trees. IBM J. Res. Development 4 (1960), 543-558.

ROTH, J. P.; KARP, R. M.

1. Minimization over Boolean graphs. IBM J. Res. Development 6 (1962), 227-238.

ROTH, J. P.; WAGNER, E. C.

1. Algebraic topological methods for the synthesis of switching systems. Part III. Minimization on nonsingular Boolean trees. IBM J. Res. Development 4 (1959), 326-344.

ROZENFEL'D, T. K.

1. Solution of Boolean equations and the decomposition of Boolean functions (Russian). Materialy Sem. Kibernet. Vyp. 68 (1974), 18-27, 28.

ROZENFEL'D, T. K.; SILAYEV, V.N.

1. Boolean equations and decompositions of Boolean functions. Engrg. Cybernet. 17 (1979), 85-92.

RUDEANU, S.

1. On Tohma's decomposition of logical functions. IEEE Trans. Electronic Comput. Ec-14 (1965), 924-931.

2. Boolean functions and equations. North-Holland, Amsterdam 1974.

3. Square roots and functional decompositions of Boolean functions. IEEE Trans. Comput. C-25 (1976), 528-532.

SASAO, T.

1. FPGA design by generalized functional decomposition. In: T. Sasao, ed.: Logic Synthesis and Optimization, Kluwer Acad. Publ., 1993; 233-258.

SASAO, T.; BUTLER, J. T.

1. On bi-decompositions of logic functions. Notes on IWLS'97, May 1997.

SAWADA, H.; SUYANA, T.; NAGOYA, A.

1. Logic synthesis for look-up table based FPGAs using functional decompositions and support minimization. Proc. ICCAD, November 1995; 353-258.

SAWADA, H.; YAMASHITA, S.; NAGOYA, A.

1. Restricted simple disjunctive decompositions based on grouping symmetric variables. Proc. Great Lakes Symp. VLSI, Urbana-Champaign Illinois, March 13-15, 1997; 39-44.

2. Restructuring logic representations with easily detectable simple disjunctive decompositions. DATE'98, Paris, February 23-26, 1998; 755-759.

SCHNEIDER, P. R.; DIETMEYER, D. L.

1. An algorithm for synthesis of multiple-output combinational logic. IEEE Trans. Comput. C-17 (1968), 117-128.

SEMON, W.

1. Characteristic numbers and their use in the decomposition of switching functions. Proc. ACM, Pittsburgh Meeting, May 1952; 273-280.

2. Synthesis of series-parallel network switching functions. Bell Syst. Techn J. 37 (1958), 877-898.

SHANNON, C.

1. The synthesis of two-terminal switching circuits. Bell Syst. Techn J. 28 (1949), 59-98.

SHEN, Y. Y.-S.; McKELLAR, A. C.

1. An algorithm for the disjunctive decomposition of switching functions. IEEE Trans. Computers C-19 (1970), 239-248.

SHEN, Y. Y.-S.; McKELLAR, A. C.; WEINER, P.

1. A fast algorithm for the disjunctive decomposition of switching functions. IEEE Trans. Computers C-20 (1971) 304-309.

SHINAR, I.; YOELI, M.

1. Group functions and multi-valued cellular cascades. Inform. and Control 15 (1969), 369-376.

SHURUPOV, A. N.

1. On functional separability of Boolean threshold functions (Russian). Diskret. Mat. 9 (1997), no.2, 59-73.

SINGER, T.

1. The decomposition chart as a theoretical aid. Harvard Comput Lab. Rep. BL-4, sect. III (1953), 1-28.

2. Some uses of truth tables. Proc. Intern. Symp. Theory Switching 29 (1959), 125-133.

SOLOV'EV, N. A.

1. Testing superpositions of Boolean functions of elementary homogeneous functions (Russian). Diskret. Mat. 8 (1996), 117-132.

STEINBACH, B.; WERESZCZYNSKI, A.

1. Synthesis of multi-level circuits using EXOR-gates. Proc. Reed- Muller'95, August 1995; 161-168.

STIEFEL, B.

1. Dekomposition Boolescher Funktionen durch Anwendung von Booleschen Differentialgleichungen. Nachrichtentechnik Elektronik 29 (1979), 333-334.

2. Über Iterierte einer Booleschen Funktion. Rostock. Math. Kolloq. 19 (1982), 119-128.

TASHKOVA, B. S.; VELINOV, Y.P.

1. Simple functional decomposition (Bulgarian). Godishnik Vissh. Uklabn. Zaved. Prilozhna Mat. 15 (1979), no.2, 59-64.

THAYSE, A.

1. A fast algorithm for the proper decomposition of Boolean functions. Philips Res. Rep. 27 (1972), 140-150.

2. Le calcul booléen des différences. Thèse, Ecole Polytechn. Lausanne, 1980.

3. Boolean calculus of differences. Lecture Notes Comput. Sci. No. 101. Springer-Verlag, Berlin 1981.

THELLIEZ, S.

1. Introduction à l'étude des structures ternaires de commutation. Gordon & Breach, Paris 1973.

TOHMA, Y.

1. Decomposition of logical functions using majority decision elements. IEEE Trans. Electronic Comput EC-15 (1964), 698-705.

TOMASHPOLSKIĬ, A. M.

1. Algorithm for the solution of logical equations for majority decomposition of two-valued logical functions (Russian). Izv. Vysh. Uchebn. Zaved. Tekhn. Kibernet. no.6 (1967), 117-122. Engl. transl.: Engrg. Cybernet. no.6 (1967), 111-115.

TRAKHTENBROT, B. A.

1. On the theory of repetition-free switching circuits (Russian). Trudy Mat. Inst. im. V. A. Steklova 51 (1958), 226-269.

TUMANYAN, G.B.

1. Minimization of indecomposable Boolean functions by means of the method of approximate functional decomposition (Russian). Dokl. Akad. Nauk SSSR 156 (1964), 525-528.

VACCA, R.

1. Decomposizioni di funzioni logiche di commutazione. Atti Conv. Naz. Logica, Torino 1964; 103-125.

VASHCHENKO, V.P.

1. Multiple functional decomposition with fixed conjugate functions. Dokl. Akad. Nauk SSSR 239 (1978), 18-21.

2. On the calculation of nontrivial decompositions of a function of the algebra of logic (Russian). Dokl. Akad. Nauk SSSR 247 (1979), 15-18.

VASHCHENKO, V. P.; FROLOV, A. B.

1. Algorithms of decomposition of logical functions and $n-$ary relations (Russian). Trudy Moskov. Energet. Inst. 412 (1979), 19-25; 154.

WALIGÓRSKI, S.

1. On superpositions of zero-one functions. Algoritmy 1 (1963), 91-98.

YAMASHITA, S.; SAWADA, H.; NAGOYA, A.

1. New methods to find optimal non-disjoint bi-decompositions. Proc. ASP-DAC'98, Yokohama, February 10-13, 1998; 59-68.

ZAKREVSKIĬ, A.D.

1. Algorithms for the partitioning of a Boolean function (Russian). Trudy Sibirsk. Fiz. Tekhn. Inst. Tomsk. Univ. 44 (1964), 15-16.

A particular line of research is the study of various expansion formulas for Boolean functions. See the survey

VINOKUROV, S.F.; PERYAZEV, N.A.

1. Polynomial decompositions of Boolean functions (Russian). Kibernet. i Sist. Analiz 1993, no.6, 34-47; 183

and the literature quoted therein, and also, by the same authors,

2. Polynomial decomposition of Boolean functions by images of homogeneous operators of nondegenerate functions (Russian). Izv. Vyssh. Uchebn. Zaved. 1996, no.1 (404), 17-21. English. transl.: Russian Math. (Iz. VUZ) 40 (1996), 15-18.

Appendix 3. Open problems

The field of lattice functions and equations and their applications is wide, therefore we begin our list of problems by suggesting one or several companion monographs to the present book. An updated monograph on truth functions and equations (TFE!) including applications, would be most welcome. As a matter of fact, even a specialized subject such as the decomposition of Boolean (truth) functions (cf. Ch.11), or the contemporary theory of MVL functions and equations (cf. the introduction to Ch.13, §1), would deserve a monograph.

Boolean geometry (cf. BFE, Chapter 13, §§2,3 and the present Ch.12, §3) is also a field developed enough to deserve a monograph.

Now we suggest several lines of research in the theory of Boolean equations:

Construct a direct theory of systems of Boolean equations, that is, without reducing the system to a single equation.

Construct a direct theory of functional Boolean equations, that is, without reducing the equation to a system of ordinary Boolean equations (cf. Problem 2.12 in BFE).

Continue the study of generalized systems of Boolean equations (cf. Problem 10.1 in BFE and the present Ch.6, §5).

As was emphasized in BFE, Chapter 15, §5 and in the present Ch.9, §4, the computer implementation of the very strong algebraic (formula-handling) methods for solving Boolean equations seems to us the major step to be done for the advancement of the theory of Boolean equations.

Many other problems arise in a natural way. Here are a few suggestions.

Study equational compactness in non-Boolean lattices.

It was seen in Ch.5 that the theory of Post functions and equations recaptures the essential features of Boolean functions and equations. Generalize as much as possible the results in Chs 8–14 to the case of Post algebras.

Construct a theory of algebraic equations in relation algebras (cf. Proposition 13.1.3).

Having in mind the isotony detectors studied in Ch.7, §2, construct theories of φ–detectors for other closure operators φ.

Solve the problem in Remark 9.2.1: investigate those classes of Boolean equations whose solutions depend on only a small subset of the coefficients.

Obtain functional characterizations as in Ch.13, §4, for classes of Boolean functions of several variables.

Construct the *complete existential theory* of conditions (M1)—(M4) in Ch.13, §5, that is, determine all the implications that exist between them (cf. Moore [1910]).

Determine classes of graphs for which a repeated application of Proposition 14.1.1 always results in the determination of a (minimal) chromatic decomposition.

Find good heuristic algorithms for the CMOS decomposition of truth functions according to the strategy suggested in the introduction to Ch.14, §3.

Bibliography

ABIAN, A.
  1970a. On the solvability of infinite systems of Boolean polynomial equations. Coll. Math. 21, 27-30.
  1970b. Generalized completeness theorem and solvability of systems of polynomial Boolean equations. Z. Math. Logik Grundlagen Math. 16, 263-264.
  1976. Boolean rings. Branden Press, Boston.

ALEKSANYAN, A. A.
  1989. Realization of Boolean functions by disjunctions of products of linear forms. (Russian). Dokl. Akad. Nauk SSSR 304, 781-784. English transl.: Soviet Math. Dokl. 39, 131-135.

ANDREOLI, G.
  1961. Formazioni algebriche booleane monotone. Ricerca (Napoli), 12, 1-9.

ANTHONISSE, J. M.
  1968. The determination of the chromatic number of a graph. Stichting Math. Centrum, Amsterdam, Adeling Math. Statistiek, S 939, March 1968.

ASHENHURST, R. L.
  1957. The decomposition of switching functions. Proc. Intern. Symp. Theory Switching, April 1957. Ann. Comput. Lab. Harvard Univ. 29, 74-116. Reprinted as appendix in: Curtis, A., A new approach to the design of switching circuits, Van Nostrand, Princeton 1962.

ASPVALL, B.; PLASS, M. F.; TARJAN, R. E.
  1979. A linear-time algorithm for testing the truth of certain quantified Boolean formulas. Inform. Process. Lett. 8, 121-123.

BAK KHYNG KHANG
  1975. The decomposition of a certain type of structures into chains (Russian). Zh. Vychisl. Mat. i Mat. Fiz. 15, 477-488.

BALAKIN, G. V.
  1973. On the distribution of the number of solutions of systems of random Boolean equations (Russian). Teor. Veroyatnost. i Primenen. 18, 627-632.

BALBES, R.; DWINGER, PH.
  1974. Distributive lattices. Univ. Missouri Press, Columbia.

BANKOVIĆ, D.
  1979. On general and reproductive solutions of arbitrary equations. Publ. Inst. Math. (Beograd) 26(40), 31-33.

1983. Solving systems of arbitrary equations. Discrete Math. 46, 305-309.

1984. The general reproductive solution of Boolean equation. Publ. Inst. Math. (Beograd) 34(48), 7-11.

1985. Some remarks on reproductive solutions. Publ. Inst. Math. (Beograd) 38(52), 17-19.

1987a. Notes on unique solutions of Boolean equations. Mat. Vesnik 39, 1-3.

1987b. The formulas of the general reproductive solution of an equation in Boolean ring with unit. Publ. Inst. Math. (Beograd) 42(56), 29-34.

1988. Formulas of the general solutions of Boolean equations. Publ. Inst. Math. (Beograd) 44(58), 9-18.

1989a. All general reproductive solutions of Boolean equations. Publ. Inst. Math. (Beograd) 46(60), 13-19.

1989b. A note on Boolean equations. Bull. Soc. Math. Belgique 41, 2, sér. B, 169-175.

1989-90. Some remarks on number of parameters of the solutions of Boolean equations. Discrete Math. 70, 229-234.

1990. All general solutions of finite equations. Publ. Inst. Math. (Beograd) 47(61), 5-12.

1992a. Certain Boolean equations. Discrete Appl. Math. 35, 21-27.

1992b. A new proof of Prešić's theorem on finite equations. Publ. Inst. Math. (Beograd) 51(65), 22-24.

1992c. A generalization of Löwenheim's theorem. Bull. Soc. Math. Belgique 44, 1, sér. B, 59-65.

1993a. Formulas of particular solutions of Boolean equations. Zb. Rad. Prir.-Mat. Fak. u Kragujevacu 14, 11-14.

1993b. A note on finite equations. Zb. Rad. Prir.-Mat. Fak. u Kragujevacu 14, 15-18.

1993c. On reproductive solutions of Boolean equations. Demonstratio Math. 26, 841-848.

1995a. All solutions of finite equations. Discrete Math. 137, 1-6.

1995b. Formulas of general reproductive solutions of Boolean equations. Fuzzy Sets and Systems 75, 203-207.

1996. Formulas of general solutions of Boolean equations. Discrete Math. 152, 25-32.

1997a. Finite equations in n unknowns. Publ. Inst. Math. (Beograd) 61(75), 1-5.

1997b. General reproductive solutions of Postian equations. Discrete Math. 169, 163-168.

1997c. Horn sentences in Post algebras. Discrete Math. 173, 269-275.

1998. Equations on multiple-valued logic. Multiple Valued Logic 3, 89-95.

2000. All reproductive general solutions of Postian equations. Rev. Roumaine Math. Pures Appl. 45, no.6 (in press)

????. All general solutions of Prešić's equation. (submitted)

2001. A note on Postian equations. Multiple-Valued Logic 6, 1-10.

BÄR, G.

1972. Zur linearen Darstellbarkeit von Ausdrücken des Aussagenkalküls. Elektron. Informationsverarbeit. Kybernetik 8, 353-378.

BÄR, G.; ROHLEDER, H.

1967. Über einen arithmetisch-aussagenlogischen Kalkül und seine Anwendung auf ganzzahlige Optimierungsprobleme. Elektron. Informationsverarbeit. Kybernetik 3, 171-195.

BATBEDAT, A.

1971. Distance booléenne sur un 3–anneau. Enseignement Math. 11e Sér. 17, 165-185.

BEAZER, R.

1974a. A characterization of complete bi-Brouwerian lattices. Coll. Math. 29, 55-59.

1974b. Some remarks on Post algebras. Coll. Math. 29, 167-178.

1974c. Functions and equations in classes of distributive lattices with pseudocomplementation. Proc. Edinburgh Math. Soc. II Ser. 19, 191-203.

1975. Post-like algebras and injective Stone algebras. Algebra Universalis 5, 16-23.

BERNSTEIN, B. A.

1924. Operations with respect to which the elements of a Boolean algebra form a group. Trans. Amer. Math. Soc. 26, 171-175.

1932. Note on the condition that a Boolean equation have a unique solution. Amer. J. Math. 54, 417-418.

BERȚI, ȘT. N.

1973. Asupra rezolvării unui sistem de ecuații booleene. Rev. Analiză Numerică Teor. Aproximației 2, 31-44.

BIRKHOFF, G.

1967. Lattice theory. Third Edition. Amer. Math. Soc., Providence. First Edition, 1940. Second Edition, 1948.

BLAKE, A.

1937. Canonical expressions in Boolean algebra. Diss., Univ. Chicago, Dept. Math. Private Printing, Univ. of Chicago Libraries.

BOCHMANN, D.

1977. Zu den Aufgaben und dem gegenwärtigen Stand des Booleschen Differentialkalküls. Wiss. Z. Techn. Hochsch. Karl-Marx-Stadt 19, 193-200.

BOCHMANN, D.; POSTHOFF, CH.

1979. Die Behandlung Boolescher Gleichungen mit Hilfe des Booleschen Differentialkalküls. Sitzungsber. Akad. Wiss. DDR 12N, 5-25.

1981. Binäre Dynamische Systeme. Springer-Verlag, Berlin.

BOICESCU, V.; FILIPOIU, A.; GEORGESCU, G.; RUDEANU, S.

1991. Lukasiewicz-Moisil algebras. North-Holland, Amsterdam.

BORDAT, J. P.

1975. Treillis de Post. Application aux fonctions et aux équations de la logique à p valeurs. Thèse, Univ. Sci. Tech. Languedoc, Montpellier.

1978. Résolution des équations de la logique à p valeurs. Rev. Roumaine Math. Pures Appl. 23, 507-531.

BOURBAKI, N.

1963. Eléments de mathématiques. Théorie des ensembles. Chapitre 8. Hermann, Paris.

BOŽIĆ, M.

1975. A note on reproductive solutions. Publ. Inst. Math. (Beograd) 19(33), 33-35.

BOZOYAN, SH. E.

1978. Some properties of the Boolean differentials and activities of the arguments of Boolean functions (Russian). Problemy Peredachi Informatsii 14, 77-89. English trans.: Problems Inform. Transmission 14, 54-62.

BREUER, M. A.; CHANG, SH.-J.; SU, S. Y. H.

1976. Identification of multiple stuck-type faults in combinational circuits. IEEE Trans. Comput. C-25, 44-54.

BROWN, F. M.

1976. Boolean equations (unpublished).

1982. Segmental solutions of Boolean equations. Discrete Appl. Math. 4, 87-96.

1990. Boolean reasoning. The logic of Boolean equations. Kluwer Acad. Publ., Boston.

BROWN, F. M.; RUDEANU, S.

1981. Consequences, consistency and independence in Boolean algebras. Notre Dame J. Formal Logic 22, 45-62.

1983. Recurrent covers and Boolean equations. Colloquia Math. Soc. János Bolyai 33 (Lattice Theory), Szeged 1980; 637-650.

1985. Triangular reproductive solutions of Boolean equations. An. Univ. Craiova Ser. Mat.-Fiz.-Chim. 13, 18-23.

1986. A functional approach to the theory of prime implicants. Publ. Inst. Math. (Beograd) 40(54), 23-32.

1988. Prime implicants of dependency functions. An. Univ. Bucureşti Ser. Mat.-Inf. 37, no.2, 16-26.

2001. Uniquely solvable Boolean and simple Boolean equations. Multiple-Valued Logic 6, 11-26.

BRYANT, R. E.

1986. Graph-based algorithms for Boolean function manipulation. IEEE Trans. Comput. C-35, 677-691.

BURRIS, S.; SANKAPPANAVAR, H. P.

1981. A course in universal algebra. Springer-Verlag, New York/ Heidelberg/Berlin.

BÜTTNER, W.

1987. Unification in finite algebras is unitary (?). Proc. CADE-9. Lecture Notes in Comput. Sci. 310, 368-377.

CAO, Z.-Q.; KIM, K. H.; ROUSH, F. W.

1984. Incline algebra and applications. Wiley, New York.

CARVALLO, M.

1967. Sur la résolution des équations de Post. C. R. Acad. Sci. Paris 265, 601-602.

1968a. Sur la résolution des équations de Post à ν valeurs. C. R. Acad. Sci. Paris 267, 628-630.

1968b. Logique à trois valeurs, logique à seuil. Gauthier-Villars, Paris.

CĂZĂNESCU, V.E.; RUDEANU, S.

1978. Independent sets and kernels in graphs and hypergraphs. Ann. Sci. Univ. Kinshasa 4, 37-66.

CERNY, E.

1977. Unique and identity solutions of Boolean equations. Digital Processes 3, 331-337.

1978. Controllability and fault observability in modular combinational circuits. IEEE Trans. Comput. C-23, 455-465.

CERNY, E.; MARIN, M. A.

1974. A computer algorithm for the synthesis of memoryless logic circuits. IEEE Trans. Comput. C-23, 455-465.

1977. An approach to unified methodology of combinational switching circuits. IEEE Trans. Comput. C-26, 745-756.

CHAJDA, J.

1973. Systems of equations over finite Boolean algebras. Arch. Math. (Brno) 9, no.4, 171-181.

CHANG, C. C.; KEISLER, J.

1973. Model theory. North-Holland, Amsterdam. Second edition, 1976. Third edition, 1999.

CHISTOV, V. P.

1994. Analytic solution of logic equations (Russian). Izv. Ross. Akad. Nauk Tekhn. Kibernet. 1994, no.2, 219-224. English translation: J. Comput. Syst. Sci. Internat. 33, no.5, 166-170.

CHVALINA, J.

1987. Characterizations of certain general and reproductive solutions of arbitrary equations. Mat. Vesnik 39, 5-12.

COHN, P. M.

1965. Universal algebra. Harper & Row, New York.

COOK, S.

1971. The complexity of theorem proving procedures. Proc. Third ACM Symp. Theory of Computing, 151-158.

CORCORAN, J.

1985. Review to Porte [1982]. Math. Rev. # 85j:03002.

COWEN, R. H.

1982. Solving algebraic problems in propositional logic by tableau. Arch. Math. Logik Grundlag. 22, 187-190.

412

CRAMA, Y.; HAMMER, P. L.; JAUMARD, B.; SIMEONE, B.

1986. Parametric representation of the solutions of a quadratic Boolean equation. Rutcor Res. Report RRR # 15-86, August 1986.

1987. Product form parametric representation of the solutions to a quadratic Boolean equation. RAIRO Rech. Opérationnelle 21, 287-306.

CRAWLEY, P.; DILWORTH, R. P.

1973. Algebraic theory of lattices. Prentice Hall, N.J.

CUNNINGHAME-GREEN, R.

1979. Minimax algebra. Lecture Notes in Economic and Math. Syst. No.166. Springer-Verlag, Berlin.

DAVEY, B. A.; PRIESTLEY, H. A.

1990. Introduction to lattices and order. Cambridge Univ. Press, Cambridge.

DAVIO, M.

1968. Flip-flop input equations. Preprint, Electronics Lab., Louvain Univ.

1970. Extremal solutions of unate Boolean equations. Philips Res. Rep. 25, 201-206.

DAVIO, M.; DESCHAMPS, J.-P.

1969. Classes of solutions of Boolean equations. Philips Res. Rep. 24, 373-378.

DAVIO, M.; DESCHAMPS, J.-P.; THAYSE, A.

1978. Discrete and switching functions. McGraw Hill, New York, and Georgi, St.Saphorin.

DAVIO, M.; RUDEANU, S.

1993. Boolean design with autodual negative gates. Rev. Roumaine Sci. Tech. Sér. Electrotech. & Energ. 39, 241-251 , 475-489.

DAVIS, A. C.

1955. A characterization of complete lattices. Pacific J. Math. 5, 311-319.

DEL PICCHIA, W.

1976. A numerical algorithm for the resolution of Boolean equations. IEEE Trans. Comput. C-23, 983-986.

DESCHAMPS, J.

1975. Fermetures i-génératrices. Application aux fonctions booléennes permutantes. Discrete Math. 13, 321-339.

1977. Obtention des ensembles principaux de décomposition simple disjointe d'une fonction booléenne comme solution d'une équation booléenne. Application aux fonctions incomplètes. Rev. Roumaine Math. Pures Appl. 22, 613-631.

1990. Fermetures génératrices et fonctions booléennes. Rev. Roumaine Math. Pures Appl. 35, 125-137.

DESCHAMPS, J.-P.

1971. Maximal classes of solutions of Boolean equations. Philips Res. Rep. 26, 249-260.

DIALLO, M. K.

1983. Contribuţii la programarea pseudobooleană. Ecuaţii booleene pătratice.

Ph. D. Thesis, Univ. of Bucharest.

DI NOLA, A.; SESSA, S.; PEDRYCZ, W.; SANCHEZ, E.

1989. Fuzzy relation equations and their applications. Kluwer Acad. Press, Dordrecht.

DIMITROV, D. P.

1980a. Directed Boolean differential operators (Bulgarian). Problemi Tekhn. Kibernet. 10, 40-47.

1980b. Two methods for computing directed Boolean differential operators (Bulgarian). Problemi Tekhn. Kibernet. 10, 48-54.

DINCĂ, A.; ȚĂNDĂREANU, N.

1981. An application of Boolean equations in numerical analysis. An. Univ. Craiova, Ser. Mat.-Fiz.-Chim. 9, 6-10.

DRABBE, J.

1969. Sur les algèbres implicatives. C. R. Acad. Sci. Paris 266, 1073.

DRECHSLER, R.; BECKER, B.

1998. Graphenbasierte Funktiondarstellung. Boolesche und pseudo-Boolesche Funktionen. B. G. Teubner, Stuttgart.

DYUKOVA, E. V.

1987. On the complexity of the realization of certain recognition procedures (Russian). Zh. Vychisl. Mat. i Mat. Fiz. 27, no.1, 114-127.

1989. Solving systems of Boolean equations of quasi-Nelson type (Russian). Voprosy Kibernet. (Moscow) no.133, 5-19.

EGIAZARYAN, È. V.

1976. A certain class of systems of Boolean equations (Russian). Zh. Vychisl. Mat. i Mat. Fiz. 16, 1073-1077, 1088.

EKIN, O.; FOLDES, S.; HAMMER, P. L.; HELLERSTEIN, L.

2000. Equational characterizations of Boolean functions classes. Discrete Math. 211, 27-51.

ÈLYASHBERG, M. E.; MOSKOVINA, L. A.; GRIBOV, L. A.

1971. The use of mathematical logic in establishing the structure of complex molecules from their vibration spectra. Zh. Priklad. Spektroskopii 15, 843-853. English transl.: J. Appl. Spectroscopy (?), 1469-1477.

EPSTEIN, G.

1960. The lattice theory of Post algebras. Trans. Amer. Math. Soc. 95, 300-317.

EVEN, S.; MEYER, A. R.

1969. Sequential Boolean equations. IEEE Trans. Comput. C-18, 230-240.

FADINI, A.

1961. Operatori che estendono alle algebre di Boole la nozione di derivata. Giorn. Mat. Battaglini 89(5), gennaio-dicembre, 42-64.

1972. Algoritmo per la construzione di una funzione booleana composta mediante un'assegnata famiglia di funzioni booleane. Ricerca (Napoli) (2) 23, gennaio-aprile, 23-30.

414

FOSTER, A. L.

1951. *p*—rings and their Boolean vector representation. Acta Math. 84, 231-261.

FÜGERT, E.

1975. Auflösen von Gleichungssysteme mittels dynamischer Operationen. Unveröff. Mitteilung, Karl-Marx-Stadt, 1975.

GAREY, M. R.; JOHNSON, D. S.

1979. Computers and intractability. W. H. Freeman, San Francisco.

GHILEZAN, C. (GILEZAN, K.)

1970. Méthode à résoudre des relations dont les résolutions appartiennent à un ensemble fini. Publ. Inst. Math. (Beograd) 10(24), 21-23.

1976. Certaines équations fonctionnelles pseudo-booléennes généralisées. Publ. Inst. Math. (Beograd) 20(34), 99-109.

1979a. Equations fonctionnelles pseudo-booléennes généralisées du deuxième ordre. Zb. Rad. Prir.-Mat. Fak. u Novom Sadu 9, 105-109.

1979b. Generalized pseudo-boolean functions on finite sets. Zb. Rad. Prir.-Mat. Fak u Novom Sadu 9, 111-113.

1980a. Some fixed point theorems in Boolean algebra. Publ. Inst. Math. (Beograd) 28(42), 77-82.

1980b. Differentials of generalized pseudo-Boolean functions. Zb. Rad. Prir.-Mat. Fak u Novom Sadu 10, 185-190.

1981a. Some properties of linear operators of discrete functions. Zb. Rad. Prir.-Mat. Fak u Novom Sadu 11, 247-252.

1981b. A note on generalized pseudo-Boolean functional equations with constant coefficients and *n* variables. Zb. Rad. Prir.-Mat. Fak. u Novom Sadu 11, 253-257.

1982. Les dérivées partielles des fonctions pseudo-booléennes généralisées. Discrete Appl. Math. 4, 37-45.

1995. Taylor formula of Boolean and pseudo-Boolean functions. Zb. Rad. Prir.-Mat. Fak. u Novom Sadu 25, 141-149.

GILEZAN, K.; UDICKI, M.

1989. Generalized pseudo-Boolean functional equations of the third order. Zb. Rad. Prir.-Mat. Fak. u Novom Sadu 19, 81-91.

GOODSTEIN, R. L.

1967. The solutions of equations in a lattice. Proc. Roy. Soc. Edinburgh , Sect. A, 67, Part III, 231-242.

GORSHKOV, S. P.

1996. On the complexity of the problem of determining the number of solutions of systems of Boolean equations (Russian). Diskret. Mat. 8, no.1, 72-85. English translation: Discrete Math. Appl.6, no.1, 77-92.

GRÄTZER, G.

1962. On Boolean functions (Notes on lattice theory. II). Rev. Roumaine Math. Pures Appl. 7, 693-697.

1964. Boolean functions on distributive lattices. Acta Math. Acad. Sci. Hungarica 15, 195-201.

1978. General lattice theory. Academic Press, New York.

1979. Universal algebra. Springer-Verlag, New York. First edition: Van Nostrand, Princeton 1968.

GRÄTZER, G.; LAKSER, H.

1969. Equationally compact semilattices. Coll. Math. 20, 27-30.

GROZEA, C.

2000. Relations between the low subrecursion classes. Preprint, University of Bucharest.

HAMMER, P. L.; RUDEANU, S.

1968. Boolean methods in operations research and related areas. Springer-Verlag, Berlin. French. translation: Méthodes booléennes en recherche opérationnelle. Dunod, Paris 1970.

HAMMER, P. L.; SIMEONE, B.

1987. Quadratic functions of binary variables. Rutcor Res. Rep. RRR 20-87.

HANSEN, P.; JAUMARD, B.

1985. Uniquely solvable quadratic Boolean equations. Discrete Appl. Math. 12, 147-154.

HELM, R.; MARRIOTT, K.; ODERSKY, M.

1991. Constrained-based query optimization for spatial databases. Proc. ACM Symp. Principles of Database Systems, 181-191.

1995. Spatial query otpimization: from Boolean constraints to range queries. J. Comput. Syst. Sci. 51, 197-210.

HENKIN, L.; MONK, J. D.; TARSKI, A.

1971. Cylindric algebras. Part I. North-Holland, Amsterdam.

HERTZ, A.

1997. On the use of Boolean methods for the computation of the stability number. Discrete Appl. Math. 76, 183-203.

IGAMBERDYEV, T. M.

1989. The number of solutions of some types of Boolean equations (Russian). Diskret. Mat. 1, no.1, 105-116.

ITOH, M.

1955. On the lattice of n−valued functions (Japanese). Kyûsyû Daigaku Kôgaku syûhô (Fukuoka) 28, 96-99, 99-100. Review in J. Symbolic Logic 22(1957), 100-101.

1956. On the general solution of the n−valued function lattice (logical) equation in one variable (Japanese). Kyûsyû Daigaku Kôgaku Syûhô 28, 239-243. Review in J. Symbolic Logic 22(1957), 100-101.

JÓNSSON, B.; TARSKI, A.

1951. Boolean algebras with operators. I. Amer. J. Math. 73, 891-939.

1952. Boolean algebras with operators. II. Amer. J. Math. 74, 177-262.

KABULOV, A. V.; BAĬZHUMANOV, A. A.

1986. Local methods for solving systems of Boolean equations (Russian).

416

Dokl. Akad. Nauk UzSSR 1986, no.3, 3-5.

KAREPOV, S. A.; LIPSKIĬ, V. B.

1974. Applications of the methods for solving Boolean equations to fault detection in combinational networks (Russian). In: Diskretnye Sistemy, sb. stateĭ. Zinatne, Riga, vol.2, 151-159.

KATERINOCHKA, N. N.; KOROLEVA, Z. E.; MADATYAN, KH. A.: PLATO-NENKO, I. M.

1988. Methods for the solution of Boolean equations (Russian). Vychisl. Tsentr Akad. Nauk SSSR. Moskva, 1988.

KEČKIĆ, J. D.; PREŠIĆ, S.

1984. Reproductivity - a general approach to equations (unpublished).

KELLY, D.

1972. A note on equationally compact lattices. Algebra Universalis 2, 80-84.

KIM, K. H.

1982. Boolean matrix theory and applications. M. Dekker, New York.

KLEENE, S. C.

1952. Introduction to metamathematics. North-Holland, Amsterdam.

KOPPELBERG, S.

1989. Handbook of Boolean algebras. Vol.1. North-Holland, Amsterdam.

KOSSOVSKIĬ, N. K.

1978. The complexity of the solvability of Boolean functional equations (Russian). Vychisl. Tekhn. i Voprosy Kibernet. no.15, 104-111.

KRNIĆ, L.

1978. O sustavima izvodnica i bazama za F_2. Mat. Vesnik 2(15) (30), 363-367.

KU, CH. T.; MASSON, G. M.

1975. The Boolean difference and multiple fault analysis. IEEE Trans. Comput. C-24, 62-71.

KUCHAREV, G.; SHMERKO, V.; YANUSHKEVICH, S.

1991. Technique of binary data parallel processing on VLSI (Russian). Vysh. Shkola Izd. Minsk.

KÜHNRICH, M.

1986a. Differentialoperatoren über Booleschen Algebren. Z. Math. Logik Grundlagen Math. 32, 271-288.

1986b. Operators on Boolean algebra. Akad. Wiss. DDR, Karl-Weierstrass-Inst. Math. Preprint P-Math-27/86.

KUNTZMANN, J.

1965. Algèbre de Boole. Dunod, Paris.

LAPSCHER, F.

1968. Application de la notion de fermeture à l'étude des fonctions booléennes. Thèse, Univ. de Grenoble.

LAUSCH, H.; NÖBAUER, W.

1973. Algebra of polynomials. North-Holland, Amsterdam.

LAVIT, C.

1974. La notion de fermeture en algèbre de Boole. Thèse, Univ. du Langue-doc, Montpellier.

1976. Classes de solutions paramétriques d'une équation booléenne. Rev. Roumaine Math. Pures Appl. 21, 1049-1052.

LEE, S. C.

1976. Vector Boolean algebra and calculus. IEEE Trans. Comput. C-25, 865-874.

LEONT'EV. V. K.; NURLYBAEV, A. N.

1975. A class of systems of Boolean equations (Russian). Zh. Vychisl. Mat. i Mat. Fiz. 15, no.6, 1568-1579. English translation: USSR Comput. Math. & Math. Phys. 15(1975), no.6, 198-210.

LEONT'EV, V. K.; TONOYAN, G. P.

1993. Approximate solutions of systems of Boolean equations (Russian). Zh. Vychisl. Mat. i Mat. Fiz. 33, 1383-1390.

LEVCHENKOV, V. S.

1999a. Boolean equations with many unknowns (Russian). Nelineĭnaya Dinamika i Upravlenie, 1999, 105-118. English translation: Comput. Math. Modeling 11(2000), 143-153.

1999b. Solutions of equations in Boolean algebra (Russian). Nelineĭnaya Dinamika i Upravlenie, 1999, 119-123. English translation: Comput. Math. Modeling 11(2000), 154-163.

1999c. Analytical solutions of Boolean equations (Russian). Dokl. Akad. Nauk 369, 325-328.

LEVIN, A. G.

1978. A method for the solution of difference logic equations (Russian). Vestnik Beloruss. Gos. Univ. Ser.I 1978, no.1, 28-31.

LEVIN, V. I.

1975a. Equations of infinite-valued logic that contain all logical operations (Russian). Teor. Konechn. Avtomatov i Prilozhen. Vyp. 5(1975), 5-14, 113.

1975b. Equations of infinite-valued logic with deviating arguments and with all possible logical operations (Russian). Teor. Konechn. Avtomatov i Prilozhen. Vyp. 5(1975), 15-31, 113.

LIANG, P.

1983. The inverse operator of Boolean difference - Boolean integration (Chinese). Chinese J. Comput. 6, 307-313.

1990. A theory of Boolean integration. Intern. J. Comput. Math. 35, 83-91.

LIU, W.-J.

1990. On some systems of simultaneous equations in a completely distributive lattice. Inf. Sci. 50, 185-196.

LIU, YANPEI

1987a. Boolean planarity characterization of graphs. Rutcor Res. Rep. # 38-87.

418

1987b. Boolean approach to planar embeddings of a graph. Rutcor Res. Rep. # 39-87.

LIU, YONGCAI
1988. Construction and counting of generalized Boolean functions. Discrete Math. 69, 313-316.

LÖWENHEIM, L.
1908. Über das Auflösungsproblem im logischen Klassenkalkul. Sitzungsber. Berl. Math. Geselschaft 7, 90-94.

1910. Über die Auflösung von Gleichungen im logischen Gebietkalkul. Math. Ann. 68, 169-207.

1913. Über Transformationen im Gebietkalkul. Math. Ann. 73, 245-272.

1919. Gebietdeterminanten. Math. Ann. 79, 222-236.

MARCZEWSKI, E.
1958. A general scheme for the notion of independence in mathematics. Bull. Acad. Polonaise Sci. Sér. Sci. Math. Astron. Phys. 6, 731-736.

1960. Independence in algebras of sets and Boolean algebras. Fund. Math. 48, 135-145.

MARENICH, E. E.
1997. Enumeration of solutions of some equations in finite lattices (Russian). Vestnik Moskov. Univ. Ser. I Mat. Mekh. 1997, no.3, 16-21, 70. English translation: Moscow Univ. Math. Bull. 52, no.3, 16-21.

MARRIOTT, K.; ODERSKY, M.
1996. Negative Boolean constraints. Theor. Comput. Sci. 160, 365-380.

MARTIN U.; NIPKOW, T.
1986. Unification in Boolean rings. Proc. CADE-8. Lecture Notes Comput. Sci. 230, 506-513.

1988. Unification in Boolean rings. J. Automat. Reason. 4, 381-396.

1989. Boolean unification - the story so far. J. Symb. Comput. 7, 275-293.

MATROSOVA, A. YU.
1975a. Solution of Boolean equations that are given in bracket form (Russian). Mat. Sb. (Tomsk), vyp.2, 95-119.

1975b. The number of computations in the solution of Boolean equations (Russian). Mat. Sb. (Tomsk) vyp. 2, 120-128.

MAYOR, G.; MARTIN, J.
1999. Locally internal aggregation functions. Intern. J. Uncertainty, Fuzziness, Knowledge-Based Syst. 7, 235-241.

McCOLL, H.
1877-80. The calculus of equivalent statements. Proc. London Math. Soc. 9(1877/78), 9-20, 177-186; 10(1878), 16-28; 11(1979/80), 113-121.

McKINSEY, J. C. C.
1936a. Reducible Boolean functions. Bull. Amer. Math. Soc. 42, 263-267.

1936b. On Boolean functions of many variables. Trans. Amer. Math. Soc. 40, 343-362.

MEILER, M.; BÄR, G.

1974. Zur Umformung eines Booleschen Ausdrucks in eine äquivalente Gleichung mit 0−1-Variablen. Elektron. Informationsverarbeit. Kybernetik 10, 341-353.

MELTER, R. A.

1988. Boolean functions which preserve levels. Mathematica (Cluj) 30(53), 145-147.

MELTER, R. A.; RUDEANU, S.

1974. Geometry of 3−rings. Colloquia Math. Soc. János Bolyai, 14 (Lattice Theory), Szeged 1974, 249-269.

1980. Generalized inverses of Boolean functions. Rev. Roumaine Math. Pures Appl. 25, 891-898.

1981. A measure of central tendency for Boolean algebras. An. Şti. Univ. "Al.I.Cuza" Iaşi 27, 411-415.

1982. Characterizations of Boolean functions. An. Şti. Univ. "Al.I.Cuza" Iaşi 28, 161-169.

1983. Functions characterized by functional Boolean equations. Colloquia Math. Soc. János Bolyai, 33 (Lattice Theory), Szeged 1980, 637-650.

1984a. Linear equations and interpolation in Boolean algebra. J. Linear Algebra Appl. 57, 31-40.

1984b. Alternative definitions of Boolean functions and relations. Archiv. Math. (Basel) 43, 16-20.

1993. Voronoi diagrams for Boolean algebras. Stud. Cerc. Mat. 45, 429-434.

MIJAJLOVIĆ, Ž.

1977. Some remarks on Boolean transformations - model theoretic approach. Publ. Inst. Math. (Beograd) 21(35), 135-140.

1980. Two remarks on Boolean algebras. Algebraic Conference, Skopje, 35-41.

MIKHAĬLOVSKIĬ, L. V.

1977. A method for solving logic problems of great dimension (Russian). Voprosy Tekhn. Diagnostiki 1977 no.17, 187-190.

MINOUX, M.

1992. The unique Horn-satisfiability problem and quadratic Boolean equations. Ann. Math. Artificial Intell. 6, 253-266.

MOISIL, GR. C.

1941. Recherches sur la théorie des chaînes. Ann. Sci. Univ. Jassy 27, 181-240.

MOORE, E. H.

1910. Introduction to a form of general analysis. New Haven Math. Colloq. 1906, Yale Univ. Press, New Haven, Connecticut.

MYCIELSKI, J.

1964. Some compactifications of general algebras. Colloq. Math. 13, 1-9.

NGOM, A.; REISCHER, C.; SIMOVICI, D. A.; STOJMENOVIĆ, I.

1997. Set-valued logic algebra: a carrier computing foundation. Multiple Valued Logic 2, 183-216.

NGUYEN, X. Q.

1976. Application of Boolean difference for fault detection in logical networks. Algorithmische Komplizierheit, Lern- und Erkennungsprozesse (Zweite Internat. Sympos., Friederich-Schiller-Univ., Jena, 1976, 61-69. Friedreich- Schiller-Univ., Jena.

NIPKOW, T.

1988. Unification in primal algebras. Proc. CAAP'88, Lecture Notes Comput. Sci. no.299, 117-131.

1990. Unification in primal algebras, their powers and their varieties. J. Assoc. Comput. Mach. 37, 742-746.

NISAN, N.; SZEGEDY, M.

1994. On the degree of Boolean functions as real polynomials. Comput. Complexity 4, 301-313.

NURLYBAEV, A. N.

1990. A class of Boolean equations (Russian). Theory of functions and problems in numerical mathematics (Russian) 35-38, 71-72, Kazakh. Gos. Univ., Alma-Ata.

OBUKHOV, V. E.; PAVLOV, V. V.

1992. Logical equations and applied problems (Russian). Naukova Dumka, Kiev.

PAPAIOANNOU, S. G.: BARRETT, W. A.

1975. The real transform of a Boolean function and its applications. Comput. Electr. Engrg. 2, 215-224.

PARFENOV, YU. M.; RYABININ, I. A.

1997. Boolean differences for monotone functions of the algebra of logic (Russian). Avtomat. i Telemekh. no.10, 193-204.

PARKER, W. L.; BERNSTEIN, B. A.

1955. On uniquely solvable Boolean equations. Univ. Calif. Publ. Math., NS, 3(1), 1-29.

PERIĆ, V.

1978. Rešenje jednog sistema skupovnih jednačina. Mat. Vesnik 2(15)(30), 273-277.

PETRESCHI, R.; SIMEONE, B.

1980. A switching algorithm for the solution of quadratic Boolean equations. Inform. Process. Lett. 11, no.4-5, 193-198. Correct. ibid. 12(1981), no.2, 109.

PETROSYAN, A. V.

1982. Some differential properties of Boolean functions (Russian). Tarnulmányok-MTA Számitátech. Automat. Kutató Intezet, Budapest, no.135, 15-37.

PIERCE, R. S.

1968. Introduction to the theory of abstract algebras. Holt, Rinehart and

Winston, New York.

PLATONENKO, I. M.

1983. On the realization of algorithms of type "Cora" by solving systems of Boolean equations of a special form (Russian). Vychisl. Tsentr Akad. Nauk SSSR, Moskva 1983.

PORTE, J.

1982. Fifty years of deduction theorems. Proc. Herbrand Symp. (Marseille 1981). Studies Logic Found. Math. 107, North-Holland, Amsterdam.

POSHERSTNIK, M. S.

1979. Solving of logical equations by the method of separation of variables (Russian). Avtomat. i Telemekh. 1979, no.2, 133-140. English translation: Automat. Remote Control 40, no.2, part 2, 260-267.

POST, E. L.

1921. Introduction to a general theory of elementary propositions. Amer. J. Math. 43, 163-185.

POSTHOFF, CH.

1978. Die Lösung und Auflösung binärer Gleichungen mit Hilfe des Booleschen Differentialkalküls. Elektron. Informationsverarbeit. Kybernet. 14, 53-80.

POVAROV, G. N.

1954. On the functional separability of Boolean functions (Russian). Dokl. Akad. Nauk SSSR 94, 801-803.

PREŠIĆ, S. B.

1968. Une classe d'équations matricielles et l'équation fonctionnelle $f^2 = f$. Publ. Inst. Math. (Beograd) 8(22), 143-148.

1971. Une méthode de résolution des équations dont toutes les solutions appartiennent à un ensemble fini donné. C. R. Acad. Sci. Paris 272, 654-657.

1972. Ein Satz über reproduktive Lösungen. Publ. Inst. Math. (Beograd) 14(28), 133-136.

1988. All reproductive solutions of finite equations. Publ. Inst. Math. (Beograd) 44(58), 3-7.

2000. A generalization of the notion of reproductivity. Publ. Inst. Math. (Beograd) 67(81), 76-84.

PRUSINKIEWICZ, P.

1976. Stany stabilne asynchronicznych liniowych układów logicznych. Arch. Automat. i Telemech. 21, no.1, 65-72.

QUINE, W. V.

1952. The problem of simplifying truth functions. Amer. Math. Monthly 59, 521-531.

1955. A way to simplify truth functions. Amer. Math. Monthly 62, 627-631.

1959. On cores and prime implicants of truth functions. Amer. Math. Monthly 66, 627-631.

RASIOWA, H.

1974. An algebraic approach to non-classical logics. North-Holland, Amsterdam.

REISCHER, C.; SIMOVICI, D. A.

1971. Associative algebraic structures in the set of Boolean functions and some applications in automata theory. IEEE Trans. Computers C-20, 298-303.

1987. New functional characterization of Boolean and Post algebras. Contributions to General Algebra 5, Proc. Salzburg Conference, May 29 - June 1, 1986. B.G. Teubner Verlag, Stuttgart.

REISCHER, C.; SIMOVICI, D. A.; STOJMENOVIĆ, I.; TOŠIĆ, R.

1997. A characterization of Boolean collections of set-valued functions. Information Sci. 99, 195-204.

ROBINSON, J. A.

1965. A machine-oriented logic based on the resolution principle. J. Assoc. Comput. Mach. 12, 23-41.

ROSENBLOOM, P. C.

1924. Post algebras. I. Postulates and the general theory. Amer. J. Math. 64, 167-188.

ROUSSEAU, G.

1970. Post algebras and pseudo-Post algebras. Fund. Math. 67, 133-145.

ROZENFEL'D, T. K.;

1974. Solution of Boolean equations and the decomposition of Boolean functions (Russian). Materialy Sem. Kibernet. 68, 18-27, 28.

ROZENFEL'D, T. K.; SILAYEV, A. N.

1979. Boolean equations and decomposition of Boolean functions (Russian). Izv. Akad. Nauk SSSR Tekhn. Kibernet. English translation: Engrg. Cybernet. 17, 85-92.

RUDEANU, S.

1964. Notes sur l'existence et l'unicité des noyaux d'un graphe. I. Rev. Française Rech. Opérationnelle 8, 345-352.

1965. On Tohma's decompositions of logical functions. IEEE Trans. Electronic Comput. EC-14, 929-931.

1966a. Notes sur l'existence et l'unicité des noyaux d'un graphe. II. Application des équations booléennes. Rev. Française Rech. Opérationnelle 10, 301-310.

1966b. On solving Boolean equations in the theory of graphs. Rev. Roumaine Math. Pures Appl. 11, 653-664.

1967. Axiomatization of certain problems of minimization. Studia Logica 20, 37-61.

1968. On functions and equations in distributive lattices. Proc. Edinburgh Math. Soc. 16 (Series II), 49-54. Correct. ibid. 17 (Series II), 105.

1969. Boolean equations for the chromatic decomposition of graphs. An. Univ. Bucureşti ser. Mat. 18, 119-126.

1973. Testing Boolean identities. Centre de rech. math., Univ. de Montréal, CRM-308, July 1973.

1974a (≡BFE). Boolean functions and equations. North-Holland, Amsterdam.

1974b. An algebraic approach to Boolean equations. IEEE Trans. Computers C-23, 206-207.

1975a. Local properties of Boolean functions. I. Injectivity. Discrete Math. 13, 143-160.

1975b. Local properties of Boolean functions. II. Isotony. Discrete Math. 13, 161-183.

1975c. On the range of a Boolean transformation. Publ. Inst. Math. (Beograd) 19(33), 139-144.

1976a. Local properties of Boolean functions. III. Extremal solutions of Boolean equations (Russian). Teoria avtomatov, M. A. Gavrilov ed., Izd. Nauka, Moskva, 25-33.

1976b. Square roots and functional decompositions of Boolean functions. IEEE Trans. Computers C-25, 528-532.

1977. Systems of linear Boolean equations. Publ. Inst. Math. (Beograd) 22(36), 231-235.

1978a. Injectivity domains of Boolean transformations. Rev. Roumaine Math. Pures Appl. 23, 113-119.

1978b. On reproductive solutions of arbitrary equations. Publ. Inst. Math. (Beograd) 24(38), 143-145.

1980. Fixpoints of lattice and Boolean transformations. An. Şti. Univ. "Al.I.Cuza" Iaşi 26, 147-153.

1983a. Isotony detectors for Boolean functions. Rev. Roumaine Math. Pures Appl. 28, 243-250.

1983b. Linear Boolean equations and generalized minterms. Discrete Math. 43, 241-248.

1990. On deduction theorems and functional Boolean equations. An. Univ. Timişoara, Ser. Şti. Natur. 28, 197-210.

1993. Unique solutions of Boolean equations. Discrete Math. 122, 381-383.

1995a. On quadratic Boolean equations. Fuzzy Sets and Systems 75, 209-213.

1995b. On certain set equations. Stud. Cerc. Mat. 47, 349-451.

1997. Gr. C. Moisil, a contributor to the early development of lattice theory. Multiple Valued Logic 2, 323-328.

1998a. On Boolean modus ponens. Mathware & Soft Comput. 5, 115-119.

1998b. On general and reproductive solutions of finite equations. Publ. Inst. Math. (Beograd) 63(77), 26-30.

RYBAKOV, V. V.

1986a. Equations in free topoboolean algebra (Russian). Algebra i Logika 25, 172-204. English translation: Algebra Logic 25, 109-127.

1986b. Equations in a free topoboolean algebra and the substitution problem (Russian). Dokl. Akad. Nauk SSSR 287, 554-557. English translation: Soviet Math. Dokl. 33, 428-431.

1992. The universal theory of the free pseudoboolean algebra $F_\omega(H)$ in the signature extended by constants for free generators. Algebra, Proc. Intern. Conf. Memory A. I. Mal'cev. Novosibirsk, 1989. Contemporary Math. 131, Part 3, 645-

424

656.

SCHAEFER, T. J.

1978. The complexity of satisfiability problems. Conference Record Tenth Annual ACM Symp. Theory of Computing (San Diego, Calif. 1978). ACM, New York 1978, 216-226.

SCHMIDT, K.

1922. The theory of functions of one Boolean variable. Trans. Amer. Math. Soc. 23, 212-222.

SCHRÖDER, E.

1890-1905. Vorlesungen über die Algebra der Logik. Leipzig; vol.1, 1890; vol.2, 1891, 1905; vol.3, 1895. Reprint Chelsea, Bronx NY, 1966.

SCHWEIGERT, D.

1975. Über idempotente Polynomfunktionen auf Verbänden. Elemente der Mathematik, vol.30/2, 30-32.

SCOGNAMIGLIO, G.

1960. Elementi uniti ed antiuniti delle funzioni monovalenti algebriche di Boole. Giorn. Mat. Battaglini 88, 135-154.

SERFATI, M.

1973a. Sur les polynômes postiens. C. R. Acad. Sci. Paris 276, 677-679.

1973b. Introduction aux algèbres de Post et à leurs applications. Inst. Statistique Univ. Paris, Cahiers du Bureau Univ. de Rech. Opérationnelle, Cahier no.21, Paris.

1977. Une méthode de résolution des équations postiennes à partir d'une solution particulière. Discrete Math. 17, 187-189.

1995. Boolean differential equations. Discrete Math. 146, 235-246.

1996. On Postian algebraic equations. Discrete Math. 152, 269-285.

1997. A note on Postian matrix theory. Intern. J. Algebra Comput. 7, 161-179.

SERIKOV, YU. A.

1972. An algebraic method for the solution of logical equations (Russian). Izv. Akad. Nauk SSSR Tekhn. Kibernet. English translation: Engrg. Cybernet. 1972, 273-282.

SHAPIRO, S. I.

1984. Solutions of logical and game-theoretic problems (Russian). Radio i Svyaz, Moskva.

SIMEONE, B.

1979. Quadratic 0−1 programming, Boolean functions and graphs. PhD Thesis, University of Waterloo. Waterloo, Ontario.

SIMOVICI, D. (SIMOVICI, D. A.); REISCHER, C.

1986. Iterative characterization of Boolean algebras. Discrete Appl. Math. 15, 111-116.

SOFRONIE, V.

1989. Formula-handling computer solution of Boolean equations. I. Ring equations. Bull. European Assoc. Theor. Comput. Sci. no.37, 184-186.

STIEFEL, B.

1982. Über Iterierte einer Booleschen Funktion. Rostock Math. Kolloq. 19,119-182.

TAKAGI, N.; NAKAMURA, Y.; NAKASHIMA, K.

1997. Set-valued logic functions monotone in the set-theoretical inclusion. Multiple Valued Logic 2, 287-304.

TAKENAKA, Y.

1978. A Boolean proof of the Richardson theorem of graph theory. Inform. and Control 39, 1-13.

ȚĂNDĂREANU, N.

1981. On generalized Boolean functions. I. Discrete Math 34, 293-299.

1982. On generalized Boolean functions. II. Discrete Math. 40, 277-284.

1983. O caracterizare algebrică a mulțimilor de funcții booleene generalizate. INFO-IASI 83, 27-29 octombrie 1983, 78-81.

1984a. Remarks on generalized Boolean functions of one variable. Rev. Roumaine Math. Pures Appl. 29, 715-718.

1984b. On generalized Boolean functions. III. The case $A = \{0,1\}$. Discrete Math. 52, 269-277.

1985a. Partial derivatives of $\{0, 1\}$−generalized Boolean functions. An. Univ. Craiova Ser. Mat.-Fiz.-Chim. 13, 63-66.

1985b. Monotonicity of the functions of one variable defined on Boolean algebras. Rev. Roumaine Math. Pures Appl. 30, 579-582.

TAPIA, M. A.; TUCKER, J. H.

1980. Complete solutions of Boolean equations. IEEE Trans. Comput. C-29, 662-665.

TARSKI, A.

1955. A lattice-theoretical fixpoint theorem and its applications. Pacific J. Math. 5, 285-309.

TESLENKO, A. A.

1974. An approach to the problem of solving logical equations (Russian). Avtomat. i Telemekh. 1974, no.12, 159-162.

THAYSE, A.

1981a. Universal algorithms for evaluating Boolean functions. Discrete Appl. Math. 3, 53-65.

1981b. Boolean calculus of differences. Lecture Notes in Comput. Sci. no.101. Springer-Verlag, Berlin.

THOMASSON, M. G.; PAGE, E. W.

1976. Boolean difference techniques in fault tree analysis. Internat. J. Comput. Inform. Sci. 5, 81-88.

TIKHONENKO, O. M.

1974. Properties of linear Boolean functions (Russian). Vestnik Beloruss. Gos. Univ. ser.I 1975, no.2, 19-23, 99.

TING, SH.-T.; ZHAO, SH.-Y.
1991. The generalized Steiner problem in Boolean space and applications. Discrete Math. 90, 75-84.

TINHOFER, G.
1972. Über die Bestimmung von Kernen in endlichen Graphen. Computing 9, 139-147.

TOHMA, Y.
1964. Decomposition of logical functions using majority decision elements. IEEE Trans. Electronic Comput. EC-13, 698-705.

TOŠIĆ, R.
1978a. Neke osobine monotonih bulovih funkcija nad konačnim bulovim algebrama. Zb. Radova Prirodno-Mat. Fak. Univ. Novom Sadu, 8, 63-68.

1978b. Constant-preserving functions over the finite Boolean algebras. Math. Balkanica 8:29, 227-234 = Zb. Radova Prirodno-Mat. Fak. Univ. Novom Sadu 10, 197-203.

1980. Jedan način predstavljanja isotonih bulovih funkcija. Zb. Radova Prirodno-Mat. Fak. Univ. Novom Sadu 10, 205-207.

TRACZYK, T.
1963. Axioms and some properties of Post algebras. Colloq. Math. 10, 193-209.

1964. An equational definition of a class of Post algebras. Bull. Acad. Polonaise Sci., Sér. Sci. Math. Astron. Phys. 12, 147-150.

TRILLAS, E.: CUBILLO, S.
1996. Modus ponens on Boolean algebras revisited. Mathware & Soft Comput. 3, 105-112.

UBAR, R. R.
1979. Detection of suspected faults in combinational circuits by solving Boolean differential equations (Russian). Avtomat. i Telemekh. 1979, no.11, 170-183. English translation: Automate Remote Control 40, no.11, part 2, 1693-1703 (1980).

USTINOV, N. A.
1980. On the numbers of solutions of a system of logical equations (Russian). Combinatorial algebraic methods in applied mathematics (Russian). Gor'kov Gos. Univ, Gorki, 206-212, 216.

VAUGHT, R.
1954. On sentences holding in direct products of relational systems. Intern. Congr. Math. Amsterdam 1954 (Noordhoff, Gronningen), 409.

VOIGT, A. H.
1890. Die Auflösung von Arteilsystemen, das Eliminationsproblem und die Kriterien des Widerspruch in der Algebra der Logik. Freiburger PhD Dissertation, Alex Danz, Leipzig.

WANG, H.
1959. Circuit synthesis by solving sequential Boolean equations. Z. Math.

Logik Grundlagen Math. 5, 291-322. Also in H. Wang, A survey of mathematical logic. Science Press, Pekin 1962, 269-305. Romanian translation: H. Wang, Studii de logică matematică, Ed. Ştiinţifică, Bucureşti 1972, 261-294.

WĘGLORZ, B.

1966. Equationally compact algebras. I. Fund. Math. 59, 289-298.

WHITEHEAD, A. N.

1901. Memoir on the algebra of symbolic logic. Amer. J. Math. 23, 139-165, 297-316.

YANUSHKEVICH, S.

1994. Development of the methods of Boolean differential calculus for arithmetical logic (Russian). Avtomat. i Telemekh. no.5, 121-137. English translation: Automat. Remote Control 55, 715-729.

1998. Logic differential calculus in multi-valued logic design. Techn. Univ. Szczecin, Szczecin.

ZAKREVSKIĬ, A. D.

1971. Algorithms for the synthesis of discrete automata (Russian). Izd. Nauka, Moskva.

1975a. Logical equations (Russian). Izd. Nauka i Tekhnika, Minsk.

1975b. The method of "reflected waves" for solving logical equations (Russian). Priklad. Aspekty Teor. Avtom., Varna. Bulgarian Acad Sci. no.2.

ZAKREVSKIĬ, A. D.; UTKIN, A. A.

1975. The solution of logical difference equation (Russian). Dokl. Akad. Nauk BSSR 19, 34-37, 91.

ZEMMER, J. L.

1956. Some remarks on p-rings and their Boolean geometry. Pacific J. 6, 193-208.

ZHURAVLEV, YU. I.; PLATONENKO, I. M.

1984. On economic multiplication of Boolean equations (Russian). Zh. Vychisl. Mat. i Mat. Fiz. 24, 164-166, 176.

Index

430

Other titles in the DMTCS series:

Combinatorics, Complexity, Logic:
Proceedings of DMTCS '96
D. S. Bridges, C. S. Calude, J. Gibbons,
S. Reeves, I. Witten (Eds)
981-3083-14-X

Formal Methods Pacific '97: Proceedings
of FMP '97
L. Groves and S. Reeves (Eds)
981-3083-31-X

The Limits of Mathematics: A Course on
Information Theory and the Limits of
Formal Reasoning
Gregory J. Chaitin
981-3083-59-X

Unconventional Models of Computation
C. S. Calude, J. Casti and M. J. Dinneen (Eds)
981-3083-69-7

Quantum Logic
K. Svozil
981-4021-07-5

International Refinement Workshop and
Formal Methods Pacific '98
J. Grundy, M. Schwenke and T. Vickers (Eds)
981-4021-16-4

Computing with Biomolecules: Theory
and Experiments
Gheorghe Paun (Ed)
981-4021-05-9

People and Ideas in Theoretical Computer
Science
C. S. Calude (Ed)
981-4021-13-X

Combinatorics, Computation and Logic:
Proceedings of DMTCS'99 and CATS'99
C. S. Calude and M. J. Dinneen (Eds)
981-4021-56-3

Polynomials: An Algorithmic Approach
M. Mignotte and D. Stefanescu
981-4021-51-2

The Unknowable
Gregory J. Chaitin
981-4021-72-5

Sequences and Their Applications:
Proceedings of SETA '98
C. Ding, T. Helleseth and H. Niederreiter (Eds)
1-85233-196-8

Finite versus Infinite: Contributions to an
Eternal Dilemma
Cristian S. Calude and Gheorghe Paun (Eds)
1-85233-251-4

Network Algebra
Gheorge Stefanescu
1-85233-195-X

Exploring Randomness
Gregory J. Chaitin
1-85233-417-7

Unconventional Models of Computation
(UMC2K)
I. Antoniou, C.S. Calude and M.J. Dineen
(Eds)
1-85233-415-0